D1161210

recent advances in phytochemistry

volume 29

Phytochemistry of Medicinal Plants

RECENT ADVANCES IN PHYTOCHEMISTRY

Proceedings of the Phytochemical Society of North America
General Editor: John T. Romeo, *University of South Florida, Tampa, Florida*

Recent Volumes in the Series:

Volume 20 The Shikimic Acid Pathway
Proceedings of the Twenty-fifth Annual Meeting of the Phytochemical Society of North America, Pacific Grove, California, June, 1985

Volume 21 The Phytochemical Effects of Environmental Compounds
Proceedings of the Twenty-sixth Annual Meeting of the Phytochemical Society of North America, College Park, Maryland, July, 1986

Volume 22 Opportunities for Phytochemistry in Plant Biotechnology
Proceedings of the Twenty-seventh Annual Meeting of the Phytochemical Society of North America, Tampa, Florida, June, 1987

Volume 23 Plant Nitrogen Metabolism
Proceedings of the Twenty-eighth Annual Meeting of the Phytochemical Society of North America, Iowa City, Iowa, June, 1988

Volume 24 Biochemistry of the Mevalonic Acid Pathway to Terpenoids
Proceedings of the Twenty-ninth Annual Meeting of the Phytochemical Society of North America, Vancouver, British Columbia, Canada, June, 1989

Volume 25 Modern Phytochemical Methods
Proceedings of the Thirtieth Annual Meeting of the Phytochemical Society of North America, Quebec City, Quebec, Canada, August, 1990

Volume 26 Phenolic Metabolism in Plants
Proceedings of the Thirty-first Annual Meeting of the Phytochemical Society of North America, Fort Collins, Colorado, June, 1991

Volume 27 Phytochemical Potential of Tropical Plants
Proceedings of the Thirty-second Annual Meeting of the Phytochemical Society of North America, Miami Beach, Florida, August, 1992

Volume 28 Genetic Engineering of Plant Secondary Metabolism
Proceedings of the Thirty-third Annual Meeting of the Phytochemical Society of North America, Pacific Grove, California, June–July, 1993

Volume 29 Phytochemistry of Medicinal Plants
Proceedings of the Thirty-fourth Annual Meeting of the Phytochemical Society of North America, August, 1994, Mexico City, Mexico

---

A Continuation Order Plan is available for this series. A continuation order will bring delivery of each new volume immediately upon publication. Volumes are billed only upon actual shipment. For further information please contact the publisher.

RS
164
.P536
1995

BELL LIBRARY TAMU-CC

recent advances in phytochemistry

volume 29

Phytochemistry of Medicinal Plants

Edited by

John T. Arnason
Université d' Ottawa
Ottawa, Ontario, Canada

Rachel Mata
Universidad Nacional Autónoma de México
Coyoacán, México

and

John T. Romeo
University of South Florida
Tampa, Florida

PLENUM PRESS • NEW YORK AND LONDON

Library of Congress Cataloging-in-Publication Data

Phytochemistry of medicinal plants / edited by John T. Arnason, Rachel Mata, and John T. Romeo.
p. cm. -- (Recent advances in phytochemistry ; v. 29)
"Proceedings of the Thirty-fourth Annual Meeting of the Phytochemical Society of North America on phytochemistry of medical plants, held August 15-19, 1994, in Mexico City, Mexico"--T.p. verso.
Includes bibliographical references and index.
ISBN 0-306-45181-6
1. Materia medica, Vegetable--Congresses. 2. Medicinal plants--Congresses. 3. Phytochemistry--Congresses. I. Arnason, J. T., 1948- . II. Mata, Rachel. III. Romeo, John T. IV. Phytochemical Society of North America. Meeting (34th : 1994 : Mexico City, Mexico) V. Series.
RS164.P536 1995
615'.321--dc20 95-39616
CIP

Cover photograph: Comparative Molecular Field Analysis (CoMFA) model of the germacranolide sesquiterpene lactone, parthenolide. Green areas indicate molecular regions where activity would be enhanced by greater steric bulk. Yellow areas indicate molecular regions where activity would be enhanced by less steric bulk. Red areas indicate molecular regions where activity would be enhanced by greater negative electrostatic potential. Blue areas indicate molecular regions where activity would be enhanced by greater positive electrostatic potential. Such models aid the rational design of natural product derivatives with enhanced activity and specificity.

Proceedings of the Thirty-Fourth Annual Meeting of the Phytochemical Society of North America on Phytochemistry of Medicinal Plants, held August 15–19, 1994, in Mexico City, Mexico

ISBN 0-306-45181-6

© 1995 Plenum Press, New York
A Division of Plenum Publishing Corporation
233 Spring Street, New York, N. Y. 10013

All rights reserved

10 9 8 7 6 5 4 3 2

No part of this book may be reproduced, stored in a retrieval system, or transmitted in any form or by any means, electronic, mechanical, photocopying, microfilming, recording, or otherwise, without written permission from the Publisher

Printed in the United States of America

PREFACE

Phytochemicals from medicinal plants are receiving ever greater attention in the scientific literature, in medicine, and in the world economy in general. For example, the global value of plant-derived pharmaceuticals will reach $500 billion in the year 2000 in the OECD countries. In the developing countries, over-the-counter remedies and "ethical phytomedicines," which are standardized toxicologically and clinically defined crude drugs, are seen as a promising low-cost alternatives in primary health care. The field also has benefited greatly in recent years from the interaction of the study of traditional ethnobotanical knowledge and the application of modern phytochemical analysis and biological activity studies to medicinal plants.

The papers on this topic assembled in the present volume were presented at the annual meeting of the Phytochemical Society of North America, held in Mexico City, August 15-19, 1994. This meeting location was chosen at the time of entry of Mexico into the North American Free Trade Agreement as another way to celebrate the closer ties between Mexico, the United States, and Canada. The meeting site was the historic Calinda Geneve Hotel in Mexico City, a most appropriate site to host a group of phytochemists, since it was the address of Russel Marker. Marker lived at the hotel, and his famous papers on steroidal saponins from *Dioscorea composita,* which launched the birth control pill, bear the address of the hotel.

The meeting was organized by Rachel Mata, José Calderón, Rogelio Pereda-Miranda, Guillermo Delgado, and José Antonio Serratos in Mexico, and by John Arnason and Dennis Awang in Canada. We are grateful to the Facultad de Química, Universidad Nacional Autónoma de México for assistance and to the many students who assisted at the meeting.

The papers presented in this volume are representative of the new trends in medicinal plants research. Wagner, Pezzuto, and Beutler *et al.* provide chapters that demonstrate the importance of such plant-based medicinals as adaptogens, immunostimulants, cancer preventatives, and anti-AIDS or anti-cancer agents. Bye *et al.*, Pereda-Miranda, Vlietinck *et al.*, Niemeyer, and Johns and Chapman provide insight into ethnobotanical traditions, and how these can lead to new phytochemical and biological activity discoveries. As is appropriate for a meeting held in Mexico, the rich tradition of medicinal plants of Latin America is emphasized. The two chapters by Wolfender and Hostettmann and Loyola-Vargas and Miranda-Ham describe state-of-the-art techniques of phytochemical analysis and root culture in medicinal plant studies, while Marles *et al.* apply the latest modeling techniques to study antimigraine activity of sesquiterpene lactones. Finally Rodriguez-Hahn and Gu *et al.* provide in-depth reviews covering the phytochemistry and biological activity of two highly interesting groups of compounds, acetogenins and neo-clerodane diterpenes.

We acknowledge the assistance of Ms. Diane Field for her skillful preparation of the camera-ready manuscript.

Rachel Mata, Universidad Nacional Autónoma de México
John Thor Arnason, Université d'Ottawa
John Romeo, University of South Florida

CONTENTS

Chapter One

IMMUNOSTIMULANTS AND ADAPTOGENS FROM PLANTS

Hildebert K. M. Wagner

University of Munich
Institute of Pharmaceutical Biology
D-80333 Munich, Karlstrasse 29
Germany

INTRODUCTION

The terms "immunostimulants" and "adaptogens" both describe drugs capable of increasing the resistance of an organism against stressors of variable origin. Both types of drugs achieve this enhancement primarily by nonspecific mechanisms of actions. Immunostimulants generally stimulate, in a non-antigen dependent manner, the function and efficiency of the nonspecific immune system in order to counteract microbial infections or immunosuppressive states. Adaptogens are believed to reinforce (increase) the non-specific power of resistance of the body against physical, chemical or biological noxious agents. With respect to the mechanisms of action immunostimulants influence primarily the humoral and cellular immune system, whereas adaptogens are thought to between the immune and endocrine system it is very often difficult or impossible to discriminate between the two mechanisms of action. Therefore, it is not surprising that both classes of drugs can influence both systems at the same time.

Experience suggests that immunostimulants are effective prophylactically as well as therapeutically, while adaptogens are of use primarily in the prevention of stress situations. One characteristic feature of both types of drugs is that, similar to vaccines, they must be applied at relatively low doses to achieve optimal effects. They are, therefore, suitable for regulative or modulating medication, i.e. for restoring body homeostasis. These characteristics, which are unknown or not observed with classical drugs, may be one of the reasons why both classes of drugs have so far failed to be fully accepted in medicine, in particular in a rational drug medication. Therefore it is necessary to search for the active principles of plant drugs applied in "traditional medicine" and to clarify their mechanisms of actions.

IMMUNOSTIMULANTS

Currently Available Drugs

Plant-derived immunostimulatory drugs available on the drug market can be subdivided into two classes: plant extracts and polysaccharides isolated from fungi.[1] The first class comprises plant drugs which are widely used in Europe and Asia in traditional medicine for self-medication as well as for prescription. Table 1 lists those plants which have been thoroughly investigated using modern immunological test models and from which some putative active principles have been isolated.

Among these plants, both the water and alcoholic extracts of *Echinacea purpurea, E. angustifolia* and *E. pallida* (herbs and roots) are most prominent.[2,3] It appears that their immunostimulating activities, as demonstrated in many *in vitro* and experimental animal studies as well as clinical trials, are associated with lipophilic compounds (alkylamides) and a polar fraction (cichoric acid and polysaccharides). This could explain the effectiveness of water extracts as well as alcoholic extracts. The most frequent major therapeutic and prophylactic applications using *Echinacea*-preparations are for: chronic and recurrent infections of respiratory organs and urogenital organs; chronic inflammations/allergies; tonsillitis and sinusitis; retarded wound healing and infected wounds; Ulcus cruris, eczemas, and psoriasis; chronic bronchitis and prostatitis; and malignant diseases (in combination with chemotherapy or irradiation).[2] Both the preventive use and the therapeutic treatment of the major indication, i.e. recurrent infections of respiratory organs and the urogenital tract, are officially accepted in the Monograph *Echinacea* by the German Drug authority (plant drug monographs filed by a commission E for phytopreparations, set up by the German Durg

Table 1. Major medicinal plants used in phytopreparations for immunostimulation[1]

Germany/Europe	China/Japan
Echinacea spp.	*Panax ginseng*
Eupatorium perf./cannabinum	*Eleutherococcus senticosus*
Thuja occidentalis	*Astragalus membranaceus*
Baptisia tinctoria	*Codonopsis pilosula*
Eleutherococcus senticosus	*Coriolus versicolor* (Krestin)
Viscum album	*Lentinus edodes* (Lentinan)
Achyrocline satureioides	*Schizophyllum commune* (Schizophyllan)
Chamomilla recutita	*Rehmania glutinosa*

Authority and containing the clinically proven indications for which a drug only should be used). The results of a placebo-controlled double blind study with 160 male and female patients suffering from influenza are shown (Figs. 1,2). These data show a significant reduction of infection-susceptibility, sniffing, and inflammatory laryngitis.[4] The treatment of leucopenia cancer patients undergoing radiotherapy with a phytopreparation containing an *Echinacea* extract in combination with a *Baptisia* extract resulted in a significant increase of the number of leukocytes[5] (Fig.3).

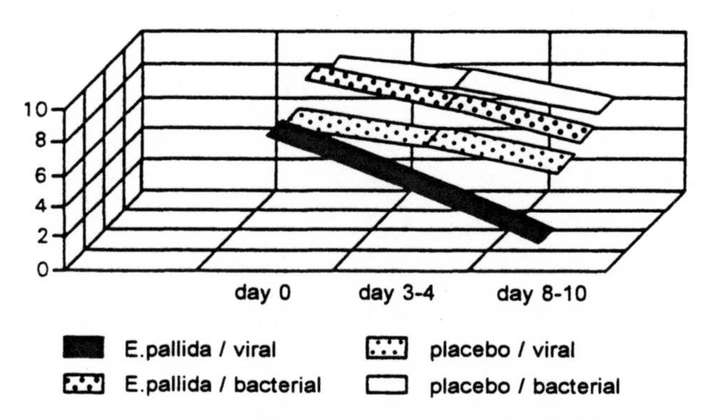

Figure 1. Result of a placebo-controlled double blind study with *Echinacea pallida* root extract (Pascotox 100®) on 160 patients (80 patients each in the verum and placebo group). Application of 90 drops/day equivalent to 900 mg drug/day[4]. Mean scores of clinical symptoms (evaluated by the physician) of patients with upper respiratory tract infections of bacterial or viral origin at baseline (day 0), at the first control 3 or 4 days after inclusion, and at the final control 8 to 10 days after inclusion. - Total score (n = 160). Clinical symptoms (total scores): $p < 0{,}0001$.

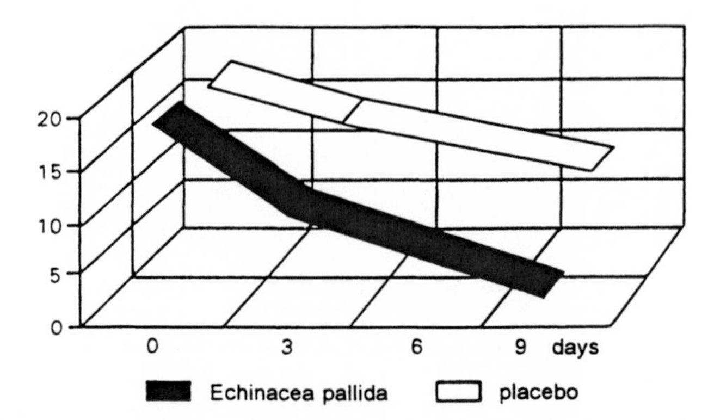

Figure 2. Result of a placebo-controlled double blind study with *Echinacea pallida* root extract (Pascotox 100®) on 160 patients (80 patients each in the verum and placebo group). Application of 90 drops/day equivalent to 900 mg drug/day[4]. Mean total scores of clinical evidences (evaluated by physicians and patients) of patients with upper respiratory tract infections of bacterial or viral origin at baseline (day 0), at the first control 3 or 4 days after inclusion, and at the final control 8 to 10 days after inclusion. Clinical evidences (total scores) : < 0,0001.

Among the isolated plant constituents available today for immunostimulatory treatment, the "antitumoral" polysaccharides Lentinan (*Lentinus edodes*) Schizophyllan (*Schizophyllum commune*) and the glycoprotein Krestin (*Coriolus versicolor*) are noteworthy.[1] Most of them are linear glucans with $1 \rightarrow 3$ linkages in the backbone and β $1 \rightarrow 6$ branches. Krestin is a β $1 \rightarrow 4$ - glucan with β 1,6 - glucopyranosidic chains at every fourth glucose unit which also carries a

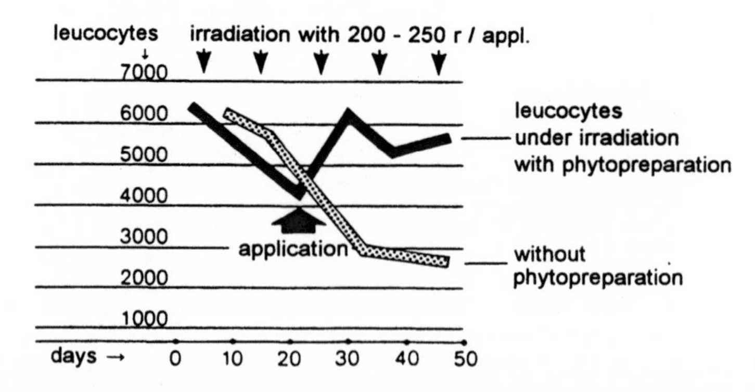

Figure 3. Results of a medication of cancer patients who were subjected to radiotherapy, with a phytocombination of the German drug market containing root extracts of *Echinacea purpurea* and *Baptisia tinctoria* (Esberitox®). Measured parameter: number of leucocytes prior to and after irradiation.

covalent bound peptide residue. The clinical usefulness of these polymers in combination with chemotherapy has been demonstrated with patients suffering from lung, gastric, colon and cervical cancer. In general, intramuscular (i.m.) injections once or twice per week with single doses of polysaccharides, ranging from 2 mg to 30 mg, were shown to be effective in prolonging considerably the survival rate of the patients.

Search for New Immunostimulating Drugs

Since the term immunostimulation is not found in the older literature, other criteria must be applied for the selection of plants. Good candidates are plants described for their antibacterial, antiviral, antifungal (antiinfectious) or antitumoral activities. Another criterion is the quantity of the administered drug needed for anti-infectious or antitumor activity. If the normally used dose is so small that a direct antimicrobial or antitumoral effect can be excluded, an immune-induced effect appears most likely.

Methods of Screening[6]

In vitro and *in vivo* test systems which allow determination of the functional state and the efficiency of the cellular and humoral nonspecific immune system are appropriate for the screening of plant constituents. These tests comprise, in particular, those *in vitro* assays utilizing granulocytes, macrophages, T-lymphocyte populations, NK-cells (Natural Killer cells) or complement (consisting of more than 15 peptides which belong to the nonspecific humoral immune system and are involved mainly in inflammatory processes), and cytokines as target cells or systems. Additionally, infection stress tests in mice with e.g. *Candida albicans*, *Staphylococcus spp.* or *Listeria euriettii* followed by determination of the survival rate can be carried out.[7] Immune-induced cytotoxicity tests using macrophages and labeled tumor cells in various modifications are also available. In recent years the flow cytometric method has been introduced with great success for monitoring the immune status of patients and for measuring phagocytosis, chemoluminescence and other cell functions during medication. Since the ability of an immunostimulant to restore an impaired (suppressed) immune system is an important criterion for its potential use, experimental animal models using cyclosporin A, corticoids or cytostatic agents as immune suppressors, should also be established.

It must be remembered, however, that results obtained in *in vivo* models in many cases have no counterpart in humans, and that the efficacy of an immunostimulant is strongly dependent on the immune status of a patient at the onset of a treatment and on the dosage and the mode of application. As far as the *in vitro* screening of extracts or isolated compounds is concerned, it also is essential to evaluate a broad range of concentrations of the immunostimulant in order to discover possible reversal effects at very low or high concentrations.

a

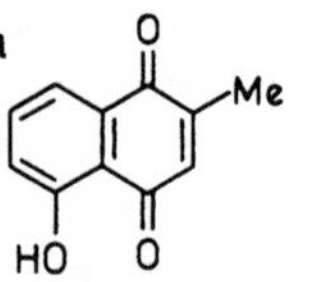

Plumbagin
(*Plumbago ceylanicum*)

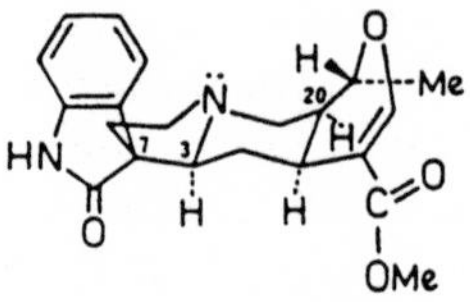

Isopterpodine
(*Uncaria tomentosa*)

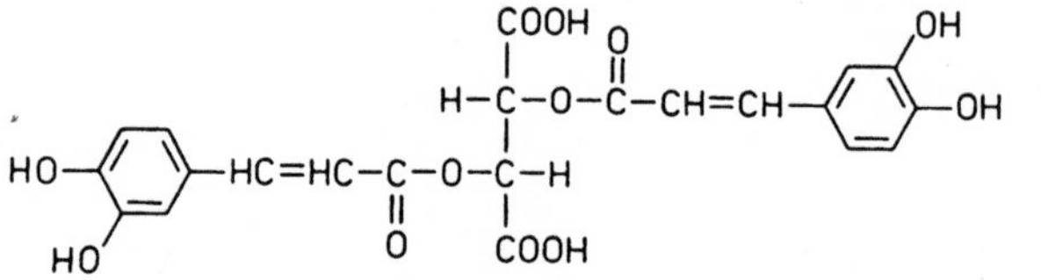

Cichoric acid
(*Echinacea purpurea*)

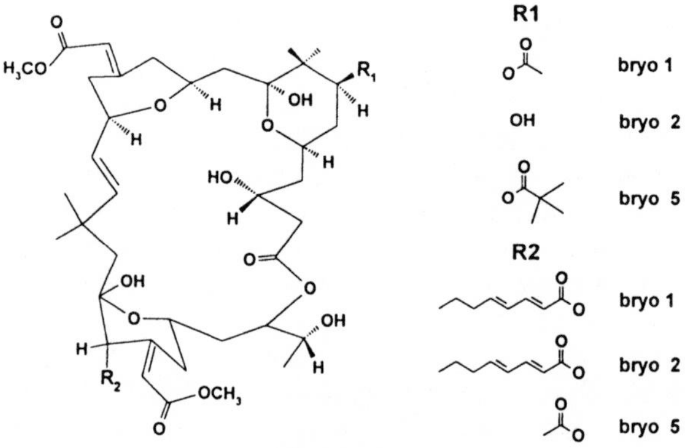

Bryostatins
(*Bugula neritina*)

Figure 4. Immunostimulating constituents of plant drugs. (a) low molecular weight compounds; (b) high molecular weight compounds.

Recent Results

As previously described, the potential immunostimulating compounds can be subdivided into low and high molecular weight compounds.[1,8] In the first group we find alkaloids (e.g. isopteropodine), terpenoids, quinones (e.g. plumbagin), macrocyclic lactones (e.g. bryostatins) as well as phenolic compounds (e.g. cichoric acid) (Fig. 4a); in the second group, polysaccharides and glycoproteins (lectins) (Fig. 4b). The most potent compounds investigated in the author's

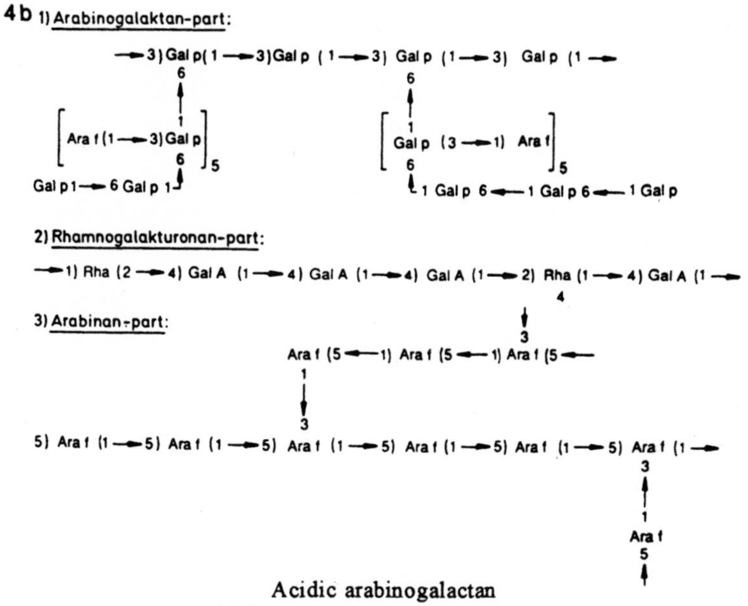

4b 1) Arabinogalaktan-part:

→3)Gal p(1→3)Gal p (1→3) Gal p (1→3) Gal p (1→
6 6
↑ ↑
[Ara f (1→3)Gal p]5 [Gal p (3→1) Ara f]5
6 6
Gal p 1→6 Gal p 1 1 Gal p 6←1 Gal p 6←1 Gal p

2) Rhamnogalakturonan-part:

→1) Rha (2→4) Gal A (1→4) Gal A (1→4) Gal A (1→2) Rha (1→4) Gal A (1→
4

3) Arabinan-part:

3
Ara f (5←1) Ara f (5←1) Ara f (5←
1
3
5) Ara f (1→5) Ara f (1→5) Ara f (1→5) Ara f (1→5) Ara f (1→5) Ara f (1→
3
1
Ara f
5

Acidic arabinogalactan
(*Echinacea purpurea* cell cultures)

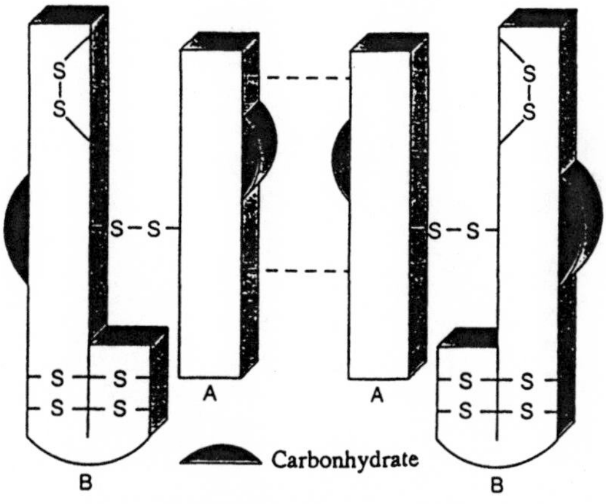

S-S S-S S-S S-S
A A
S S S S
S S S S
Carbonhydrate
B B

Molecular model of ß-galactose specific lectin I
of *Viscum album* (as dimer).

laboratory in various *in vitro* and *in vivo* models include: oxindol alkaloids of *Uncaria tomentosa*; naphthoquinones (e.g. plumbagin, lapachol); cichoric acid; alkylamides; acidic polysaccharides (pectins) of higher plants, algae, and fungi; and glycoproteins (e.g. lectins of *Viscum album* and *Urtica dioica*). The following observations are noted.

While investigating well known cytostatic or cytotoxic alkaloids we noticed that most of the compounds exhibiting the expected immunosuppressive effects at high doses possessed immunostimulating activities when applied at very low doses.[9] In the light of these findings it can be hypothesized that many plant derived anti-cancer drugs such as mistletoe (*Viscum album*), the South American lapacho (*Tabebuia avellanedae*), the extract of *Dionaea muscipula*, and others exert their antitumoral activities by a total or partial immune-induced mechanism of action. This mechanism of action has been confirmed in cancer patients for the β-galactose specific lectin I of mistletoe (Fig. 4b), which induces optimal immune response at a concentration of 1 ng/kg.[10]

The anti-neoplastic Bryostatins (Fig. 4a) from the marine organism *Budula neritina*, at present number 2 for clinical trials, were found in our assays to be powerful immunostimulating agents at concentrations of 10 ng/ml-1 μg[11,12] (Fig 5). Since the Bryostatins, in contrast to phorbol esters, lack complete tumor promoting potential and mimic many effects of the multipotential recombinant human granulocyte-macrophage colony stimulating factor (hGM-CSF), they may be candidates for a new promising concept of tumor therapy involving treatment with alternating dosages of Bryostatins.

Many polysaccharides of higher plants, e.g. those of *Echinacea spp., Achyrocline saturoioides, Chamomilla recutita, Sedum telephium* or *Urtica dioica*

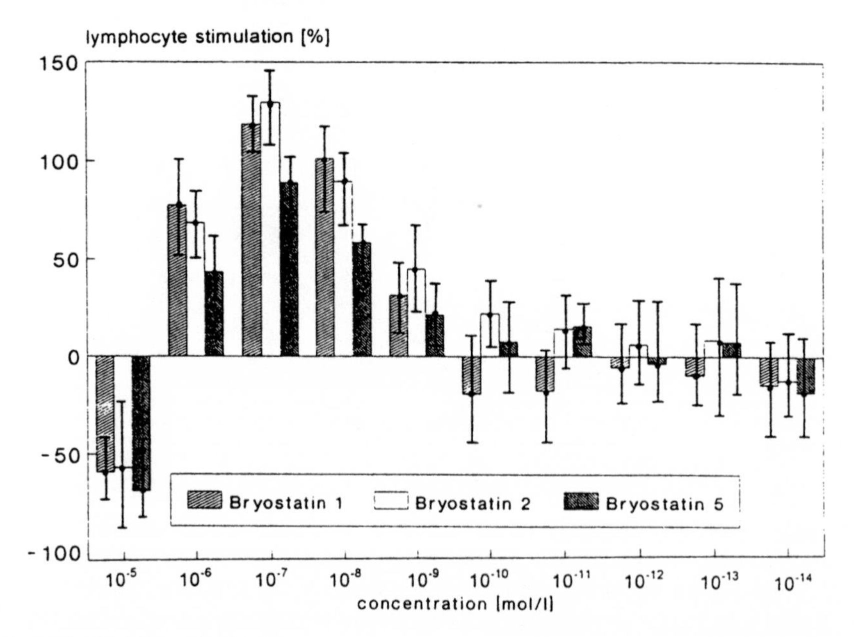

Figure 5. Dose dependent modulation of T-lymphocytes proliferation by various bryostatins after preincubation with Concanavalin A.

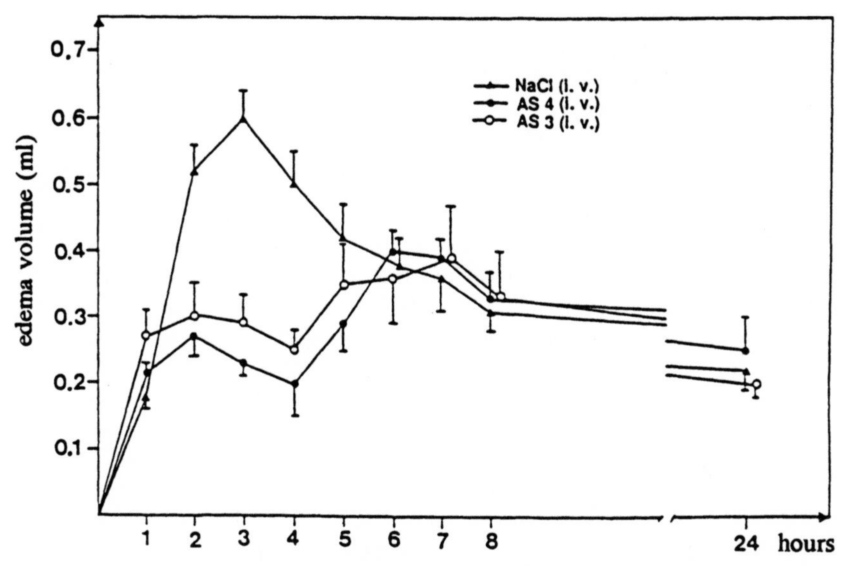

Figure 6. Inhibition of carrageenan induced rat paw edema by two acidic polysaccharides of *Achyrocline* satureioides (AS4, AS3). Conc. 3 mg/kg i.v.

were found to be powerful stimulators of various immune functions: phagocytosis, T-lymphocyte proliferation, release of cytokines (interferons, interleukins, TNF) and activation of complement factors.[1] An immunologically active arabinogalactan, produced by *Echinacea purp.* cells culture,[13,14,15] is scheduled for Phase II clinical studies. While such activities may explain the anti-infectious and antitumoral potentials of these polysaccharides, their influence on the complement cascades may also explain the additional anti-inflammatory activities reported for these plant extracts. Some polysaccharides, such as those from *Sedum telephium* and *Urtica dioica,* showed remarkable antiinflammatory activities, as measured in the carrageenan rat paw edema assay[1] (Fig. 6). While carrageenan (polysaccharide from red algae) is used for the generation of inflammations and edemas in experimental animal studies, some polysaccharides of higher plants were found to have the opposite antiinflammatory, effect.

ADAPTOGENS[16]

The term adaptogen was coined by the Russian scientist Lazarev in 1947 when he discovered the adaptogenic activity of Dibazol. It was later delineated by Brekhman[17] as follows: 1. an adaptogen shows a non-specific activity, i.e. it increases the power of resistance against physical, chemical or biological noxious agents; 2. an adaptogen must have a normalizing influence independent of the nature of the pathological state and; 3. an adaptogen must be innocuous and must not influence normal body functions more than required. Adaptogens can be best understood on the basis of Selye's General Adaptation Syndrome (GAS), which divided GAS into three individual states[18,19](Fig. 7).

The General Adaptation Syndrome

State of alarm defines an immediate response of an organism to stress. It results in a stimulation of the hypothalamus-pituitary-adrenal gland axis which, in turn, affects catabolic processes. The catecholamine and corticosterone level is increased. Cholesterol and ascorbic acid in the adrenals are decreased as are the weights of thymus, spleen and lymphatic glands. It has been suggested that, in analogy to microorganisms, so-called stress proteins are also produced by the body.[16]

State of resistance is elicited by the repeated or chronic exposure of a stressor signal. The catabolic state of the alarm phase is gradually replaced by anabolic functions. The organism then develops a certain habituation or adaptation, depending upon the nature of the stressor.[16]

State of exhaustion depends upon the individual magnitude of adaptation energy. The resistance power is exhausted and this is followed by a severe damage of organs or a total break-down[16].

Single administrations of an adaptogen cause alteration of endocrine functions of the pituitary-adrenal gland axis, e.g. an increase in the serum levels of ACTH and corticosterone.[16] Experimental animal studies show that subchronic pretreatment with adaptogens, however, causes a normalization of stress hormone levels and a generally decreased stress predisposition in behavioral tests.[16] In addition, there are several communication links between the endocrine, nervous

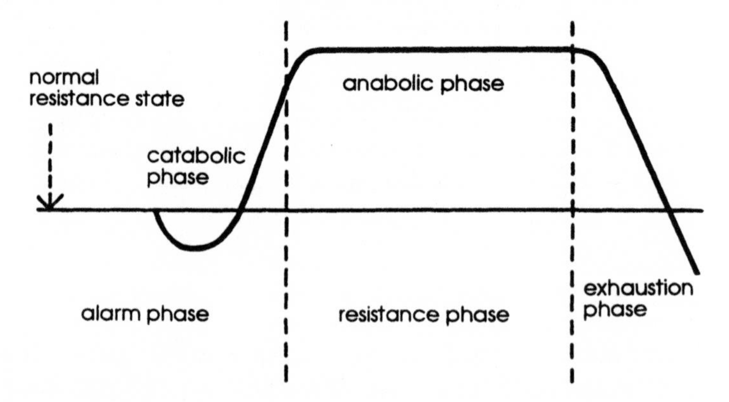

Figure 7. Specific resistance during the three states of the General Adaptation Syndome (GAS) [Seyle's Stress Model].

and immune systems, which have not yet been fully elucidated.[16] In this context it must be stressed that, similar to the immunostimulants, the result of an adaptogen application depends on the resistance state and the adaptation energy of an individual. Furthermore, the administered dose and the mode of application may influence the efficacy of an adaptogen.

With regards to Seyle's stress model it must be the general aim of an adaptogen application to reduce stress reactions during the alarm state of the stress response, or at least delay the establishment of the state of exhaustion, thus providing a certain level of protection against long-term stress.[18] In Brekhman's view the long-term effect of an adaptogen can be described as a potentiation or prolongation of physiological adaptation. He reasons that this effect can be attributed to attempts of the organism to protect energy resources from depletion and to accelerate the biosynthesis of proteins and nucleic acids.[17]

Drugs with Putative Adaptogen Activities[16]

As the term adaptogen is relatively new, it is not found in old treatises of drugs. I present here a retrospective classification of drugs as adaptogens based on empirical medical knowledge and, in some cases, on experimental data derived from *in vitro* and *in vivo* tests. Adaptogenic plant drugs belong to chemically diverse classes of compounds. They differ markedly in their chemical composition. The most important ones described in the literature are listed in Table 2.

Panax Ginseng[16]. In light of current knowledge the adaptogenic activity of ginseng can be ascribed to the so-called ginsenosides or panaxosides, glycosides of the tetracyclic dammaran type (Fig. 8). The adaptogenic efficacy of ginseng, i.e. the ability to increase the power of resistance, has been demonstrated in animal experiments with a variety of stressors. Antistress effects have been observed

Table 2. Major adaptogenic plants described in the literature[16]

Albizzia julibrissin	*Aralia elata (Miq.)*
Aralia manshurica	*Aralia schmidtii*
Cicer arietinum	*Codonopsis pilosula*
Echinopanax elatus	*Eleutherococcus senticosus*
Eucommia ulmoides	*Hoppea dichotoma*
Leuzea carthamoides	*Ocimum sanctum*
Panax ginseng	*Panax quiquefolium*
Rhodiola crenulata	*Rhodiola rosea*
Schizandra chinensis	*Tinospora cordifolia*
Trichopus zeylanicus	*Withania somnifera*

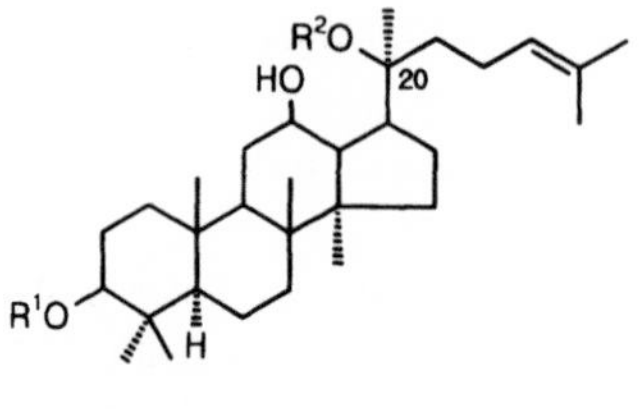

20 (S)-Protopanaxadiol

Ginsenoside	R^1	R^2
Rb_1	Glc^2-Glc	Glc^6-Glc
Rb_2	Glc^2-Glc	Glc^6-Ara (p)
Rc	Glc^2-Glc	Glc^6-Ara (f)
Rd	Glc^2-Glc	Glc

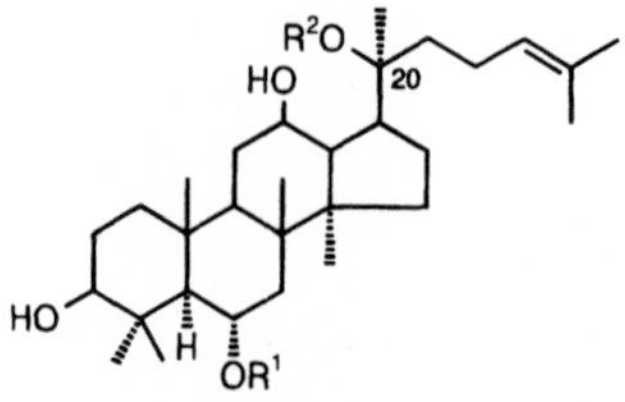

20 (S)-Protopanaxatriol

Ginsenoside	R^1	R^2
Re	Glc^2-Rha	Glc
Rf	Glc^2-Glc	H
Rg_1	Glc	Glc
Rg_2	Glc^2-Rha	H

Panax ginseng - Ginsenosides

Figure 8. Examples of some major compounds found in adaptogenic plant extracts.

from: aqueous extracts (against emotional stress - open field test, forced exercise stress, and hanging stress); root powder suspended in distilled water (against emotional stress - open field test, thirsty rat conflict test); and ginsenoside fraction (against cold stress, radio activity, ethanol treatment, heat stress, and hypoxia stress). Endocrinological investigations have shown that all preparations markedly elevated ACTH and corticosterone serum levels. Animal studies revealed antifatigue effects after administration of ginsenosides. In passive avoidance response tests with mice a marked improvement in learning capacities in the presence of stress was observed. Ginsenosides were found to enhance nerve growth factor-mediated outgrowth of neurites from cultures of embryonic brain cortex. Human studies revealed all typical symptoms of elevated corticoid levels (e.g. nervousness, sleeplessness) after overdoses of ginseng.

Eleutherococcus senticosus[16,21]. The main constituents of Siberian Ginseng differ markedly from those of the ginseng root. They can be classified into phenylpropane derivatives (e.g. syringin = eleutheroside B) (Fig. 8), lignane derivatives (e.g. syringaresinol-diglucoside = eleutheroside E(D)), coumarins, polysaccharides, and some less important minor components. Interesting antistress effects of *Eleutherococcus* have been observed from: root ethanolic extracts (against alloxane treatment, cytostatica treatment, cold stress and $NaClO_4$ treatment); root aqueous extracts (against both acute and chronic stress); and both eleutheroside B and eleutheroside E (against immobilization, acute, and chronic stress).

Recent experiments with primary cultures of rat pituitaries have shown a significant liberation of ACTH following addition of aqueous *Eleutherococcus*

8

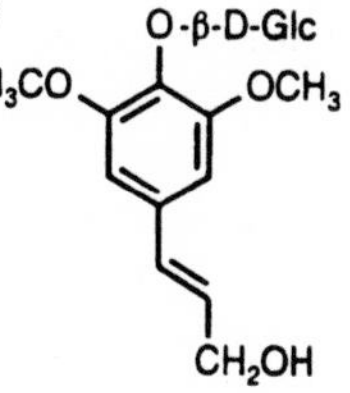

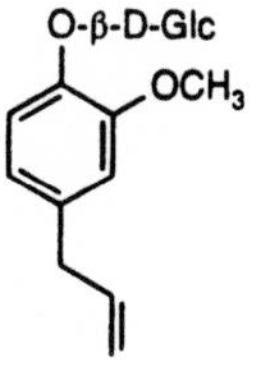

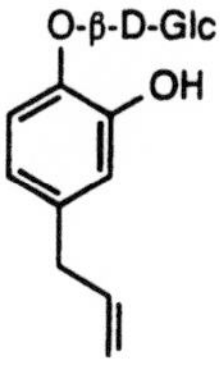

Eleutheroside B = Syringin
Eleutherococcus senticosus

Eugenol-1-O-ß-D-glucoside

4-Allyl-catechol-1-O-ß-D-glucoside

Ocimum sanctum

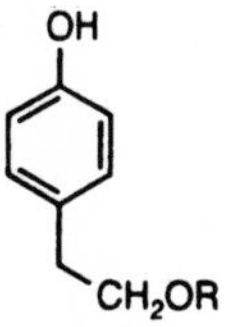

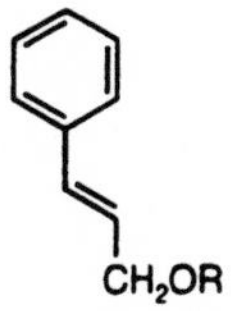

p-Tyrosol: R = H
Salidrosid: R = ß-D-Glc

Cinnamon alcohol: R = H
Rosavidin: R = ß-D-Glc-α-L-Ara

Rhodiola rosea

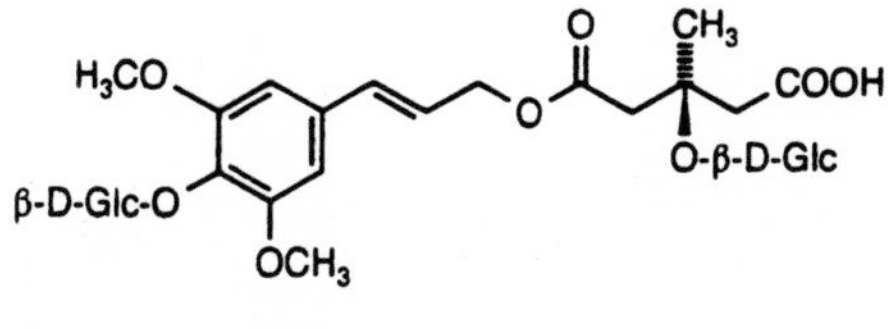

Codonopsis pilosula Tanshenoside I

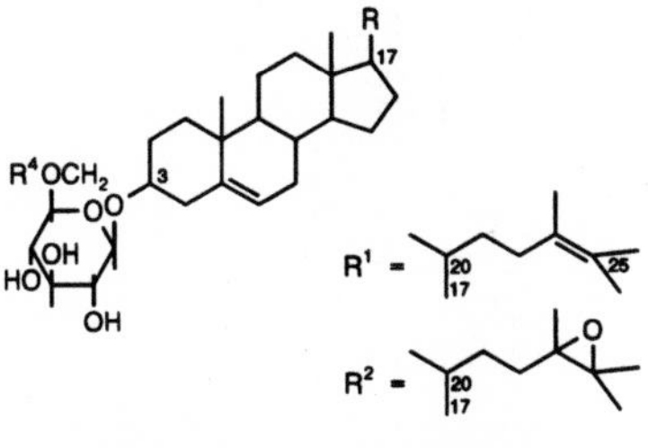

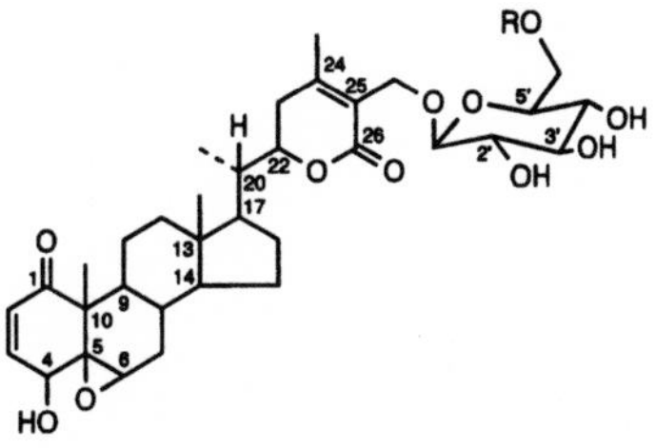

Sitoindoside VII : R = R^1 R^4 = palmitoyl
Sitoindoside VIII: R = R^2 R^4 = palmitoyl

Sitoindoside IX : R = H
Sitoindoside X : R = palmitoyl

Withania somnifera - Sitoindosides

extracts at doses of 0.1 mg/ml[22] (Fig. 9). Basal levels of luteinizing hormone secretion also were elevated significantly. *In vivo* experiments with rats reveal that a single intraperitoneal dose of an aqueous extract standardized for eleutheroside B and E at a dose of 3 mg/kg significantly enhances the liberation of corticosterone while a subchronic administration of the same extract (i.p. 3 mg/kg or p.o. 500 mg/kg) does not lead to any significant alterations in either ACTH or corticosterone levels in body or organ weights after seven weeks.[22] A remarkable finding in these experiment was the observation that elevations of corticosterone serum levels induced by mild stress were suppressed significantly in animals treated either subchronically intraperitoneally or per os. These animals also were less stress-prone in behavioral test models.

Investigations carried out mainly by Brekhman have shown that *Eleutherococcus* improves the physical performance of humans and mice.[17] They also found characteristic antistress effects in mice when they administered eleutheroside B. The *Eleutherococcus* plant drug is an example of the bifunctionality of a drug. A double blind study with 36 human subjects showed that *Eleutherococcus* improves non-specific immune reactivities as determined by quantitative flow cytometry. The immunocompetent cells, in particular those of T-lymphocytes and natural killer cells, were found to be markedly increased after administration of the extracts for four weeks.[20] It is likely that these effects are generated by the *Eleutherococcus* polysaccharides, and not by the low molecular

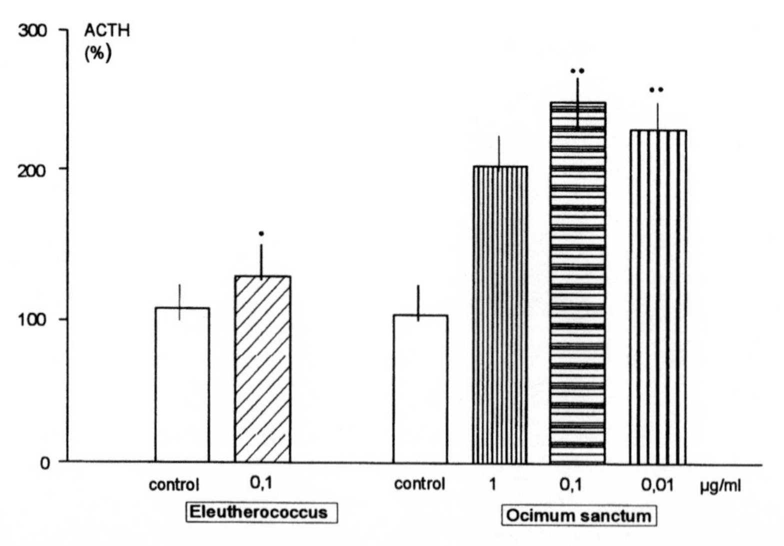

Figure 9. Induction of ACTH secretion in primary cultures of pituitary cells after incubation with aqueous extracts of *Eleutherococcus senticosus* and *Ocimum sanctum*. Stress injection = physiol. salin. injection.

weight compounds, because only the polysaccharides, not the phenylpropane derivatives, were found also to exert immune stimulating activity.

Other Adaptogenic Drugs[16]. Among the other plant drugs, four deserve to be mentioned: *Ocimum sanctum, Codonopsis pilosula* (Dangshen), *Rhodiola rosea* and *Withania somnifera*.

Ocimum sanctum contains a volatile oil consisting of a high percentage of phenyl propane derivatives such as methyleugenol, a great variety of flavonoids, and phenylpropane-glucosides[16] (Fig. 8). The "antistress" activity of the extracts has been evaluated in numerous *in vitro* and *in vivo* assays. As far as the influence on the endocrine system is concerned, a significant liberation of ACTH was observed in primary cultures of pituitary cells after incubation with aqueous extracts[22] (Fig. 9). The various measured effects on the CNS suggest a possible dopaminergic influence. *In vivo* investigations carried out with various stress factors (CCl_4, cold stress, immobilization stress, swim test) showed a remarkably enhanced resistance of animals having received the extract. An immunostimulatory effect in albino rats after oral doses of water and alcoholic extracts was also observed.

Codonopsis pilosula contains several major substances.[16] They include phenylpropane glycosides (syringin and tanghenoside I) (Fig. 8), lignans (e.g. pinoresinol), triterpenes, alkaloids, and the recently found polyacetylene compounds. As with *Ocimum*, a marked secretion of ACTH in primary cultures of rat pituitary cells has been found, and in mice the corticosteroid level increased after p.o. (per oral), i.p. (intra peritoneal) or i.v. (intra venous) administration of an alcoholic extract. Characteristic antistress activities, however, were observed only in some experiments.

Rhodiola rosea is used by native peoples of Siberia and Mongolia to prevent fatigue and a general disinclination to work.[16] Apart from salidroside (p-tyrosol glucoside), cinnamon alcohol glycosides are believed to be the major active components (Fig. 8). Rosavidin, the cinnamyl-O-(6'-O-L-arabinosyl)-D-glucoside deserves special mention. Other constituents are p-tyrosol and cinnamon alcohol, volatile oils, flavonoids, anthraglycosides and triterpenes. Increased resistance against electro trauma and other stressors has been observed after oral administration of extracts, in particular, salidroside and various cinnamylglycosides. *In vitro* ACTH-release in pituitary cells and various stimulatory effects on the CNS (improved learning behavior and memory in mice models) were found.

Withania somnifera roots are said to act as a tonic to protect the body against diseases by "maintaining a healthy balance of body powers".[16] In addition to several alkaloids the roots contain the steroid lactone Withaferin A, and related Withanolides (Fig. 8). Measurements of physical perseverance in swim tests in mice pretreated with "Withania" extract showed almost a doubling of staying time. Anabolic as well as immunomodulating activities were demonstrated. The anti-depressant effect also observed may be due to a reduction in stress efficacy or influences on monoamine metabolism in the brain. The marked improvement observed in short-term and long-term memory, after peroral administration of withanolides, also suggests that these steroids are, at least in part, responsible for the adaptogenic activity of the extract.

Verification Methods of Adaptogens[16]

Since adaptogens are expected to cause nonspecific resistance against any kinds of stressors, experimental animals are usually pretreated with the putative adaptogen and then exposed to stressor stimuli. Alterations in general resistance against stressors observed in these animals are compared with control groups. Enhanced power of resistance can manifest itself by a variety of phenomena, including: prolonged maintenance of body temperature following cold temperature stress; improvement in coordinate functions; improved cognitive abilities; increase in locomotor and explorative activities; improvement in emotional behavior; prevention of stomach ulcers by aspirin, cold stress, or immobilization; decrease in milk-induced leucocytis; and improvement of general immune defenses.

The relationship between stress and the resulting pituitary release of ACTH or adrenergic corticoid production is schematically depicted in Fig. 10. Elevated levels of ACTH and corticosterones observed after administration of test drugs are usually taken as endocrinological evidence for the adaptogenic activities of the test drug. At the same time measurements of ACTH and corticosterone levels in stress models can be valuable parameters allowing evaluation and monitoring of the stress disposition of animals, particularly if combined with stress behavioral studies. In addition to tests measuring altered performance, anabolic efficacy tests and tests measuring alterations in brain metabolism are of great value because they give a better view of the total potential of the adaptogens.

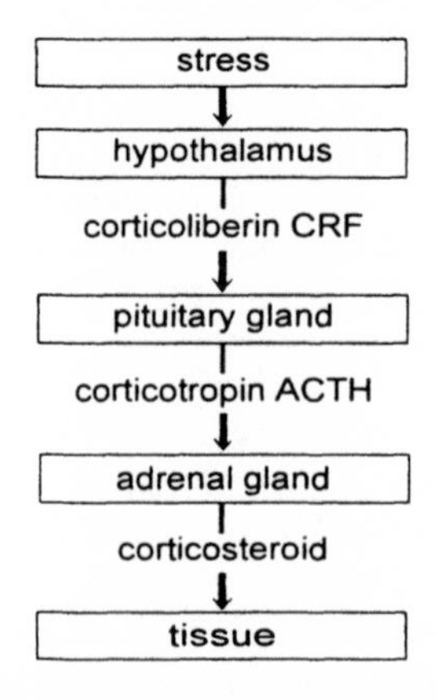

Figure 10. Reaction elicited by stress: Stress and the hypothalamus pituitary adrenal axis.

SUMMARY AND CONCLUSION

In summarizing the present state of the research on immunostimulating drugs, we can conclude that at present the application of plant extracts can be recommended only for the therapeutic treatment of certain afflictions. Restrictions are needed mainly due to the lack of chemically or biologically standardized phytopreparations. A rational therapy for severe diseases requires chemically well defined pure compounds, whether used alone or in combination with other chemotherapeutics. With the exceptions of those diseases for which vaccine treatment is obligatory, immunostimulating drugs may be sufficient for preventive purposes and adjuvant therapy. As far as the adaptogens are concerned, more information on the active principles of the plant extracts and additional efficacy studies are necessary to rationalize their applications. It is expected that the major areas of use will be for prevention purposes.

REFERENCES

1. WAGNER, H. 1990. Search for plant derived natural products with immunostimulatory activity (recent results). Pure & Appl. Chem. 62: 1217-1222.
2. BAUER. R., WAGNER, H. 1990. Echinacea. Handbuch für Ärzte, Apotheker und andere Naturwissenschaftler. Wiss. Verlagsgesellschaft Stuttgart.
3. BAUER, R., WAGNER, H. 1991. Echinacea species as potential immunostimulatory Drugs. In: Economic and Medicinal Plant Res., vol. 5 (H. Wagner, H. N. Farnsworth, eds.). Academic Press, London pp. 253-321.
4. BRÄUNING, B., KNICK, E. 1993. Therapeutische Erfahrungen mit *Echinacea pallida* bei grippalen Infekten. Naturheilpraxis 1: 72 -75.
5. BENDEL, R., BENDEL, V., RENNER, K., CARSTENS, V., STOLZE, K. 1989. Zusatzbehandlung mit Esberitox N bei Patientinnen mit chemo-strahlen-therapeutischer Behandlung eines fortgeschrittenen Mamma-Karzinoms. Onkologie 12: 32-38.
6. WAGNER, H., JURCIC, K. 1991. Assays for immunomodulation and effects on mediators of inflammation. Methods in Plant Biochemistry (ed. K.Hostettmann) Academic Press (London) 6: 195-217.
7. STIMPEL, H., PROKSCH, A., WAGNER, H., LOHMANN-MATTHES, M.L. 1984. Macrophage activation and induction of macrophage cytotoxicity by purified polysaccharide fractions from the plant *Echinacea purpurea*. Infect. Immun. 46: 845-849.
8. WAGNER, H., PROKSCH, A. 1985. Immunstimulatory drugs of fungi and higher plants. In: Economic and Medicinal Plant Res. (H. Wagner, N. Farnsworth, eds.). Academic Press, London, p. 113-153.
9. WAGNER, H., KREHER, B., JURCIC, K. 1988. In vitro stimulation of human granulocytes and lymphocytes by pico- and femtogramm quantities of cytostatic agents. Arzneim. Forsch. Drug Res. 38: 272-275.
10. HAJTO, T., HOSTANKA, K., GABIUS, H.J. 1989. Modulatory potency of the ß-galactoside-specific Lectin from mistletoe extract (Iscador) on the host defense system in vivo in rabbits and patients. Canc. Res. 49: 4803-4808.
11. PETTIT, G.R. 1991. Bryostatins. In: Progress in the Chemistry of Organic Natural Products Vol 57 (W. Herz, G.W. Kirby, W. Steglich, T. Tamm, eds.). Springer Verlag, Wien-New York, pp. 153-195.

12. EISEMANN, K., TOTOLA, A., JURCIC, K., PETTIT, G.R., WAGNER, H. Bryostatins 1, 2 and 6 as activators of human granulocytes and lymphocytes - in vitro - and in vivo-Studies. Pharm. Pharmacol. Letters (in press).
13. WAGNER, H., STUPPNER, H., SCHÄFER, W., ZENK, M. 1988. Immunologically active polysaccharides of *Echinacea purpurea* cell cultures. Phytochemistry 27: 119-126.
14. LUETTIG, B., STEINMÜLLER, C., GIFFORD, G.E., WAGNER, H., LOHMANN-MATTHES, M.-L. 1989. Macrophage activation by the polysaccharide arabinogalactan from the plant cell cultures of *Echinacea purpurea*. J. Nat. Cancer Inst. 81: 669-675.
15. ROESLER, J., STEINMÜLLER, CH., NIDERLEIN, A., IMMENDÖRFER, A., WAGNER, H., LOHMANN-MATTHES, M.-L. 1991. Application of purified polysaccharides from cell cultures of the plant *Echinacea* to mice mediates protection against systemic infections with Listeria monocytogenes and Candida albicans. Int. J. Immunopharm. 13: 27-37.
16. WAGNER, H., NÖRR, H., WINTERHOFF, H. 1994. Plant Adaptogens. Phytomedicine, 1: 63-76.
17. BREKHMAN, I.I. 1980. Man and Biologically active substances, the effect of Drugs, Diet and Pollution and Health. Pergamon Press Ltd., Oxford.
18. SEYLE, H. 1936. A syndrom produced by diverse nocuous agents. Nature 138: 32.
19. SEYLE, H. 1937. Studies on Adaptation. Endocrinology 21: 169-188.
20. BOHN, B, NEBE, C.T., BIRR, C. 1987. Flow Cytometric studies with *Eleutherococcus senticosus* extract as an immunmodulatory agent, Arzneimittel-Forsch. (Drug Res.) 37: 1193-1196.
21. FARNSWORTH, N.R., KINGHORN, A.D., SOEJARTO, D.D., WALLER, D.P. 1985. Siberian Ginseng (*Eleutherococcus senticosus*): Current Status as an Adaptogen. In: Economic and Medicinal Plant Research, Vol 1. (H. Wagner, H., N.R. Farnsworth, eds.), Academic Press, London, pp. 155-215.
22. WINTERHOFF, H., GUMBINGER, H.G., VAHLENSIECK, U., STREUER, M., NÖRR, H., WAGNER, H. 1993. Effects of *Eleutherococcus senticosus* on the pituitary-adrenal system of rats. Pharm. Pharmacol. Lett. 3: 95-98.

Chapter Two

NATURAL PRODUCT CANCER CHEMOPREVENTIVE AGENTS

John M. Pezzuto

Department of Medicinal Chemistry and Pharmacognosy
Program for Collaborative Research in the Pharmaceutical Sciences
College of Pharmacy and
Department of Surgical Oncology
College of Medicine
University of Illinois at Chicago
Chicago, Illinois 60612

INTRODUCTION

At the current time, cancer claims the lives of approximately seven million people worldwide on an annual basis. In the United States alone, there are approximately one million new cases diagnosed each year, and approximately one-half

Phytochemistry of Medicinal Plants
Edited by John T. Arnason et al., Plenum Press, New York, 1995

million succumb to the disease. However, various cancer causes and methods of prevention are now obvious, and this knowledge should be brought to bear by members of an enlightened society. As an example, over 100,000 individuals in the United States die per year due to the manifestations of lung cancer, and a large percentage of these deaths could undoubtedly be negated by abolishing the smoking of cigarettes. In fact, the National Cancer Institute has devised a campaign in which the goal is to reduce the 1985 cancer mortality rate by 50% by the year 2000.[1] The approach to achieving this goal is understandably comprehensive. In addition to primary prevention strategies (e.g., cessation of cigarette smoking, reduction of exposure to chemical carcinogens), elements such as early diagnosis, dietary modification, and cancer training programs will need to be emphasized.[2]

An adjunct approach to reducing the incidence of cancer is chemoprevention.[3-5] Cancer chemoprevention is a term that was coined by Dr. Michael B. Sporn as part of his classical work dealing with retinoids and cancer prevention. It may be defined in general terms as the prevention of cancer in human populations by ingestion of chemical agents that inhibit the process of carcinogenesis. Various groups of compounds have been classified as cancer chemopreventive agents,[6] largely based on the results of animal studies. Of key importance, of course, is the potential of these agents to affect the incidence of cancer in human populations.

Cancer chemoprevention in humans remains largely speculative. However, epidemiological data have suggested inverse correlations between human cancers and various dietary constituents. Particularly strong epidemiological evidence has been provided suggesting an inverse correlation between lung cancer and consumption of carotene-rich foods among smokers.[7,8] One study involved over 250,000 subjects,[9] and similar conclusions regarding smoking and carotene consumption have been reached in cohort and case-control studies. These types of observations are not limited to carotenoids in that other epidemiological evidence suggests an inverse correlation between vitamin C and esophageal and stomach cancers, selenium and various types of cancer, vitamin E and lung cancer, protease inhibitors and breast, colon, prostate and oral cancers, and folic acid and cervical dysplasia.[10-14]

While the health benefits of cancer chemoprevention are now generally recognized with human populations, large-scale rational drug discovery has not been undertaken. Thus, it is reasonable to ask "Can additional cancer chemopreventive agents be discovered that will be useful adjuncts or replacements for the agents in the current armamentarium?" The focus of this manuscript will be to describe our efforts in this area of drug discovery and to briefly overview some recent results.

NATURAL PRODUCT CANCER CHEMOPREVENTIVE DRUG DISCOVERY

Certain aspects of natural product drug discovery programs conducted at the University of Illinois at Chicago have been described in the literature[15-22] as has the overall philosophical approach of natural product drug discovery.[23] For individuals who have not previously been exposed to natural product drug discovery programs, the rationale of such an endeavor may not be obvious. For example, one may ask, "Why should a plant produce a drug that is useful for the treatment or prevention of human cancer?" An attempt to answer such a question is beyond the scope of this discussion, but it seems sufficient to point out that numerous drugs used in modern day medical practice are either chemical modifications of natural products or natural products themselves.[24] Based on this information and the structural diversity of chemical entities that have been identified, it is likely that additional compounds possessing chemopreventive activity (which have yet to be uncovered and characterized) are present in plant materials, and our goal is the discovery and characterization of these agents.

The experimental approach regarded as most practical for natural product drug discovery is bioassay-directed fractionation. In this endeavor, starting materials are selected (e.g., random collection, information derived from ethnomedical systems of medicine, or literature surveillance),[25] and extracts are prepared that are suitable for biological evaluation. The materials are then tested in a bioassay system, and substances demonstrating a positive response are considered as active leads. After a number of active leads are identified, decisions are made to fractionate the most promising materials. Each fraction is monitored for potential to mediate a positive response in the bioassay test system, and this process continues until a pure active substance is obtained. The resulting substance is then subjected to procedures of structure elucidation. Once an active isolate is obtained, more thorough biological evaluation procedures are often performed and, based on the accumulated data, the material is a candidate for more advanced testing and development. This is the general approach we employ for the discovery and development of cancer chemopreventive drug discovery.

PROGRAMMATIC RESEARCH: A KEY TO SUCCESS IN DRUG DISCOVERY

As illustrated in Figure 1, we have established a research program comprised of five separate projects. Each project is directed by a leading expert in the field. Expertise is represented in the areas of agriculture, botany, biochemistry, carcinogenesis, computer-assisted literature surveillance, database management, epidemiology, nutrition, pharmacology, phytochemistry, structure elucidation, and synthetic organic chemistry. Such a multidisciplinary approach is required to realize progress in this area of research, and each project contributes to the program in an obligatory and synergistic manner. An overview of the respective projects follows.

Project 1 (Provision of Source Materials, Dr. Norman R. Farnsworth, Project Leader, Drs. C.W.W. Beecher, H.H.S Fong, and D.D. Soejarto, Co-Investigators). Utilizing the resources of the University of Illinois Pharmacognosy Field Station and field collection, this project provides sufficient quantities of

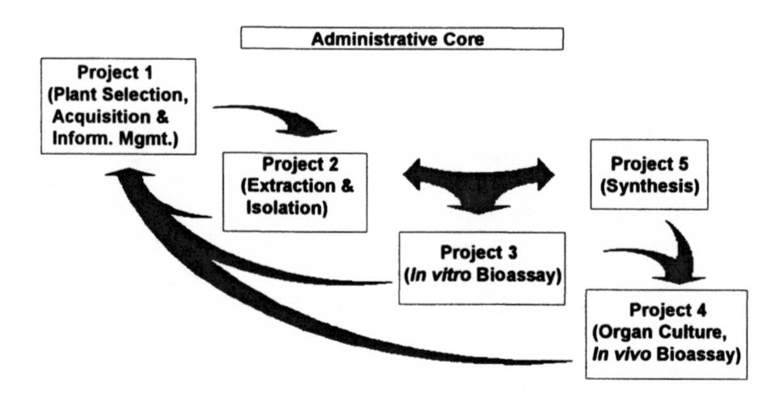

Figure 1. Interrelationships of projects aimed toward the discovery of natural product cancer chemopreventive agents.

plant materials for the overall program. An extensive network of international collaborators has been established to provide plant material, and the botanical characteristics of the resulting materials are unequivocally established and archived. This sometimes involves the use of the herbarium of Chicago's Field Museum of Natural History. As part of the selection process, this project has developed and maintains an information database of chemopreventive literature and all potential cancer chemopreventive compounds. These datasets are used to provide dereplication support on all samples found active. Finally, this project collects all of the data generated by the program into a single database that is available to all investigators.

Project 2 (Isolation and Identification, Dr. A. Douglas Kinghorn, Project Leader). Based on the results of bioassays (conducted in projects 3 and 4), plant materials are fractionated to permit the isolation of active principles. The structures of these isolates are elucidated by comtemporary spectroscopy or other relevant methods. This bioassay-directed approach has been shown previously to limit the scope of such investigations to biologically active chemicals. Experimental procedures are also interfaced with computer-assisted dereplication, and the use of LC/MS is becoming increasingly important. Isolates are sometimes provided to project 5 for the preparation of derivatives that are used in the investigation of structure-activity relationships.

Project 3 (*In Vitro* Evaluation, Dr. John M. Pezzuto, Project Leader). Employing short-term bioassays, the potential of plant extracts, fractions or compounds (supplied by projects 2 or 5) to mediate responses consistent with cancer chemoprevention is determined. The assays are selected to correlate with the studies conducted in project 4, and include procedures relevant to inhibition of tumor initiation, promotion and progression. Appropriate assay systems are selected to guide fractionation procedures, and the experimental procedures also are used to assess the mechanism of novel isolates.

Project 4 (Carcinogenesis Inhibition, Dr. Richard C. Moon, Project Leader, Dr. R.G. Mehta, Co-Investigator). Plant extracts, fractions or compounds (supplied by projects 2 or 5) are evaluated for potential to inhibit 7,12-dimethylbenz(a)anthracene (DMBA)-induced nodule-like alveolar lesions in mammary gland organ culture. Additionally, plant extracts or compounds are assessed for potential to inhibit tumorigenesis in two primary animal models (mouse skin, rat mammary). The data obtained in mammary organ culture and short-term assay procedures are of utmost importance in deciding (a) which plants to subject to bioassay-directed fractionation procedures, or (b) which plant materials or compounds to evaluate in full-term tumorigenesis studies. Evaluation of tumor inhibition is obviously necessary to establish the efficacy of any test substance. This activity also is of importance in verifying the effectiveness of our selection and bioassay procedures (projects 1 and 3).

Project 5 (Synthesis and Modification, Dr. Robert M. Moriarty, Project Leader). One objective of project 5 is the preparation of a series of compounds to investigate the structural requirements of cancer chemoprevention. As novel chemopreventive agents are uncovered, derivatives are prepared for similar mechanistic evaluations that are performed in projects 3 and 4. Another important aspect of the project is the synthetic production of large quantities of novel chemopreventive agents for structural confirmation or evaluation of tumor inhibition with laboratory animals.

In addition to these five projects, the program contains an administrative core that performs organizational and managerial tasks. One important member of the administrative core is Dr. Samath Hedayat. Dr. Hedayat manages all biostatistical aspects of the program.

OVERALL EXPERIMENTAL APPROACH FOR THE DISCOVERY OF CANCER CHEMOPREVENTIVE AGENTS

In the area of antitumor drug discovery, a large number of *in vitro* test systems and the issues that need to be considered when attempting to interrelate *in vitro* test results with *in vivo* efficacy studies have been described.[23] In general, recent advances in molecular biology have led to a greater understanding of the molecular basis of human disease states. As a correlate, *in vitro* systems that monitor a response that is either closely related to or identical with the molecular event yielding the disease condition can be devised. As summarized by Angerhofer and Pezzuto,[26] these systems can be employed to rapidly and inexpensively assess the biological potential of a large number of test materials.

In the area of cancer chemoprevention, definition of the "best" bioassay system remains subjective. However, we have devised a series of assays and an experimental approach to permit the discovery of chemopreventive agents from

Table 1. Isolation of biologically active substances from plant material using inhibition of MNU-induced mammary cancer as a model system for bioassay-directed fractionation

Sample	Test Material	No. Animals[a]	Time Required
Original extract	2 fractions	270	8 months
Column 1	10 fractions	1,100	8 months
Column 2	10 fractions	1,100	8 months
Column 3	10 fractions	1,100	8 months
Column 4	10 fractions	1,100	8 months
Column 5	10 fractions	1,100	8 months
Isolates	4 compounds	500	8 months
Total		6,270	56 months

[a]Based on the results of preliminary work and evaluation of test substances at approximately 0.8x maximum tolerated dose.

natural sources. When designing this program, it first seemed logical to consider the use of animal models wherein a chemically-induced tumor can be reproducibly generated and wherein blocking agents can be studied under well-defined conditions. However, using the standard approach of bioassay-directed fractionation, a "typical" fractionation procedure using *N*-methyl-*N*-nitrosourea (MNU)-induced mammary cancer as a model system would require the use of over 6,000 rats, more than 1 kg of nontoxic test material, and a time period of over four years for each plant (see Table 1). Clearly, under normal circumstances this is not acceptable.

An alternate system for detecting a physiological response that is indicative of cancer chemoprevention involves organ cultures of mammary glands. In response to carcinogen treatment, the mammary gland undergoes a preneoplastic change, and transplantation of cells prepared from these lesions form adenocarcinomas in syngeneic mice. As summarized in Table 2, there is an excellent correlation between positive inhibitors in the mammary organ culture test system and positive inhibitors of carcinogen-induced mammary tumors in rodents. In fact, several agents which have shown positive *in vivo* results were selected initially on the basis of the results obtained with mammary gland organ cultures. As described below, this point also is exemplified by the current program project. Both brassinin and *Mundulea sericea* were selected due to their impressive activities in mammary organ cultures, and as evident from the currently available results both substances have shown effective *in vivo* activities.

Thus, there is little doubt that organ culture provides an excellent tool to select potential chemopreventive agents effectively. Of particular note, in contrast

Table 2. Effectiveness of natural product chemopreventive agents in mammary gland organ cultures and comparison with *in vivo* rat mammary carcinogenesis and antimutagenesis

Compound	*In Vivo* Mammary Carcinogenesis	DMBA-Induced Mammary Lesions in Organ Culture	Antimutagenic Activity
Esculetin	ND[a]	Effective	Effective
Ajoene	Ineffective	Ineffective	Effective
Brassinin	Effective	Effective	ND
β-Carotene	Effective	Effective	Effective
Catechin	ND	Effective	Effective
Curcumin	Effective	Effective	Effective
Diallyl disulfide	Effective	Ineffective	ND
Erythoxydiol X	ND	Effective	Effective
β-Glycyrrhetinic acid	Effective	Effective	Effective
Limonene	Effective	Effective	ND
Nordihydroguaiaretic acid	Marginally effective	Effective	Effective
Purpurin	In Progress	Effective	Effective
Silymarin	Ineffective	Effective	ND
β-Sitosterol	Ineffective	Effective	Effective
Taurine	ND	Ineffective	ND
α-Tocopherol acetate	Marginally effective	Marginally effective	Effective

[a]ND = Not Determined

to full-term carcinogenesis inhibition systems, evaluations can be performed in the mammary organ culture test system with only a few milligrams of test substance. Moreover, results can be obtained in approximately one month, the expense is relatively low, and the required number of laboratory animals is kept to a minimum. Nonetheless, as summarized in Table 3, mammary organ culture is not generally suitable for bioassay-directed fractionation procedures, due to time, expense, and the number of mice required.

Thus, short-term procedures suitable for the discovery and characterization of cancer chemopreventive agents are necessary but not prevalent. A few examples follow. Cassady *et al.*[27] investigated the effects of chemopreventive agents on carcinogen binding and metabolism with cultured CHO cells, Muto *et al.*[28] studied the inhibition of TPA-induced early antigen of Epstein-Barr virus in culture, and Bertram *et al.*[29] reported inhibition of C3H/10T½ cell transformation by enhancing gap junction communication. The experimental approach we have developed is summarized in Table 4. In brief, plant extracts (projects 1 and 2) are

Table 3. Isolation of biologically active substances from plant materials using mammary organ culture as a model system for bioassay-directed fractionation

Sample	Test Material	No. Animals[a]	Time Required
Original extract	2 fractions	180	2 months
Column 1	10 fractions	780	2 months
Column 2	10 fractions	780	2 months
Column 3	10 fractions	780	2 months
Column 4	10 fractions	780	2 months
Column 5	10 fractions	780	2 months
Isolates	4 compounds	330	2 months
Total		4,410	14 months

[a]Based on the use of 15 glands per test substance (5 concentrations).

screened through a battery of *in vitro* systems (project 3) and active leads are then evaluated in the mammary organ culture system described above (project 4). Active leads in both the mammary organ culture and an *in vitro* test system are then subjected to bioassay-directed fractionation (project 2) utilizing an *in vitro* test system as a monitor. Once pure active principles are identified, they are tested in mammary organ culture for efficacy. Based on these data, compounds are considered for large-scale isolation (projects 1 and 2), synthesis or synthetic

Table 4. Experimental approach for the discovery of cancer chemopreventive agents from natural product source materials

1. Select and procure plant material.
2. Prepare extracts.
3. Conduct short-term *in vitro* bioassays.
4. Evaluate *in vitro* active leads as inhibitors of DMBA-induced lesions in mammary organ culture.
5. Active in 3 and 4: Isolate active principles utilizing select short-term *in vitro* tests.
6. Test resulting active principles in mammary organ culture.
7. Active in 6: Explore structural aspects of activity and chemical synthesis.
8. Test resulting active principles in full-term tumorigenesis systems.
9. Active in 8: Consider as candidates for development.

manipulation (project 5), mechanistic evaluation (projects 3, 4 or outside collaborations), or more advanced animal testing (project 4), assuming sufficient material is available.

This experimental approach recognizes the necessity of utilizing *in vitro* bioassay procedures but is designed to limit the isolation of "false positives" (i.e., compounds capable of mediating a positive response with an *in vitro* system but not an *in vivo* system). The approach has been validated utilizing a number of *in vitro* assays as monitors.[30] Plant antimutagenic compounds are one good example.[19] In this case, there is little probability that the random discovery of an agent demonstrating *in vitro* antimutagenic activity would lead to the provision of a useful cancer chemopreventive agent. However, correlation of short-term *in vitro* activity with a response that is more physiologically indicative of cancer chemoprevention, *viz.*, inhibition of carcinogen-induced lesions in mammary organ culture, should yield useful compounds. As described recently, certain plant-derived antimutagens also demonstrate activity in the organ culture system. Based on the evaluation of approximately 70 compounds (see Table 4 and Shamon *et al.*[31]), a correlation of approximately 80% has been established (i.e., active in both test systems or inactive in both test systems). Due to the magnitude of this correlation, we believe it is valid to use inhibition of DMBA-induced mutagenicity as a monitor to conduct bioassay-directed fractionation of plant materials that are active *in both* test systems. Further, we are confident that the overall strategy of correlating activities will continue to yield compounds that will demonstrate the desired chemopreventive activity.

BIOASSAY PROCEDURES CURRENTLY USED FOR ACTIVITY-DIRECTED ISOLATION OF CANCER CHEMOPREVENTIVE AGENTS

In designing our program for activity-directed isolation of cancer chemopreventive agents, several factors were taken into account: (a) Known principles of carcinogen action, metabolism and detoxification, (b) the need for establishing new methodology for the discovery of novel chemoprevention agents, (c) our area of expertise, (d) interrelationships with projects in the program, and (e) the level of effort that can be committed to this element of the program. First consider the general stages of chemically-induced carcinogenesis. As illustrated in Figure 2, agents can be broadly classified as inhibitors of carcinogen formation (e.g., ascorbic acid, tocopherols, phenols), inhibitors of initiation (e.g., phenols, flavones, aromatic isothiocyanates, diallyl disulfide, ellagic acid, antioxidants, glutathione, S_2O_3), and inhibitors of post-initiation events (e.g., β-carotene, retinoids, tamoxifen, dehydroepiandrosterone, terpenes, protease inhibitors, prostaglandin inhibitors, Ca^{2+}, nerolidol). In addition, certain agents may demonstrate pleiotropic mechanisms of action, and combinations of agents may demonstrate chemopreventive activity in a synergistic manner.[32] We have elected

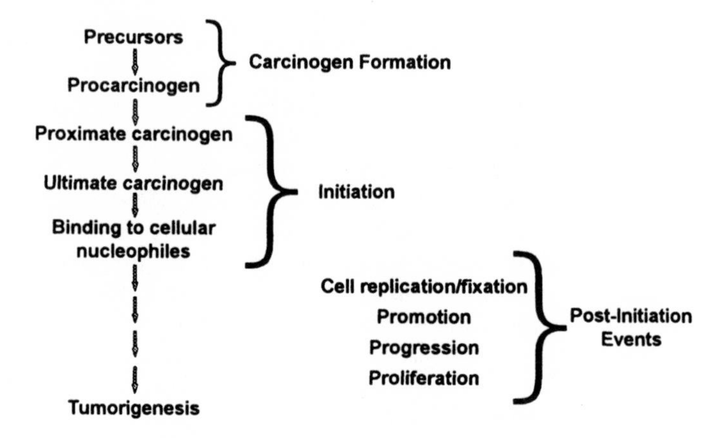

Figure 2. Stages of chemical carcinogenesis.

to conduct a broad-based drug discovery program in which assays have been devised to monitor inhibitors of initiation (i.e., DNA damage) and post-initiation events (i.e., promotion and progression). This rather comprehensive range of endpoints has been permitted by improvements in methodology. Nearly all of the procedures currently in use have been adapted to a 96-well microtiter plate format, and semi-automated data processing has been implemented. The ability to evaluate a greater number of samples in a greater range of assays should distinctly improve the chances of discovering drugs of potential clinical value. A brief description of the specific assays and the rationale involved in their selection follows.

Assays Based on Factors Associated with Tumor Initiation

Antimutagenicity. As alluded to above, we have performed an extensive series of studies to investigate the relationship between antimutagenic activity and chemopreventive activity observed in the mammary organ culture system. It is certain that a large number of compounds classified as antimutagens, isolated solely on the basis of an antimutagenic response observed with an *in vitro* test system, could not demonstrate chemopreventive activity. Since the process of antimutagenicity is obviously much less complex than cancer chemoprevention, the basis of this statement is quite obvious. However, with reasonable certainty, we have shown that inhibition of DMBA-induced mutagenicity with *Salmonella typhimurium* strain TM677 correlates with inhibition of DMBA-induced preneoplastic lesion formation in mammary organ culture. This is not to suggest that a majority of antimutagens will demonstrate such activity, but it is reasonable to suggest the opposite, i.e., that a majority of compounds inhibiting lesion formation will also demonstrate antimutagenicity in this test system. Therefore, it follows that it is sensible to conduct bioassay-directed fractionation using antimutagenicity as a monitor with plant extracts that have been shown to be active in the mammary organ culture test system.

Induction of Quinone Reductase with Cultured Hepa 1c1c7 Cells. The second assay we utilize that yields results indicative of cancer chemoprevention at the stage of initiation is the induction of quinone reductase in cultured Hepa 1c1c7 cells. This system recently has been used by Talalay and coworkers for the isolation of sulforaphane from broccoli,[33] and it was established as a useful system for evaluation of extracts.[34] Like other *in vitro* assay systems used in our laboratory, induction of quinone reductase in cell culture correlates with *in vivo* activity that is consistent with cancer chemoprevention. While it is not generally clear that induction of phase I enzymes will result in chemoprevention, since carcinogen activation is also possible, induction of phase II enzymes is more logically associated with protection. Through the courtesy of Dr. James P. Whitlock, Stanford University, we have obtained Hepa 1c1c7 cells, and testing of enzyme induction is now routinely conducted in our laboratory. In addition, to help establish selectivity in inducing phase II enzymes rather than dual induction of both phase I and II enzymes, we use BP^rc1 and $TAOc1BP^rc1$ cells. BP^rc1 cells have normal TCDD receptor but appear to be unable to translocate the bound receptor complex to the nucleus. On the other hand, $TAOc1BP^rc1$ cells contain TCDD receptor that is either reduced or of altered affinity, but translocation to the nucleus is possible.[35,36] Therefore, induction of quinone reductase in the Hepa 1c1c7 cell line, but not in the BP^rc1 or $TAOc1BP^rc1$ cell lines, implies a receptor-mediated mechanism associated with a dual inducer. Conversely, activity in all three cell lines implies selectivity, and this is the major objective in our drug discovery program.

Assays Based on Factors Associated with Tumor Promotion

Protein Kinase C. Considerable effort has been directed toward examining the potential of test substances to inhibit the promotion of tumorigenesis. First, we have taken advantage of advances that have been made regarding phorbol ester mode of action and receptor sites.[37,38] TPA binds to and activates protein kinase C,[37-40] and, consistent with the importance of this mechanism, dioctanoylglycerol (an analog of the "natural" activator) has been established as a Stage II tumor promoter.[41] Thus, we have examined the potential of test substances to modulate the binding of ligands (PDBu and staurosporine) to protein kinase C as well as effects on catalytic activity.

Ornithine Decarboxylase. One well-established consequence of tumor promoter action is the induction of ornithine decarboxylase activity. This provides another level of potential control through which a chemopreventive agent may function. For example, retinoids are established as inhibitors of TPA-induced ornithine decarboxylase activity, and this may relate to mechanism of action. For monitoring this activity, we have routinely employed cultured mouse 308 cells (originally supplied through the courtesy of Dr. Stuart Yuspa, NIH). Results obtained with this test system are not redundant with those obtained in the protein kinase C assays. For example, rotenoids derived from *Mundulea sericea* inhibit ODC induction with extreme efficacy but no inhibition of protein kinase C is observed (see below).

Arachidonic Acid Metabolism. In addition to the promotion of tumors that presumably involves the activation of protein kinase C, either directly by phorbol esters or indirectly by the products of oncogenes, it is apparent that certain oxidants also may play a critical role in the promotion of carcinogenesis (for a detailed review see Cerutti[42]). Included in this category of tumor promoters are UV light and chemicals such as hydrogen peroxide, peroxyacetic acid, chlorobenzoic acid, benzoylperoxide, decanoylperoxide, cumene-hydroperoxide, *p*-nitroperbenzoic acid, and IO^{4-}. One effect of these tumor promoters is the activation of phospholipase A_2 which catalyzes the production of arachidonic acid. In turn, arachidonic acid serves as a substrate for cyclooxygenase which catalyzes the production of prostaglandins, thromboxane and prostacyclin. The precise role of these substances in tumor promotion is not clear, but a correlation of elevated prostaglandins in carcinogenesis has been established,[43] and certain nonsteroidal anti-inflammatory agents (e.g., indomethacin and piroxicam) demonstrate chemopreventive activity.[44] We have identified several active leads as inhibitors of cyclooxygenase, and these leads are distinct from those identified on the basis of other assays related to tumor promotion inhibition. Inclusion of the cyclooxygenase assay is also sensible from a mechanistic viewpoint since inhibition of TPA-induced ODC activity by indomethacin is overcome by concurrent application of PGE_2.[45]

Assays Based on Factors Associated with Tumor Progression

Differentiation of HL-60 Cells. Terminal differentiation can be induced in HL-60 cells by treatment with a variety of clinically interesting agents such as 1α,25-dihydroxyvitamin D_3 [1α,25-$(OH)_2D_3$] (or analogs thereof), DMSO, actinomycin D, retinoic acid, etc.[46] Treatment of HL-60 cells with such agents results in the induction of granulocytes, monocytes, eosinophils, or macrophage-like cells, and in certain cases, this induction process is mediated without toxicity. We have selected HL-60 differentiation as a target in our drug discovery program. This type of response is unique and not monitored in any of our other test systems. Thus, these determinations broaden the base of our test battery.

One obvious clinical implication of agents found to be active in this test system is the treatment of leukemia since terminal monocytic differentiation suggests a mechanism of controlling the proliferation of progenitor cells. However, as judged by $1\alpha,25\text{-}(OH)_2D_3$, one of the most potent compounds known to mediate HL-60 cell differentiation, agents found to be active in this test system should be of interest for the control of other cancer types, in particular, cancer of the breast.[47-49]

We have established this assay and routinely evaluate extracts and compounds for potential to induce nonspecific esterase (indicative of monocytes and macrophage-like cells), esterase (monocytes and other cell-types), and ability to produce superoxide when challenged with phorbol ester as judged by nitroblue tetrazolium (NBT) reduction (indicative of monocytes and granulocytes). The potential to inhibit thymidine incorporation is also monitored to assess specificity and antiproliferative activity. Cells are enumerated to assess cytotoxicity, and results are considered relative to $1\alpha,25\text{-}(OH)_2D_3$, as a positive control.

Assays Designed for the Discovery of Antihormones. Although many aspects of this overall program are directed toward chemoprevention of breast cancer, no component thus far described takes into account a strategy to exploit the hormone-dependence of this disease or certain other human cancers (e.g., cervical, ovarian). Thus, we have recently added a component to focus on this class of chemopreventive agents. The first assay simply detects displacement of estrogen binding to the estrogen receptor. Fundamentally, this is based on action that can be mediated by tamoxifen. Although controversial,[50] tamoxifen has been entered into a breast cancer reduction intervention trial that will involve approximately 16,000 subjects at a total cost of approximately $100 million.[51] It is apparent that a clear definition of the mechanism of tamoxifen relevant to antitumor activity is not available. However, the majority of *in vitro* studies have been performed with cultured MCF-7 cells, an estrogen receptor positive (ER^+) cell line wherein tamoxifen demonstrates antiproliferative activity, and tamoxifen and metabolites (such as 4-OH-tamoxifen) bind to the estrogen receptor, even though the binding differs from estradiol as shown by conformation and proteolytic susceptibility studies.[52,53] Agonistic activity of 4-OH-tamoxifen has been attributed to binding at the N-terminal region of the receptor.[54] When complexed with tamoxifen, the receptor binds to DNA as judged by *in vitro* studies, but this compound does not induce a conformational change.[55] However, consistent with this interaction being of physiologic importance, tamoxifen demonstrates complete antiestrogenic activity in estrogen-responsive breast cancer cells wherein estrogen inhibits c-erbB-2 proto-oncogene expression.[56] Additionally, with MCF-7 cells, tamoxifen down-regulates a 52 kDa protein induced by estrogen,[57] 4-OH-tamoxifen blocks estrogen-induced down-regulation of estrogen receptor, and inhibition of cell proliferation in the low concentration range (10^{-10}-10^{-8} M) can be completely negated by the addition of estradiol.[58] Thus, we surmise displacement of estrogen binding to its receptor is a valid parameter to monitor.

One shortcoming of this procedure, as with all receptor binding assay methods, is that compounds capable of displacing estrogen may demonstrate either antagonistic (antiestrogenic) or agonistic (estrogenic) activity. Thus, a rapid follow-up screen is required, and we use cultured Ishikawa cells (supplied through the courtesy of Dr. C. Richard Lyttle, Wyeth-Ayerst Research) for this purpose. Treatment of this cell line with estrogen or estrogenic compounds has been shown to result in significant elevation of alkaline phosphatase that can be monitored easily with a chromogenic substrate using a 96-well plate format.[59-61] Further, this estrogenic response is effectively blocked by the addition of antiestrogens to the medium. Thus, the assay is suitable for large-scale assessment of either antagonistic or agonistic activity.

A final assay in this area has yet to be implemented but we are planning to search for novel inhibitors of aromatase. The significance of aromatase and its relevance as a therapeutic target has been considered in detail, as summarized in a supplemental issue of *Cancer Research*.[62] Aromatase catalyzes the conversion of androstenedione to estrone, and this is considered an important source of estrogens in postmenopausal women. Therefore, since a significant number of human breast tumors are hormone-dependent, inhibition of aromatase is a reasonable intervention strategy.

Examples of aromatase inhibitors that have been examined in clinical trials include 4-hydroxyandrostenedione, fadrozole and aminoglutethimide.[63] R76713 is a non-steroidal inhibitor that is specific for aromatase and is approximately 1000-fold more active than aminoglutethimide.[64,65] Since this compound induced regression of DMBA-induced mammary tumors in rats, specific follow-up strategies for active isolates procured during the course of the current investigation are obvious.

Although certain clinical success has been realized with these compounds, adverse side effects have been noted, and other limitations exist such as metabolic conversion of inhibitors and interference with aldosterone biosynthesis. Thus, there is a need for the discovery of new chemotypes, and natural product drug discovery is a reasonable approach. Since aromatase converts androgens to estrogens with the liberation of formic acid and water, inhibition of the reaction can be readily with suitable radiolabelled substrates.

EXAMPLES OF RECENT RESULTS

Brassinin and Related Compounds

One substance of considerable interest in our program is brassinin. Brassinin has been identified as a phytoalexin; it originates biosynthetically by conversion of tryptophan to glucobrassinin with indole-3-methylisothiocyanate as a probable intermediate. In a sense, brassinin can therefore be considered a hybrid molecule containing an indole ring and a methyldithiocarbamate group, and thus can be structurally associated with other indoles and isothiocyanates from cruciferous vegetables that play important roles in chemoprevention. Based on the structure of brassinin and its possible biosynthetic relationship to other indolic dietary anti-carcinogenic compounds such as indole-3-carbinol, indole-3-acetonitrile and 3,3'-diindoylmethane, this compound has been of key importance in the direction taken by our synthetic program. Brassinin was synthesized and proved to be highly active in various anticarcinogenic screens. The effect of brassinin on the development of DMBA-induced mammary lesions in organ culture was most notable. At 10^{-5}M, it inhibited development of mammary lesions by more than 60%. A linear increase in the percent inhibition was observed with increasing concentrations.[66,67]

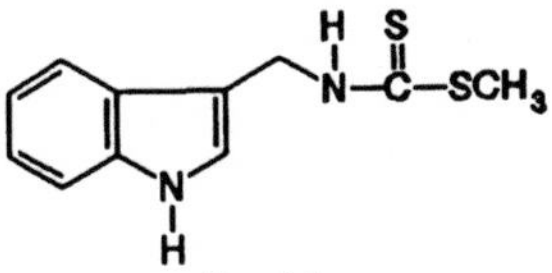

Brassinin

Based on these results, two courses of action were taken: (a) Synthesis of larger quantities of brassinin for more advanced testing, and (b) preparation of a series of synthetic brassinin derivatives to establish efficacy and structure-activity relationships. In order to make larger quantities of brassinin a novel scalable and economically feasible synthesis was developed. A two-step process was standardized in which the oximino derivative of indole-3-carboxaldehyde is reduced with Raney Nickel and the amine thus formed is converted to the methyl dithiocarbamate product, brassinin. With this material, we established that brassinin inhibits tumors in the two-stage skin carcinogenesis model with CD-1 mice and the DMBA-induced mammary tumor model with Sprague-Dawley rats. Both incidence and multiplicity were inhibited in both model systems.

Several brassinin analogs also were synthesized. Methyl analogs (2-, 4-, 5-, 6- and 7-methylbrassinin) were synthesized from 2-methylindole-3-carboxaldehyde. C-3 cyanation was carried out using a triphenyl phosphine/thiocyanogen reagent. Reduction and methyldithiocarbamylation yielded respective substituted brassinins. Other procedures for the synthesis of cyclo-, spiro- and dihydrobrassinin were also devised.

All the analogs were evaluated for effectiveness against DMBA-induced mammary lesions in organ cultures. Cyclobrassinin inhibited mammary lesion formation more effectively than brassinin. Additionally, spirobrassinin and *N*-ethyl-2,3-dihydrobrassinin inhibited mammary lesion formation by more than 60%. 2-Methylbrassinin was totally inactive, whereas some other analogs inhibited mammary lesion development to some extent.[68]

In a second test system, quinone reductase activity was monitored utilizing Hepa 1c1c7 cells. Complete dose-response determinations were conducted and the following results obtained: *N*-ethyl-2,3-dihydrobrassinin > cyclobrassinin > brassinin ≈ 5-chlorobrassinin ≈ 1-methoxybrassinin > 2-, 4-, 5-, or 7-methylbrassinin > cyclohomobrassinin (not active).[69] These data suggest the compounds are inducers of phase II enzymes.

In addition, we have begun to evaluate effects of brassinin on the induction of glutathione *S*-transferase and quinone reductase activity in mammary gland organ culture. In a three day experiment, brassinin and cyclobrassinin induce glutathione *S*-transferase activity in a dose-dependent manner. Little or no difference between brassinin or cyclobrassinin was observed with respect to glutathione *S*-transferase activity. However, cyclobrassinin at a concentration of 50 μg/ml dramatically enhanced quinone reductase activity by ~30-fold. Quinone reductase activity was also elevated by spirobrassinin and brassinin, but not to the extent induced by cyclobrassinin. Methyl analogs were inactive for the induction of both glutathione *S*-transferase and quinone reductase activities. These results provide evidence for the possible mechanism of the anti-initiating activity of brassinin demonstrated against DMBA-induced mammary tumors in rats.

Overall, the results we have obtained with cell culture systems and the mammary organ culture test system are remarkably consistent. In addition to the superior activity of cyclobrassinin, the activity of *N*-ethyl-2,3-dihydrobrassinin was notable. These results have prompted the synthesis of other *N*-ethyl-brassinin derivatives that will be tested for potential to induce quinone reductase. The studies are of great importance since we are currently at the stage of producing brassinin on the scale of 500 g in order to conduct additional full-term tumor inhibition studies. Based on mechanistic insight, this effort may focus on a brassinin derivative rather than the parent compound itself.

Isolates from Macleaya cordata

Macleaya cordata (Papaveraceae) was selected based both on a report of the leaves demonstrating antimutagenic activity[70] and our computerized selection method.[71] Although a rather extensive literature on the biology and chemistry of this plant was found, the reported antimutagenic activity was not explainable on the basis of the known chemical constituents, even when the profile was extended to the entire genus. Thus, samples of the aerial parts and the stems were collected in the Hyogo Prefecture, Japan, in August, 1991, by Dr. Kazuko Kawanishi, Kobe Women's College of Pharmacy, Kobe, Japan. Identification of the original and recollected samples was made by Prof. C. Miyoko of Kobe Women's College of Pharmacy, and voucher specimens were deposited at the herbarium of that institution.

An ethyl acetate extract of the plant was found to inhibit significantly phorbol ester interaction with partially purified protein kinase C (IC_{50} = 2.8 μg/ml). The extract was evaluated further at concentrations of 1, 10 and 50 μg/ml in mammary organ culture. It inhibited mammary lesion formation by 44% at a concentration of 10 μg/ml; at 50 μg/ml, it was toxic. Based on these data, the plant was selected for bioassay-directed fractionation, and PDBu binding was used as a monitor for activity. These procedures afforded a benzophenanthridine alkaloid, angoline (previously known also as 8-methoxy-dihydrochelerythrine and 9-methoxychelerythrine), as the active constituent of the plant crude extract. Angoline, originally isolated as a constituent of *Fagara angolensis*, was identified primarily by comparison of its spectroscopic and physical data with published values. However, in our investigation, complete ^{13}C-NMR assignments were made for the first time for angoline, using a combination of modern one- and two-dimensional NMR techniques. In addition, the stereochemistry of the C-8 attached methoxy group was confirmed by appropriate nuclear Overhauser enhancement (NOE) NMR observations.

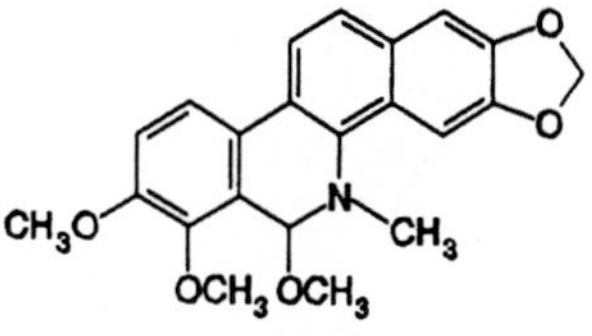

CH_3O N CH_3 OCH_3 OCH_3

Angoline

A related compound, 8-methoxydihydrosanguinarine, was isolated as an inactive constituent. As a result of this finding, 17 benzophenanthridine alkaloids and structurally-related compounds were examined for potential to inhibit PDBu or staurosporine binding, induction of ODC activity, and modulation of protein kinase C-mediated catalysis.[72] Surprisingly, of the compounds studied, greatest inhibition of PDBu binding was observed with the original isolate, angoline. Chelerythrine, a compound reported to inhibit protein kinase C,[73] does not interact with the phorbol ester binding site, nor does sanguinarine or 8-methoxydihydrosanguinarine. These results suggest that all three methoxy groups associated with the structure of angoline are required for binding to the regulatory domain.

Isolates from Mundulea sericea

Mundulea sericea (Leguminosae) was randomly selected from the repository of plant materials available in the Program for Collaborative Research in the Pharmaceutical Sciences, Department of Medicinal Chemistry and Pharmacognosy, College of Pharmacy, University of Illinois at Chicago. This plant material was originally collected in Kenya in 1976 as part of an earlier NCI sponsored antitumor agents from plants project. A voucher specimen was deposited at the Medicinal Plant Resources Laboratory, Plant Genetics and Germplasm Institute, Agriculture Research Center-East, Beltsville, MD. A check of the NAPRALERT database indicated a previous biological literature, but no prior cancer chemopreventive studies had been carried out. Ethnomedically, reports indicate the plant has been used for minor ailments. The chemical profile was not useful in providing any indication of its potential as a carcinogenesis inhibitor.

A sample of the bark of *Mundulea sericea* was extracted with methanol, then defatted with petroleum ether, and partitioned between water and ethyl acetate. The ethyl acetate extract was found to significantly inhibit TPA-induced ODC induction (IC_{50} = 0.016 μg/ml) without affecting protein kinase C activity.

The extract was also tested in mammary gland organ cultures. Results showed that this extract dramatically inhibited mammary lesion formation. Nearly 100% inhibition was observed at 10μg/ml concentration. This was the most effective extract tested during the past two years of experiments on this project. Based on this profile of activity, the plant was selected for bioassay-directed fractionation.

Activity-monitored purification using inhibition of ODC induction as a monitor has afforded an unusually varied group of active compounds. These compounds represent altogether five structural classes, namely, chalcones (4-hydroxylonchocarpin, munsericin), flavanones (lupinifolin, lupinifolinol, mundulin, mundulinol), isoflavonoids (munetone, mundulone), rotenoids (deguelin, 13α-hydroxydeguelin, 13α-hydroxytephrosin, tephrosin), and a triterpenoid (oleanolic alcohol). All compounds were active to some degree in the ODC induction assay, with the four rotenoids being particularly potent in this regard.[74]

Three of the ODC-active compounds isolated from *M. sericea* bark were novel, and required detailed analysis of their spectral parameters in order to discern their structures. The compounds concerned, 13α-hydroxydeguelin, 13α-hydroxytephrosin, and munsericin, were structurally determined with reference to the closely related reference compounds, deguelin, tephrosin, and musericin, respectively, which also occurred as *M. sericea* constituents. In each case, a variety of NMR and other spectroscopic techniques were employed in combination to solve these structural problems.

Thus, *Mundulea sericea* was found to yield a number of active isolates, notably deguelin and tephrosin. These compounds are incredibly potent inhibitors of phorbol ester-induced ornithine decarboxylase in mouse 308 cells, demonstrating IC_{50} values of <1 ng/ml. By comparison, a positive control compound 13-*cis*-retinoic acid, demonstrates an IC_{50} value of ~0.1 μg/ml.[75]

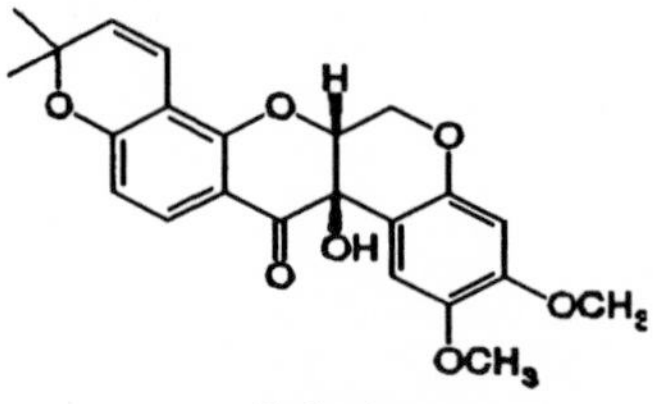

Tephrosin 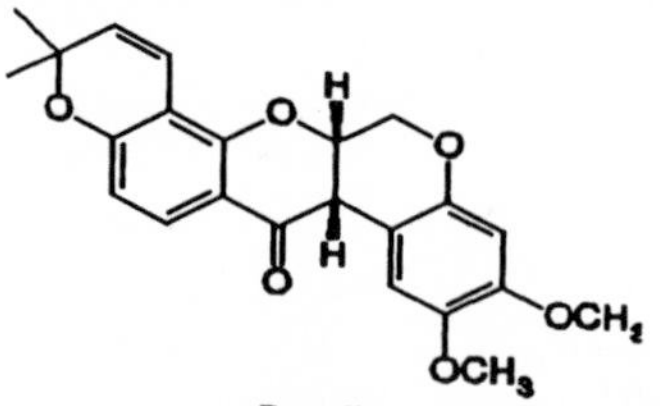 Deguelin

Deguelin was further evaluated at five different concentrations for its dose responsive activity in mammary gland organ cultures. Results showed that deguelin inhibited development of mammary lesions by 70% at 10^{-5}M concentration. In addition, an extract of *Mundulea sericea* was evaluated for its activity as a chemopreventive agent against skin carcinogenesis in CD-1 mice. Results showed that an 1% extract dramatically inhibited the occurrence of skin tumors. As compared to 95% tumor incidence in control animals, the group treated with extract had only 15% incidence. The tumor numbers showed even more dramatic results. Instead of 595 tumors in the control animals, the extract-treated animals had a total of only 4 tumors. Since the extract contains several components such as deguelin, tephrosin and flavonoids, additional work is required. Nonetheless, it is clear that the approach chosen to select agents for further evaluation based on *in vitro* assays and mammary gland organ cultures is very effective.

CONCLUSIONS

At the present time, we are fractionating approximately 20 plants by monitoring activity in the assay systems described above, and various lead compounds are emerging. The long-term objective of this research is to identify chemopreventive candidates that may be entered into clinical intervention trials. In contemplating the requirements of attaining this long-term goal, it is worth considering cancer chemopreventive agents currently under investigation in clinical intervention trials. A number of chemopreventive agents in this category are summarized in Table 5. One of these agents, *N*-(4-hydroxyphenyl)retinamide (4-HPR), was developed largely by the efforts of investigators affiliated with the University of Illinois at Chicago, and a brief historical overview is presented below.

The first definitive experiments to show the chemopreventive effect of retinoids, specifically 13-*cis*-retinoic acid-mediated inhibition of urinary bladder cancer[77] and retinyl acetate-mediated inhibition of mammary cancer,[78] provided the rationale of evaluating synthetic retinoids as possible chemopreventive agents. Sporn and colleagues used hamster tracheal organ culture to evaluate numerous retinoids for the identification of analogs that would reverse keratinization caused by vitamin A deficiency.[79] In addition to several other retinoids, 4-HPR was active in tracheal organ culture and selected for further *in vivo* experimentation.

Moon *et al.*[80] reported that 4-HPR was less toxic than other retinoids and was not stored in the liver. It inhibited both DMBA- and MNU-induced mammary carcinogenesis in a dose-dependent manner. In addition to 4-HPR, several other retinoids were studied for their efficacy in rat mammary carcinogenesis model, but 4-HPR proved superior to all other retinoids evaluated to date. It was also shown that 4-HPR inhibited mammary carcinogenesis in a synergistic manner with ovariectomy.[81] However, combination of tamoxifen and 4-HPR resulted in additive rather than synergistic inhibition of carcinogenesis. Subsequently it was observed that 4-HPR inhibited chemically-induced carcinogenesis of a variety of other target organs, including urinary bladder, two-stage skin carcinogenesis, DEN-induced lung cancers, MNU-induced prostate cancer in rats and carcinogen-induced pancreatic cancer in hamsters.[82] The results provided by the laboratory of Moon and collaborators constituted the basis of initiating a clinical trial with 4-HPR for women at a high risk of developing cancer in contralateral breasts. This study is being conducted in Milan, Italy, by Drs. Veronesi, Costa and

Table 5. Examples of chemoprevention intervention studies[a]

Target site organ	Target/risk group	Inhibitory agents
Cervix	Cervical dysplasia	trans-Retinoic acid
Cervix	Cervical dysplasia	Folic acid
Colon	Familial polyposis	Vitamins C and E and fiber
Colon	Familial polyposis	Calcium
Colon	Adenomatous polyps	β-Carotene and vitamins C and E
Colon	Adenomatous polyps	Piroxicam
Lung	Chronic smokers	Folic acid and vitamin B12
Lung	Asbestosis	β-Carotene and retinol
Lung	Cigarette smokers	β-Carotene and retinol
Lung	Smoking males	β-Carotene
Lung	Asbestosis	β-Carotene
Skin	Albinos	β-Carotene
Skin	Basal cell carcinoma	β-Carotene and vitamins C and E
Skin	Basal cell carcinoma	β-Carotene
Skin	Actinic keratoses	Retinol
Skin	Basal cell carcinoma	Retinol and 13-cis-retinoic acid
Breast	Adenocarcinoma	N-(4-hydroxyphenyl)retinamide
All sites	American physicians	β-Carotene

[a]Adapted from Boone et al.[76]

colleagues, with inclusion of 2900 women in the trial. Results from this trial are still accumulating.

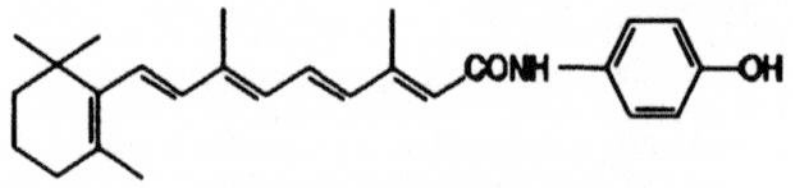

CONH — OH

N-(4-Hydroxyphenyl)retinamide

This brief overview indicates types of considerations that need to be taken into account regarding clinical trials. It is likely that newly discovered agents would be subjected to similar criteria of activity. For natural product drugs, the following are desirable characteristics: (a) a novel chemical structure or a chemical structure that would not have been predicted to mediate the observed biological response, (b) an understanding of structure-activity relationships such that the

Table 6. Putative intermediate endpoints[a]

- Genetic
  - Oncogene activation/suppression
  - Micronuclei
  - Quantitative DNA analysis, DNA ploidy
- Biochemical
  - Ornithine decarboxylase
  - Prostaglandin synthetase
- Cellular
  - Sputum metaplasia/dysplasia
  - Cervical dysplasia
  - Gastric metaplasia
  - Colonic cell proliferation
- Precancerous lesions
  - Colon polyps
  - Bladder papillomas
  - Oral leukoplakia
  - Dysplastic nevi

[a]Adapted from Malone.[83]

most promising agent of a series can be selected for more advanced testing, (c) a basic understanding of mechanism of action, (d) a unique mechanism of action or a notable potency. In addition, however, a strong indication of chemopreventive activity needs to be obtained in an *in vivo* system, as judged by biomarkers. Several intermediate endpoints (biomarkers) are summarized in Table 6. Modulation of these endpoints can be monitored much more quickly and less expensively than tumor incidence in human populations, and efficacy in affecting biomarkers can provide a strong rationale for entering the stage of preclinical development.

To conclude, cancer chemoprevention is a field of great promise. It is likely that humans exist in a state of homeostasis, and certain dietary components have contributed to our overall state of well-being including cancer prevention. This has recently been advocated by certain dietary campaigns inaugurated by the American Cancer Society, the National Cancer Institute, and others. Although it is not reasonable to assume cancer chemopreventive agents will safeguard humans from known carcinogenic risks such as cigarette smoking, it is reasonable to anticipate that these agents will play an increasing role in cancer prevention strategies. Our efforts in the area of natural product drug discovery support this undertaking.

ACKNOWLEDGMENTS

This work is conducted under the auspices of a program project grant (PO1 CA48112) funded by the National Cancer Institute. Each component is of integral importance for the success of the program. The projects are ably conducted by the following principal and co-investigators: Drs. N.R. Farnsworth, H.H.S. Fong, C.W.W. Beecher, D.D. Soejarto, A.D. Kinghorn, R.C. Moon, R.G. Mehta, R.M. Moriarty, and S. Hedayat. The author is grateful to the following students and postdoctoral associates for their tireless efforts in evaluating the biologic potential of chemopreventive agents discovered as part of this program project: C. Gerhäuser, M. You, E. Pisha, M.-S. Jang, L. Shamon, S.K. Lee, and N. Suh. C.G is supported in part by the Feodor Lynen program of the Alexander von Humboldt Foundation.

REFERENCES

1. GREENWALD, P., SONDIK, E., LYNCH, B.S. 1986. Diet and chemoprevention in NCI's research strategy to achieve national cancer control objectives. Annu. Rev. Public Health 7:267-291.
2. GREENWALD, P., CULLEN, J.W., McKENNA, J.W. 1987. Cancer prevention and control from research through applications. J. Natl. Cancer Inst. 79:389-400.
3. BERTRAM, J.S., KOLONEL, L.N., MEYSKENS, Jr., F.L. 1987. Rationale and strategies for chemoprevention of cancer in humans. Cancer Res. 47:3012-3031.
4. MALONE, W.F., KELLOFF, G.J., BOONE, C., NIXON, D.W. 1989. Chemoprevention and modern cancer prevention. Prevent. Med. 18:553-561.
5. GREENWALD, P., NIXON, D.W., MALONE, W.F., KELLOFF, G.J., STERN, H.R., LITKIN, K.M. 1990. Concepts in cancer chemoprevention research. Cancer 65:1483-1490.
6. COSTA, A., SANTORO, G., ASSIMAKOPOULOS, G. 1990. Cancer chemoprevention: A review of ongoing clinical trials. Acta Oncol. 29:657-663.
7. BJELKE, E. 1975. Dietary vitamin A and human lung cancer. Int. J. Cancer 15:561-565.
8. SHEKELLE, R.B., LIU, S., RAYNOR, Jr., W.J., LEPPER, M., MAKZA, C., ROSSOF, A.H., PAUL, O., SHRYOCK, A.M., STAMILER, J. 1981. Dietary vitamin A and risk of cancer in the Western Electric study. Lancet 2:1185-1189.
9. HIRAYAMA, T. 1979. Diet and cancer. Nutr. Cancer 1:67-81.
10. BLOCK, G. 1992. Vitamin C status and cancer. Epidemiologic evidence of reduced risk. Ann. N.Y. Acad. Sci. 669:281-292.
11. HOCMAN, G. 1992. Chemoprevention of cancer: Protease inhibitors. Int. J. Biochem. 24:1365-1375.
12. TROLL, W. 1989. Protease inhibitors interfere with the necessary factors of carcinogenesis. Environ. Health Perspect. 81:59-62.
13. KNEKT, P. 1992. Vitamin E and cancer: Epidemiology. Ann. N.Y. Acad. Sci. 669:269-279.

14. BUTTERWORTH, Jr., C.E. 1992. Effect of folate on cervical cancer. Ann. N.Y. Acad. Sci. 669:293-299.
15. CORDELL, G.A., PEZZUTO, J.M., HAMBURGER, M.O., GUNAWARDANA, Y.A.G.P., SHIEH, H.-L., McPHERSON, D.D. 1989. Recent studies on anticancer agents. In: Progress in Chemistry of Medicinal Plants in Asia. Proceedings of the Sixth Asian Symposium on Medicinal Plants and Spices. Bandung, Indonesia. pp 243-254.
16. CORDELL, G.A., FARNSWORTH, N.R., BEECHER, C.W.W., SOEJARTO, D.D., KINGHORN, A.D., PEZZUTO, J.M., WALL, M.E., WANI, M.C., COBB, R.R., HARRIS, T.J.R., O'NEILL, M.J., TAIT, R.M. 1994. Novel strategies for the discovery of plant-derived anticancer agents. In: Second Symposium on Anticancer Drug Discovery and Development, Traverse City, Michigan, June 27-29, 1991. Kluwer Adademic Publishers, Boston. pp. 63-83.
17. TAN, G.T., PEZZUTO, J.M., KINGHORN, A.D. 1992. Screening of natural products as HIV-1 and HIV-2 reverse transcriptase (RT) inhibitors. In: Natural Products as Antiviral Agents. (C.K. Chu, H. Cutler, eds.). Plenum Press, N.Y. pp 195-222.
18. CORDELL, G.A., KINGHORN, A.D., PEZZUTO, J.M. 1993. Separation, structure elucidation and bioassay of cytotoxic natural products. In: Bioactive Natural Products: Detection, Isolation and Structural Identification. (S.M. Colegate, R.J. Molyneux, eds.). CRC Press, Boca Raton. pp 195-219.
19. SHAMON, L., PEZZUTO, J.M. 1994. Plant antimutagens: A review and strategy for the identification of therapeutically useful agents. In: Economic and Medicinal Plant Research. Vol. 6. (N.R. Farnsworth, H. Wagner, eds.). Academic Press, London. pp 235-297.
20. ANGERHOFER, C.K., KÖNIG, G.M., WRIGHT, A.D., STICHER, O., MILHOUS, W.K., CORDELL, G.A., FARNSWORTH, N.R., PEZZUTO, J.M. 1993. Selective screening of natural products: A resource for the discovery of novel antimalarial compounds. In: New Trends in Natural Products Chemistry. (Atta-ur-Rahman, ed.). pp 431-443.
21. CORDELL, G.A., FARNSWORTH, N.R., BEECHER, C.W.W., SOEJARTO, D.D., KINGHORN, A.D., PEZZUTO, J.M., WALL, M.E., WANI, M.C., BROWN, D.M., O'NEILL, M.J., LEWIS, J.A., TAIT, R.M., HARRIS, T.J.R. 1993. Novel strategies for the discovery of plant-derived anticancer agents. In: Human Medicinal Agents from Plants (ACS Symposium Series No. 534). (A.D. Kinghorn, M.F. Balandrin, eds.) A.C.S. Wash. D.C. pp 191-204.
22. PEZZUTO, J.M. 1993. Cancer chemopreventive agents: From plant materials to clinical intervention trials. In: Human Medicinal Agents from Plants (ACS Symposium Series No. 534). (A.D. Kinghorn, M.F. Balandrin, eds.). A.C.S. Wash. D.C. pp 205-217.
23. SUFFNESS, S.M., PEZZUTO, J.M. 1991. Assays for cytotoxicity and antitumor activity. Chapter 4. In: Methods of Plant Biochemistry, Vol. 9. (K. Hostettmann ed.). Academic Press, London. pp 71-133.
24. KINGHORN, A.D., BALANDRIN, M.F. (eds.). 1993. Human Medicinal Agents from Plants (ACS Symposium Series No. 534). A.C.S. Wash. D.C. pp 356.
25. CORDELL, G.A., BEECHER, C.W.W., PEZZUTO, J.M. 1991. Can ethnopharmacology contribute to the development of new anticancer drugs? J. Ethnopharmacol. 32:117-133.
26. ANGERHOFER, C.K., PEZZUTO, J.M. 1993. Applications of biotechnology for drug discovery and evaluation. In: Biotechnology and Pharmacy. (J.M. Pezzuto, M.E. Johnson, H.R. Manasse, Jr., eds.). Chapman and Hall, N.Y. pp 312-365.
27. CASSADY, J.M., BAIRD, W.M., CHANG, C.-J. 1990. Natural products as a source of potential cancer chemotherapeutic and chemopreventive Agents. J. Nat. Prod. 53:23-41.
28. MUTO, Y., NINOMIYA, M., FUJIKI, H. 1990. Present status of research on cancer chemoprevention in Japan. Jpn. J. Clin. Oncol. 20:219-224.
29. BERTRAM, J.S., HOSSAIN, M.Z., PUNG, A, RUNDHAUG, J.E. 1989. Development of *in vitro* systems for chemoprevention research. Prev. Med. 18:562-575.

30. BEECHER, C.W.W., FARNSWORTH, N.R., FONG, H.H.S., KINGHORN, A.D., MEHTA, R.G., MOON, R.C., MORIARTY, R.M., PEZZUTO, J.M., SOEJARTO, D.D., WALL, M.E., WANI, M. 1993. Identification and Characterization of Natural Inhibitors of Carcinogenesis. Proc. Am. Assoc. Cancer Res. 34:559.
31. SHAMON, L.A., CHEN, C., MEHTA, R.G., STEELE, V., MOON, R.C., PEZZUTO, J.M. A correlative approach for the discovery of antimutagens that demonstrate cancer chemopreventive activity. Anticancer Res. 14: in press.
32. IP, C., GANTHER, H. E. 1991. Combination of blocking agents and suppressing agents in cancer prevention. Carcinogenesis 12:365-367.
33. ZHANG, Y., TALALAY, P., CHO, C.-G., POSNER, G.H. 1992. A major inducer of anticarcinogenic protective enzymes from broccoli: Isolation and elucidation of structure. Proc. Natl. Acad. Sci. USA 89:2399-2403.
34. PROCHASKA, H.J., SANTAMARIA, A.B., TALALAY, P. 1992. Rapid detection of inducers of enzymes that protect against carcinogens. Proc. Natl. Acad. Sci. USA 89:2394-2398.
35. De LONG, M., SANTAMARIA, A.B., TALALAY, P. 1987. Role of cytochrome P1-450 in the induction of NAD(P)H:quinone reductase in a murine hepatoma cell line and its mutants. Carcinogenesis 8:1549-1553.
36. MILLER, A.G., ISRAEL, D., WHITLOCK, Jr., J.P. 1983. Biochemical and genetic analysis of variant mouse hepatoma cells defective in induction of benzo(a)pyrene-metabolizing enzyme activity. J. Biol. Chem. 258:3523-3527.
37. TAKAI, Y., KIKKAWA, U., KAIBUCHI, K., NISHIZUKA, Y. 1984. Membrane phospholipid metabolism and signal transduction for protein phosphorylation. Adv. Cyclic Nucleotide Prot. Phosphoryl. Res. 18:119-158.
38. NISHIZUKA, Y. 1984. The role of protein kinase C in cell surface signal transduction and tumour promotion. Nature 308:693-698.
39. CASTAGNA, M., TAKAI, Y., KAIBUCHI, K., SANO, K., KIKKAWA, U., NISHIZUKA, Y. 1982. Direct activation of calcium-activated phospholipid-dependent protein kinase by tumor-promoting phorbol esters. J. Biol. Chem. 257:7847-7851.
40. BELL, R.M. 1986. Protein kinase C activation by diacylglycerol second messengers. Cell 45:631-632.
41. VERMA, A.K. 1988. The protein kinase C activator L-α-dioctanoylglycerol: A potent stage II mouse skin tumor promoter. Cancer Res. 48:1736-1739.
42. CERUTTI, P.A. 1988. Oxidant tumor promoter. In: UCLA Symposia on Molecular and Cellular Biology New Series, Vol. 58, Growth Factors, Tumor Promoters, and Cancer Genes. (N.H. Colburn, H.L. Moses, E.J. Stanbridge, eds.). Alan R. Liss, Inc., N.Y. pp 239-247.
43. LUPULESCU, A. 1978. Enhancement of carcinogenesis by prostaglandins. Nature 272:634-636.
44. REDDY, B.S. 1992. Inhibitors of the arachidonic acid cascade and their chemoprevention of colon carcinogenesis. In: Cancer Chemoprevention. (L. Wattenberg, M. Lipkin, C.W. Boone, G.L. Kelloff, eds.). CRC Press, Boca Raton. pp 153-164.
45. VERMA, A.K., ASHENDEL, C.L., BOUTWELL, R.K. 1980. Inhibition by prostaglandin synthesis inhibitors of the induction of epidermal ornithine decarboxylase activity, the accumulation of prostaglandins, and tumor promotion caused by 12-*O*-tetradecanoylphorbol-13-acetate. Cancer Res. 40:308-315.
46. COLLINS, S.J. 1987. The HL-60 promyelocytic leukemia cell line: Proliferation, differentiation, and cellular oncogene expression. Blood 70:1233-1244.
47. EISMAM, J.A., KOGA, M., SUTHERLAND, R.L., BARKLA, D.H., TUTTON, P.J.M. 1989. 1,25-Dihydroxyvitamin D_3 and the regulation of human cancer cell replication. P.S.E.B.M. 191:221-226.

48. COLSTON, K.W., BERGER, U., COOMBES, R.C. 1989. Possible role for vitamin D in controlling breast cancer cell proliferation. Lancet, January 28, 1989, pp 188-191.
49. COLSTON, K., WILKINSON, J.R., COOMBES, R.C. 1986. 1,25-Dihydroxyvitamin D_3 binding in estrogen-responsive rat breast tumor. Endocrinol. 119:397-403.
50. HAN, X.L., LIEHR, J.G. 1992. Induction of covalent DNA adducts in rodents by tamoxifen. Cancer Res. 52:1360-1363.
51. SMIGEL, K. 1992. Breast cancer prevention trial takes off. J. Natl. Cancer Inst. 84:669-670.
52. SASSON, S., NOTIDES, A.C. 1988. Mechanism of the estrogen receptor interaction with 4-hydroxytamoxifen. Mol. Endocrinol. 2:307-312.
53. GEIER, A., BEERY, R., HAIMSOHN, M., LUNENFELD, B. 1987. Analysis of the 4-hydroxytamoxifen (4-OHTAM) bound nuclear estrogen receptor from MCF-7 cells by limited proteolysis. J. Steroid Biochem. 28:471-478.
54. METZGER. D., LOSSON, R., BORNERT, J.M., LEMOINE, Y., CHAMBON, P. 1992. Promoter specificity of the two transcriptional activation functions of the human oestrogen receptor in yeast. Nucleic Acids Res. 20:2813-2817.
55. FAWELL, S.E., WHITE, R., HOARE, S., SYDENHAM, M., PAGE, M., PARKER, M.G. 1990. Inhibition of estrogen receptor-DNA binding by the "pure" antiestrogen ICI 164,384 appears to be mediated by impaired receptor dimerization. Proc. Natl. Acad. Sci. USA 87:6883-6887.
56. ANTONIOTTI, S., MAGGIORA, P., DATI, C., DEBORTOLI, M. 1992. Tamoxifen upregulates c-erbB-2 expression in oestrogen-responsive breast cancer cells *in vitro*. Eur. J. Cancer. 28:318-321.
57. LYKKESFELDT, A.E., SORENSEN, E.K. 1992. Effect of estrogen and antiestrogens on cell proliferation and synthesis of secreted proteins in the human breast cancer cell line MCF-7 and a tamoxifen resistant variant subline, AL-1. Acta Oncol. 31:131-138.
58. SUTHERLAND, R.L., WATTS, C.K., HALL, R.E., RUENITZ, P.C. 1987. Mechanisms of growth inhibition by nonsteroidal antioestrogens in human breast cancer cells. J. Steroid Biochem. 27:891-897.
59. LITTLEFIELD, B.A., GURPIDE, R., MARKIEWICZ, L., MCKINLEY, B., HOCHBERG, R.B. 1990. A simple and sensitive microtiter plate estrogen bioassay based on stimulation of alkaline phosphatase in Ishikawa cells: Estrogenic action of Δ^5 adrenal steroids. Endocrinol. 127:2757-2762.
60. ALBERT, J.L., SUNDSTROM, S.A., LYTTLE, C.R. 1990. Estrogen regulation of placental alkaline phosphatase gene expression in a human endometrial adenocarcinoma cell line. Cancer Res. 50:3306-3310.
61. HOLINKA, C.F., HATA, H., KURAMOTO, H., GURPIDE, E. 1986. Effects of steroid hormones and antisteroids on alkaline phosphatase activity in human endometrial cancer cells (Ishikawa line). Cancer Res. 46:2771-2774.
62. Anon. 1982. Aromatase: New Perspectives for Breast Cancer. Cancer Res. (Suppl.) 42:3261s-3469s.
63. COOMBES, R.C., EVANS, T.R.J. 1991. Aromatase Inhibitors 2. In: The Medical Management of Breast Cancer. (T. Powles, I.E. Smith, eds.). Martin Dunitz, London, pp 81-93.
64. WOUTERS, W., DE COSTER, R., KREKELS, M., VAN DUN, J., BEERERNS, D., HAELTERMAN, C., RAEYMAEKERS, A., FREYNE, E., VAN GELDER, J., VENET, M., JANSSEN, P.A.J. 1989. R76713, A new specific non-steroidal aromatase inhibitor. J. Steroid Biochem. 32:781-788.
65. De COSTER, R., WOUTERS, W., BOWDEN, C.R., VANDEN BOSSCHE, H., BRUYNSEELS, J., TUMAN, R.W., VAN GINCKEL, R., SNOECK, E., VAN PEER, A., JANSSEN, P.A.J. 1990. New non-steroidal aromatase inhibitors: Focus on R76713. J. Steroid Biochem. Mol. Biol. 37:335-341.

66. MEHTA, R.G., CONSTANTINOU, A., MORIARTY, R., PEZZUTO, J.M., MOON, R.C. 1993. Brassinin: A novel chemopreventive agent. Proc. Am. Assoc. Cancer Res. 34: 127.
67. MEHTA, R.G., MORIARTY, R.M., LIU, J., CONSTANTINOU, A., THOMAS, C., HAWTHORNE, M., YOU, M., GERHÄUSER, C., PEZZUTO, J.M., MOON, R.C. Synthesis and cancer chemopreventive activity of brassinin, a phytoalexin from cabbage. Carcinogenesis, in press.
68. MEHTA, R.G., LIU, J., CONSTANTINOU, A., HAWTHORNE, M., PEZZUTO, J.M., MOON, R.C., MORIARTY, R.M. Structure-activity relationships of brassinin in preventing the development of carcinogen-induced mammary lesions in mammary gland organ culture. Anticancer Res. 14:1209-1213.
69. YOU, M., GERHÄUSER, C., LIU, J., MORIARTY, R.M., MEHTA, R.G., MOON, R.C., PEZZUTO, J.M. 1994. Induction of quinone reductase activity mediated by brassinin and its derivatives. Proc. Am. Assoc. Cancer Res. 35:627.
70. ISHII, R., YOSHIKAWA, K., MINAKATA, H., KOMURA, H., KADA, T. 1984. Specificity of bio-antimutagens in plant kingdom. Agr. Biol. Chem. 48:2587-2591.
71. BEECHER, C.W.W., PEZZUTO, J.M., FARNSWORTH, N.R. 1993. Meta-analysis of published literature as a means of chemopreventive drug discovery. Proc. Am. Assoc. Cancer Res. 34:127.
72. KINGHORN, A.D., MEHTA, R.G., MOON, R.C., PEZZUTO, J.M. 1993. Bioassay-directed isolation of natural inhibitors of carcinogenesis. Proc. Am. Assoc. Cancer Res. 34:558.
73. HERBERT, J.M., AUGEREAU, J.M., GLEYE, J., MAFFRAND, J.P. 1990. Chelerythrine is a potent and specific inhibitor of protein kinase C. Biochem. Biophys. Res. Commun. 172:993-999.
74. LUYENGI, L., LEE, I.-S., MAR, W., FONG, H.H.S., PEZZUTO, J.M., KINGHORN, A.D. Rotenoids and chalcones from *Mundulea sericea* that inhibit phorbol ester-induced ornithine decarboxylase activity. Phytochemistry 36:1523-1526.
75. PEZZUTO, J.M., MAR, W., LUYENGI, L., KINGHORN, A.D., FONG, H.H.S., HALLINE, A.G., MEHTA, R.G., MOON, R.C. 1994. Rotenoid-mediated inhibition of ornithine decarboxylase activity as a mechanism of chemopreventive activity. Proc. Am. Assoc. Cancer Res. 35:627.
76. BOONE, C.W., KELLOFF, G.J., MALONE, W.E. 1990. Identification of candidate cancer chemopreventive agents and their evaluation in animal models and human intervention trials: A review. Cancer Res. 50:2-9.
77. SPORN, M.B., SQUIRE, R.A., BROWN, C.C., SMITH, J.M., WENK, M.L., SPRINGER, S. 1977. 13-*cis*-Retinoic acid: Inhibition of bladder carcinogenesis in rats. Science 195:487-489.
78. MOON, R.C., GRUBBS, C.J., SPORN, M.B., GOODMAN, D.G. 1977. Retinyl acetate inhibits mammary carcinogenesis induced by *N*-methyl-*N*-nitrosourea. Nature 267:620-621.
79. SPORN, M.B., CLAMON, G.H., DUNLOP, N.M., NEWTON, D.L., SMITH, J.M., SAFFIOTTI, U. 1975. Activity of vitamin A analogs in cell cultures of mouse epidermis and organ culture of hamster trachea. Nature 253:47-50.
80. MOON, R.C., THOMPSON, H.J., BECCI, P.J., GRUBBS, C.J., GANDER, R.J., NEWTON, D.L., SMITH, J.M., PHILLIPS, S.L., HENDERSON, W.R., MULLEN, L.T., BROWN, C.C., SPORN, M.B. 1979. *N*-(4-Hydroxyphenyl)retinamide, a new retinoid for prevention of breast cancer in rat. Cancer Res. 39:1339-1346.
81. McCORMICK, D.L., MEHTA, R.G., THOMPSON, C.A., DINGER, N., CALDWELL, J.L., MOON, R.C. 1982. Enhanced inhibition of mammary carcinogenesis by combination of *N*-(4-hydroxyphenyl)retinamide and ovariectomy. Cancer Res. 42:508-512.

82. MOON, R.C., MEHTA, R.G., RAO, K.V.N. 1993. Retinoids and cancer in experimental animals. In: The Retinoids: Biology, Chemistry and Medicine, Second edition. (M.B. Sporn, A.B. Roberts, D.S. Goodman, eds.). pp 573-592.
83. MALONE, W. F. 1991. Studies evaluating antioxidants and β-carotene as chemopreventives. Am. J. Clin. Nutr. 53:305S-313S.

Chapter Three

ANTIVIRAL AND ANTITUMOR PLANT METABOLITES

John A. Beutler, John H. Cardellina II, James B. McMahon,
Robert H. Shoemaker, and Michael R. Boyd

Laboratory of Drug Discovery Research and Development
Developmental Therapeutics Program, Division of Cancer Treatment
National Cancer Institute, NCI-FCRDC
Frederick, Maryland 21702-1201

INTRODUCTION

The U.S. National Cancer Institute (NCI) currently has in place a large screening program which supports researchers worldwide in the search for new effective drugs for cancer and AIDS.[1-3] While a majority of screening in this

Phytochemistry of Medicinal Plants
Edited by John T. Arnason et al., Plenum Press, New York, 1995

program is focused upon pure compounds submitted by non-NCI investigators for NCI testing, a portion of the screening capacity is allocated to the initial testing of natural product extracts from the NCI repository.

A major focus of our NCI intramural laboratory is the use of these primary NCI screens to attempt to discover and characterize, chemically and biologically, new potential drug development leads from NCI repository extracts. The NCI repository is also used extensively by many scientists and organizations outside of NCI employing their own diverse screening systems and research strategies. However, the scope of the present review is restricted to specific current examples from our laboratory which illustrate particular approaches to application of the NCI primary screens in plant natural products research.

The NCI repository contains extracts from diverse microbial, marine invertebrate, algal and plant sources. Terrestrial plants make up 48 percent of the extracts tested thus far against the human immunodeficiency virus (HIV) and 60 percent of the extracts tested thus far in the cancer screen. NCI contractors from the New York Botanical Garden, Missouri Botanical Garden, and University of Illinois at Chicago funnel approximately 5,000 new plant samples per year into the NCI repository. In addition, other plants and plant extracts are obtained for screening through collaborative arrangements with other organizations or individuals. Most of the plant specimens are collected in tropical regions, specifically, Central and South America, tropical Africa, and Southeast Asia.

Plant samples average from 500 g to 1 kg dry weight. Individual plant parts such as leaves, bark, wood, flowers or roots are collected separately where possible and dried in the field. After shipment to the U.S., each plant specimen is frozen at -20°C, ground, and processed to yield first an organic and then an aqueous extract, which are tested separately. Extracts which test positive are confirmed with a second test, then prioritized for fractionation and isolation of the active constituents. At the present time, approximately 15,000 plant specimens (30,000 extracts) have been tested against HIV, and about 19,000 plant specimens have been tested in the primary cancer screen.

SCREENING MODELS

Cancer

The cancer screening system presently uses a panel of 60 human tumor cell lines organized into subpanels representing different types of tumors (leukemia, colon, breast, prostate, melanoma, brain, renal, lung, ovarian). The screening format has been described in detail.[2,4] Briefly, cells are grown in 96-well microtiter plates, with a two-day exposure to a range of concentrations of the test sample. Cell number is assessed using a sulforhodamine stain as an endpoint. From the resulting dose-response curves, three response parameters are calculated for each cell line: a growth inhibitory dose for 50 percent reduction to growth

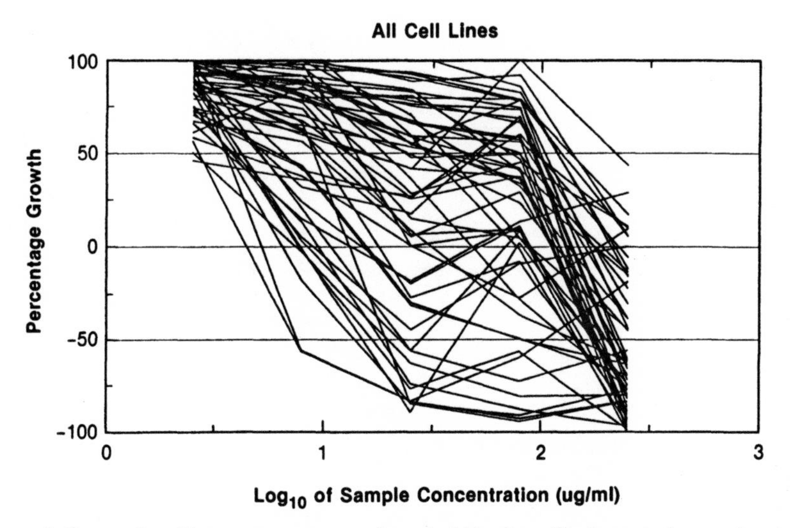

Figure 1. Composite of dose-response curves from *Archidendron ellipticum* crude organic extract tested in the NCI 60 cell line screen.

(GI_{50}), a total growth inhibition level (TGI), where the number of cells remaining is equal to the starting number, and a lethal concentration for 50 percent mortality (LC_{50}), where half of the initial cell number remains viable (Fig. 1). The derivations and definitions of these response parameters are reviewed in detail elsewhere.[5,6]

The 60-cell assay yields 60 dose response curves for each test. One useful way of viewing the data is to plot the relative sensitivities with respect to the particular dose parameter (GI_{50}, TGI, or LC_{50}) as a mean bar graph in which the sensitive cell lines are depicted as bars to the right of the calculated mean for all cells, and resistant lines are depicted as bars to the left of the mean (Fig. 2). The pattern of selectivity depicted in this fashion can be analyzed by pattern-matching algorithms. It has been found that certain classes of cytotoxic compounds which have similar mechanisms of action tend to generate similar patterns of selectivity in this screen.[6-8]

HIV

Extracts are tested routinely against HIV type 1 (HIV-1) in a cytopathicity assay which has been described in detail.[1,9-11] Briefly, the CEM-SS lymphocyte cell line is infected with the RF-1 strain of HIV-1. Extracts which block the cell-killing effect of the virus are detected after a six-day incubation using cellular metabolic reduction of the tetrazolium compound XTT to a colored formazan. Eight concentrations of test extract are used to generate dose-response curves

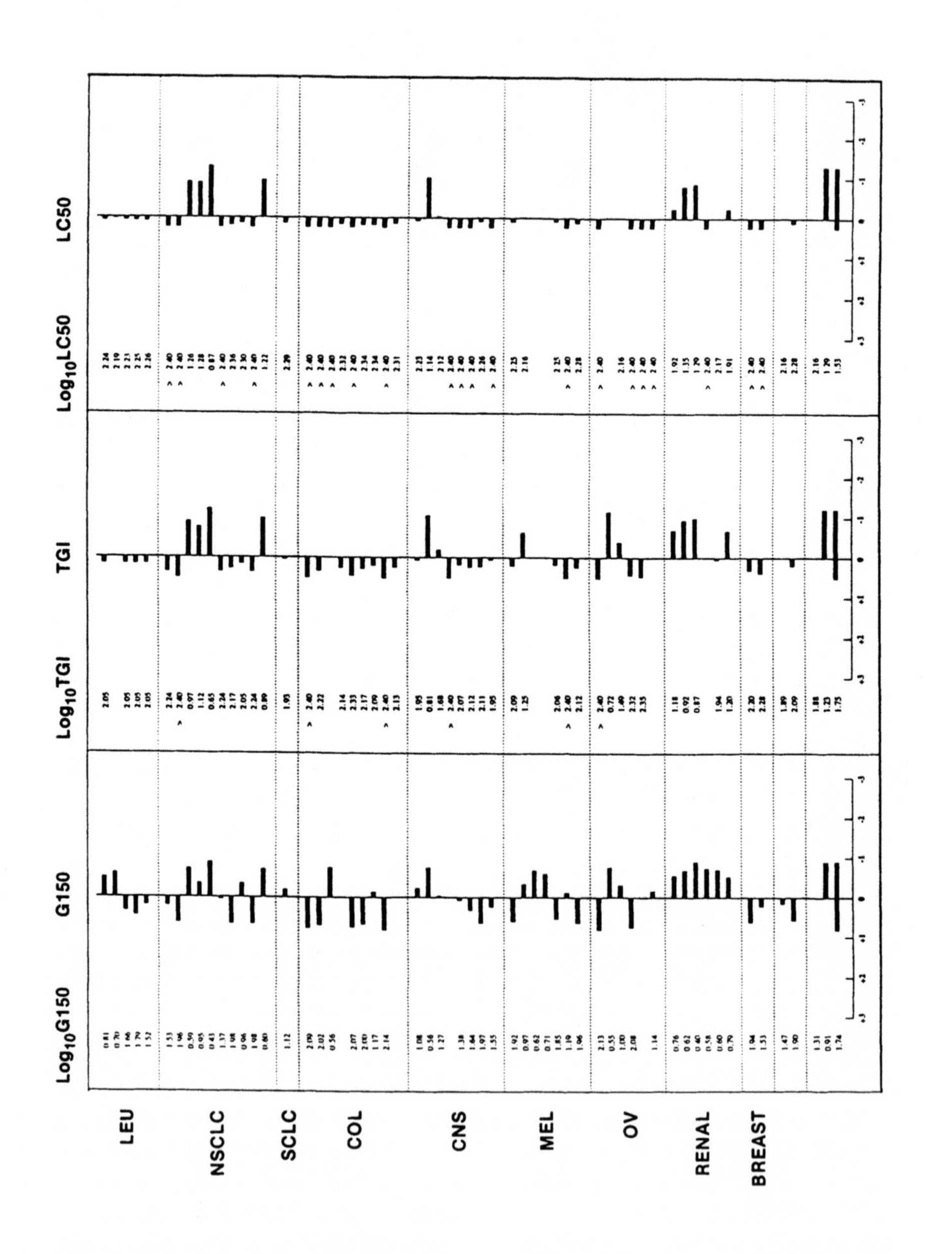

Log10GI50
GI50
Log10TGI
TGI
Log10LC50
LC50
LEU
NSCLC
SCLC
COL
CNS
MEL
OV
RENAL
BREAST

Figure 2. Mean bar graph presentation of the same data as in figure 1, concentration units in μg/ml. Abbreviations: GI50: Growth inhibitory dose 50 percent, TGI: Total growth inhibition dose level, LC50: Lethal concentration 50 percent. Human cell line panel codes: LEU: leukemia cell lines, NSCLC: Non-small cell lung cancer cell lines, SCLC: Small cell lung cancer cell lines, COL: colon cancer lung cell lines, CNS: Central nervous system cancer cell lines, MEL: Melanoma cell lines, OV: Ovarian cancer cell lines, RENAL: Renal cancer cell lines, BREAST: Breast cancer cell lines.

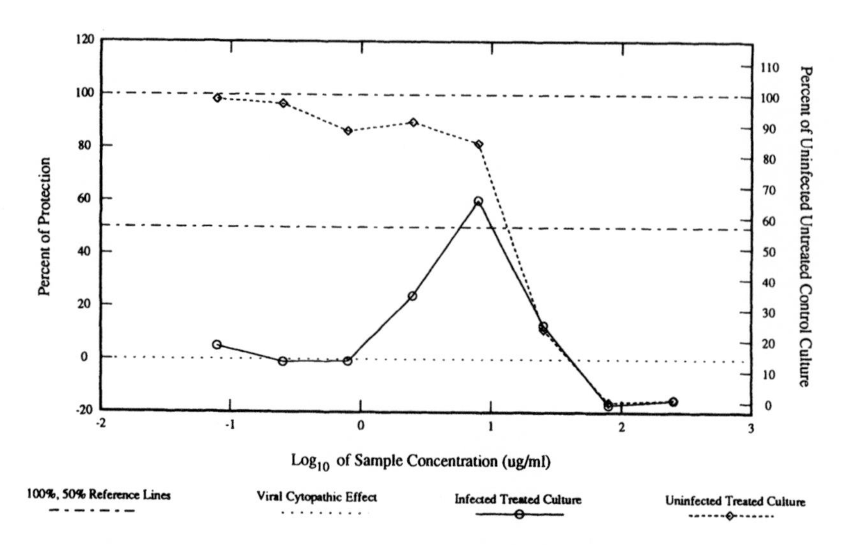

Figure 3. Anti HIV screening profile of compound **1**, O-demethylbuchenavianine.

(Fig. 3) in which protection is measured at an effective concentration 50 percent (EC_{50}), and extract toxicity to the CEM cells is interpreted as an inhibitory concentration 50 percent (IC_{50}). The ratio between these two parameters is also reported as an *in vitro* therapeutic index (TI).

DEREPLICATION OF "NUISANCE COMPOUNDS"

Especially in the anti-HIV screening, many common plant metabolites show some degree of activity. The major recurrent classes of plant compounds active in the primary HIV assay are tannins, acidic polysaccharides, and phorbol esters. Polysaccharides have been found only in the aqueous extracts, while tannins occur in both organic and aqueous extracts. Phorbol esters appear to be limited to the organic extracts, although trace amounts are sometimes observed in an aqueous extract when the companion organic extract is very potently active.

The broad distribution of polysaccharides and tannins in plants has required that the large number of initial HIV leads be carefully evaluated before devoting time and effort to bioassay-guided fractionation. For aqueous extracts, our current dereplication protocol involves precipitation of aqueous extracts with 50% ethanol. The supernatant, the precipitate, and the parent crude extract are then tested in the anti-HIV assay. If all of the bioactivity resides only in the precipitate, this is taken as presumptive evidence for the presence of polysaccharides or other large biopolymers, and the extract is dropped from further consideration. If the supernatant has activity, it is often due to tannins. These tannins are eliminated by passing the ethanolic supernatant through a small column of polyamide gel and eluting successively with water, 50% methanol, and methanol. Each fraction is evaporated and tested in the anti-HIV assay. If no activity is eluted from the polyamide column, this is interpreted to mean that all of the anti-HIV activity in the supernatant was due to tannins, which do not elute under these conditions, and the extract is then dropped from further consideration. Approximately 42 percent of active plant aqueous extracts survive ethanol precipitation, and about 1 percent of those survive the polyamide evaluation. Further evaluation of surviving extracts involves a panel of solid phase extraction cartridges.[12]

For organic anti-HIV extracts, a solvent-solvent partition scheme is used to investigate the polarity of the active species. The polar fractions in this scheme may contain tannins, and these active fractions are passed through polyamide columns in the same fashion as for aqueous supernatants.

The third category of "nuisance compounds", the phorbol esters, are a quite different story. The distribution of phorbol esters is limited to the Euphorbiaceae and Thymelaceae. Many compounds in this class are capable of binding to and activating protein kinase C. This mechanism appears to be responsible for the anti-HIV activity of phorbol esters,[13] as well as for many other kinds of activity in biological systems. Since most of the NCI's plant samples were collected in the tropics, where the Euphorbiaceae comprise a large and prominent fraction of the flora, the occurence of phorbol esters poses a significant challenge. About eight percent of the plant specimens in the NCI repository are from this family.

We have been able to use taxonomy as one criterion in the dereplication of extracts from the Euphorbiaceae. The family is divided into five subfamilies, according to the most recent taxonomic monograph.[14] We have found that bioactive phorbol esters, as measured by [^{3}H]-phorbol dibutyrate binding to rat brain membranes,[15] appear to be limited to only two of these subfamilies, the Euphorbiodeae and the Crotonoideae (see below). This distribution correlates well with the chemical literature on phorbol ester occurence. Thus, if a collection belongs to one of the other three subfamilies, its chances of containing phorbol esters are considered very low, assuming that the plant identification is correct.

Each HIV-positive organic extract is routinely tested in the phorbol dibutyrate (PDBu) binding assay. We have found several cases where extracts from other plant families are active, and these have been evaluated further. For the Euphorbiaceae and Thymelaeaceae, a positive PDBu test is followed by a rapid diol column elution procedure developed in our laboratory (Fig. 4). The anti-HIV and PDBu activities are compared for each of the five fractions generated, and ^{1}H-nmr spectra are acquired. In addition, the dose-response curve from the anti-HIV assay is examined carefully, since phorbol esters produce a characteristic cytostatic effect on the CEM cells which gives the dose-response curve a unique, recognizable shape. The combination

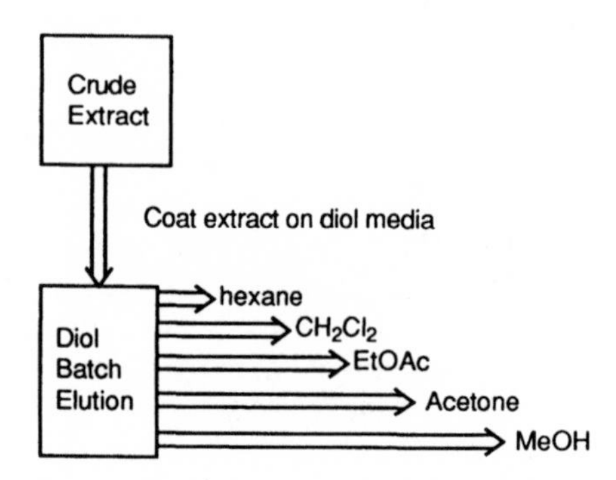

Figure 4. Scheme for diol column dereplication of PDBU-active crude organic extracts.

of evidence from all of these methods is used to prioritize extracts for further work. Most known inhibitors of PDBu binding (phorbol esters, bryostatin, teleocidins, lyngbyatoxin, aplysiatoxins) elute from the diol column in the two least polar fractions, and extracts with a more polar activity profile are given higher priority for investigation. Several leads from this protocol are currently under investigation.

HIV-INHIBITORY PLANT METABOLITES

Buchenavia Alkaloids

An organic extract of *Buchenavia capitata* (Vahl) Eichl. (Combretaceae) showed activity in the anti-HIV assay. Inspection of the phytochemical literature showed that an interesting group of piperidine-flavone alkaloids had been isolated by Ahond and coworkers[16] from this species. Testing of an alkaloid partition fraction of our extract revealed that the bioactivity resided therein. Bioassay-guided fractionation of the extract resulted in the isolation of O-demethylbuchenavianine (**1**) as the primary active compound. Other related alkaloids were isolated but had lesser protective action against HIV.

The cytotoxic concentration of **1** was sufficiently close to the antiviral concentration (i.e., a narrow therapeutic index) that the compound was not considered further as a viable anti-HIV drug development lead. We also investigated antitumor activity of **1** in the NCI screen, but there was only modest differential cytotoxicity. Consideration of the similar structure of rohitukine and its analogs[17] leads us to speculate that the anti-HIV and antitumor activity of **1** may involve tyrosine kinase,[17] although we have not directly tested that hypothesis.

Prostratin

An organic extract of *Homalanthus nutans* (Forster) Pax (Euphorbiaceae) was strongly active in the anti-HIV screen. While nonpolar partition fractions of this extract appeared to contain typical phorbol esters with lipophilic side chains, a more polar fraction yielded the known compound prostratin (**2**, 12-deoxyphorbol-13-acetate), which had activity in both the anti-HIV and PDBu binding assays. Prostratin showed the characteristic cytostatic effect for phorbols mentioned above. Confirmatory assays showed that the compound protected against virus-mediated cytopathicity in CEM cells. Other viral strains and host cells were tested as well. Prostratin was active against HIV-1 strain RF-1 in CEM-SS cells and C-1866 cells, but inactive in the MT-2, C-344, or LDV-7 lines. It was also active against HIV-2, and drug-resistant HIV-1 strains. The compound also inhibited HIV replication in macrophages and monocytes.[18] While most phorbols are tumor promoters, prostratin was inactive in a classical Berenblum tumor-promotion experiment, where mouse skin was initiated with dimethylbenzanthracene before prostratin application. Futhermore, prostratin was found to *block* the tumor promoting effect of TPA in the same type of assay.[19-21] Nevertheless, the drug development potential for prostratin remains uncertain.

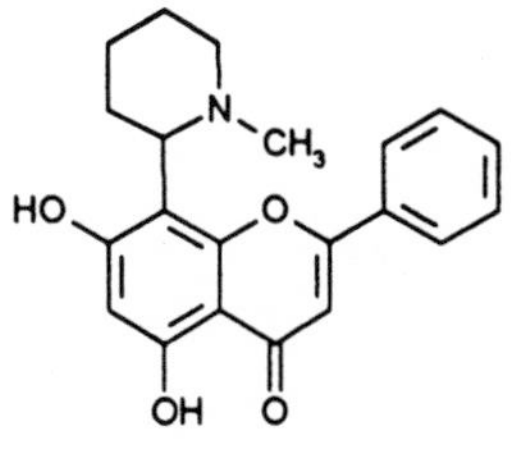

1. O-demethylbuchenavianine

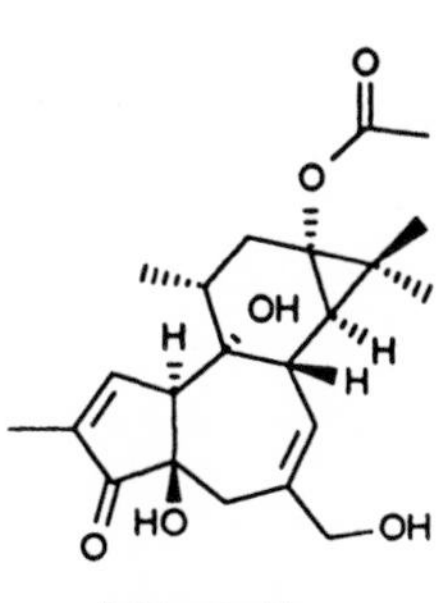

2. Prostratin

The ethnobotanical use of *Homalanthus nutans* in Samoa suggested its potential in therapy.[22] If prostratin was present in the healer preparation in substantial amounts, its use in Samoa might provide preliminary evidence for human toxicity or safety. We undertook a chemical examination of an ethnomedical preparation of *H. nutans* to determine if prostratin were present. Samoan healers, or *taulasea*, typically prepare aqueous extracts by steeping the stem bark in boiling water. In cooperation with Dr. Paul Cox of Brigham Young University, we obtained a sample of such a preparation for analysis. Our preliminary, unpublished results showed that the amount of prostratin in a preparation was very small (37 μg/liter), and that other phorbol esters, as well as tannins, were present in higher concentrations, and likely contributed to the *in vitro* anti-HIV activity of the extract. We are continuing to pursue analysis of a larger number of healer preparations to further clarify this issue.

Phorbol Ester Distribution Studies

The Euphorbiaceae family is well-represented in the tropics where many of the NCI contract collections have been made. Since many of these plant extracts might be suspected of containing phorbol esters which could be active in the anti-HIV screen, we explored the distribution of phorbol esters using the PDBu binding assay as a primary tool.[23]

To date, 403 samples from 79 genera of Euphorbiaceae, representing 193 species, have been tested in the PDBu binding assay. When viewed with respect to the subfamily from which the samples derive (Table 1), it is clear that PDBu

Table 1. Distribution of phorbol ester binding activity in the plant family Euphorbiaceae

Subfamily	Genera tested	Species tested	Samples tested	PDBu active
I. Subfamily Phyllanthoideae	22	71	141	0
II. Subfamily Oldfieldioideae	1	1	4	0
III. Subfamily Acalyphoideae	26	52	99	0
IV. Subfamily Crotonoideae	18	34	85	43
V. Subfamily Euphorbioideae	12	35	74	45

activity is found only in the Crotonoideae and the Euphorbioideae, as noted above. Based upon this information we now can steer our efforts away from probable phorbol ester-containing samples and instead focus upon anti-HIV extracts more likely to yield novel chemistry. In addition, we have identified a number of previously unstudied genera which presumably contain phorbol esters as the anti-HIV principles.

Maprounea

One HIV lead we investigated early in these studies was *Maprounea africana* Muell.-Arg. (Euphorbiaceae, subfamily Euphorbioideae), which showed both HIV- and PDBu-inhibitory activity in the crude organic extract. A series of cytotoxic triterpenes had been reported from this species by Wall and coworkers several years previously,[24] and we obtained reference samples of these compounds from the original investigators. Several of the compounds showed anti-HIV and PDBu activity in our assays, prompting a reinvestigation of our own extracts of *M. africana* and the related *M. membranacea* in order to obtain larger quantities of the supposedly active triterpenes for further biological characterization.

The anti-HIV and PDBu activity fractionated in a consistent manner, but the bioactivity diverged away from the fractions which had characteristic triterpene nmr signals. Eventually, we were able to conclude that the triterpenes did not have the desired activity, and that small amounts of very potent phorbol esters were the major active principles. Gas chromatography-mass spectral analysis of the peracetylated hydrolysate of the active fraction supported this conclusion. Several novel triterpenes **(3-6)** in the maprounic acid/aleuritolic acid class were isolated in the course of the pursuit of the bioactivity. The details of this work have recently been submitted for publication.[25]

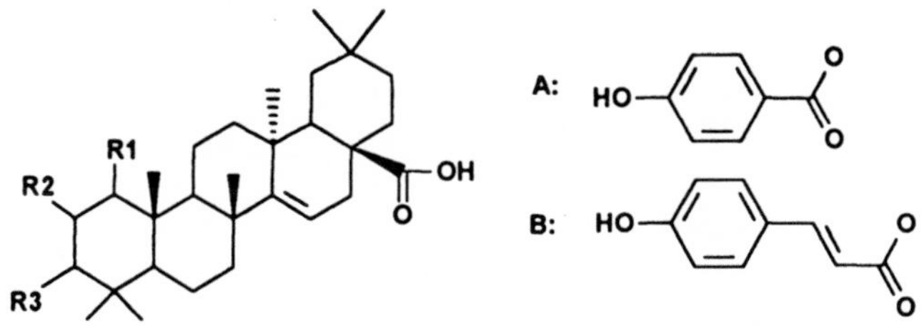

3. R_1=H, R_2=α-OH, R_3=β-A

4. R_1=H, R_2=α-A, R_3=β-OH,

5. R_1=R_2=H, R_3=β-B

6. R_1=H, R_2=α-A, R_3=α-OH

ANTITUMOR PLANT METABOLITES

As mentioned above, the NCI primary antitumor screen is designed to identify extracts or compounds which have a selective effect on particular cell lines or groups of cell lines. While conceptually it is a pragmatic, mechanism-independent assay, the patterns of selectivity toward different cell lines can nonetheless be used to identify extracts with particular mechanisms of action, or leads with entirely novel mechanisms of cytotoxicity.

Compounds Affecting Tubulin

Given the recent interest in taxol as a clinically promising antitumor agent, and the established clinical use of vincristine and vinblastine, there has been some continuing interest in elucidating new types of compounds which interfere with tubulin polymerization and function. It was shown that compounds which share this general type of mechanism of action (e.g. taxol, the vinca alkaloids, rotenone, rhizoxin) also have similar profiles of differential cytotoxicity in the NCI cancer screen.[26]

We have extended this idea to the identification of crude extracts with screening profiles which match the "seed" patterns for known tubulin-interactive agents. Using a panel of seven representative tubulin-interactive drugs (taxol, colchicine, vincristine, rotenone, rhizoxin, maytansine, and dolastatin 10), extracts were sought which gave the best matches to these related patterns of cytotoxicity. A number of extracts which matched were eliminated due to the liklihood that they contained other well-studied compounds. For example, extracts of several *Maytenus* species gave excellent matches to the seed compounds, but it was considered highly probable that they contained maytansinoids as the active principles. Similarly, extracts of *Lonchocarpus* and *Derris* were rejected due to the probable occurence of rotenoids.

This taxonomic winnowing was not entirely effective, unfortunately, since known compounds can occur in new taxa. We have found rotenone and several of its congeners in the extracts of *Dalea cylindrica* Hook. (Leguminosae), and in *Phyllanthus mirabilis* Muell.-Arg. (Euphorbiaceae). While rotenone had not previously been found in the family Euphorbiaceae, it is apparently quite common in the legumes.

Centaureidin

Of greater interest was the result from an extract of *Polymnia fruticosa* Benth. (Compositae), where cytotoxicity and inhibitory activity of the polymerization of tubulin was traced to the flavone centaureidin (**7**).[27] This compound, repeatedly noted in various genera of the Compositae, had been reported by Kupchan's group to have cytotoxic properties,[28] an effect confirmed by a second group.[29] In collaboration with Hamel's group in Bethesda, we found that the cytotoxic effect of centaureidin correlated well with the appearance of mitotic figures, indicating that an antimitotic effect was produced in whole cells. In addition, centaureidin had a strong inhibitory effect on *in vitro* tubulin polymerization.[27]

Flavone SAR

The interesting actions of centaureidin prompted us to evaluate structurally related flavones in the 60-cell cancer screen and the mechanism-based tubulin assays. We chose to focus our attentions initially on flavones with 3-methoxy substitution. While a number of the flavones we chose to examine were cytotoxic, only one other compound (**8**) yielded a pattern which matched that of centaureidin and was also active in the tubulin assays. This result has been recently confirmed by others.[30] We are continuing to explore the structure-activity relationships in this series of compounds. *In vivo* testing of centaureidin awaits an adequate supply of compound, since we were able to isolate it in only relatively low yield from *P. fruticosa*.[27]

Chalcones

The organic extract of *Calythropsis aurea* C.A.Gardner (Myrtaceae) showed a modest correlation to the tubulin "seed" compounds, and was fractionated to yield two novel chalcones, calythropsin (**9**) and dihydrocalythropsin (**10**) as the active principles.[31] Calythropsin was moderately cytotoxic in the 60-cell assay, having a mean GI_{50} value of 0.66 μg/ml, however, the correlation of its screening profile to tubulin "seed" compounds was poor (correlation coefficients of 0.36 to 0.57), and the compound was inactive in the *in vitro* tubulin polymerization assay. However, a further experiment using L1210 cells showed the presence of some increased mitotic figures at the cytotoxic IC_{50} (7 μg/ml) for those cells. Calythropsin is thus, at best, only a very weak antimitotic agent.

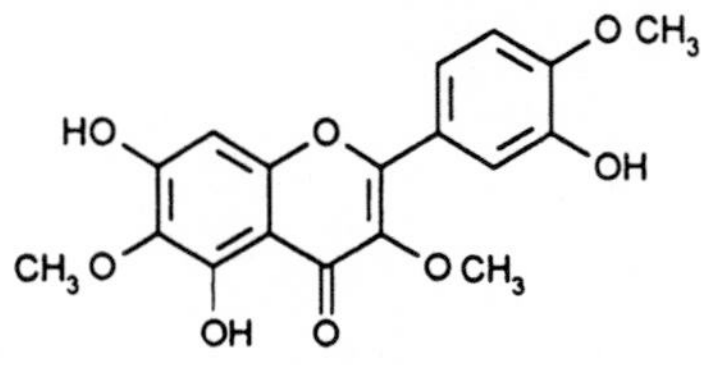

7. Centaureidin

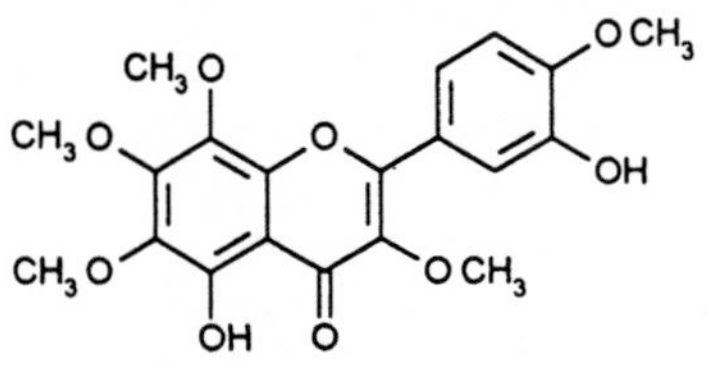

8. 5,3'-dihydroxy-3,6,7,8,4'-pentahydroxyflavone

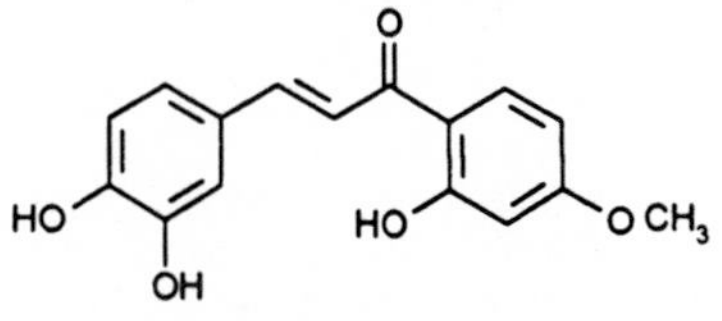

9. Calythropsin

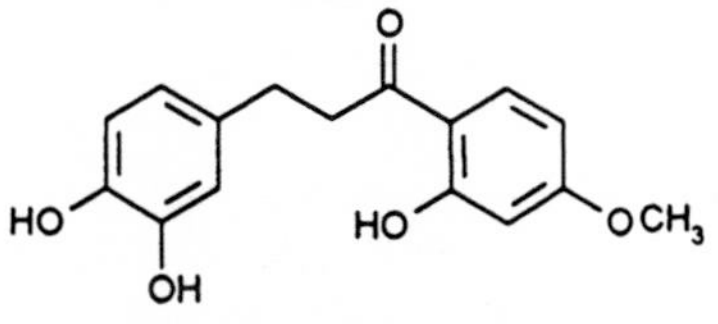

10. Dihydrocalythropsin

Renal Selective Saponins from Legumes

A quite different type of project involves the investigation of extracts with patterns of cytotoxic selectivity which are *not* related to any known mechanism of action. One such project involves the organic extract of *Archidendron ellipticum* (Bl.) Nielsen (Leguminosae), which inhibited the growth of all six of the renal cancer cell lines in the 60-cell panel (Fig. 1). While the renal cells were only about one-half to one log unit more sensitive than the average, the fact that all of the renal cell lines were sensitive was somewhat unusual.

Bioassay-guided fractionation rapidly showed that the active principle was a complex mixture of saponins. However, the purification of the individual saponins to homogeneity proved to be a very difficult task. Countercurrent methods were tried, but the recovery of bioactivity was poor. The key to successful separation proved to be utilization of wide-pore bonded-phase media. In particular, a 300 Å (mean pore diameter) amino-bonded phase chromatography medium, eluted with acetonitrile-water mixtures gave relatively high resolution of the saponin mixture (Fig. 5). Examination of the fractions by nmr and wide-pore C4 hplc showed that the peaks were still relatively complex mixtures. Further

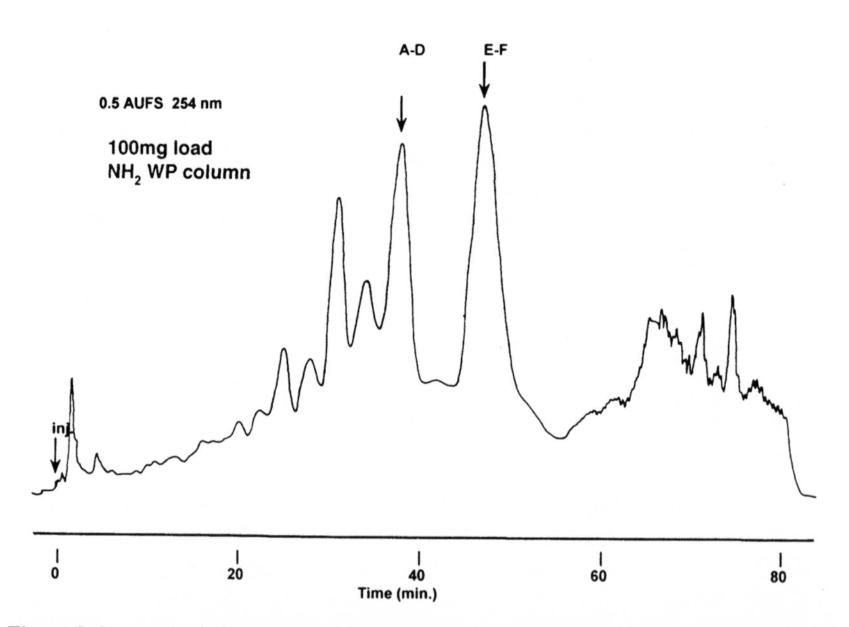

Figure 5. Preparative hplc separation of crude saponin mixture by wide-pore amino bonded phase column. Arrows point to peaks containing elliptosides A-D and E-H. Eluant acetonitrile-H_2O 80% isocratic 25 ml/min., linear gradient to 60% acetonitrile-H_2O at 45 min. Column YMC PBMN Polyamine 20 x 250 mm, 5μ, 300 Å.

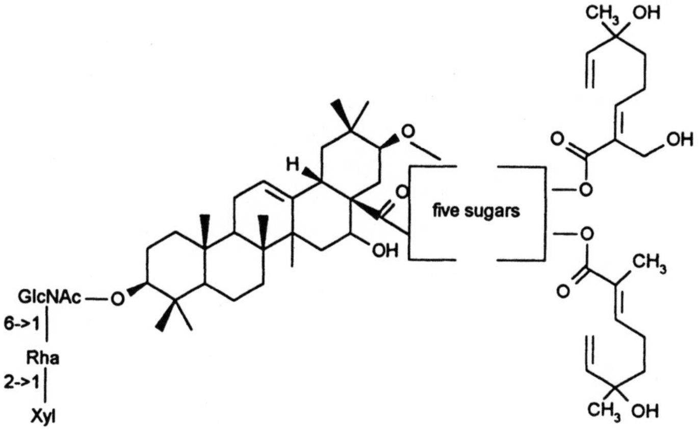

Figure 6. Elliptoside A partial structure.

purification using wide-pore C4 reversed-phase hplc was required to give pure saponins.

The saponins, named elliptosides A-J, are bisdesmosides of acacic acid, with a trisaccharide unit attached at position C-3, and an ester linkage from C-28 to a tetrasaccharide unit and further functionalized at C-21 (Fig. 6). Basic hydrolysis of the saponin mixture liberates two fragments missing the C-21 and C-28 portions of the molecule and differing in the internal sugar, which is either glucose or N-acetylglucosamine. More vigorous hydrolysis liberates acacic acid as the primary genin. The details of the structure and the sites of monoterpene attachment are under investigation.

While the saponins as a class are relatively abundant in the extract, individual members are present in much smaller amounts, which has limited the biological work on the pure compounds. However, the mixture of saponins has proven to have interesting biological activity in experimental therapeutic models, and the limited amount of testing with the major pure compound elliptoside A has been consistent with the activity of the mixture. While it would be preferable to test renal cells lines *in vivo*, our initial testing has employed the more tractable melanoma cell line LOX IMVI, which is also sensitive to the saponins (figure 1). A minimal challenge model, where both drug and tumor cells were injected i.p. into nude mice, yielded complete protection from the tumors at several doses which were not toxic to the animals (i.e., < 2 mg/kg). More stringent trials where drug is injected i.v. or s.c. are ongoing, with promising results to date. Thus, if the structure determination and supply problems can be solved, the elliptosides may prove to be an interesting antitumor lead for further investigation.

CONCLUSIONS

Higher plants have provided interesting new leads active in the anti-HIV or antitumor screens currently in operation at the NCI. Suitable procedures for the elimination of recurring classes of compounds such as phorbol esters, acidic polysaccharides, and tannins have been developed which can rapidly and efficiently identify these types of compounds. Some rather common, known classes of compounds (flavones, saponins) have been found to contain some heretofore unknown profiles of biological activity of potential interest for future investigation.

ACKNOWLEDGMENTS

We thank Glenn N. Gray and Lewis Pannell for mass spectroscopy, Mark Tischler for his work on *Maprounea* and rotenoids, Kirk Gustafson for his work on prostratin, Anne Monks, Dominic Scudiero and Tanya Prather for cytotoxicity assays, Robert Moran, Robert Gulakowski, Owen H. Weislow, and David Clanton for anti-HIV assays, Ernest Hamel for tubulin assays, K.H. Lee for the generous supply of compound **8**, and Gordon Cragg and Kenneth Snader of the NCI Natural Products Branch for their work in obtaining the plant collections, and Thomas McCloud for making the plant extracts.

REFERENCES

1. BOYD, M. R. 1988. Strategies for the identification of new agents for the treatment of AIDS: A national program to facilitate the discovery and preclinical development of new drug candidates for clinical evaluation. In: AIDS, Etiology, Diagnosis, Treatment and Prevention, (V.T. Devita, S. Hellman, S.A. Rosenberg,, eds.). J.B. Lippincott, Philadelphia. pp. 305-319.
2. BOYD, M. R. 1989. Status of the NCI preclinical antitumor drug discovery screen. In: PPO Updates, vol.3, no.10 (V.T. Devita, S. Hellman, S.A. Rosenberg, eds.). J.B.Lippincott, Philadelphia. pp. 1-12.
3. BOYD, M. R. 1993. The future of new drug development. In: Current Therapy in Oncology, (J.E. Niederhuber, ed.). B.C. Decker, Inc., Philadelphia. pp. 11-22.
4. MONKS, A., SCUDIERO, D., SKEHAN, P., SHOEMAKER, R., PAULL, K., VISTICA, D., HOSE, C., LANGLEY, J., CRONISE, P., VAIGRO-WOLFF, A., GRAY-GOODRICH, M., CAMPBELL, H., BOYD, M. 1991. Feasibility of a High-Flux Anticancer Drug Screen Utilizing a Diverse Panel of Human Tumor Cell Lines in Culture. J. Natl. Cancer Inst. 83: 757-766.
5. BOYD, M. R., PAULL, K. D., RUBINSTEIN, L. R. 1992. Data display and analysis strategies for the NCI disease-oriented *in vitro* antitumor drug screen. In: Cytotoxic Anticancer Drugs: Models and Concepts for Drug Discovery and Development, (F.A. Valeriote, T. Corbett, L. Baker, eds.). Kluwer Academic Publishers, Amsterdam. pp. 11-34.
6. BOYD, M. R . and PAULL, K. D. 1995. Some practical considerations and applications of the NCI *in vitro* drug discovery screen. Drug Develop. Res. (In Press).

7. PAULL, K. P., HAMEL, E., MALSPEIS, L. 1994. Prediction of biochemical mechanism of action from the *in vitro* antitumor screen of the National Cancer Institute. In: Cancer Chemotherapeutic Agents, ACS Books, Washington, D.C..
8. PAULL, K. D., SHOEMAKER, R. H., HODES, L., MONKS, A., SCUDIERO, D.A., RUBINSTEIN, L., PLOWMAN, J., BOYD, M. R. 1989. Display and analysis of patterns of differential activity of drugs against human tumor cell lines: Development of mean graph and COMPARE Algorithm. J. Natl. Cancer Inst. 81: 1088-1092.
9. GULAKOWSKI, R. J., MCMAHON, J. B., STALEY, P. G., MORAN, R. A., BOYD, M. R. 1991. A semiautomated multiparameter approach for anti-HIV drug screening. J. Virol. Meth. 33: 87-100.
10. BADER, J. P. 1991. The AIDS antiviral screening programme of the National Cancer Institute. Biotechnology International 1992, Century Press, London. pp. 271-275.
11. WEISLOW, O. S., KISER, R., FINE, D. L., BADER, J., SHOEMAKER, R. H., BOYD, M. R. 1989. New soluble-Formazan assay for HIV-1 cytopathic effects: Application to high-flux screening of synthetic and natural products for AIDS-antiviral activity. J. Natl. Cancer Inst. 81: 577-586.
12. CARDELLINA, J. H., II, MUNRO, M. H. G., FULLER, R. W., MANFREDI, K. P., MCKEE, T. C., TISCHLER, M., BOKESCH, H. R., GUSTAFSON, K. R., BEUTLER, J. A., BOYD, M. R. 1993. A chemical screening strategy for the dereplication and prioritization of HIV-inhibitory aqueous natural products extracts. J. Nat. Prod. 56: 1123-1129.
13. FIRESTEIN, G. S., REIFLER, D., RICHMAN, D., GRUBER, H. E. 1988. Rapid and reversible modulation of T4 (CD4) on monocytoid cells by phorbol myristate acetate: Effect on HIV susceptibility. Cell. Immunol. 113: 63-69.
14. WEBSTER, G. L. 1975. Conspectus of a new classification of the Euphorbiaceae. Taxon. 24: 593-601.
15. DEVRIES, D. J., HERALD, C. L., PETTIT, G. R., BLUMBERG, P. M. 1988. Demonstration of sub-nanomolar affinity of bryostatin 1 for the phorbol ester receptor in rat brain. Biochem.Pharmacol.. 37: 4069-4073.
16. AHOND, A., FOURNET, A., MORETTI, C., PHILOGèNE, E., POUPAT, C., THOISON, O., POTIER, P. 1984. Premiers alcaloïdes vrais isolés de Combrétacées, *Buchenavia macrophylla* Eichl. et *Buchenavia capitata* Eichl.. Bull. Soc. Chim. Fr. 2: 41-45.
17. HARMON, A. D., WEISS, U., SILVERTON, J. V. 1979. The Structure of rohitukine, the main alkaloid of *Amoora rohituka* (Syn. *Aphanamixis polystachya*) (Meliaceae). Tet. Lett. 721-724.
18. GUSTAFSON, K. R., CARDELLINA, J. H., II, MCMAHON, J. B., GULAKOWSKI, R. J., ISHITOYA, J., SZALLASI, Z., LEWIN, N. E., BLUMBERG, P. M., WEISLOW, O. S., BEUTLER, J. A., BUCKHEIT, R. W., CRAGG, G. M., COX, P. A., BADER, J. P., BOYD, M. R. 1992. A nonpromoting phorbol from the Samoan medicinal plant *Homalanthus nutans* inhibits cell killing by HIV-1. J. Med. Chem. 35: 1978-1986.
19. SZALLASI, Z., KRSMANOVIC, L., BLUMBERG, P. M. 1993. Nonpromoting 12-deoxyphorbol 13-esters inhibit phorbol 12-myristate 13-acetate induced tumor promotion in CD-1 mouse skin. Cancer Res. 53: 2507-2512.
20. KAZANIETZ, M. G., KRAUSZ, K. W., BLUMBERG, P. M. 1992. Differential irreversible insertion of protein kinase C into phospholipid vesicles by phorbol esters and related activators. J. Biol. Chem. 267: 20878-20886.
21. SZALLASI, Z., KRAUSZ, K. W., BLUMBERG, P. M. 1992. Non-promoting 12-deoxyphorbol 13-esters as potent inhibitors of phorbol 12-myristate 13-acetate-induced acute and chronic biological responses in CD-1 mouse skin. Carcinogenesis. 13: 2161-2167.
22. COX, P. A., BALICK, M. J. 1994. The ethnobotanical approach to drug discovery. Scientific American. 270: 82-87.

23. BEUTLER, J. A., ALVARADO, A. B., MCCLOUD, T. G., CRAGG, G. M. 1989. Distribution of phorbol ester bioactivity in the Euphorbiaceae. Phytother. Res. 3: 188-192.
24. WANI, M. C., SCHAUMBERG, J. P., TAYLOR, H. L., THOMPSON, J. B., WALL, M. E. 1983. Plant antitumor agents, 19. Novel triterpenes from *Maprounea africana*. J. Nat. Prod. 46: 537-543.
25. BEUTLER, J. A., KASHMAN, Y., TISCHLER, M., CARDELLINA, J. H. II, GRAY, G. N., CURRENS, M. J., WALL, M. E., WANI, M. C., BOYD, M. R.. A reinvestigation of Maprounea Triterpenes. J. Nat. Prod., submitted, 1995.
26. PAULL, K. D., LIN, C. M., MALSPEIS, L., HAMEL, E. 1992. Identification of novel antimitotic agents acting at the tubulin level by computer-assisted evaluation of differential cytotoxicity data. Cancer Res. 52: 3892-3900.
27. BEUTLER, J. A., CARDELLINA, J. H., II, LIN, C. M., HAMEL, E., CRAGG, G. M., BOYD, M. R. 1993. Centaureidin, a cytotoxic flavone from *Polymnia fruticosa*, inhibits tubulin polymerization. Bioorg. Med. Chem. Lett. 3: 581-584.
28. KUPCHAN, S. M., BAUERSCHMIDT, E. 1971. Cytotoxic flavonols from *Baccharis sarothroides*. Phytochemistry. 10: 664-666.
29. JEFFERIES, P. R., KNOX, J. R., PRICE, K. R., SCAF, B. 1974. Constituents of the tumor-inhibitory extract of *Olearia muelleri*. Austr. J. Chem. 27: 221-225.
30. LICHIUS, J. J., THOISON, O., MONTAGNAC, A., PAÏS, M., GUÉRITTE-VOGELEIN, F., SÉVENET, T., COSSON, J.-P., HADI, A. H. A. 1994. Antimitotic and cytotoxic flavonols from *Zieridium pseudobtusifolium* and *Achronycha porteri*. J. Nat. Prod. 57: 1012-1016.
31. BEUTLER, J. A., CARDELLINA, J. H., II, GRAY, G. N., PRATHER, T. R., SHOEMAKER, R. H., BOYD, M. R., LIN, C. M., HAMEL, E., CRAGG, G.M . 1993. Two new cytotoxic chalcones from *Calythropsis aurea*. J. Nat. Prod. 56: 1718-1722.

Chapter Four

BIOLOGICAL DIVERSITY OF MEDICINAL PLANTS IN MEXICO

Robert Bye[1], Edelmira Linares[1], and Eric Estrada[2]

[1] Instituto de Biología
Universidad Nacional Autónoma de México
04510 México, DF, México
[2] Departamento de Fitotecnia
Universidad Autónoma Chapingo
Chapingo, México, México

INTRODUCTION

Biological diversity can be described as the product of the richness or variety of entities (usually species) and the variance of that richness or its importance value.[1] Diversity can be extended to include ecological and biogeographic heterogeneity, and various indices can be used to measure it.[2] Biological diversity also can be appreciated by the number of endemic species whose distributions are restricted to a confined geographic area. In recent years, public attention has been given to diversity at the world level[3] as well as in the Western Hemisphere[4] and Mexico.[5]

Preliminary evaluation indicates that Mexico is the third ranked megadiversity country after Brazil and Colombia, followed by Madagascar, Zaire and Indonesia,[6] even though its 1,972,544 km^2 territory places it as the 14th largest

Phytochemistry of Medicinal Plants
Edited by John T. Arnason et al., Plenum Press, New York, 1995

Table 1. Plant diversity by vegetation types expressed in terms of richness (number of species and percentage of flora) and endemism (percentage of species).[7,73] The ecological zones are: ASA—arid and semi-arid; SHTm—subhumbid temperate; HTm—humid temperate; HTr—humid tropical; SHTr—subhumid tropical

VEGETATION TYPE	ECOLOGICAL ZONE	% AREA	#/% SPECIES	% ENDEMICS
desert shrub& grassland	ASA	50	6,000 / 20	60
conifer-oak forest	SHTm	21	7,000 / 24	70
cloud forest	HTm	1	3,000 / 10	30
tropical evergreen forest	HTr	11	5,000 / 17	5
tropical deciduous forests	SHTr	17	6,000 / 20	40

country in world. The juxtaposition of this country's medley of topography with the confluence of the Neotropical Kingdom with warm tropical floristic elements and the Holarctic Kingdom with cold temperate floristic components amplifies the variety of organisms and their interactions.

Although the vascular plant flora is not thoroughly explored, it is estimated that it consists of at least 21,600 species.[7] The diversity within Mexico's flora can be appreciated by considering the distribution of richness and endemism by vegetation types (Table 1). One of the most striking observations from this summary is that the conifer and oak forest, which is usually associated with the mountains and the subhumid temperate zones, has the greatest number of species (about 24%) and the highest number of endemics (7,000) even though it covers only about 21% of country's land surface. The other vegetation types of importance are the desert shrub and grassland vegetation and the tropical deciduous forests, which have about 6,000 species each and cover 60% and 17% of the Mexican territory, respectively.

In addition to biological diversity, cultural diversity of Mexico is high, placing it third in the Western Hemisphere after Brazil and Colombia. Over 8 million indigenous people (or about 7.5% of the population in 1980) speak a native language. Of the original 120 language families that were spoken in Mexico at the time of European contact, 54 persist today.

A third type of diversity, ethnobotanical diversity, is based upon the interaction of biological and cultural diversities.[7] Although only one-fifth of Mexican indigenous societies has recorded ethnobotanical information,[8,9] Mexico is recognized as a pivotal geographic point for plant-human interaction. For example, one of the three primary centers of the origin of agriculture is located here.[10] Mexican wild and cultivated plants are recognized as a major source of germplasm for world exploitation.[11] Nonetheless, one should be cautious in assuming a positive correlation between high taxonomic richness and high importance value of plants. For example, the century plant or *Agave*, a multiple use plant that was critical to human survival in Mexico for many centuries and which has 136 species distributed primarily in Mexico,[12] produces "aguamiel" (the sugar rich sap) and "pulque" (the fermented juice used as medicine and beverage). This represents one of the most intricate relationships between humans and plants. Nonetheless, this significantly important ethnobotanical interaction that evolved in the arid southern Mexican Central Plateau is not correlated geographically with the three areas of high species richness (in the Sierra Madre Occidental, the southern Sierra Madre Oriental and the southern Valley of Tehuacan).[13] A greater degree of species richness does not necessarily indicate higher ethnobotanical diversity. Although medicinal plants have not been studied comparably as to their richness and variation, the *Agave*-man symbiosis suggests that the more phytochemically interesting plant-human interactions may be found in environmentally marginal or stressful areas rather than in species-dense regions.

RICHNESS

The richness of Mexican medicinal plants can be estimated from the number of species utilized. Since the first contact with Spaniards, efforts have been made to provide a basic listing of plant species that afford remedial properties. Such endeavors, using different sources and conducted on diverse scales, continue today. Even less well known is the variation within species. On one hand, there is little documentation of variation in phytochemistry of Mexican plants in response to different environmental factors such as altitude, day length, fertilizers and other components, although it is assumed that plasticity in the production of biodynamic principles follows the general patterns known for plants in other parts of the world.[14] On the other hand, the existence of genetically distinct chemical races in Mexican plants is suspected based upon preliminary research.

Inventory of Medicinal Plants

The richness of medicinal plants in Mexico has been estimated using various sources. To date, however, there is no systematic survey of medicinal plants grown and consumed in Mexico. Programs at various universities and national agencies (e.g., Secretaría de Salud, Instituto Mexicano de Seguro Social, Instituto Nacional Indigenista) are compiling inventories based upon bibliography, specimens, and field surveys. Attempts to survey the medicinal flora have been documented since the early colonial period. A selection of the critical publications are discussed below.

During the first half century of the colonial period, three major works were compiled in central Mexico. Martin de la Cruz[15] prepared an herbal with a Nahuatl text and colored illustrations. Known as the *Badianus Manuscript*, for Juan Badiano who translated the text into Latin, it was discovered in the Vatican Library in 1929 and subsequently published. The original document, which was prepared in 1552, contains 263 plants named in Nahuatl along with remedial animals, stones and soil. Present-day analysis of 185 plants recognizes 83 plants to species, 53 to genus, and 21 to family, all of which are distributed in 61 botanical families.[16] The most important vascular plant families in order of greatest number of species are Asteraceae, Fabaceae, Convolvulaceae, Lamiaceae and Solanaceae. Many of the ethnomedical concepts are native in origin; thus this document probably is the closest of all postConquest material to prehispanic view of illnesses and curatives.

In order to aid the conversion of the natives to the Catholicism, Fray Bernardino de Sahagún[17] worked from 1558 to 1582 on various manuscripts that were condensed into the *Florentine Codex* which first became available to Western scholars in 1829.[18] Fray Sahagún's technique of systematic questioning and recording in the native language with illustrations distinguishes his work as the first major ethnographic study of the New World. Certain sections were translated into Spanish so that the Church authorities could use the information to identify pagan rituals and plants of the devil. Of the 11 books that survive today, Book 10 (The People) and Book 11 (Earthly Things) list 724 useful plants with native names. The most important anthropocentric use category is that of medicinal with 266 species.[19]

In addition to the inventories made by the native health practitioner and the church ethnographer, the representative of the Spanish royalty, Francisco Hernández (King Philip II's physician), also made a systematic study of the curative flora of New Spain as part of his *Natural History of New Spain*.[20] Between 1571 and 1576, he travelled in a geographic area that was larger than that visited by Sahagún and characterized the properties of the Mexican plants in European terms. Although indigenous information is included, this document is least reflective of the native knowledge and employment of medicinal plants during the early colonial period. The Latin text later was complemented by illustrations and the data were incorporated into the works of other authors.[21] Of the 3,076 plants listed by common names, only 667 have been identified to species, 249 to genus and 98 to family.[22]

These three references document the importance of various medicinal plants during the early colonial period and probably reflect to a certain degree their significance prior to the arrival of Hernán Cortés. Although many of the same plants are found in all three references, the ethnobotanical data do not always coincide. In some cases (e.g., jimson weed, *Datura stramonium* L. (Solanaceae) called "tlápatl"), the names, uses, plant parts utilized, and forms of preparation and administration are similar, while in others (e.g., the hand flower tree, *Chiranthodendron pentadactylon* Larr. (Sterculiaceae) known as "macpalxóchitl") there is little correspondence.[23]

As part of the Spanish colonial interest in recording and exploiting the natural resources of New Spain, a systematic program of questioners known as the *Relaciones Geográficas* were answered by church authorities throughout the New Spain. Particular attention was given to the medicinal plants of the region, especially antidotes for poison arrows and venomous animals. No contemporary analysis of the plants in these surveys has been made.

Based on the accumulated knowledge of the missionaries in New Spain, Esteyneffer[24] compiled the *Florilegio Medicinal* as a medical guide for his colleagues who were saving souls and bodies throughout the Spanish colony. A preliminary study of this provincial manual first published in 1712 reveals that about 273 plants are mentioned.[24]

Later, other books recorded the plants used in medical practice in the colony.[25] The reawakening of scientific interest and the desire of Charles III to promote the medical sciences (which included botany) in New Spain formed the basis of the Royal Scientific Expedition (1875-1803). The Expedition's base was the Royal Botanical Garden of Mexico that was established in 1788 in order to promote the experimentation and use of Mexican medicinal plants.[26] The botanical publications (*Flora Mexicana* and *Plantae Novae Hispaniae*) did not become available until 1894 and contain little information on medicinal plants. In 1801, Juan Navarro[27] assembled information about medicinal plants of central Mexico along with color illustrations of the plants that are incorporated in the recently discovered *Historia Natural o Jardín Americano*.

After the release from colonialism and stabilization of the country, the Mexican government sponsored a concentrated effort to document, analyze and promote medicinal plants in Mexico. Between 1888 and 1915 the Instituto Médico Nacional (IMN) published results of interdisciplinary studies in such publications as "El Estudio" (4 volumes), "Anales del Instituto Médico Nacional" (12 volumes), 14 monographs, and 5 books.[28] At the same time, the United States Department of Agriculture and the Smithsonian Institution were conducting field studies on medicinal and other plants in Mexico. Rose[29] was so impressed with the advancement of the Mexican program that he decided not to publish all their information so that the Mexican program would have full credit. Unfortunately, the Mexican Revolution terminated the program and the remnants of the specimens and data were used to create the Instituto de Biología de la Universidad Nacional de México in 1929.

Much of the information generated by IMN was used in the formulation of the Mexican Pharmacopeia and in the classic work on medicinal plants of Mexico written by Maximino Martínez.[30] Martínez' *Plantas Medicinales de México* first appeared in 1933. Considered the bible of Mexican medicinal plants, it contains information on 883 species and is divided into three parts. The first section contains data derived from the beginning of this century while the last two portions have only fragmentary data. Unfortunately, there are no voucher specimens with which one can confirm the botanical identity of the entries. Given the poor state of knowledge of the Mexican flora during that period, the species to which curative powers are attributed are not reliable. Popular guides to Mexican medicinal plants and herbolaria that were published subsequently often copy and miscopy data from Martínez' pioneering work.

The next major compilation was published in 1976 by the Instituto Mexicano sobre el Estudio de Plantas Medicinales (IMEPLAM), a multidisciplinary group from universities and governmental institutions that promoted the evaluation and use of Mexican medicinal plants. Both the species inventory[31] and the summary of the uses and state of experimental knowledge[32] are based primarily on 35 references from historical documents, Instituto Médico Nacional, compilations based upon IMN studies and Martínez' book. Recorded were 2,196 species of vascular plants in 900 genera and 161 families. The initial compilation of botanical, nomenclatural, phytogeographical, ecological, historical, chemical, pharmacological, and bibliographic data to create the medicinal flora of Mexico was undertaken by the Instituto Mexicano de Seguro Social (IMSS).[33]

As part of a national evaluation of the genetic resources in Mexico, a list of medicinal plants was generated based upon publications, state inventories and student theses. This study[14] revealed 3,352 vascular plant species distributed in 1,214 genera and 166 families in Mexico. Hence, 15% of the Mexican flora has been employed for remedial purposes. The five families with the greatest number of species are: Asteraceae (383), Fabaceae (324), Euphorbiaceae (137), Lamiaceae (92), and Solanaceae (92). The geographic distribution revealed few studies from the states of Baja California Norte, Baja California Sur, Sinaloa, Coahuila, Nayarit, Zacatecas, Colima, Aguascalientes and Campeche. Most surveys are concentrated in central Mexico, where major universities and institutions have been located throughout Mexican history. The ecological arid and semi-arid zones have the least number of studies.

In order to produce catalogs of regional medicinal plants, many institutions have compiled state inventories. These initial works are unreliable as they are often based on limited information and they report plants as being medicinal based on the species' presence within the state's boundary although there are no documented uses for remedial purposes in the state. The states with highest percentage of locally documented medicinal plants are in order: Quintana Roo (99% of 373 species),[34] Yucatan (60% of 623 species),[35] Veracruz (28% of 548 species),[36] Durango (26% of 255 species), [37] and Sonora (18% of 450 species).[38]

The most recent effort to produce systematically an inventory of Mexican medicinal plants was carried out by Instituto Nacional Indigenista (INI) using native participants as well as bibliographic and herbarium sources.[39] Based upon four phases of development, the project documented the plants used in 35 contemporary rural communities[40]. The bibliography[41] consists of 2049 references on medicinal plants used in Mexico and contains chemical studies on 394 species, chemical and pharmacological studies on 280 species, chemical, pharmacological, and active principal studies on 88 species, chemical, pharmacological and toxicological studies on 177 species, and chemical, pharmacological, toxicological and active principal studies on 69 species. Combining these sources,[42] 3103 species in 183 families are currently used in Mexico. Based upon the 1000 principal medicinal plants, 45.2% are herbs, 28.2% shrubs, 27.7% trees and 5.7% vines.[42] These vegetal remedies grow, in order of importance though not limited exclusively to, in dry tropical forest (44.4%), oak forest (44.1%), pine forest (37.5%), desert (34.7%) and humid tropical forest (32.5%).[42]

Infraspecific Variation

Although phytochemical variation in medicinal plants based upon genetic forms has been demonstrated in many European plants,[43] few detailed studies have been made to verify the existence of chemovars in Mexican plants. Two exceptions are described here.

Mexican hyssops of the genus *Agastache* section *Brittonastrum* (Lamiaceae) has its center of diversity in Mexico and adjacent southwestern USA where 14 species grow in the pine-oak forests.[44] *Agastache mexicana* (HBK.) Lint & Epling) is highly prized as a calmative tea where it grows naturally in the Transvolcanic Mountains of central Mexico. Two subspecies are readily recognized by flower color and distinctive aroma. "Toronjil rojo" (*A. mexicana* subsp. *mexicana*) has reddish flowers and an anise odor, while "toronjil blanco" (*A. mexicana* subsp. *xolocotziana* Bye, Ramamoorthy & Linares) has white flowers and a lemon fragrance. The red and white hyssops are mixed with "toronjil blue" (*Dracocephalum moldavica* L., Lamiaceae) and drunk as a tea to calm the nerves and irregular heartbeats.[45] Filtered infusions of each plant (25 gm dry biomass in 300 ml of double distilled water) have been used to immerse isolated tissue of aorta, trachea, ileum, bladder and uterus from laboratory rats.[46] While the infusions produced contraction in aorta and bladder muscles and no response in tracheal muscle, there were notable differences with the others. *Agastache mexicana* subsp. *mexicana* induced intense relaxation of the intestinal and utreal muscles while *A. mexicana* subsp. *xolocotziana* produced intense contraction. Preliminary chemical studies[47] also confirm the differences between these two subspecies. The red Mexican hyssop has 6 flavonoids while the white hyssop has 12, of which they share only 3. Of the 46 components of the essential oil in the red Mexican hyssop and 31 in the white hyssop, only 14 are common to both taxa. The four principal essential oils (comprising over 30%) of the red Mexican hyssop are α-terpinene, α-thujene, α-pinene and camphene, while the four major constituents (accounting for 80%) of the white Mexican hyssop are isopulegone, pulegone, α-terpinenol and ρ-cimene. Of the 14 shared compounds, pulegone and isopulegone are the most abundant. Only thujene, menthone, estragol and methyl-anthranilate are found in "toronjil rojo" while only bornyl acetate and linalyl-oxide are present in "toronjil blanco".

The center of diversity of *Datura* (Solanaceae) is Mexico where all 14 species of these annual or perennial herbs are native. Jimson weed (*Datura stramonium*), a cosmopolitan weed, produces well known physiological effects in traditional medicine that are attributed to its alkaloids which are of pharmaceutical interest. Mika[48] has demonstrated the existence of chemovars which have a total alkaloid content as low as 0.086% dry weight to as high as 0.248%. The hyoscyamine varies from 0.051 to 0.215% while scopolamine ranges from 0.033 to 0.092%.

IMPORTANCE

One measure of the importance of medicinal plants is their direct utilization by people. The consensus or repetitive use of remedies by a group of people has been considered an important measure of consistency while studying cultural dynamics and searching for biological active and efficacious medicinal plants.[49,50] Often, considered an error in sampling, the use of different species of plants that are thought to be the same medicinal plant complex may be based upon human experience that spans time and geographic space. The consistency of specific plants in herbal remedies in the market nonetheless suggests the existence of fundamental plants for classes of ailments. The continued use of plants over time, as well as the cross-cultural acceptance of those effective plants by people with differing ethnomedical concepts, is a measure of the importance value of vegetal remedies. It also serves as a guide for selecting samples for phytochemical and pharmacological analyses. The therapeutic effectiveness of plants is another measure of importance, and although there are no systematic studies of this nature in Mexico, selected examples provide insight as to the variance of this component of biological diversity.

A nation-wide consensus of the important medicinal plants of Mexico requires an infrastructure covering all sections of the country and the human population. Although limited to selected medical practitioners in rural areas, the IMSS's 1983-1984 national survey suggests productive tendencies. Made in collaboration with the rural development program (COMPLAMAR) and 3,500 IMSS's rural clinics) 13,034 traditional medical practitioners were interviewed.[51] Each person designated the ten most important medicinal plants based only on common names, resulting in 6,175 plant names. Because no specimens were collected, the scientific name most commonly associated with the particular common name was applied. Of the ten most important plants, half were introduced (*Mentha piperita* L., Lamiaceae; *Ruta chalepensis* L., Rutaceae; *Matricaria recutita* L., Asteraceae; *Ocimum basilicum* L., Lamiaceae; *Aloe barbadensis* Mill., Liliaceae) and the other half were native (*Zea mays* L., Poaceae; *Artemisia ludoviciana* Nutt., Asteraceae; *Chenopodium ambrosioides* L., Chenopodiaceae; *Calamintha macrostema* Benth., Lamiaceae; *Heterotheca inuloides* Cass., Asteraceae).

A survey of selected medicinal plants throughout Mexico revealed the existence of medicinal plant complexes in which taxonomically distinct plants share a common name, traditional remedial use, and certain morphological and chemical properties.[52] Each medicinal plant complex had species that extended the length of Mexico but usually only one species was universally recognized as the best form. The dominant taxon was not only employed within its geographic distribution but also available through the marketing system that extended beyond its range. If the dominant taxon, which is considered to the be best, is not available, people would use local forms (i.e., other species) whose employment was usually limited to their native range. For example, the dominant species of the "matarique" medicinal complex, which is valued for treating diabetes, kidney ailments and pains, is *Psacalium decompositum* (Gray) H. Robins. & Brett. (Asteraceae). It grows in the pine-oak forests of the northern Sierra Madre Occidental but is commercialized throughout Mexico and adjacent USA. The other species that are employed for similar purposes, when *P. decompositum* is not available, also belong to the Asteraceae. They are *P. peltatum* (HBK.) Cass., *P. sinuatum* (Cerv.) H. Robins. & Brett., *P. palmeri* (Greene) H. Robins. & Brett., and *Acourtia thurberi* (Gray) Reveal & King (Asteraceae). They, however, are employed only within each plant's geographical distribution.

In central Mexico, remedial infusions often consist of mixtures of medicinal plants. A survey of these mixtures in the Sonora Market (Mexico's principal medicinal plant market that is located in Mexico City[53]) for treatment of different ailments suggests there is a consistency in the use of specific plants for treating classes of ailments. For each class, there are fundamental plants that are always included. Then, depending upon the variant of the illness, the degree of severity, and the time of the year, supplementary herbs are added. Often taken into consideration is the appropriate balance of the mixture from the perspective of the "hot-cold" classification of illness and the corresponding plants that counteract it. For instance, an herbal remedy for diarrhea (primarily a cold illness) (Table 2) usually includes five fundamental plants (mostly hot plants) and one or more supplementary plants that are added in order to provide the correct balance depending upon the type of diarrhea (e.g., bloody, mucus, chunky, watery, etc.).

Cross-cultural comparison of medicinal plants of two ethnic groups with access to the same vegetal resources but who do not share ethnomedical concepts can provide an indication of the probable efficacy of species with similar uses. For example, in the state of Chihuahua, the Tarahumara Indians inhabit the pine-oak forest as well as the dry tropical forest in the west. The major urban centers with markets that sell medicinal plants are found along the central north-south axis. There is little communication between the native peoples of the mountains and canyons who retain their indigenous world view and the urban Mexicans who have access to institutional Western medicine. Of the hundreds of vegetal remedies sold in the market, 47 plants originate from the western mountains of Chihuahua. Of these 30 species are used in a similar fashion by both the Tarahumara and the mainstream Mexicans.[54] The continued employment by both groups assumes that they are effective in the same human physiological system and hence produce favorable biodynamic effects independent of mental expectations.

The therapeutic effectiveness of medicinal plants, evaluated upon the cultural basis of disease etiology and its cure, can provide meaningful insight into the phytochemical properties of plants and human perception and exploitation of

Table 2. Fundamental medicinal plants for the treatment of diarrhea based upon remedial mixtures sold in Mexican markets

Fundamental plants		
Name	part utilized	hot-cold classification
Psidium guajava L. guayaba	leaves	hot
Artemisia mexican Willd. estafiate	plant	hot
Waltheria americana L. tapacola	plant	hot
Teloxys graveolens (Willd.) Weber epazote de zorrillo	plant	hot
Tagetes erecta L. cempasuchil	flowers	temperate
Supplementary plants		
Name	part utilized	hot-cold classification
Hintonia latiflora (Sessé & Mociño ex DC.) Bullock quina	bark	hot
Satureja oaxacana (Fernald) Briq. menta verde	leaves	hot
Hedeoma piperitum Benth. tabaquillo	plant	hot
Tecoma stans (L.) HBK. tronadora	leaves, flowers	hot
Buddleia scordioides HBK. hierba de perro	leaves	hot
Matricaria recutita L. manzanilla	flowers	hot
Peumus boldus Mol. boldo	leaves	temperate
Krameria secundiflora DC. clameria	roots	temperate
Senna skinneri (Benth.) Irwin & Barneby paraca	bark	temperate
Agastache mexicana (HBK.) Lint & Epling toronjil	plant	temperate
Heterotheca inuloides Cass. árnica	flowers	cool
Equisetum myriochaetum Schlecht. & Cham. cola de caballo	plant	cool

them. There is growing interest world wide in promoting the assessment of herbal medicines.[55] Although there are insufficient data to evaluate Mexican plants at this time, the pharmacological activities of known phytochemicals of 16 of 25 native vegetal remedies from the early colonial period coincide with the expected therapeutic benefits in the native healing system.[56]

Bioassays also provide an indication of the probable effectiveness of medicinal plants. The importance of the cultural context for evaluating the effectiveness of a given plant remedy is supported by a more detailed study of 20 Aztec plants used in treating headaches.[57] Based upon the pharmacological properties of chemical constituents of the plants and Aztec etiological beliefs and Western biomedical standards, either 90% or 30%, respectively, of these herbs would be considered effective.[57]

Bioassays not only indicate possible effectiveness but also are useful in fractionation of extracts while in pursuit of active principles. Many ethnopharmacological and phytochemical studies of Mexican medicinal plants incorporate bioassays as part of the sampling routine. Summaries of such broad screens of Mexican medicinal plants can be found in Domínguez and Alcorn,[58] Jiu,[59] and Rojas *et al.*[60]

CHANGES IN IMPORTANCE

Cultural interactions with vegetal remedies may alter the importance of and, hence, the diversity of medicinal plants. The suppression of indigenous beliefs and practices often drove traditional medicine and its associated plants underground. Efforts of the Mexican government to provide modern health care to all sectors of the population have revealed unexpected levels of resistance to modern treatments. This situation is due, in part, to the persistence of centuries' old traditional medical practices. There has been sensitivity on part of institutionalized medicine in Mexico to recognize and tolerate this "invisible medicine" which successfully attends a large percentage of the rural and urban Mexican population.[61] Examples of the changes of importance can be seen in continuity and disjunction, discontinuity, and synchronism.

The continued use of medicinal plants represents **continuity** in which the element and the context in which it operates have not been altered over time.[62] On the other hand, certain elements of the past may be employed today but not in the same context or with the same conceptual background; this situation is **disjunction**. This permanence of vegetal elements but in altered cultural contexts can be found in several Mexican plants.[63] "Nantzinxócotl" or "nanche" as it is known today (*Byrsonimia crassifolia* (L.) HBK., Malpighiaceae) was present in the Aztec markets of the 16th century in the form of a powdered bark or an edible but bland fruit that was used to cure sores, aid digestion, assist in childbirth, and to treat swollen legs. Today, "nanche" is represented only in the context of fresh and preserved fruits that are exported from the lower tropical zones to the higher markets of central Mexico; the knowledge and commercialization of its medicinal properties have been lost.

The loss of importance of certain vegetal remedies represents **discontinuity**, which may occur due to socioeconomic determinants among others (e.g., extinction of plant, lack of efficacy, etc.). For example, "goma de Sonora" (an exduate from *Coursetia glandulosa* Gray (Fabaceae) that is produced by the scale insect *Tachardiella fulgens* (Homoptera:Coccidae)) was gathered in northwestern Mexico and commercialized by the Jesuits who prized it for treating stomach ailments. When the Jesuits were expelled from Mexico in 1775, the highly esteemed orange gum disappeared from world commerce. Only indigenous groups in this region, such as the Tarahumara who call it "arí", use it today.

One characteristic of acculturation is **synchronism** of elements from the socioeconomically dominant culture that substitutes similar components in the subjected society. The synchrony of medicinal plants is common in central Mexico where European ideas and plants have gradually substituted those of the Aztecs. Shared concepts of independent origins, such as the "hot - cold" spectrum of illnesses classification,[64] may have contributed to the ready acceptance by the Aztecs of European medicinal plants. For example, "hot" plants are used to treat "cold" ailments. In preHispanic times, the "cold" illnesses attributed to Tlaloc (the god of rain) were treated ritually with such hot plants as "iztauhyatl" (wormwood or *Artemisia ludoviciana* Nutt. subsp. *mexicana* (Willd.) Keck, Asteraceae) and "yauhtli" (sweet marigold or *Tagetes lucida* Cav., Asteraceae). Today, these have been substituted by introduced "hot" plants such as rue (*Ruta chalepensis*) and rosemary (*Rosmarinus officinalis* L., Lamiaceae).[65] The gradual substitution may have occurred in order to disguise the plants' ritual connection with Tlaloc or also because of unavailability of native species as the frontier of New Spain moved northward beyond the original Aztec Empire.

CONSERVATION

Biological surveys and inventories are the first step to successful, long-term conservation.[66] This initial process should determine the genetic strains, species, ecological assemblages, distribution, abundance, patterns in the landscape, role in ecological processes, utility (potential or proven) for human benefit and response to human or natural disturbances. The need to focus on the conservation of medicinal plants on a world basis has drawn considerable attention.[67] As of 1981, about 30% of the natural vegetation of Mexico had been altered by humans for agricultural, grazing and other purposes (Table 3). Habitat alteration has increased since then so this approximation is an underestimate of the actual situation that threatens the destruction of natural populations of medicinal plants. Shade-requiring plants such as "chuchupate" (*Ligusticum porteri* C. & R., Apiaceae) of the pine-oak forests of Chihuahua have disappeared entirely as clear cutting for pulpwood harvest advances.[68]

Table 3. Land use patterns in Mexico based upon 1981 agricultural and forestry census by ecological zones.[7,73] The ecological zones are: ASA—arid and semi-arid; SHTm—subhumbid temperate; HTm—humid temperate; HTr—humid tropical; SHTr—subhumid tropical

ECOLOGICAL ZONE	% FOREST	AGRICULTURE	CATTLE	OTHER USES	NO-FOREST VEGETATION
ASA	5	9	11	2	72
SHTm	63	15	11		11
HTm	64	8	22		6
HTr	58	13	19	6	4
SHTr	45	24	9	6	4

Over-collecting of certain plants has lead to the local extinction of certain species. In the temperate forests of central Mexico, "valeriana" (*Valeriana ceratophylla* HBK. and *V. edulis* Nutt. subsp. *procera* (HBK.) Meyer, Valerianaceae) has disappeared as a consequence of increased pressure to gather wild plants to satisfy European markets' demand for this sedative root. The popularity of "cancerina" (*Hemiangium excelsum* (HBK.) A.C. Sm., Hippocrateaceae) in treating ulcers throughout Mexico has increased the demand for root of this vine of the dry tropical forest; it is especially vulnerable because the plant remains green during the dry season when most of leaves fall off trees.

Plants with roots are especially susceptible to local extinction. Trees that yield medicinal bark may also become locally extinct if massive bark harvesting girdles the tree or if the stems are cut for massive stripping elsewhere. The regeneration of bark and persistence of harvested trees is possible, but only if traditional appropriate technology is applied such as in the case of the dry tropical forest tree "cuachalalate" (*Amphypterigium adstringens* (Schlecht.) Schiede, Julianiaceae) whose bark is prized for its gastrointestinal ulcer healing properties.[69]

Currently there is no legal regulation for the collecting of wild medicinal plants, if the collectors do so for personal consumption. Commercial collecting is regulated by official norms of the Secretaría de Agricultura y Recursos Hidraúlicos and the Secretaría de Desarrollo Social y Ecología. Nonetheless, reckless hoarding of high-demand medicinal plant occurs in many parts of Mexico in order to satisfy foreign commercial interests.

Conservation of medicinal plants calls for both *in situ* and *ex situ* efforts. Maintenance and restoration of original habitats where medicinal plants grow naturally is most desirable. Traditional harvesting schedules and techniques may assure the permanence and, in some cases, enhancement of populations of some species.

Ex situ conservation programs carried out by the government or universities are scarce. There are a few germplasm banks in Mexico but all are dedicated to major food plants. There is no officially designated Mexican seed bank for medicinal plants. Botanical gardens could play an important role in the study, education and conservation of medicinal plants. Of the 35 botanical gardens registered in Mexico, two are dedicated to the propagation and exhibition of this class of plants.[70]

Meanwhile, farmers and medicinal plant suppliers have taken the initiative at the local level to relieve the collection pressure on natural populations by cultivating medicinal plants. This activity is part of the domestication process that existed in prehispanic times and continues to influence plant diversity in Mexico.[71] Various medicinal plants exploited for their roots (e.g.,"jalapa" or *Ipomoea purga* (Wender.) Hayne, Convolvulaceae), woody stems and leaves (e.g., "laurel" or *Litsea glaucescens* HBK., Lauraceae), and leaves and flowers (e.g., "árnica" or *Heterotheca inuloides*; *Agastache mexicana*; "epazote" or *Teloxys ambrosioides* (L.) Weber, Chenopodiaceae) are cultivated on small farms in order to satisfy the local and national markets.[68,72]

CONCLUSIONS

Mexico is favorably positioned so that its rich flora is derived of both temperate and tropical elements. Cultural diversity has persisted over the centuries so that over 3,350 species of 21,600 vascular plants form part of the medicinal flora. Although native, church, and government authorities began to inventory remedial plants shortly after the Spanish Conquest of Mexico, most of the plants in these documents remain unidentified today. The effort to provide a preliminary list of Mexico's medicinal plants continues along with an interest in measuring their importance. The frequency of use, their inclusion as fundamental elements in remedial mixtures, the recognition and substitution within medicinal plant complexes, and the assessment of their effectiveness are some of the avenues being explored to evaluate their prominence.

Changes have occurred in the use of medicinal plants. Because many native plants were associated with ritual, they were disguised and substituted with introduced European plants. Some plants have continued over time to be used in the same manner and cultural context while others, although still used, are disjunct from their original context. In some cases, medicinal plants have been abandoned or the knowledge about them lost.

The demand for medicinal plants continues in Mexico. A few cases of local extinction of plant populations are known to be caused by overcollecting. Perhaps the greatest threat is the alteration of the natural habitat. Some highly valued plants have been brought into cultivation by local farmers interested in the conservation and commercialization of remedial herbs.

ACKNOWLEDGMENTS

F. Basurto and M. Trejo assisted in the compilation of data used in part of this paper.

REFERENCES

1. PIELOU, E.C. 1975. Ecological Diversity, John Wiley, New York. p. 165.
2. MAGURRAN, E. 1988. Ecological Diversity and Its Measurment, Princeton University Press, Princeton, NJ. p. 179.
3. WILSON, E.O., PETER, F.M. (eds.) 1988. Biodiversity, National Academy Press, Washington, DC. p. 521.
4. HALFFTER, G., (ed.) 1992. La Diversidad Biológica de Iberoamerica I, Instituto de Ecología, Xalapa, Veracruz. p.389.
5. RAMAMOORTHY, T.P., BYE, R., LOT, A., FA, J. (eds.) 1993. Biological Diversity of Mexico: Origins and Distribution, Oxford University Press, New York. p. 812.
6. MITTERMEIER, R.A. 1988. Primate diversity and the tropical forest, In: Biodiversity, (E.O. Wilson, F.M. Peter, eds.), National Academy Press, Washington, DC, pp. 145-154.
7. RZEDOWSKI, J. 1993. Diversity and origins of the phanerogamic flora of Mexico, In: Biological Diversity of Mexico: Origins and Distribution, (T.P. Ramamoorthy, R. Bye, A. Lot, J. Fa, eds.), Oxford University Press, New York, pp. 129-144.
8. CABALLERO, J. 1987. Etnobotánica y desarrollo: la busqueda de nuevos recursos vegetales, In: Memorias, IV Congreso Latinoamericano de Botánica: Simposio de Etnobotánica, Instituo Colombiano para el Fomento de la Educación Superior, pp. 79-96.
9. TOLEDO, V.M. 1987. La etnobotánica en Latinoamerica: vicisitudes, contextos, disafios, In: Memorias, IV Congreso Latinoamericano de Botánica: Simposio de Etnobotánica, Instituo Colombiano para el Fomento de la Educación Superior, Bogotá, Colombia, pp. 13-34.
10. HARLAN, J.R. 1975. Crops and Man. American Society of Agronomy, Crop Science Society of America. Madison, WI. p. 295.
11. VAVILOV, N.I. 1951. The Origin, Variation, Immunity and Breeding of Culitvated Plants. Chronica Botanica 13: 1-366.
12. GENTRY, H.S. 1982. Agaves of Continental North America, University of Arizona, Tucson. p. 670.
13. REICHENBACHER, F.W. 1985. Conservation of southwestern agaves. Desert Plants 7: 88, 103-106.
14. BYE, R., ESTRADA LUGO, E., LINARES MAZARI, E. 1991. Recursos genéticos en plantas medicinales de México, In: Avances en el Estudio de los Recursos Fitogenéticos de México, (R. Ortega Paczka, G. Palomino Hasbach, F. Castillo González, V.A. González Hernández, M. Livera Munoz, eds.), Sociedad Mexicana de Fitogenética, A.C., Chapingo, México, pp. 341-359.
15. DE LA CRUZ, M. 1991. Libellus de Medicinalibus Indorum Herbis, Manuscrito azteca de 1552, según traducción latina de Juan Badiano [Facsimile y Versión española con estudios y comentarios por diversos autores.], Fondo de Cultura Económica y Instituto Mexicano del Seguro Social, México. facsimile (64 folios) + p. 258.
16. VALDÉS GUTIÉRREZ, J., FLORES OLVERA, H., OCHOTERENA-BOOTH, H. 1992. La botánica en el Códice de la Curz, In: Estudios Actuales sobre El Libellus de Medicinalibus Indorum Herbis, (J. Kumate, ed.), Secretaría de Salud, México, pp. 129-180.
17. SAHAGÚN, B. DE. 1979. Códice Florentino, Archivo General de la Nación, México, Tomo I (p. 353), Tomo II (p. 375), Tomo III.
18. MARTÍNEZ, J.L. 1989. El "Códice Florentino" y la "Historia General" de Sahagún, Archivo General de la Nacion, México. p. 157.
19. ESTRADA LUGO, E.I.J. 1989. El Codice Florentino. Su Información Etnobotánica, Colegio de Posgraduados, Chapingo, México. p. 399.

20. HERNÁNDEZ, F. 1959. Historia de las plantas de Nueva España, In: Historia Natural de Nueva España Volumenes I y II, Obras Completas Tomos II y III, Universidad Nacional Autónoma de México, México. p. 476. + p. 554.
21. XIMÉNEZ, F. 1615. Quatro Libros de la naturalez y Virtudes de las Plantas, Casa de la viuda de Diego López Dávalos, México. p.203.
22. VALDÉS, J., FLORES, H. 1985. Historia de las plantas de Nueva España, In: Comisión Editora de las Obras de Francisco Hernández, Comentarios a la Obra de Francisco Hernández, Obras Completas Tomo VII, Universidad Nacional Autónoma de México, México, pp. 7-222.
23. BYE, R., LINARES, E. 1987. Usos pasados y presentes de algunas plantas medicinales econtradas en los mercados mexicanos. América Indígena 47: 199-230.
24. ESTEYNEFFER, J. DE. 1978. Florilegio Medicinal de Todas las Enfermedades [Notes by Ma. del Carmen Anzures y Bolaños], Academia Nacional de la Medicina, México. p. 973.
25. LOZOYA, X. 1984. Bibilografía Básica sobre Herbolaria Medicinal de México, Secretaría de Desarrollo Urbano y Ecología, México. p. 86.
26. MORENO, R. 1988. La Primera Cátedra de Botánica en México: 1788, Sociedad Mexicana de Historia de la Ciencia y de la Tecnología and Sociedad Botánica de México, México. p. 145.
27. NAVARRO, J. 1992. Historia Natural o Jardín Americano, Universidad Nacional Autónoma de México, Instituto Mexicana del Seguro Social, and Instituto de Seguridad y Servicios Sociales de los Trabajadores del Estado, México. p. 314.
28. FERNÁNDEZ DEL CASTILLO, F. 1961. Historia Bibliográfica del Instituto Médico Nacional de México (1888-1915), Imprenta Universitaria, México. p. 207.
29. ROSE, J.N. 1899. Notes on useful plants of Mexico. Contributions from the U.S. National Herbarium 5: 209-259.
30. MARTÍNEZ, M. 1990. Las Plantas Medicinales de México, Ediciones Botas, México. p. 657.
31. DÍAZ, J.L., (ed.) 1976. Indice y Sinonimia de las Plantas Medicinales de México, Instituto Mexicano para el Estudio de las Plantas Medicinales, México. p. 358.
32. DÍAZ, J.L., (ed.) 1976. Usos de las Plantas Medicinales de México, Instituto Mexicano para el Estudio de las Plantas Medicinales, México. p. 329.
33. LOZOYA, X., LOZOYA, M. 1982. Flora Medicinal de México, Primera Parte: Plantas Indigenas, Instituto Mexicano del Seguro Social, México. p. 309.
34. PULIDO SALAS, MA. T., SERRALTA PERAZA, L. 1993. Lista Anotada de las Plantas Medicinales de Uso Actual en el Estado de Quintana Roo, México, Centro de Investigaciones de Quintana Roo, Chetumal, Quintana Roo. v + p. 105.
35. MENDIETA, R.M., DEL AMO R., S. 1981. Plantas Medicinales del Estado de Yucatán, Instituto Nacional de Investigaciones sobre Recursos Bióticos, Xalapa, Veracruz. xxv + p. 428.
36. DEL AMO R., S. 1979. Plantas Medicinales del Estado de Veracruz, Instituto Nacional de Investigaciones sobre Recursos Bióticos, Xalapa, Veracruz. ix + p. 279.
37. GONZÁLEZ ELIZONDO, M. 1984. Las Plantas Medicinales de Durango: Inventario Básico. Cuadernos de Investigación Tecnológica [Instituto Politécnico Nacional - CIIDIR, Vicente Guerrero, Durango] 1(2): 1-115.
38. LÓPEZ ESTUDILLO, R., HINOJOSA GARCÍA, A. 1988. Catálogo de Plantas Medicinales Sonorenses, Universidad de Sonora, Hermosillo, Sonora. vi + p. 134.
39. AGUILAR, A., ARGUETA, A., CANO, L. (eds.) 1994. Flora Medicinal Indígena de México, Instituto Nacional Indegenista, México. p. 1591.
40. ARGUETA, A., CANO, L. 1993. El Atlas de las Plantas de la Medicina Tradicional Mexicana, In: La Investigación Científica de la Herbolaria Medicinal Mexicana, (M. Juan, A. Bondani, J. Sanfilippo B., E. Berumen, eds.), Secretaria de Salud, México, pp. 103-113.

41. ARGUETA, A., ZOLLA, C. (eds.) 1994. Nueva Bibliografía de la Medicina Tradicional Mexicana, Instituto Nacional Indigenista, México. p. 450.
42. ARGUETA VILLAMAR, A., CANO ASSELEIH, L.M., RODARTE, M.E. (eds.) 1994. Atlas de las Plantas Medicinales de la Medicina Tradicional Mexicana, Instituto Nacional Indigenista, México. p. 1786.
43. TÉTÉNYI, P. 1970. Intraspecific Chemical Taxa of Medicinal Plants, Chemical Publishing Co., New York. p. 225.
44. SANDERS, R.W. 1987. Taxonomy of *Agastache* Section *Brittonastrum* (Lamiaceae-Nepeteae). Systematic Botany Monographs 15: 1-92.
45. LINARES MAZARI, E., FLORES PEÑAFIEL, B., BYE, R. 1988. Selección de Plantas Medicinales de México, Editorial Limusa, México. p. 125.
46. GALINDO MANRIQUE, Y. 1982. Estudio farmacológico de algunoas plantas medicinals reportadas popularmente por la población mexicana para el tratameinto de padecimientos cardiovasculares, Undergraduate biology thesis, Escuela Nacional de Estudios Profesionales-Iztacala, Universidad Nacional Autónoma de México. México. p.100.
47. ESPIRITU CRUZ, L.P. 1991. Estudio quimiotaxonómico comparativo entre el toronjil rojo y el toronjil blanco, Undergraduate chemical-pharmaceutical biology thesis, Universidad Femenina de México, México. p. 65.
48. MIKA, E.S. 1962. Selected aspects of the effect of environment and heredity on the chemical composition of seeds plants. Lloydia 25: 291-295.
49. JOHNS, T., KOKWARO, J.O., KIMANANI, E.K. 1990. Herbal remedies of the Luo of Siaya District, Kenya: establishing quantitative critera for consensus. Economic Botany 44: 369-381.
50. TROTTER, R.T., LOGAN, M.H. 1986. Informant consensus: a new approach for identifying potentially effective medicinal plants, In: Plants in Indigenous Medicine and Diet: Behavioral Approaches, (N.L. Etkin, ed.), Redgrave Publishing Company, Bedford Hills, NY, pp. 91-112.
51. LOZOYA, X., VELÁZQUEZ D., G., FLORES A., A. 1988. La medicina tradicional en México - Experiencia del programa IMSS-COMPLAMAR 1982-1987, Instituto Mexicano del Seguro Social, México.
52. LINARES, E., BYE, R. 1987. A study of four medicinal plant complexes of Mexico and adjacent United States. Journal of Ethnopharmacology 19: 153-183.
53. BYE, R., LINARES, E. 1983. The role of plants found in the Mexican markets and their importance In ethnobotanical studies. Journal of Ethnobiology 3: 1-13.
54. BYE, R. 1986. Medicinal plants of the Sierra Madre: comparative study of Tarahumara and Mexican market plants. Economic Botany 40: 103-124.
55. AKERELE, O. 1993. Summary of the WHO guidelines for the assessment of herbal medicines. HerbalGram 28: 13-20.
56. ORTIZ DE MONTELLANO, B.R. 1986. Aztec medicinal herbs: evaluation of therapeutic effectivenss, In: Plants in Indigenous Medicine and Diet: Behavioral Approaches, (N.L. Etkin, ed.), Redgrave Publishing Company, Bedford Hills, NY, pp. 113-127.
57. ORTIZ DE MONTELLANO, B.R. 1975. Empiracal Aztec medicine. Science 188: 215-220.
58. DOMÍNGUEZ, X.A., ALCORN, J.B. 1985. Screening of medicinal plants used by Huastec Mayans of northeastern Mexico. Journal of Ethnopharmacology 13: 139-156.
59. JIU, J. 1966. A survey of some medical plants of Mexico for selected biological activities. Lloydia 29: 250-259.
60. ROJAS, A., HERNÁNDEZ, L., PEREDA-MIRANDA, R., MATA, R. 1992. Screening for antimicrobial activity of crude drug extracts and pure natural products from Mexican medicinal plants. Journal of Ethnopharmacology 35: 275-283.
61. LOZOYA, X., ZOLLA, C., (eds.) 1986. La Medicina Invisible, Introducción al Estudio de la Medicina Tradicional de México, Folios Ediciones, México. p. 303.

62. KUBLER, G. 1961. On the colonial extinction of the motifs of pre-Columbian art, In: Essays in precolumbian Art and Archaeology, (S.K. Lothrop, ed.), Harvard Univeristy Press, Cambridge, pp. 14-34.
63. BYE, R., LINARES, E. 1990. Mexican market plants of the 16th century. I. Plants recorded in Historia Natural de Nueva España. Journal of Ethnobiology 10: 151-168.
64. LÓPEZ AUSTIN, A. 1984. El Cuerpo Humano, Las Concepciones de los Antiguos Nahuas, Universidad Nacional Autónoma de México, Instituto de Investigaciones Antropológicas, México. Vol. 1 (p. 490), Vol. 2 (p. 334).
65. ORTIZ DE MONTELLANO, B.R. 1990. Aztec Medicine, Health, and Nutrition, Rutgers University Press, New Brunswick, NJ. p. 308.
66. RAVEN, P. 1992. Conserving Biodiversity - A Research Agenda for Development Agencies, National Academy Press, Washington, DC. p. 116.
67. AKERELE, O., HEYWOOD, V., SYNGE, H., eds. 1991. The Conservation of Medicinal Plants, Cambrdige University Press, Cambridge. p. 362.
68. BYE, R., MERAZ CRUZ, N., HERNÁNDEZ ZACARIAS, C.C. 1987. Conservation and development of food and medicinal plants in the Sierra Tarahumara, Chihuahua, Mexico, In: Strategies for Classification and Management of Native Vegetation for Food Production in Arid Zones, (E.F. Aldon, C.E. Gonzales Vicente, W.H. Moir, eds.), U.S. Department of Agriculture, Forest Service, Rocky Mountain Forest and Range Experiment Station, General Technical Report RM-150, Fort Collins, CO, pp. 66-70.
69. SOLARES ARENAS, F., BOYAS DELGADO, J.C., DÍAZ BALDERAS, V. 1992. Avances del estudio sobre efecto del descortezamiento en la capacidad de regeneración de corteza de cuachalalate (*Amphypterigium adstringens* Schiede & Schectl.) en el estado de Morelos, In: Avances de Investigación del INIFAP en Selvas Bajas Caducifolias (SBC) del Estado de Morelos, Secretaría de Agricultura y Recursos Hidraúlicos, Instituto Nacional de Investigaciones Forestales y Agropecuarias, Centro de Investigación Regional del Centro, Zacatepec, Morelos, pp. 91-98.
70. HERRERA, E., GARCÍA-MENDOZA, A., LINARES, E. 1993. Directorio de los Jardines Botánicos de México, Asociación Mexicana de Jardines Botánicos, México. p. 63.
71. BYE, R. 1993. The role of humans in the diversification of plants in México, In: Biological Diversity in México: Origins and Distribution, (T.P. Ramamoorthy, R. Bye, A. Lot, J. Fa, eds.), Oxford University Press, New York, pp. 707-731.
72. LINAJES, A., RICO-GRAY, V., CARRIÓN, G. 1994. Traditional production system of the root of jalapa, *Ipomoea purga* (Convolvulaceae), in central Veracruz, Mexico. Economic Botany 48: 84-89.
73. TOLEDO, V.M., ORDÓÑEZ, M.J. 1993. The biodiversity scenario of Mexico: a review of terrestrial habitats, In: Biological Diversity of Mexico: Origins and Distribution, (T.P. Ramamoorthy, R. Bye, A. Lot, J. Fa, eds.), Oxford University Press, New York, pp. 757-777.

Chapter Five

BIOACTIVE NATURAL PRODUCTS FROM TRADITIONALLY USED MEXICAN PLANTS

Rogelio Pereda-Miranda

Departamento de Farmacia
Facultad de Química
Universidad Nacional Autónoma de México
Coyoacán 04510, México, D.F., México

INTRODUCTION

Drug discovery based on the investigation of terrestrial plants for their therapeutic potential has been a goal of mankind since prehistoric times and is a universal feature of human cultures. Traditional methods of trial and error and use of lepto-organic and taxonomic clues have now given way to modern phytochemical and pharmacological methods for drug development. Plant de-

Phytochemistry of Medicinal Plants
Edited by John T. Arnason et al., Plenum Press, New York, 1995

rived natural products are a source of economically important chemicals, clinically useful drugs, leads for synthesis of derivatives and analogues and tools for biochemical investigations of mamalian receptors. They can also be useful "over the counter" remedies for self administration, and therefore could represent effective strategies of health assistance programmes in developing countries.

Worldwide, the traditional science found in ethnomedical records has been a major guide for scientific discovery of bioactive compounds (the so-called "ethnobotanical hypothesis"). In Mexico, the rich traditions of Aztec and Maya civilizations provide a pharmacopoeia of several thousand medicinal plants, and several thousand more may be contemplated if the many other cultural traditions of Mesoamerica are considered. The majority of these known medicinal and economically important plants remain unexplored by modern science suggesting that we may be far less knowledgeable that our ancestors in the understanding and exploration of biologically active agents traditionally employed by man.

Reference to the traditional use of plants in Mexico has been helpful in designing an efficient search plan for sampling plant material at UNAM to investigate the chemical composition and biological activities of some traditionally used Mexican plants.[1] Recent advances in our program are outlined in this chapter and emphasize the important continuity between traditional and modern science in investigations of selected species from the Lamiaceae and Convolvulaceae.

Many bioactive secondary chemicals may have evolved as antiherbivore or antimicrobial defences in plants and subsequently adapted for their physiological activity in human medicine. For this reason the line between therapy and pathology is not always clearly drawn. The final section in this chapter illustrates how modern phytochemistry and toxicology can more clearly distinguish between therapeuthic and potentially dangerous drugs and allow selection of an elite group of "ethical phytomedicines" that are both safe and effective.

ANTIMICROBIAL AND CYTOTOXIC CONSTITUENTS FROM HYPTIS SPECIES (LAMIACEAE)

Ethnobotany

Mexican Lamiaceae are comprised of 26 genera, represented by over 512 species which are predominantly mountainous. They are richly distributed in desert and arid vegetation zones, but rather poorly in the tropical lowlands of Mexico. *Salvia* (312 species) *Scutellaria* (37 species) and *Hyptis* (32 species) are the largest genera and together constitute about 78% of all the Mexican Lamiaceae.[2]

Lamiaceae (Labiatae) species have provided important resources for the Old and New World, and their uses in medicine and as condiments in regional cuisine are of central importance. Folk medicine has for a long time placed emphasis on extracts or essential oils of the Lamiaceae used for antibacterial, antiviral, carminative and spasmolytic activities.[3] The chemical components of Lamiaceae oils and their exploitation recently have been reviewed.[4] In America, some introduced Lamiaceae species from the Old World are important economically (e.g., spearmint and peppermint) as condiments, spices, perfumes and remedies. They also have played a relevant role in local economies based on exchange of vegetal remedies and aromatic herbs in medicinal plant markets.[5] The family has provided important remedies especially to treat neurological, respiratory, cardiac and gastrointestinal disorders.[6] A recent review on the economic botany of American Lamiaceae summarized the data on traditional uses of this family by indigenous and mestizo cultures of Mexico, the Caribbean, Central and South America.[6] A large number of medicinal species are frequently used as gastrointestinal remedies, and these usages are integrated into complex cultural systems.[7,8] The genus *Hyptis* in particular has yielded a great number of species which are important in Mexican folk medicine. While this genus has a considerable variety of traditional uses,[6] it is noteworthy that almost all Mexican medicinal *Hyptis* species are used as multipurpose remedies in the treatment of gastrointestinal disturbances and skin infections, as well as for treating rheumatism, cramps and muscular pains.

The employment of some members of *Hyptis* goes back to ancient Mesoamerican civilizations. In the 16th century manuscript "*Historia Plantarum Novæ Hispaniæ*", a post-conquest account of the prehispanic herbolaria written by the Spanish physician Francisco Hernández,[9] "huitsiquia" or "xoxouhcapatli" is mentioned as an effective medicinal plant for treating dysentery and tumors.[10] Rheumatism, sores[11] and eye infections also were alleviated with a decoction of this herb.[10] Recently it has been suggested that these documented uses of "huitsiquia" by the Purepecha in the western central state of Michoacán, also known as "xoxouhcapatli" by the Aztecs in central Mexico, represent a medicinal Labiatae probably a *Hyptis* species.[12]

Hyptis albida H.B.K. is a shrub or small tree, 1-6 m high with a lanate pubescence. The species is restricted to the Pacific drainage of Mexico and ranges from southwestern Sonora to central Guerrero.[13] The plant is generally known as "salvia" or "orégano" and these popular names are related to its use as a condiment in the regional cuisine of the Pacific coast of Mexico.[14] "Salvia blanca" (white salvia) is a descriptive variant with reference to its gray wool-like appearance. The most common use of this herb is medicinal, for which it is widely recognized as a remedy for treating gastrointestinal upsets.[15] The leaves are used to alleviate earache and applied in the form of a decoction in the treatment of rheumatic pain.[16] The plant is also reputed to act as a potent insect repellent.[15]

Hyptis mutabilis (Rich.) Briq. is a widely distributed weed in the tropical Americas,[17] where it is used as an important indigenous drug in various countries for treating gastrointestinal ailments.[6] In Mexico, it ranges from southern Nayarit to Chiapas in the Pacific drainage, and is found as well as in Veracruz and Tabasco in the coast Gulf of Mexico. "Hierba del burro" is the most frequently encountered common name.[14] The infusion of leaves is used as a carminative in order to alleviate stomach pains and flatulence.[18] The tea is said to be an effective antipyretic.

Hyptis pectinata (L.) Poit., popularly known by the Mexican rural population as "hierba del burro" and with the Maya name of "xolte-xnuk",[14] is an herbaceus plant with a pantropical distribution[13] and a wide medicinal usage in Central America.[6] Its distribution in Mexico includes the intertropical and tropical regions, where it is valued for its medicinal qualities and for its smell and taste in the regional cuisine. Formulations of the plant are used in popular medicine as a domestic remedy in the treatment of fevers, certain skin diseases, gastric disturbances, rhinopharyngitis, lung congestions and rheumatism.[6,19] *Hyptis suaveolens* (L.) Poit. is another pantropical Lamiaceae of great importance in popular medicine of the Americas.[17] The uses range from the treatment of rheumatism and other painful conditions to the therapy of stomach problems and gastrointestinal parasites. The infusion prepared using leaves or the inflorescence is employed for its stimulant, carminative, diuretic and antipyretic properties.[20] The decoction of the whole plant is used to alleviate diarrhea[21] and various kidney ailments. This medicinal plant, together with *Salvia hispanica* L. and *Salvia polystachya* Ort., is part of the so-called "chia"[14] complex in Mexico. The pre-conquest Aztecs cultivated "chian" or "chiantzotzolli" extensively to use its seeds against urinary obstruction[6] and in the treatment of fevers.[11] In modern Mexico and Central America, the seeds are used to prepared refreshing drinks, and also medically to cure constipation.[16] The root is prepared as a bitter tea by the rural population of Michoacán and Guerrero in the Balsas River drainage and drunk before breakfast to cure malarial fevers.[22]

Hyptis verticillata Jacq., known by the rural population of Mexico as "hierba martina"and "hierba negra",[14] is a bitter-aromatic herb common in the tropical regions of southern Mexico, Central America, the Caribbean and northern South America. The infusion prepared with fresh leaves is a remedy for diverse ailments, but is particularly used as drinking water in the treatment of gastro-intestinal disorders such as stomach aches, indigestion, and colics. It is also valued for its carminative and cathartic properties. The bitter infusion prepared with the flowered plant is drunk as a hot tea in order to treat headaches, gastro-intestinal infections and parasites.[23] The whole plant is boiled and rubbed on the skin for rheumatism, skin infections, such as wounds and insect bites,[6] and also used as a bath for undiagnosed ailments.[24] The leaves are used in topical treatments of a variety of warts. Ritual cleansing ceremonies ("limpias") are performed using this plant where its penetrating persistent odor is of great psycotherapeutic significance in these practices by the Chinantec and Totonac Indians "curanderos" (folk healers).[24,25] Roots are cut into small pieces, mixed with animal feed, and used as a dewormer by the lowland Maya people.[26]

Other members of the genus *Hyptis* have only minor ethnobotanical importance in Mexico, since their uses in traditional medicine are confined exclusively to their geographical distribution. For example, *Hyptis emoryi* Torr.

occurs in the Sonoran desert, where it has been used by the Seri Indian as a medicine to treat colds, internal illnesses and ear ailments.

Screening for Antimicrobial and Cytotoxic Activities

The screening of plant extracts and natural products for antimicrobial activity has revealed their potential as sources of new anti-infective agents.[27] As part of a research program directed towards isolation of bioactive compounds from Mexican medicinal plants,[28] preliminary antimicrobial screenning of crude extracts prepared from five *Hyptis* species was undertaken. The "rationale" for selection of plant material was based on folk medicine knowlegde. Table 1 shows antimicrobial activity of the MeOH extracts tested against *Candida albicans* and selected Gram-positive and Gram-negative bacteria. Relative activities are expressed by the diameter of the developed inhibiton zones compared to those of widely used antibiotics (streptomycin and nystatin). The extracts of *Hyptis mutabilis*, *Hyptis pectinata* and *Hyptis suaveolens* were active only against the Gram-positive bacteria. However, the extracts of *Hyptis albida* and *Hyptis verticillata* significantly inhibited the growth of all tested microorganisms.

The lipophilic fractions of *Hyptis albida*, which are particularly active against all the microorganisms tested, afforded rosmarinic acid (Labiatae-tannin) in high yields (0.2% dry wt), as well as several flavonoids and triterpenoids.[15,29] The antibacterial and antiviral activity of rosmarinic acid is well known.[3] Among the flavonoids, cirsimaritin (**1**) showed the widest spectrum of antimicrobial activity: against *Staphylococcus aureus* [MIC (minimum inhibitory concentration) = 31.25 μg/ml]; *Bacillus subtilis* (MIC = 15 μg/ml); *Pseudomonas aerugi-*

Table 1. Antimicrobial activity of crude extracts from selected *Hyptis* species

Plant material	Part used[a]/ extract yield (%)	Inhibition zone (mm)[b] 1	2	3	4	5
Hyptis albida	LF (13.6)	6	8	5	5	4
Hyptis mutabilis	PL (8.7)	7	11	-[c]	-	-
Hyptis pectinata	PL (7.3)	6	10	-	-	-
Hyptis suaveolens	LF (9.4)	6	9	-	-	-
Hyptis verticillata	LF (18.6)	6	10	6	7	7
Standard[d]	-	7	12	7	9	9

[a](LF) leaves; (PL) whole plant. Yield = w/w in terms of dry starting material.
[b] The crude extracts were assays as aqueous supensions in 1% Tween-80 at a concentration of 20 mg of extract per ml. Holes having a diameter of 11 mm were made on agar plates containing 1 ml of an overnight broth culture (10^6 bacteria/ml) and filled with 100 μl of the test solution. Microorganism: (1) *Staphylococcus aureus*; (2) *Bacillus subtilis*; (3) *Escherichia coli*; (4) *Pseudomonas aeruginosa*; (5) *Candida albicans*.
[c]No zone of inhibition was observed.
[d]Antibiotic standards: streptomycin sulfate (1 mg/ml) for bacteria and nystatin (3 mg/ml) for *C. albicans*.

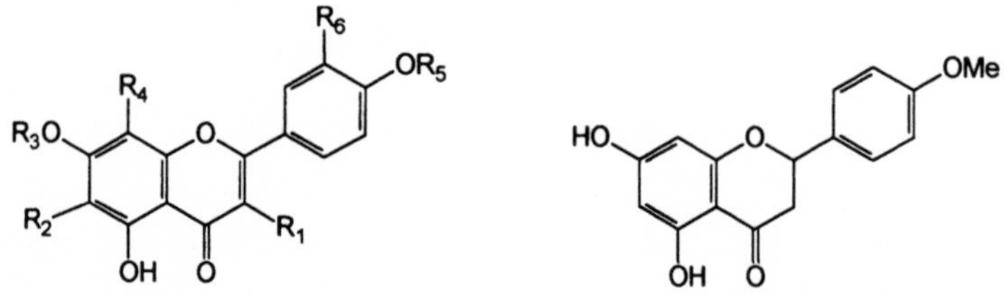

Cirsimaritin (**1**) $R_1 = R_4 = R_5 = R_6 = H$; $R_2 = OMe$; $R_3 = Me$

Isosakuranetin (**4**)

Ermanin (**2**) $R_1 = OMe$; $R_2 = R_3 = R_4 = R_6 = H$; $R_5 = Me$

Nevadensin A (**3**) $R_1 = R_3 = R_6 = H$; $R_2 = R_4 = OMe$; $R_5 = Me$

Gardenin (**5**) $R_1 = R_6 = H$; $R_2 = R_4 = OMe$; $R_3 = R_5 = Me$

5-Hydroxy-4',3,6,7,8-pentamethoxyflavone (**6**) $R_1 = R_2 = R_4 = OMe$; $R_3 = Me$; $R_5 = Me$; $R_6 = H$

Sideritoflavone (**7**) $R_1 = R_4 = R_5 = H$; $R_2 = OMe$; $R_3 = Me$; $R_6 = OH$

Figure 1. Bioactive flavonoids from some medicinal *Hyptis* species.

nosa (MIC = 31.25 μg/ml); and *Candida albicans* (MIC = 50 μg/ml). Ermanin (**2**), nevadensin A (**3**) and isosakuranetin (**4**) exhibited marginal activity against the Gram-positive bacteria tested (MIC = 50-70 μg/ml).[28] These preliminary screening results support the documented use of *Hyptis* species as antimicrobial agents in traditional medicine.

Plants from the genus *Hyptis* have been studied fairly extensively from a phytochemical point of view. However, only a few chemical studies of medicinal *Hyptis* species have been conducted using a biological activity-driven procurement of cytotoxic principles. It was found that the ethanolic extract of *Hyptis tomentosa* Poit. was active against P-388 lymphocytic leukemia and KB cell culture systems.[30] By activity-guided fractionation of this extract, the primary cytotoxic component was identified as deoxydopophyllotoxin, together with two mild cytotoxic flavones, gardenin B (**5**) and 5-hydroxy-4',3,6,7,8-pentamethoxyflavone (**6**).[30] The sideritoflavone (**7**) was one of the cytotoxic constituents present in the aerial parts of *Hyptis verticillata*.[23] Structures of the bioactive flavonoids from *Hyptis* species are shown (Fig. 1).

The $CHCl_3$ extract of the aboveground parts of *Hyptis emoryi* Torr. showed activity in the Walker carcinoma 256 (intramuscular) tumor system, and betulic acid (**8**) was identified as the active principle (52 % T/C at 300 mg/kg).[31] This triterpene acid has been isolated also from roots of *Hyptis suaveolens* and aerial parts of *Hyptis albida*[15] and *Hyptis pectinata*.[19] It has been suggested that the cytotoxic activity detected during the screening procedures and fractionation of

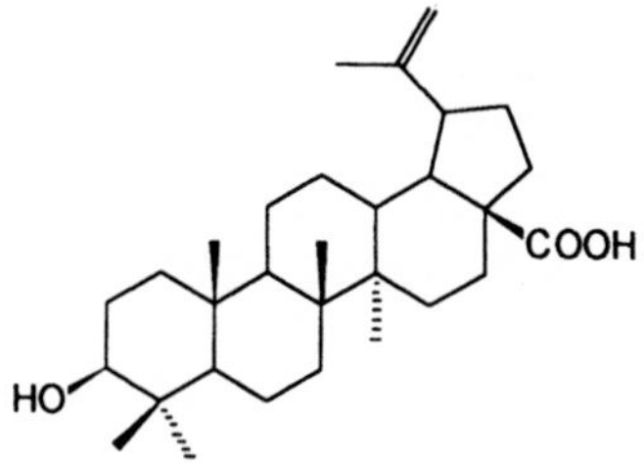

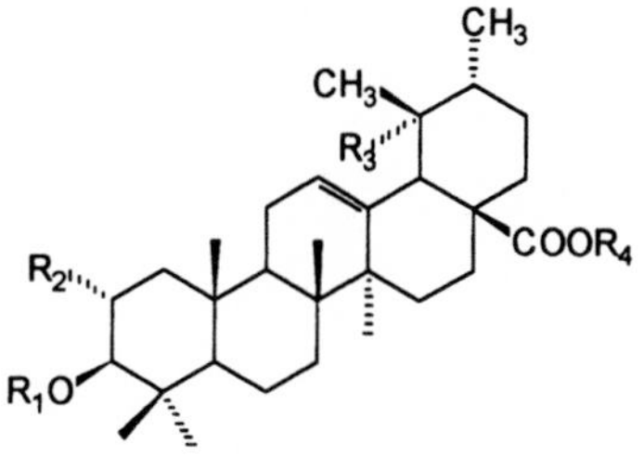

Betulinic acid (**8**)

Ursolic acid (**9**) $R_1 = R_2 = R_3 = R_4 = H$

Pomolic acid (**10**) $R_1 = R_2 = R_4 = H$; $R_3 = OH$

2α-Hydroxyursolic acid (**11**) $R_1 = R_3 = R_4 = H$; $R_2 = OH$

Acetylursolic acid (**12**) $R_1 = MeCO$; $R_2 = R_3 = R_4 = H$

Methyl ursolate (**13**) $R_1 = R_2 = R_3 = H$; $R_4 = Me$

Methyl acetylursolate (**14**) $R_1 = MeCO$; $R_2 = R_3 = H$; $R_4 = Me$

Figure 2. Cytotoxic pentacyclic triterpenoids from some medicinal *Hyptis* species.

organic solvent-soluble extracts derived from several traditional remedies of the Lamiaceae family is associated with the presence in large quatities of moderately cytotoxic pentacyclic triterpenoids, among these ursolic (**9**), pomolic (**10**) and 2α-hydroxyursolic (**11**) acids.[32] Structures of the cytotoxic triterpenes **8-14** isolated from some medicinal *Hyptis* species, and cytotoxicity data in KB and P-388 cell culture systems are shown (Fig. 2, Table 2).

The biological test data obtained during a screening for cytotoxicity of MeOH extracts prepared from selected *Hyptis* species are shown (Table 3). All these crude extracts demonstrated general nonspecific cytotoxic activity against a panel of cell lines comprising a number of human cancer cell types {breast, colon, fibrosarcoma, lung, melanoma, KB and KB-VI (a multidrug resistant cell line derived from KB)}, and murine lymphocytic leukemia (P-388). These results were useful for ranking potentially valuable plant material to pursue in bioactivity-guided fractionation of extracts for isolation of the active principles. The extracts of *Hyptis verticillata* and *Hyptis pectinata* showed the strongest cytotoxicity, and these candidate plants were collected in sufficient quantity for phytochemical studies. In the following paragraphs the steps involved with the isolation of the bioactive constituents will be briefly described.

Table 2. Cytotoxicity of triterpenes **8-14** in cell culture systems

Compound	ED_{50} (μg/ml)		Source[a]	Reference
	KB	P-388		
8	>15	5.3	HA,HE,HS,HP	15,19,30
9	6.6	9.0	HA,HM,HV	15,18,23
10	8.7	2.9	HM	18
11	5.7	3.2	HM	18
12	3.7	3.9	HM, HV	18,23
13	>5	>5	HM	18
14	>5	>5	HM	18

[a]HA = *Hyptis albida*, HE = *Hyptis emoryi*, HM = *Hyptis mutabilis*, HP = *Hyptis verticillata*, HS = *Hyptis suaveolens*, HV = *Hyptis verticillata.*

5,6-Dihydro-α-Pyrones from Hyptis Pectinata

Steam distillation of the fresh leaves of *Hyptis pectinata* afforded an essential oil (1.3 %), which exhibited a significant broad-spectrum of antimicrobial activity at a concentration of 10 μg/ml and, therefore, corroborated its use as an antimicrobial agent in traditional medicine. GC-MS analysis of this extract indicates the presence of a large amount of thymol (32.2 %), which probably accounts for its antiseptic properties. Solvent-soluble extracts ($CHCl_3$) derived from the deffated aerial parts also were found to demonstrate a strong inhibitory activity (7-10 mm zone of growth inhibition) against Gram-positive bacteria in qualitative biological assays. In addition, these extracts were cytotoxic (ED_{50} 2.2 μg/ml) when tested in the *in vitro* P-388 murine lymphocytic leukemia assay system. Therefore, a phytochemical study was initiated, and the antimicrobial and cytotoxic activities were

Table 3. Cytotoxic activity of crude extracts from selected *Hyptis* species

Plant material	Cell line[a] ED_{50} (μg/ml)								
	BC1	HT	Lu1	Mel2	Col2	KB	KB-V	A431	P388
H. albida	>20	8.9	11.1	8.9	>20	11.7	>20	9.2	>5
H. mutabilis	>20	9.0	8.9	>20	5.9	11.5	>20	>20	>5
H. pectinata	6.8	7.9	8.3	4.3	6.6	6.7	6.8	7.2	2.2
H. suaveolens	12.8	8.9	11.2	8.9	13.3	12.4	>20	>20	>5
H. verticillata	3.9	2.0	0.8	4.6	0.6	0.2	0.4	2.9	0.3

[a]Abbreviations: BC1, human breast cancer; HT (HT-1080), human fibrosarcoma; Lu1, human lung cancer, Mel2, human melanoma; Col2, human colon cancer; KB, human nasopharyngeal carcinoma; KB-V, vinblastine resistant KB; A431, human epidermoid carcinoma; P-388, murine lymphocytic leukemia.

traced, using bioassay-directed fractionation, to a mixture of pectinolides A-C (**15-17**), three novel 5,6-dihydro-α-pyrones (Fig. 3). The plant also was shown to contain hyptolide (**18**), a natural α-pyrone previously isolated from this species,[33] and the cytotoxic pentacyclic triterpenoids betulinic (**8**) and ursolic (**9**) acids.[19] The structure and stereochemical elucidation of the bioactive isolates from *Hyptis pectinata* was achieved as follows. Pectinolide A (**15**) was found to have a molecular formula of $C_{16}H_{22}O_6$, based on high-resolution mass spectral data (HRMS). The presence of an α,β-unsaturated δ-lactone was inferred from the observed UV and IR spectral maxima at 208 nm and 1730 cm^{-1}, respectively. The ^{1}H-NMR spectrum indicated the presence of two acetoxy substituents (δ 2.05 and 2.10). The characteristic resonances of the vinylic protons on C-3 (δ 6.24) and C-4 (δ 6.96) in the lactone ring, as part of an ABX spin-system with H-5 (δ 5.19), indicated the substitutiton at C-5 by one of the acetyloxy functionalities. The discernible 10.5 Hz coupling constant for the two sets of olefinic protons at C-1' (δ 5.73) and C-2' (δ 5.62) indicated the *cis* configuration of the exocyclic double bond. The ^{13}C-NMR spectrum of **15**, assisted by off-resonance, APT, and ^{1}H-^{13}C HETCOR techniques, confirmed the presence of a 3-(acetyloxy)-1-heptenyl moiety at C-6. Pectinolides B (**16**) and C (**17**) were found to be monodeacetylated isomeric forms of compound **15** with a molecular formula of $C_{14}H_{20}O_5$, as determined by HRMS. On acetylation, both compounds afforded the same derivative which was identical (GC-MS, NMR) to pectinolide A (**15**). ^{1}H- and ^{13}C-NMR spectra of **16** and **17** were comparable to those obtained for **15**. After measuring the ^{1}H-^{1}H COSY and ^{1}H-^{13}C HETCOR, the OH group was placed at C-3' in pectinolide B (**16**). The H-3'signal had been shifted significantly upfield ($\delta_{(H\text{-}3')16}$ 4.39 - $\delta_{(H\text{-}3')15}$ 5.35 = -0.96), and there was a

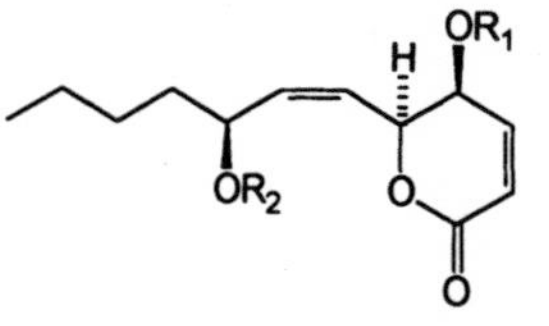

Pectinolide A (**15**) $R_1 = R_2 = Ac$

Pectinolide B (**16**) $R_1 = Ac; R_2 = H$

Pectinolide C (**17**) $R_1 = H; R_2 = Ac$

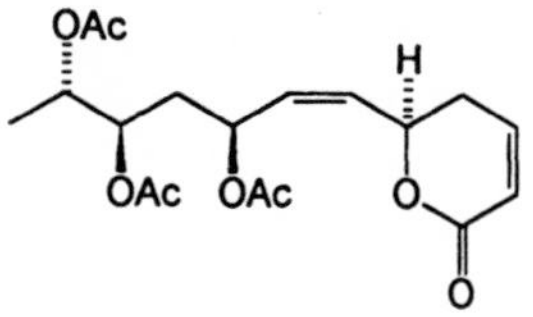

Hiptolide (**18**)

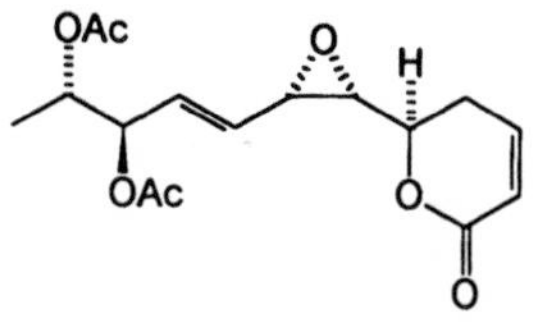

Deacetylepiolguine (**19**)

Figure 3. Cytotoxic 5,6-dihydro-α-pyrones from *Hyptis* species.

diamagnetic shift observed for C-3' ($\Delta\delta$ = -1.1). Similarly, the data obtained for pectinolide C (**17**) supported the placement of the OH group at C-5. The signal for H-5 was shifted 1.07 ppm upfield, and the C-5, 1.12 ppm upfield, relative to the same resonances in compound **15**.

The absolute configuration for the chiral centers was determined as follows. The pseudo-equatorial orientation of the side chain at C-6 and the axial configuration for the substituent at C-5 were established by the H-5, H-6 coupling constant value (J = 3 Hz). This spectroscopic evidence was used to assign an *S* configuration to the chiral center at C-6, together with the observation of a positive circular dichroism[33] (CD) absorption maximun at 265 nm, which is consistent with model 5,6-dihydro-α-pyrones.[34] Ozonolysis of pectinolides A (**15**) and C (**17**) yielded the same major degradation product, 2-acetyloxyhexanoic acid, which was saponified to afford (+)-2*S*-hydroxyhexanoic acid, the ozonolysis product of pectinolide B (**16**).[19] Accordingly, these chiroptical and chemical results provided conclusive evidences for the formulation of pectinolide A (**15**) as 6*S*-{(3*S*-acetyloxy)-1Z-heptenyl}-5*S*-(acetyloxy)-5,6-dihydro-2H-pyron-2-one.

Table 4 summarizes the antibacterial activity of the four 5,6-dihydro-α-pyrones (**15-18**) isolated from *Hyptis pectinata*. Pectinolide A (**15**), of all compounds tested, showed the highest antimicrobial activity against the Gram positive bacteria. *Staphylococcus aureus* and *Bacillus subtilis* were sensitive to **15** in the concentration range of 6.25-12.5 μg/ml. However, the natural monodeacetylated derivatives, pectinolides B (**16**) and C (**17**), were much less active, with an MIC of 12.5-25 μg/ml against *Bacillus subtilis* and a value of 100 μg/ml against *Staphylococcus aureus*. In contrast, hyptolide (**18**) was inactive at concentrations below 100 μg/ml. These results suggest that the presence of an acetyloxy group at C-5 and the possible hydrophophic properties conferred by the 3-acetyloxy-1-heptenyl residue at C-6 on the 5,6-dihydro-α-pyrone nucleus play important roles in the antimicrobial activity of pectinolides.[19]

The cytotoxic potential of pectinolides A-C (**15-17**) was evaluated with a number of cultured cell lines. The three isolates from the antimicrobial fraction demostrated general nonspecific cytotoxic activity, (ED_{50} < 4 μg/ml, Table 5). The intensities of the responses displayed by these compounds were similar to each other and approximately 2- to 5-fold more intense than those demostrated by hyptolide (**18**) and deacetylepiolguine (**19**), a structurally related δ-lactone isolated from *Hyptis oblongifolia*.[34]

Pectinolide A (**15**) formed conjugate addition products with nucleophilic substances such as methanol. Therefore, it is evident that the α,β-unsaturated δ-lactone nucleus was the site responsible for the antimicrobial and cytotoxic potentials of pectinolides A-C, and it appears that an acetyloxy group at C-5 may increase these biological activities.[19] A number of 6-substituted 5,6-dihydro-α-pyrones closely related to the bioactive isolates from *Hyptis pectinata* have displayed interesting physiological properties, e.g., smooth muscle relaxant, local anaesthetic, anti-inflammatory, antipyretic, antifungal and antimicrobial activi-

Table 4. Antibacterial activity of compounds **15-18**

Compound	Zone of inhibition per microorganism[a], mm (MIC, μg/ml)[b]				
	1	2	3	4	5
15	10 (6.25)	8 (12.5)	-[c] (200)	- (200)	- (250)
16	6 (25)	4 (100)	- (300)	- (300)	- (300)
17	8 (12.5)	4 (100)	- (>500)	- (>500)	- (>500)
18	5 (100)	- (>100)	- (>500)	- (>500)	- (>500)
Standard[d]	13 (3.12)	9 (6.25)	NT[e]	NT	7 (NT)

[a]For microorganisms see Table 1.
[b]Determination of the minimun inhibitory concentration (MIC) was accomplished by the two-fold serial dilution technique in nutrient broth with test sample concentrations ranging from 100-0.2 μg/ml.
[c]No zone of inhibition observed below 100 μg/ml.
[d]For antibiotic standards see Table 1.
[e]Not tested.

ties.[33] Therefore, pectinolides A-C (**15-17**) could be considered as potential parent compounds for development of new classes of phamacodinamic agents.

Lignans from *Hyptis verticillata*

Prior to the commencement of our study, phytochemical work on *Hyptis verticillata* revealed the presence of two aryltetralin lignans, β-peltatin (**20**) and 4'-demethyldeoxypodophyllotoxin (**21**), which account for the antimitotic action of this plant material.[35] We found that the MeOH extract of this plant has, in addition to the significant cytotoxicity observed with cultured mammalian cells

Table 5. Cytotoxicity data for compounds **15-19**

Sample	Cell line[a] ED_{50} (μg/ml)								
	BC1	HT1	Lu1	Mel2	Col2	KB	KB-V	A431	P-388
15	1.0	1.7	0.9	0.7	1.0	1.8	1.8	1.4	0.9
16	2.5	2.3	3.8	2.2	1.1	1.4	2.0	0.6	0.1
17	2.0	1.8	2.3	3.3	1.6	1.7	3.2	0.8	2.2
18	2.4	9.6	4.9	12.2	4.7	3.6	4.5	3.7	1.6
19	4.8	5.7	3.3	4.1	3.0	2.9	3.4	1.9	0.4

[a]For cell lines abbreviations see Table 3.

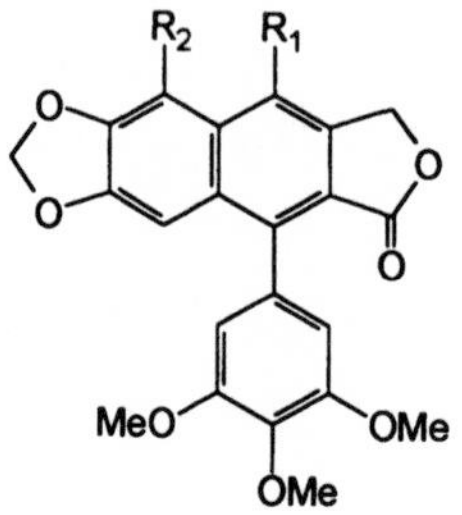

Dehydro-β-peltatin methyl ether (**22**) R_1 = H; R_2 = OMe

5-Methoxydehydropodophyllotoxin (**23**) R_1 = OH; R_2 = OMe

Dehydropodophyllotoxin (**24**) R_1 = OH; R_2 = H

Deoxydehydropodophyllotoxin (**25**) R_1 = R_2 = H

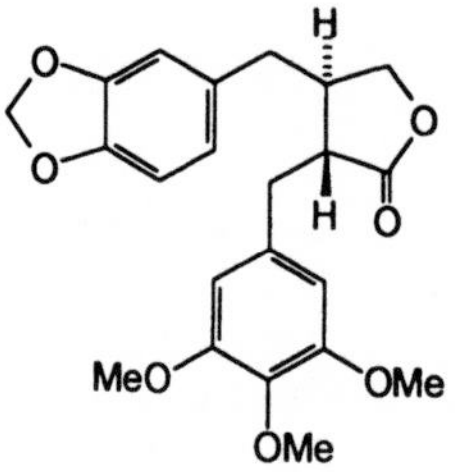

Yatein (**26**)

Figure 4. Cytotoxic arylnaphtalene and dibenzylbutyrolactone lignans from *Hyptis verticillata.*

(Table 3), a significant activity in the brine shrimp lethality test[36] (BST; LC_{50} 13.4 μg/ml). The cytotoxic activity was associated with a $CHCl_3$ soluble extract, and it was monitored throughout the phytochemical investigation using lethality to brine shrimp and P-388 cell line. When this $CHCl_3$ extract was subjected to Si gel column chromatography, one of the eight combined fractions was found to be bioactive (BST, LC_{50} 7.9 μg/ml; P-388, ED_{50} 0.1 μg/ml). Bioassay-directed fractionation of this sample yielded compounds **21-29** (Figs. 4 and 5). The toxicity to brine shrimp and the cytotoxic activities of the bioactive fraction (Pool IV) and all active isolates from *Hyptis verticillata* are summarized (Table 6).

Dehydro-β-peltatin methyl ether (**22**) was reported for the first time as a natural product. Compound **23** exhibited a molecular formula of $C_{23}H_{20}O_9$, based on its HRMS. The UV absorption maxima at 268, 325 and 357 nm were a clear indication of the presence of a naphthalene nucleus. The ^{1}H-NMR spectrum of **23** was comparable to that obtained for **22** the only differences being the absence of the low-field singlet at δ 8.16, which was ascribed to the proton on C-4 of **22**, and the presence of a phenolic-OH proton resonance at δ 9.58. Verification of the proposed C-4 position for the OH group on lignan **23** was provided by ^{1}H-NMR nOe experiments. As expected, enhancement was observed for the methylene protons of the lactone ring (δ 5.34) on irradiation of the signal belonging to the MeO group on the naphthalene nucleus (δ 4.29). Assignment of the ^{13}C-NMR of compounds **22** and **23** was achieved by selective INEPT (Insensitive Nuclei Enhancement by Polarization Transfer) experiments. These spectral data provided

β-Peltatin (**20**) R_1 = H; R_2 = OH; R_3 = Me

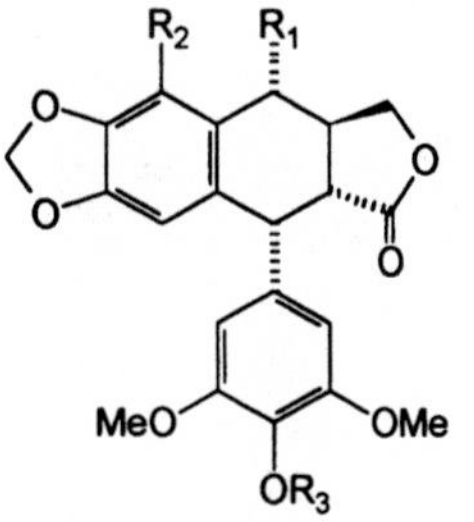

4'-Demethyldeoxypodophyllotoxin (**21**) R_1 = R_2 = R_3 = H

β-Apopicropodophyllin (**29**) R_1 = R_2 = H; R_3 = Me; Δ^2

Podophyllotoxin (**30**) R_1= OH; R_2 = H; R_3 = Me

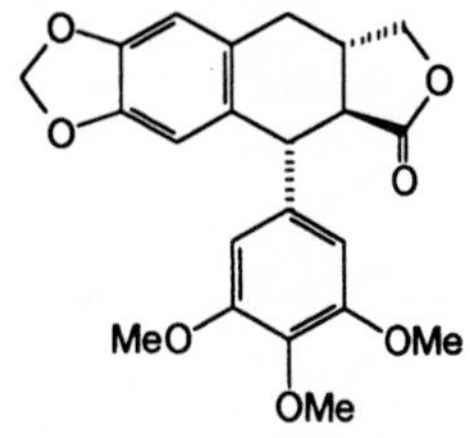

Isodeoxypodophyllotoxin (**27**)

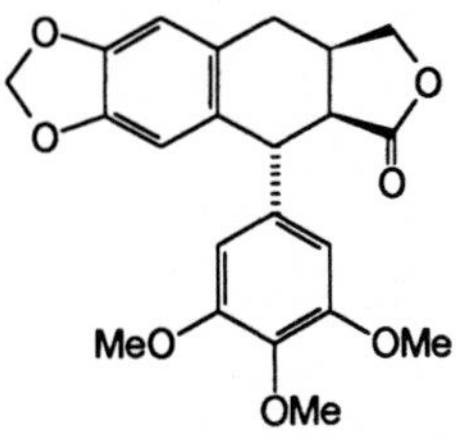

Deoxypicropodophyllin (**28**)

Figure 5. Cytotoxic podophyllotoxin type lignans from *Hyptis verticillata.*

conclusive evidence for the formulation of **23** as 5-methoxydehydropodophyllotoxin.[23]

The identification of known lignans, 4'-demethyldeoxypodophyllotoxin (**21**), dehydropodophyllotoxin (**24**), deoxydehydropodophyllotoxin (**25**), (-)-yatein (**26**), isodeoxypodophyllotoxin (**27**), deoxypicropodophyllin (**28**) and β-apopicropodophyllin (**29**) was achieved by comparison of physical and spectroscopic data (UV, IR, ^{1}H-NMR) with literature values.[23]

The cytotoxic potential of isolates **21-29** was evaluated with a number of human culture cell lines. Dibenzylbutyrolactone (**26**) and aryltetralin lignans (**21, 27-29**) demonstrated general nonspecific cytotoxic activity (ED_{50} < 10^{-2} μg/ml) comparable to that of podophyllotoxin (**30**, Table 6). The intensities of the responses displayed by **21** and **29** were similar to each other and approximately 10- to 100-fold more intense than those demonstrated by the iso- and picroisomers, compounds **27** and **28**. As expected, the arylnaphthalanes **22-25** were 100- to 1000-fold less active that podophyllotoxin (**30**) because of the planar aromatized naphthalane ring which alters the conformation of the molecule significantly relative to **30**. Compounds **22** and **23** showed only marginal cytotoxicity with the human cancer cell lines, and the most intense activity was observed with the murine lymphocytic leukemia (P-388) in cell culture (Table 6).[23]

The aryltetralins (**21, 29**), as well as (-)-yatein (**26**), were found to show strong inhibitory activity against *Candida albicans* at the concentration of 0.2-1.0

Table 6. Selected bioactivities of compounds **21-30**

Sample	BST[a]	Cell line[b] ED_{50} (μg/ml)				
		P-388	KB	A-549	MCF-7	HT-29
Pool IV[c]	7.9	0.1	0.09	-[d]	-	-
21	0.2	0.005	0.01	3.03×10^{-3}	3.40×10^{-3}	2.85×10^{-2}
22	434.7	1.8	2.2	2.19	2.94	3.35
23	>500	4.0	6.0	3.18	2.05	2.06
24	255.0	>5	5.0	2.32	2.99	3.31
25	>500	>5	11.4	17.5	>20	>20
26	2.8	0.4	0.08	$<10^{-4}$	$<10^{-4}$	$<10^{-4}$
27	>500	>20	6.7	15.7	13.5	10.3
28	141.5	0.1	0.1	2.36×10^{-2}	1.15×10^{-2}	2.88×10^{-2}
29	0.2	0.002	0.05	$<10^{-4}$	$<10^{-4}$	$<10^{-4}$
30	0.2	0.003	0.08	$<10^{-4}$	$<10^{-4}$	$<10^{-4}$

[a]Brine shrimp lethality test, LC_{50} μg/ml.
[b]Abbreviations: P-388, murine lymphocytic leukemia; KB, human nasopharyngeal carcinoma; A-549, human lung carcinoma; MCF-7, human breast carcinoma; HT-29, human colon carcinoma.
[c]Bioactive fraction obtained by column chromatography of the crude extract.
[d]Not tested.

μg/ml when examined by the standard dilution technique. This activity, which could account for the antiseptic properties of the infusion prepared from *Hyptis verticillata*, is presumably due to the established antimitotic activity of the podophyllotoxin lignan derivatives, and is consistent with the demonstrable reversal astrocyte formation (ASK activity) observed for the cytotoxic isolates **21-29**. The dose required (0.032 μg/ml) for **21** and **29** to effect 100% reversal conversion of cultured ASK cells was comparable to that of podophyllotoxin (**30**). Compound **26** was less active, with an antimitotic activity of 0.16 μg/ml. The concentration needed for the same response in the ASK system for the aryl-naphthalene lignanes (**22-25**) and the iso- and picroisomers (**27**, **28**) was in the interval of 0.8-20.0 μg/ml.[23]

In conclusion, the high content of cytotoxic lignans in *Hyptis verticillata* can be easily associated with some its medicinal attributes. The cathartic, anthelminthic and anticancer properties of infusions prepared with the whole plant, as well as its traditional use in dermatology for the treatment of warts, resemble the biological action of the resin produced from an alcoholic extract of the roots and rhizomes of *Podophyllum* species (Berberidaceae), sources of podophyllotoxin type compounds.[37] Clearly these lignans posses a diverse spectrum of biological activities of great interest for developement of new and effective pharmacological agents.

BIOLOGICALLY ACTIVE GLYCOLIPIDS FROM THE RESINS OF *IPOMOEA* SPECIES (CONVOLVULACEAE)

Ethnobotany

The most conspicuous anatomical features of species belonging to any genus of erect, trailing, or twining herbs and shrubs of the morning-glory family (Convolvulaceae) is the occurrence of rows of secretory cells with resinous contents in foliar tissues, especially in roots and rhizomes. These glycoresins represent an important chemotaxonomic marker of this family, and are responsible for the purgative properties of some medicinal members of the Convolvulaceae family, e.g., *Convolvulus*, *Ipomoea*, *Merremia* and *Operculina*.[38]

Ipomoea species constitute a group of economically important plants in modern Mexico whose use goes back to prehispanic times. One of the most important members of this genus is *Ipomoea batatas* (L.) Lam., the sweet potato. Some varieties of this species are appreciated for their large thick and nutritious tuberous root, popularly known in Mexico as "camote". It is thought that this edible especies was derived from *Ipomoea tiliacea* (Willd.) Choisy, as a result of selection by the native inhabitants of tropical Americas. Before the introduction of the potato into Europe, the sweet potato was regularly imported as a wholesome article of diet, and was grown in Spain and Portugal, to which it had been brought from the West Indies. The potato which Shakespeare mentions in the *Merry Wives of Windsor* is the sweet potato, and not the common potato (*Solanum tuberosum* L.).

To review the uses of *Ipomoea* species and their import in Mexican popular medicine would require a separate study. Here, only a few interesting and well documented examples of traditional usages of selected species will be presented.

The rhizomes of the Jalap bindweed, *Ipomoea purga* (Wender) Hayne (*Ipomoea jalapa* Pursh), with purgative properties were introduced in Europe about 1565, and since then they have represented one of the crude drugs with cathartic and purgative action of major trade between the New and Old World. The drug jalap is prepared from a resin which abounds in the roots of the jalap bindweed, known in Mexico as "raíz de Jalapa" or "raíz de Veracruz". It derives its local names from the City of Xalapa in the state of Veracruz, where it is abundant.[39] The dried tubercules yield a nearly tasteless purgative with a slight smoky odour, often given to children as an anthelminthic on account of the small purgative dose required.[39] Jalap resin occurs as a yellowish brown mass or powder, and is a cathartic possessing hydragogue activity, i.e a drastic laxative that causes abundant watery evacuations. It is reputed to be an excellent purge for rheumatism, and is used in constipation, pain and colic in the bowels and general intestinal torpor, being combined, in compound powder, with other laxatives, and with carminatives such as ginger and cloves. The dry root of *Ipomoea orizabensis* (Pelletan) Steud. ex Ledanois (*Ipomoea tyrianthina* Lindley), known as "jalapa de Orizaba", has been exported as a substitute or adulterant of the true jalap.[39]

In central Mexico, the root fragments of *Ipomoea stans* Cav., known as "tumbavaqueros" are used as a cathartic medicine and for the treatment of heart diseases. The decoction of the root (ca 3% dry weight) is used for neurological disorders such as epilepsy and chorea; it is also registered that it can be used for kidney inflamation and excess bile, although large doses are said to be dangerous. In the State of Puebla, where "tumbavaqueros" is known by its indian name of "soyoquilitil" by the population of Pahuatlan, the leaves are boiled and the liquid drunk by mothers to promote milk production and by bruised individuals to reduced inflammation.[40] Other uses have been listed such as, an antispasmodic, a sedative, and use in treating nephritis, ophthalmia and paralysis.[41]

The leaves of *Ipomoea batatas* are applied as a poultice for the treatment of inflammatory tumors and their deccoction used in baths and gargles for tumors of the mouth and throat. The aerial parts of *Ipomoea pes-caprae* (L.) Sweet in the form of a decoction or a poultice are applied to ulcerating tumors and used for the treatment of rheumatic pain. The infusion is drunk to treat kidney ailments, functional digestive disorders and internal pain.[41]

Screening for Cytotoxic Activity

In the course of our screening studies for bioactive natural products, a total of 12 crude extracts corresponding to six different *Ipomoea* species were investigated. Table 7 shows the species and the part of the plant tested, the type of solvent used for extraction, as well as the cytotoxicity ED_{50} values obtained for each extract against a panel of human tumor cells in culture. Samples active only at concentrations below 20 μg/ml are considered promising materials for further phytochemical studies. The extracts prepared from the aerial parts of *Ipomoea pes-caprae* and *Ipomoea batatas* did not demonstrate any such activity. Among the crude extracts showing important activities, the $CHCl_3$ extract of *Ipomoea tricolor* displayed a general cytotoxicity. In contrast, the $CHCl_3$ extracts of *Ipomoea purga* showed a significant inhibitory effect ($ED_{50} < 4$ μg/ml) against the human epidermoid carcinoma and breast cancer cell cultures. The $CHCl_3$ extract of *Ipomoea purga* and *Ipomoea tyrianthina* also presented non-specific activity. We suggest that the stronger activity observed for the chloroform extract in comparison to the methanol soluble ones, is due to the lipophilic affinity of actives to target cell membranes.

In our tests, the crude resins obtained from *Ipomoea* species were shown to display a significant antagonism to phorbol ester binding with protein kinase C (PKC).[42,43] No activity was detected for the MeOH extract of *Ipomoea pes-caprae*. However, concentration-dependent inhibition of phorbol 12,13-dibutyrate (PDBu) binding to partially purified PKC were observed for all the cytotoxic crude extracts. The corresponding IC_{50} values were calculated from dose response experiments for the extracts with greater than 60% displacement at 0.2 mg/ml (Table 8). Since PKC has been shown to be a receptor that is activated by tumor-promoting phorbol esters, susbtances that antagonize PKC activity logically may be considered as putative antitumor agents. Therefore, the cytotoxic

Table 7. Cytotoxic activity of selected *Ipomoea* species

Species[a]	Part used[b]/ Yield (%)	Extract	Human cell lines[c] ED_{50} (μg/ml)			
			KB	A-549	MCF-7	HT-29
Ib	AP (14.2)	MeOH	>20	>20	>20	>20
	RT (21.2)	MeOH	11.7	11.1	>20	8.9
Ipc	AP (18.4)	$CHCl_3$	>20	>20	>20	>20
	AP (20.3)	MeOH	>20	>20	>20	>20
Ip	RT (14.6)	$CHCl_3$	3.7	7.7	2.4	5.9
	RT (24.8)	MeOH	>20	8.5	>20	9.5
Is	RT (19.5)	$CHCl_3$	4.2	6.4	3.5	6.0
	RT (17.5)	MeOH	12.0	8.6	>20	>20
It	AP (6.6)	$CHCl_3$	5.8	5.3	5.7	8.9
	AP (8.9)	MeOH	11.7	11.1	>20	>20
Ity	RT (18.2)	$CHCl_3$	6.1	2.4	8.4	6.4
	RT (20.7)	MeOH	11.7	12.1	>20	>20

[a]Species abbreviations: Ib = *Ipomoea batatas*, Ipc = *Ipomoea pes-caprae*, Ip = *Ipomoea purga*, Is = *Ipomoea stans*, It = *Ipomoea tricolor*, Ity = *Ipomoea tyrianthina*.
[b](AP) aerial parts; (RT) root. Extract yield = w/w in terms of dry starting material.
[c]For cell lines abbreviations see Table 6.

$CHCl_3$ extract prepared from the roots of *Ipomoea purga* was selected as a candidate for further phytochemical work.

A yellowish brown resinous mass was obtained by macerating powdered jalap with $CHCl_3$. Preliminary inspection of the residue by TLC showed two major spots on SiO_2 eluted with $CHCl_3$-MeOH-H_2O (6:4:1) with Rfs of about 0.55 and 0.50. These two clearly defined zones were separately eluted by preparative column chromatography of the crude glycoresin. The cytotoxic activity associated with these resins was monitored throughout the phytochemical investigation, using the P-388 and KB cell lines. Preliminary characterization indicated that each sample was an oligosaccharide glycosidically linked to a hydroxylated long-chain fatty acid, combined with the carbonyl group to form a macrolactone. As previously described for other *Ipomoea* species,[44,45] other short-chain fatty acids esterify the oligosaccharide core at different positions. The major difference between the two samples appeared to be the extent of esterification, with the more polar zone containing fewer ester linkages. HPLC analysis indicated that neither sample was pure. Instead, each one represented a mixture of several related compounds that apparently were isomers involving different esters or sites of esterification. In order to characterize the oligossacharide core of the jalap glycoresins, both samples were subjected to alkaline hydrolysis and afforded the same glycosidic acid (**31**).

Table 8. Phorbol dibutyrate receptor binding assay data for selected *Ipomoea* species

Plant material	Part used/Extract[a]	PDBu displacement[b]	IC_{50} (μg/ml)
Ipomoea batatas	AP-MeOH	0%	-
	RT-MeOH	58%	-
Ipomoea pes-caprae	AP-$CHCl_3$	46%	-
	AP-MeOH	0%	-
Ipomoea purga	RT-$CHCl_3$	73%	82.6
	RT-MeOH	64%	96.4
Ipomoea stans	RT-$CHCl_3$	68%	93.4
	RT-MeOH	58%	–
Ipomoea tricolor	AP-$CHCl_3$	80%	76.6
	AP-MeOH	55%	-
Ipomoea tyrianthina	RT-$CHCl_3$	62%	99.5
	RT-MeOH	55%	-

[a]For abbreviations of part used and extract yield (%) see Table 7.
[b]Specific displacement of [^{3}H]-PDBu at concentration of 0.2 mg/ml.

The structure of derivative **31** was elucidated by spectral analysis including a combination of two-dimensional NMR techniques (DQF-COSY, HETCOR, COLOC and HMBC), and it corresponded to the previously described scammonic acid A, one of the glycosidic acids derived by alkaline hydrolisis of the ether-soluble glycoresins obtained from scammony roots (*Convolvulus scammonia* L.).[44]

A major compound (**32**) was isolated from the bioactive jalap glycoresin and its purification was successfully achieved by a high resolution preparative HPLC methodology. The structure of **32** was determined on the basis of the similar chemical and spectral evidences previously reported for scammonin I.[44] Compound **32** displayed a marginal cytotoxicity (Table 9) which was completely lost in its glycosidic acid derivative (**31**). The molecular structures of the cytotoxic glycolipid **32** from *Ipomoea purga* and its corresponding glycosidic acid, single oligossacharide derivative (**31**) obtained from the saponification of the whole crude glycoresin prepared from jalap, are shown (Fig. 6).

Glycoresins Constituents as Phytogrowth Inhibitors

The allelopathic potential of plants in the genus *Ipomoea* (Convolvulaceae), especially the suppressive effects on weed growth of sweet potatoes, *Ipomoea batatas*, has been demonstrated.[46] These species produce aggresive competitive effects, due to their strong propagative power, complemented with a high allelopathic interference. In tropical and temperate zones of Mexico, polycultures represent a traditional and efficient management of resources in agroe-

Table 9. Cytotoxicity data of compounds **31-34**

Compound	Cell line[a] ED_{50} (μg/ml)				
	A-549	MCF-7	HT-29	KB	P-388
31	3.25	3.21	3.00	2.5	2.8
32	>20	>20	>20	>20	>20
33	3.18	2.05	2.06	>20	2.2
34	>20	>20	>20	>20	>20

[a]For cell lines abbreviations see Table 6.

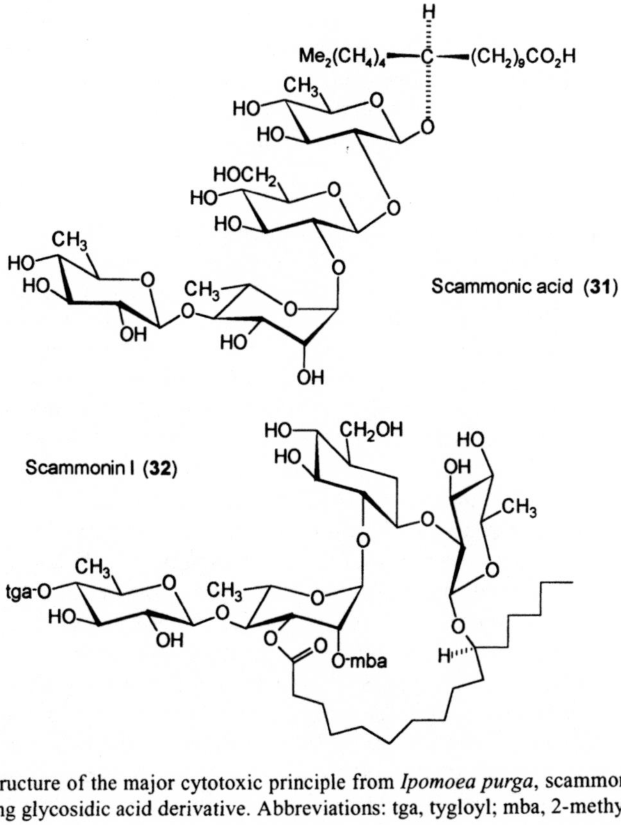

Scammonic acid (**31**)

Scammonin I (**32**)

Figure 6. Structure of the major cytotoxic principle from *Ipomoea purga*, scammonin I, and its corresponding glycosidic acid derivative. Abbreviations: tga, tygloyl; mba, 2-methylbutyroyl.

cosystems that provide a natural control of weeds. In such systems, the cultivation of semi-domesticated legumes, e.g. *Sorghum vulgare* and *Stizolobium pruriens*, as well as *Ipomoea* species, is a common practice to minimize the growth of companion weeds.[45]

Ipomoea tricolor Cav. (*Ipomoea violacea* L.) is used extensively as a cover crop, especially during August to October, the fallow period in the sugar-cane fields of the state of Morelos in the southeastern intertropical region of Mexico. The allelopathic potential of this species was previously demonstrated by measuring the inhibitory activity of aqueous and organic-solvent-soluble extracts (hexane, $CHCl_3$, and MeOH) on seed germination and seedling growth of *Amaranthus leucocarpus* Watts. and *Echinochloa crus-galli* (L.) Beauv. Bioassays detected significant activity in the $CHCl_3$ solution. When this bioactive extract was subjected to column chromatography, the phytogrowth-inhibitory activity was concentrated in the crude glycoresin. Further chromatographic analysis by HPLC yielded the major component in pure form, which was given the trivial name of tricolorin A (**33**).[45]

The molecular formula of tricolorin A (**33**) was determined as $C_{50}H_{86}O_{21}$ by elemental analysis. Its negative ion fab-ms exhibited a quasi-molecular ion peak at m/z 1021 $\{M-H\}^-$. On alkaline hydrolisis, **33** liberated an organic acid fraction together with an H_2O-soluble glycosidic acid, designated as tricoloric acid (**34**). Analysis of the organic acid fraction by GC-MS afforded a single peak which was identified as 2*S*-methylbutyric acid. The acid-catalyzed hydrolysis of **33** gave a mixture of sugars which were characterized as rhamnose, fucose, and glucose in a ratio of 2:1:1.

The ^1H and ^{13}C-NMR data and the linkages of the sugar moieties were assigned by use of a combination of one and two-dimensional NMR techniques (SINEPT, DQF-COSY, HOHAHA, ROESY, HETCOR and COLOC). The selective INEPT technique, applied to tricolorin A (**33**), confirmed the nature of saccharide substitution. Irradiation of the anomeric proton H-1 of fucose at δ 4.68 ($^3J_{CH}$ = 6 Hz) selectively enhanced C-11 (δ 80.76) of the aglycone. When H-2 of fucose (δ 4.78) was irradiated ($^3J_{CH}$ = 6 Hz), a clear enhancement of the anomeric carbon C-1 of glucose resulted (δ 99.76). Irradiation of the anomeric proton of one of the rhamnose units at δ 5.58 ($^3J_{CH}$ = 4 Hz) enhanced the glucose C-2 (δ 80.60). Finally, irradiation of the inner rhamnose H-3 at δ 4.81 ($^3J_{CH}$ = 6 Hz) resulted in the polarization transfer to the anomeric carbon C-1' (δ 104.49) of the terminal rhamnose. The application of a COLOC NMR experiment on tricolorin A further confirmed the sequence of the sugar moiety. ^{13}C-^1H long-range cross peaks were observed and unambiguously assigned as those between C-1 of fucose and H-11 of the aglycone, C-1 of glucose and fucose H-2, C-1 of inner rhamnose and glucose H-2, and C-1' of terminal rhamnose and inner rhamnose H-3. Through-space ^1H-^1H nOe responses for the anomeric protons in **33** were used to confirm the glycosidation secuence as well as the anomeric configuration of each sugar. In order to establish the site of lactonization and the location of the two additional ester linkages at the oligosaccharide core, the signals for the sugar moiety in the ^1H-NMR spectra of **33** and **34** were compared. Significant downfield shifts owing to acylation were observed for glucose H-3 ($\Delta\delta$ 1.66) and inner rhamnose protons H-2 ($\Delta\delta$ 0.82) and H-4 ($\Delta\delta$ 1.22). Subsequently, in a selective INEPT experiment, magnetization transfer from rhamnose H-4 at δ 5.74 ($^3J_{CH}$ = 4 Hz) enhanced the carbonyl ester resosance at δ 175.42. Therefore, one of the methyl butyric acid residues is concluded to be esterified at the geminal hydroxyl group at this position. In the negative fabsms of **33** and **34**, besides the common fragments peaks observed at m/z 271 and 417, tricolorin A showed a peak at m/z 561 in place of that detected at m/z 579 in the spectrum of **34**. The difference of 18 mass units suggested that the ester linkange of jalapinolic acid is placed at C-3 of glucose. The molecular structure of bioactive tricolorin A (**33**) and its saponification derivative (**34**) are shown (Fig. 7).

Bioassays showed that radicle elongation of the two weed seedlings tested was strongly inhibited by tricolorin A, with IC_{50} values ranging from 12 to 37 μM. The usual concentration threshold required for most natural phytogrowth inhibitors is in the 100-1000 μg/ml range. Additional biological evaluation procedures were performed. Quantitative antimicrobial assays against *Staphylococcus aureus* determined a MIC of 1.8 μg/ml for compound **33**. Competitive antagonism of PDBu binding to partially purified PKC was analyzed. Tricolorin A demonstrated a dose-dependent inhibition of the specific binding of {^{3}H}PDBu with an IC_{50} of 43.2 μM.[45] Finally, this compound displayed only marginal cytotoxicity with ten human cancer cell lines and murine lymphocytic leukemia in cell culture. As summarized in Table 9, the most intense cytotoxic activity (ED_{50} 2.2 μg/ml) was observed with human breast cancer and P-388 cells.[45] Tricoloric acid (**34**) was inactive against all the cell lines tested. These results were similar to those obtained for **32** and its saponification derivative (**31**). It suggests that the biological activity must be associated with the macrocycle structure of these glycolipids, major constituents of the Convolvulaceae glycoresins, which also is consistent with the purgative effect confined to the whole molecule. The laxative activity of the glycoresins presumably is bound to the intact complex mixture of constituents, since the glycosidic acids derived from them have shown no pharmacological activity.[38]

It can be infered that the *Ipomoea* glycoresins are potent inhibitors of plant growth and are primarily responsible for the allelopathic interference exhibited by these species. The active principles may also be involved in the chemical ecology of the plant family Convolvulaceae, as demonstrated by the broad range of biological activities displayed by tricolorin A (**33**).

HEPATOTOXIC PYRROLIZIDINE ALKALOIDS IN *PACKERA CANDIDISSIMA* (ASTERACEAE)

Ethnobotany

Packera candidissima (Greene) Weber & Löve is a low growing, gray-green perennial herb with yellow flowers. It is found frequently as extensive

Tricolorin A (**33**)

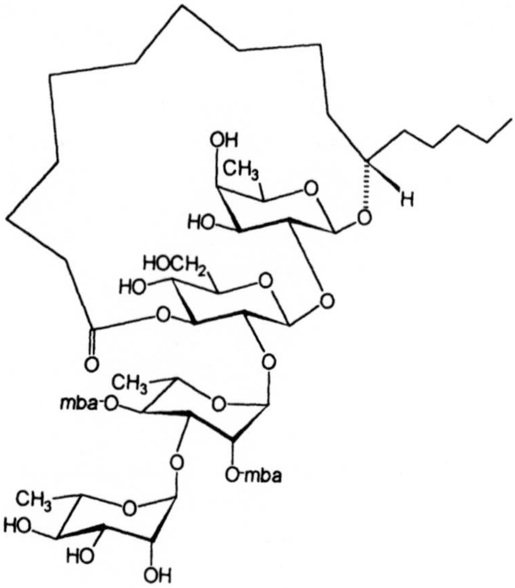

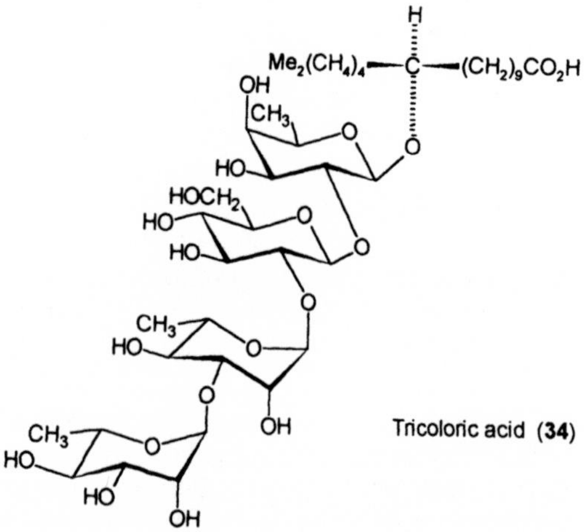

Tricoloric acid (**34**)

Figure 7. Structure of tricolorin A, major phytogrowth inhibitor from *Ipomoea tricolor*, and its corresponding glycosidic acid derivative. Abbreviation: mba, 2-methylbutyroyl.

populations in openings of recently disturbed pine-oak forest in the Sierra Madre Occidental of northwestern Mexico, principally in the state of Chihuahua. Traditional Asteraceae classification places this taxon in the genus *Senecio* as *Senecio candidissimus* Greene. A more radical taxonomic perspective is that of Weber and Löve who recognize *Packera* as a monophyletic American line that deviated from the European *Senecio*. This view is based upon a different basic chromosome number ($x = 8$ rather than $x = 10$ as in the basal genus *Senecio*), and a distinctive set of morphological characteristics, *e.g.* prolonged rhizomes, particular basal leaf arrangement and form and typical tomentum.[47] Only four species in this recently recognized taxon have been analysed chemically: *Packera anonyma* (Wood) Weber & Löve;[48] *P. clevelandii* (Greene) Weber & Löve;[49] *P. multilobata* (T. & G. ex Gray) Weber & Löve;[50] and *P. toluccana* (DC.) Weber & Löve.[51] The studies reveal the presence of furanoeremophilanes and 12-membered macrocyclic pyrrolizidine alkaloids, both as free bases and N-oxides, which are characteristic of the tribe Senecioneae. Of these species, only two are reportedly used for medicinal purposes. In the United States, *P. multilobata* is employed by the Ramah Navaho of New Mexico for ceremonial purposes and the alleviation of menstrual pains. In Mexico, *P. toluccana* has been used in the treatment of tetanus.[47]

Mexicans living in cities (Chihuahua City and Hidalgo de Parral) east of the Sierra Madre Occidental employ the dried leaves of *P. candidissima*, popularly known as "chucaca" or "lechugilla", for various medicinal purposes. The quantity of leaves added to a liter of boiling water varies from three plants to the amount gathered by three fingers. The bitter infusion is drunk as a hot tea or as drinking water in order to treat kidney ailments and ulcers. The whole plant is boiled and drunk as a purgative. Also, the leaves in the form of a decoction or a poultice are applied to boils, skin eruptions and superficial infections.[47] Mexicans living in rural areas also appreciate "chucaca" for medicinal properties. As a tea, it is drunk by those with bladder and kidney trouble, while the fresh plant is ground with olive oil and applied as a poultice which is changed several time a day for boils, tumors and infections. The mestizos use the decoction of this medicinal herb as a vaginal wash in the treatment of venereal diseases.[47] The Tarahumara Indians of the Sierra Madre Occidental collect this perennial herb of the pine-oak forest throughout the year and use it in a similar manner. Known in the "rarámuri" language as "chucá" or "chucaca" or in Mexican Spanish as "lechugilla de la sierra" or "hierba de milagro", the plant is prepared by the Indians as a bitter tea using a few leaves and roots to alleviate chest (or heart), stomach and kidney pains as well as for treatment of ulcers and diarrhea. A stronger infusion is said to be a good purgative. The dried plant is prepared as a wash or poultice for curing sores and boils, especially running sores called "chaanare". The Tarahumara women use a decoction of this herb as a vaginal douche. Pieces of the root also are placed in dental caries to alleviate toothaches.[47] "Chucaca" has been probably in use for at least 200 years, and its present employment as a traditional remedy is not only valued by both the urban Mexicans and the Tarahumara in northern Mexico, but also by various vendors in Chihuahua markets. They export this plant to the southwestern part of the United States (Texas, New Mexico and Arizona) where it is employed for comparable purposes by the Hispanic population under the name of "té de milagros" (miracle tea).[47] Thus, it is a major medicinal herb in current use in northern Mexico and southwestern United States.

Phytochemical Aspects

The main goal of our chemical study of *Packera candidissima* was the determination of its harmful potential by quantifying the total content of hepatotoxic pyrrolizidine alkaloids.[47] Acid-base partition and reductive processes led to isolation of free pyrrolizidine alkaloids and N-oxides, respectively. Free bases comprised 44.2 % of the total alkaloid content (0.76% w/w) obtained from the root. Aerial parts contained lower amount of free alkaloids (0.36% w/w). The yield of free pyrrolizidine alkaloids and their N-oxides obtained in terms of starting crude material are summarized (Table 10).

In order to identify individual components of these mixtures, capillary gas chromatography coupled to mass spectrometry was used to analyze the alkaloid extracts prepared by both reductive and non-reductive procedures. The analysis of the crude alkaloid extracts prepared by reductive and non-reductive procedures demonstrated no qualitative difference in their contents. The quantification of senecionine (**35**), integerrimine (**36**), retrorsine (**37**), and usaramine (**38**) from the aerial parts was carried out. Analysis of the root alkaloidal fraction indicated senkirkine (**39**) as the major component (Table 11). Molecular structures of the alkaloids **35-39** are shown (Fig. 8).

Despite the frequent demonstration of pyrrolizidine alkaloid poisoning by medicinal plants,[52] and the vast scientific literature on the toxicology of these alkaloids in humans,[53] there is still a lack of recognition in Mexico and the United States of the potential problems associated with the use of pyrrolizidine-containing herbs in traditional remedies. Of particular public health significance is the use of "lechugilla de la sierra" as a medicinal tea. The content of free bases and N-oxide pyrrolizidine alkaloids in the root and aerial parts of *Packera candidissima* is one of the highest values reported in the literature.[52,53] Thus, infusions of only 1.3 g of the root or 2.7 g of the aerial parts might provide enough pryrrolizidine alkaloids (10 mg) to produce chronic or acute veno-occlusive deseases.[47]

Table 10. Amounts of pyrrolizidine alkaloids determined in *Packera candidissima*

Sample	Root		Aerial part	
	Quantity (g)	Yield (%)[a]	Quantity (g)	Yield (%)[a]
Crude material	6.50	100	4.54	100
MeOH extract	1.00	13.4	1.00	22.0
Crude alkaloids	0.05	0.76	0.017	0.36
Free bases	0.02	0.33	0.003	0.06
N-oxides	0.03	0.43	0.013	0.30

[a]Yield = w/w in terms of dry starting material.

Table 11. Composition of the alkaloid mixture prepared from aerial parts and root of *Packera candidissima*

Compound	Trivial name	Rt (min)[a]	m/z (M^+)[b]	Concentration (%)[c]	
				Aerial parts	Root
1	Senecionine	10.5	335	23.80	10.34
2	Integerrimine	11.2	335	68.16	12.15
3	Retrorsine	12.5	351	7.09	-
4	Usaramine	12.9	351	0.95	-
5	Senkirkine	12.0	365	-	53.48
	Unidentified	14.5	-	-	8.70
	Unidentified	15.7	-	-	15.33

[a]Capillary gas chromatography retention time.
[b]Electron impact mass spectrometry (70 ev).
[c]Percentage in terms of total alkaloid content.

Despite potential toxicity, the high levels of both free bases and their N-oxides in *P. candidissima* are probably related to some of its curative properties, due to the cytotoxic and antimitotic properties of pyrrolizidine alkaloids. The local anaesthetic action of some pyrrolizidine derivatives also have received attention and their pharmacology has been reviewed.[53] It is noteworthy that various *Senecio* have been used in the treatment of cancer by many cultures.[54] Indeed, *Senecio vulgaris* L., known to contain senecionine (**35**), has been used in England since the 4th century A.D. for tumors of the feet and sinews, and since the 10th century

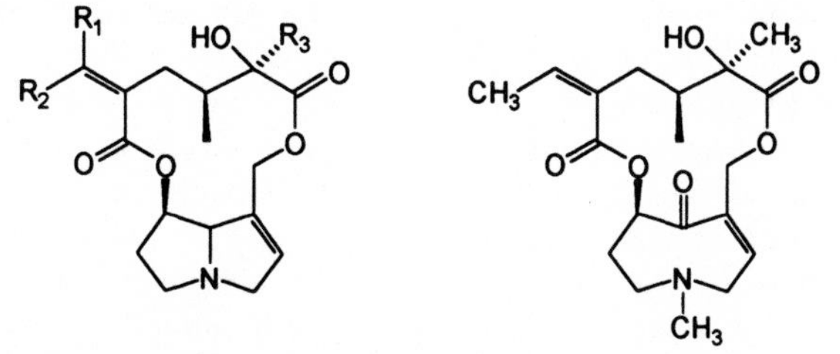

Senecionine (**35**) R_1 = H; R_2 = R_3 = Me

Senkirkine (**39**)

Integerrimine (**36**) R_1 = R_3 = Me; R_2 = H

Retrorsine (**37**) R_1 = H; R_2 = Me; R_3 = CH_2OH

Usaramine (**38**) R_1 = Me; R_2 = H; R_3 = CH_2OH

Figure 8. Hepatotoxic pyrrolizidine alkaloids from *Packera candidissima*.

for cancer. Senecionine (**35**) and its N-oxide were found to be the active principles in *Senecio triangularis* Hook extracts against Walker 256 (intramuscular) tumors.[53] Pyrrolizidine alkaloids **35-39** also were the constituents of the alkaloidal fraction from *Packera anonyma* (*Senecio anonymus*), and found to be active against A204 rhabdomyosarcoma cells.[48] Despite their interest as antitumor agents, none of the naturally occuring *Senecio* pyrrolizidine alkaloids has undergone clinical trials as an anticancer drug due to their hepatotoxicity which severely represents an obstacle to therapeutic usefulness.[53]

The consumption of *Packera candidissima* should be suspended. Appropriate health and regulation authorities in both Mexico and the United States should warn the vendors, "curanderos" and the consumers of herbal products of the dangers involved with the drinking and topical application of this herb. Members of the medical profession who attend patients who use herbal products should be trained in the detection of symptons of pyrrolizidine alkaloid poisoning such as veno-occlusive disease of the liver. Human pathologists and coroners can contribute to this poorly documented health hazard by inquiring into the background of individuals who had cirrhosis of the liver, primary liver cancer, and lung and heart damage in respect to use of herbal medications.

CONCLUSIONS

Herbal medicines traditionally have contributed to the identifcation of natural products for potential development as drugs. We have provided several illustrations form the Lamiaceae and Convolvulaceae where biological activity-guided fractionation of plant extracts has led to isolation of active principles. In our phytochemical investigations, the "ethnobotanical hypothesis" that interdisciplinary scientific exploration of plants traditionally employed or observed by man may lead to the discovery of bioactive natural products has been upheld. Modern analytical methods are critical, as seen in the example of *Packera candidissima* (Asteraceae), for making sophisticated scientific and ethical decisions on which plants should be eliminated from development as clinical drugs or over the counter phytomedicines. The persistence of medicinal plants in Mexican markets and their employment in modern Mexico by people with different cultural backgrounds in both urban and rural communities can be considered as a testimony of their efficacy. This cultural heritage constitutes a rich source of ethnomedical information for selecting effective plants as candidates for scientific evaluation.

ACKNOWLEDGMENTS

This work was supported, in part, by grants from the Dirección General de Asuntos del Personal Académico (DGAPA IN202493; IN203394; IN203494), Universidad Nacional Autónoma de México (UNAM). It was carried out through efforts of several graduate students (M. Bah, L. Hernández, M. Novelo and M.J. Villavicencio) and collaborators in Mexico (Drs. A.L. Anaya, R. Bye and R. Mata, UNAM) and the United States (Drs. A.D. Kinghorn and J.M. Pezzuto, University of Illinois at Chicago), whose names appear on the respective papers which detail the studies. The author is indebted to the staff of the NMR laboratories of the Instituto de Química (UNAM), primarily to F. del Río, M.I. Chávez and A. Gutiérrez, for their technical assistance and to Dr. J.L. McLaughlin (Purdue University, U.S.A.) who provided facilities to conduct some of the cytotoxicity evaluations. The critical comments and suggestions of Dr. J.T. Arnason during the preparation of this manuscript are acknowledged with appreciation. Most importantly, I thank my former professor, Dr. Rachel Mata, for her constant and unfailing support and especially for injecting a breath of fresh air into the study of natural products from Mexican medicinal plants.

REFERENCES

1. MATA, R. 1993. Chemical studies and biological aspects of some Mexican plants used in traditional medicine. In: Phytochemical Potential of Tropical Plants. (K. R. Downum, J. T. Romeo, H. A. Stafford, eds.). Plenum Press, New York. pp. 41-64.
2. RAMAMOORTHY, T. P., ELLIOTT, M. 1993. Mexican Lamiaceae: Diversity, distribution, endemism, and evolution. In: Biological Diversity of Mexico: Origins and Distribution. (T. P. Ramamoorthy, R. Bye, A. Lot, J. Fa, eds.), Oxford University Press, New York, pp. 513-539.
3. WAGNER, H. 1977. Pharmaceutical and economic use of the Labiatae and Rutaceae families. Rev. Latinoamer. Quim. 8: 16-25.
4. LAWRENCE, B. M. 1992. Chemical components of Labiatae oils and their exploitation. In: Advances in Labiatae Science. (R.M. Harley, T. Reynolds, eds.), Royal Botanic Gardens, Kew, pp. 399-436.
5. LINARES, E., BYE, R., FLORES, B. 1984. Tés curativos de México. FONART/SEP-Cultura, México, pp. 1-95.
6. HEINRICH, M. 1992. Economic Botany of American Labiatae. In: Advances in Labiatae Science. (R.M. Harley, T. Reynolds, eds.), Royal Botanic Gardens, Kew, pp. 475-488.
7. ALCORN, J. B. 1984. Huastec Mayan Ethnobotany. University of Texas Press, Austin, p. 982.
8. HEINRICH, M., RIMPLER, H. BARRERA, N. A. 1992. Indigenous phytotherapy of gastrointestinal disorders in a lowland Mixe community (Oaxaca, Mexico): Ethnopharmacologic evaluation. J. Ethnopharmacol. 36: 63-80.
9. HERNANDEZ, F. 1942. Historia de las Plantas de Nueva España por Francisco Hernández, Médico e Historiador de su Majestad don Felipe II, Rey de España y de la Indias, y Protomédico de Todo el Nuevo Mundo. Instituto de Biología, Universidad Nacional Autónoma de México, México. Vol. 1., p. 318.
10. HERNANDEZ, F. 1959. Obras Completas. Historia Natural de Nueva España I. Universidad Nacional Autónoma de México, México, vol. II, p. 397.
11. NAVARRO, J. 1992. Historia Natural o Jardín Americano (Manuscrito de 1801). UNAM-IMSS-ISSSTE, México, p. 112.
12. HERNANDEZ, F. 1984. Obras Completas. Comentarios a la Obra de Francisco Hernández. Universidad Nacional Autónoma de México, México, vol. VII, p. 187.

13. EPLING, C. 1949. Revisión del género *Hyptis* (Labiatae). Revista del Museo de la Plata 8: 153-497.
14. MARTINEZ, M. 1979. Catálogo de Nombres Vulgares y Científicos de Plantas Mexicanas. Fondo de Cultura Económica, México, pp. 1133-1134.
15. PEREDA-MIRANDA, R., DELGADO, G. 1990. Triterpenoids and flavonoids from *Hyptis albida*. J. Nat. Prod. 53: 182-185.
16. MARTINEZ, M. 1989. Las Plantas Medicinales de México. 6th ed. Ed. Botas, México, pp. 412-488.
17. REIS ALTSCHUL, S. von. 1973. Drugs and Foods from Little-Known Plants. Notes in Harvard University Herbaria. Harvard University Press, Cambridge, pp. 263-265.
18. PEREDA-MIRANDA, R., GASCON-FIGUEROA, M. 1988. Chemistry of *Hyptis mutabilis*: New pentacyclic triterpenoids. J. Nat. Prod. 51: 996-998.
19. PEREDA-MIRANDA, R., HERNANDEZ, L., VILLAVICENCIO, M. J., NOVELO, M., IBARRA, P., CHAI, H., PEZZUTO, J. M. 1993. Structure and stereochemistry of pectinolides A-C, novel antimicrobial and cytotoxic 5,6-dihydro-α-pyrones from *Hyptis pectinata*. J. Nat. Prod. 56: 583- 593.
20. ROIG Y MESA, J. T. 1974. Plantas Medicinales, Aromáticas o Venenosas de Cuba. Instituto de Libro, La Habana, Cuba, pp. 596-597.
21. DEL AMO, R. S. 1979. Plantas Medicinales del Estado de Veracruz. Instituto Nacional de Investigaciones sobre Recursos Bióticos, México, p. 110.
22. SOTO NUÑEZ, J. C. 1987. Las Plantas Medicinales y su Uso Tradicional en la Cuenca del Rio Balsas; Estados de Michoacán y Guerrero, México. B.S. Thesis. Universidad Nacional Autónoma de México, México, p. 56.
23. NOVELO, M., CRUZ, J. G., HERNANDEZ, L., PEREDA-MIRANDA, R., CHAI, H., MAR, W., PEZZUTO, J. M. 1993. Cytotoxic constituents from *Hyptis verticillata*. J. Nat. Prod. 56: 1728-1736.
24. MARTINEZ ALFARO, M. A. 1984. Medicinal plants used in a Totonac community of the Sierra Norte de Puebla: Tuzamapan de Galeana, Puebla, Mexico. J. Ethnopharmacol. 11: 203-221.
25. HEINRICH, M., VELAZCO, O., RAMOS, F. 1990. Ethnobotanical report on the treatment of snake-bites in Oaxaca, Mexico. Curare 13: 11-16.
26. ARNASON, T., UCK, F., LAMBERT, J., HEBDA, R. 1980. Maya medicinal plants of San Jose Succotz, Belize. J. Ethnopharmacol. 2: 345-364.
27. MITSCHER, L. A., DRAKE, S., GOLLAPUDI, S. R., OKWUTE, S. K. 1987. A modern look at folkloric use of anti-infective agents. J. Nat. Prod. 50: 1025-1040.
28. ROJAS, A., HERNANDEZ, L., PEREDA-MIRANDA, R., MATA, R. 1992. Screening for antimicrobial activity of crude drug extracts and pure natural products from Mexican medicinal plants. J. Ethnopharmacol. 35: 275-283.
29. PEREDA-MIRANDA, R., IBARRA, P., HERNANDEZ, L., NOVELO, M. 1990. Bioactive constituents from *Hyptis* species. Planta Medica 56: 560-561.
30. KINGSTON, D. G. I., RAO, M. M., ZUCKER, W. V. 1979. Plant Anticancer Agents. IX. Constituents of *Hyptis tomentosa*. J. Nat. Prod. 42: 496-499.
31. SHETH, K., JOLAD, S., WIEDHOPF, R., COLE, J. R. 1972. Tumor-inhibitory agent from Hyptis emoryi (Labiatae). J. Pharm. Sci. 61:1819.
32. PEREDA-MIRANDA, R., HERNANDEZ, L., LOPEZ, R. 1992. A novel abietane-type diterpene from *Salvia albocaerulea*. Planta Medica 58: 223-224.
33. DAVIES-COLEMAN, M. T., RIVETT, D. E. A. 1989. Naturally occurring 6-substituted 5,6-dihydro-α-pyrones. In: Progress in the Chemistry of Organic Natural Products. (W. Herz, H. Grisebach, G. W. Kirby, C. Tamm, eds.), Springer-Verlag, New York. Fortschr. Chem. organ. Naturstoffe pp. 1-35.

34. PEREDA-MIRANDA, R., GARCIA, M., DELGADO, G. 1990. Structure and stereochemistry of four α-pyrones from *Hyptis oblongifolia*. Phytochemistry 29: 2971-2974.
35. GERMAN, V. F. 1971. Isolation and characterization of cytotoxic principles from *Hyptis verticillata* Jacq. J. Pharm. Sci. 60: 649-650.
36. SAM, T. W. 1993. Toxicity testing using the brine shrimp: *Artemia salina*. In: Bioactive Natural Products. Detection, Isolation, and Structural Determination. (S. M. Colegate, R. J. Molyneux, eds.), CRC Press, Boca Raton, Florida, pp. 441-456.
37. AYRES, D. C., LOIKE, J. D. 1990. Lignans. Chemical, Biological and Clinical Properties. Cambridge University Press, Great Britain, pp. 85-112.
38. WAGNER, H. 1973. The chemistry of resin glycosides of the Convolvulaceae family. In: Medicine and Natural Sciences, Chemistry in Botanical Classification. (G. Bendz, J. Santesson, eds.), Academic Press, New York, pp. 235-240.
39. LINAJES, A., RICO-GRAY, V., CARRION, G. 1994. Traditional production system of the root of jalap, *Ipomoea purga* (Convolvulaceae), in Central Vercruz, Mexico. Economic Bot. 48: 84-89.
40. CASTRO RAMIREZ, A. E. 1988. Estudio Comparativo del Conocimiento sobre Plantas Medicinales Utilizadas por Dos Grupos Etnicos del Muncipio de Pahuatlán, Puebla. San Juan Iztacala, México. B.S. Thesis. Escuela Nacional de Estudios Profesionales Iztacala. Universidad Nacional Autónoma de México, pp. 60-61.
41. DIAZ, J. L. 1976. Usos de las Plantas Medicinales de México. Monografías Científicas II. Instituto Mexicano para el Estudio de las Plantas Medicinales, A.C., México, pp. 66-67.
42. BEUTLER, J. A., ALVARADO, A. B., McCLOUD, T. G., CRAGG, G. M. 1989. Distribution of phorbol ester bioactivity in the Euphorbiaceae. Phytother. Res. 3: 188-192.
43. EVANS, F. J. 1991. Natural products as probes for new drug target identification. J. Ethnopharmacol. 32: 91-101.
44. NODA, N., KOGETSU, H., KAWASAKI, T., MIYAHARA, K. 1990. Scamonins I and II, the resin glycosides of radix scammoniae from *Convolvulus scammonia*. Phytochemistry 29: 3565-3569.
45. PEREDA-MIRANDA, R., MATA, R., ANAYA, A. L., WICKRAMARATNE, D. B., PEZZUTO, J. M., KINGHORN, A. D. 1993. Tricolorin A, major phytogrowth inhibitor from *Ipomoea tricolor*. J. Nat. Prod. 56: 571-582.
46. PETERSON, J. K., HARRISON H. F., Jr. 1991. Isolation of substance from sweet potato (*Ipomoea batatas*) periderm tissue that inhibits seed germination. J. Chem. Ecol. 17: 943-951.
47. BAH, M., BYE, R., PEREDA-MIRANDA., R. 1994. Hepatotoxic pyrrolizidine alkaloids in the Mexican medicinal plant *Packera candidissima* (Asteraceae: Senecioneae). J. Ethnopharmacol. 43: 19-30.
48. ZALKOW, L. H., ASIBAL, C. F., GLINSKI, J. A., BONETTI, S. J., GELBAUM, L. T., VANDERVEER, D., POWIS, G. 1988. Macrocyclic pyrrolizidine alkaloids from *Senecio anonymus*. Separation of a complex alkaloid extract using droplet counter-current chromatography. J. Nat. Prod. 51: 690-702.
49. BOHLMANN, F., ZDERO, C., KING, R. M., ROBINSON, H. 1981. The first acetylenic monoterpene and other constituents from *Senecio clevelandii*. Phytochemistry 20: 2425-2427.
50. McCOY, J. W., ROBY, M. R., STERMITZ, F. R. 1983. Analysis of plant alkaloid mixtures by ammonia chemical ionization mass spectrometry. J. Nat. Prod. 46: 894-900.
51. PEREZ, A. L., VIDALES, P., CARDENAS, J., ROMO DE VIVAR, A. 1991. Eremophilanolides from *Senecio toluccanus* var. *modestus*. Phytochemistry 30: 905-908.
52. HUXTABLE, R .J. 1989. Human health implications of pyrrolizidine alkaloids and herbs containing them. In: Toxicants of Plant Origen. Alkaloids, Vol. I. (P. R. Cheeke, ed.), CRC Press, Boca Raton, FL. pp. 41-86.

53. MATTOCKS, A. R. 1986. Chemistry and Toxicology of Pyrrolizidine Alkaloids. Academic Press, London, pp. 158-315.
54. HARTWELL, J. L. 1968. Plants used against cancer. A survey. LLoydia 31:7 1-170.

Chapter Six

BIOASSAY-GUIDED ISOLATION AND STRUCTURE ELUCIDATION OF PHARMACOLOGICALLY ACTIVE PLANT SUBSTANCES

A. J. Vlietinck, L. A. C. Pieters and D. A. Vander Berghe

Department of Pharmaceutical Sciences
University of Antwerp (U.I.A.)
B-2610, Antwerp, Belgium

INTRODUCTION

One of the successful methodologies for the investigation of traditional medicines as sources of new drugs includes the pharmacological screening of plant preparations followed by a bioassay-guided fractionation leading to isolation of pure active plant constituents. Ideally, this methodology entails the *in vivo* testing of the traditional drug for the claimed pharmacological activity. After experimental confirmation of this activity, a corresponding *in vitro* method is developed, which can then be used for the monitoring of activity during purifica-

Phytochemistry of Medicinal Plants
Edited by John T. Arnason et al., Plenum Press, New York, 1995

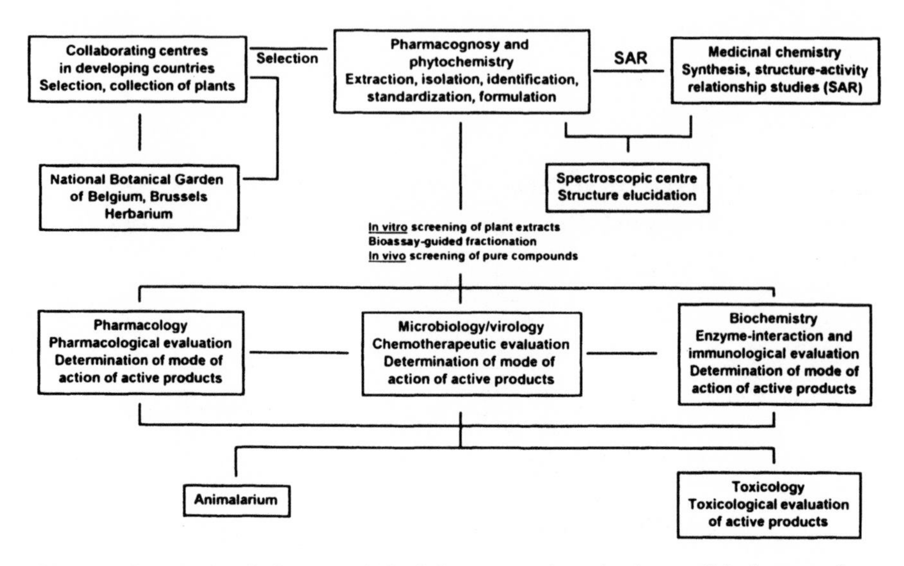

Figure 1. Organizational chart - multidisciplinary team investigating medicinal agents from higher plants.

tion of the active plant constituents. Once these active plant substance(s) have been identified, standardization of a plant preparation can be developed, or structure activity relationship studies can be started by partial or total synthesis of the active plant substance(s).

Such a research program is best carried out by a multidisciplinary team consisting of at least a pharmacognosist and a microbiologist, pharmacologist or biochemist, depending on the kind of test models used in the screening battery. As shown in the organizational chart (Fig. 1) the team should collaborate with a center where the selection and collection of the plants to be tested is carried out, and with a medicinal or organic chemist who is responsible for the synthesis and structure activity relationship studies of the lead compounds. In this paper, the strategy for finding new leads from plants used in traditional medicine will be illustrated with several examples of plants with wound healing and antiviral activity.

SANGRE de DRAGO

Wound Healing Properties

Deep-red blood-like sap from various *Croton* spp. (Euphorbiaceae), most commonly *Croton lechleri* L., *Croton draconoides* (Muell.) Arg. and *Croton erythrochilus* (Muell.) Arg, is known as Dragon's blood or Sangre de Drago (SdD)

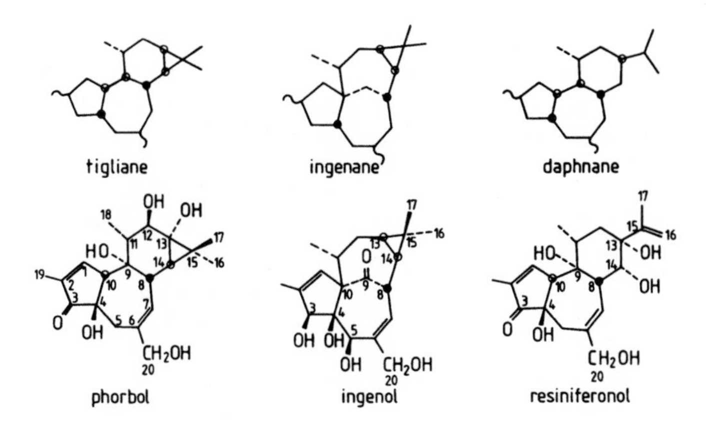

Figure 2. Diterpene parent hydrocarbons tigliane, ingenane and daphnane, and prototype parent alcohols phorbol, ingenol and resiniferonol.

in Spanish. It is widely used throughout South America as a folk medicine for the treatment of wounds, inflammation and cancer.[1,2] The sap is collected from the bark of felled trees, and as a consequence the popular widespread use of the sap has threatened the existence of the plants.

Efforts have been made by our team to identify the wound healing principles and to examine their mechanism(s) of action. Since it is known that many species of the Euphorbiaceae contain tumor promoting diterpene esters, which are derived from the tetracyclic or tricyclic diterpene parent alcohols phorbol, ingenol and resineferol,[3,4] it was decided to examine whether or not such compounds were present in Sangre de Drago, before starting a bioassay-guided isolation (Fig 2). By TLC and NMR it was found that Sangre de Drago contains no detectable amounts of tumor-promoting diterpene esters. Therefore, there appears to be no reason to restrict or discourage the use of this popular South American medicine.[5]

Process of Wound Healing

The process of wound healing begins immediately following surface lesion or when skin protein is exposed to radiation, chemical damage or extreme temperatures. It can be divided into four overlapping stages, including coagulation, inflammation, formation of granulation tissue, and matrix formation and remodeling.[6,7] Tissue injury results in the release of blood components into the wound site, activation of the clotting cascade and coagulation, including the formation of thrombin, which stimulates the release of α-granules from aggregated platelets. These granules contain locally acting growth factors. Growth factors may be defined as polypeptides that stimulate cell proliferation through binding to specific high-affinity cell membrane receptors.[8] They presumably diffuse short distances through intercellular spaces and act locally. The combination, concentration, and timing of growth factor release and activation at the site of injury regulate the complex process of wound healing. Growth factors released in the traumatized area promote cell migration into the wound area (chemotaxis), stimulate the growth of epithelial cells and fibroblasts (mitogenesis), initiate the formulation of new blood vessels from endothelial cells (angiogenesis), and stimulate matrix formation and remodeling of the affected region.

After an injury site has been sterilized during inflammation, granulation tissue, consisting of a dense array of fibroblasts, macrophages and neovasculature embedded in a loosely woven matrix of collagen, fibronectin and hyaluronic acid, is formed. In response to certain growth factors, the fibroblasts proliferate and migrate into the wound site where they display different phenotypes. Fibroblasts first assume a migratory phenotype, then a collagen-producing phenotype, and finally a contractile phenotype. In the contractile state, the so-called myofibroblasts align themselves along the radial axis of the newly deposited extracellular matrix within the wound, and form cell-cell and cell-matrix links to generate a contractive force that aids wound closure. Endothelial cells in the wound proliferate and form new blood vessels to supply the injured site with nutrients and oxygen. Within hours of injury, reepithelialization begins to restore the integrity of the damaged surface.

Reepithelialization begins with the migration of epithelial cells from the edges of the tissue across the wound. Within 24 hours, epithelial cells at the original edge of the wound begin proliferating, thereby generating more cells for migration. Once reepithelialization is complete, the epithelial cells revert to their non-migrating phenotype and become attached to the basement membrane through hemidesmosomes. The final phase of wound healing is the replacement of granulation tissue with connective tissue consisting of a framework of collagen and elastin fibers providing tissue strength and elastic properties, respectively. This framework then becomes saturated with proteoglycans and glycoproteins. Remodeling involves the synthesis of new collagen and the degradation of old collagen.[9]

Bioassay-Guided Isolation of the Active Component(s)

Since endothelium plays a crucial role in the process of wound healing, and an *in vivo* guiding test was excluded for practical and ethical reasons, an *in vitro* test system for the stimulation of endothelial cells was selected.[10] Endothelial cells were obtained from human umbilical cord vein (HUVEC) as described by Jaffe et al.[11] The cells were cultured in medium 199 (Gibco), supplemented with 30% heat-inactivated human adult serum (HAS), 100 U/ml penicillin and 20 μg/ml gentamicin, at 37°C, under a humidified atmosphere containing 5% $C0_2$. The cells were fed every 5 days with a complete change of fresh culture medium until confluence was reached.

For subculture, HUVEC were harvested with trypsin-EDTA solution (Gibco) and split at a ratio of 1:3 for inoculation into new culture flasks. Cultures between the second and fourth passage were used in the experiments. In all stimulation experiments, cells were subcultured in microtiter plates containing media which did not allow normal growth so that potential stimulation could be easily detected. In such media, cells remained alive for several days. In the presence of only 5% HAS, without any additional stimulating agent, HUVEC did not multiply in medium 199 (negative control experiment). On the other hand, starting with 4 x 10^3 cells per well of the microtiter plate, a confluent monolayer containing about 2 x 10^4 cells per well was usually obtained after 6 days in medium 199 with 30% HAS (positive control experiment). For the stimulation experiments, HUVEC were inoculated, after trypsinization, into 96-well microtiter plates (Costar) at a density of 4x10^3 cells per well (0.3 cm^2/well) in medium 199 supplemented with 20% HAS and 100 U/ml penicillin. After 8 hours, cells were carefully washed twice with medium 199 and then exposed to the same medium containing 2% HAS with or without solutions to be tested for stimulation of endothelial cells. Cell growth was evaluated microscopically every other day during 6 days.[5,12]

Bioassay-guided fractionation of Dragon's blood using the *in vitro* test system for the stimulation of HUVEC, has resulted in the isolation of a dihydrobenzofuran lignan, 3',4,0-dimethylcedrusin (DMC) as the biologically active principle[13] (Fig. 3). A related compound, 4-0-methylcedrusin (MC) and the alkaloid taspine (T), also isolated from Dragon's blood, were not active in the same assay as shown in Table 1. Whereas 3',4-0-dimethylcedrusin, at 25 and 5 μg/ml, stimulated HUVEC to the same extent as 30% HAS, 4-0-methylcedrusin did not stimulate HUVEC, and taspine was cytotoxic down to a concentration of about 0.5 μg/ml.

For a more precise evaluation of the effects on HUVEC observed for 3',4-0-dimethylcedrusin (stimulation) and taspine (cytotoxicity), an *in vitro* thymidine incorporation assay was performed.[5,12] An increase in cell density

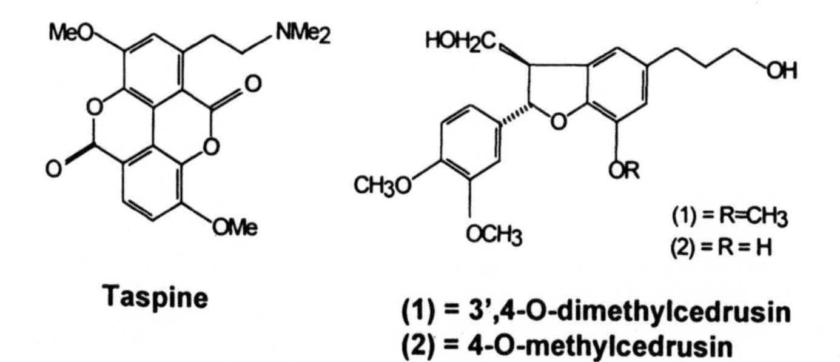

Figure 3. Chemical structures of taspine (T) and 3',4-0-dimethylcedrusin (R = CH_3) (DMC) and 4-0-methylcedrusin (R = H).

Table 1. Results of the stimulation of HUVEC with natural products

Compound	Concentration (μg/ml)						
	250	50	25	5	2.5	0.5	0.25
Et_2O extract of SdD	+	+	=				
DMC	T	+	++	++	=		
MC	T	=					
T	nt	nt	T	T	T	-	=

++ = similar to the positive control experiment (30% HAS)
+ = better than the negative control experiment (5% HAS)
= = similar to the negative control experiment (no cell growth)
- = worse than the negative control experiment (slightly toxic)
T = toxic to the cells
nt = not tested
SdD = *Sangre de Drago*
DMC = 3',4-0-dimethylcedrusin
MC = 4-0-methylcedrusin
T = taspine

should normally correspond to an increase in DNA synthesis, characterized by a higher thymidine incorporation into cellular DNA of the cultured cells. Therefore, the incorporation of tritiated thymidine can be used as a sensitive index of cell proliferation. Results are shown for 3',4-0-dimethylcedrusin (Fig.4).

Figure 4 shows the relative number of cells at the end of the experiment as a function of the concentration of 3',4-0-dimethycedrusin, compared to the negative control experiment at 100%. The maximum number of cells obtained was between 50 and 100 μg/ml, which was higher than observed in the previous experiment (Table 1). This is because in the thymidine incorporation assay only 2% HAS was used, instead of 5% as in the cell growth assay, thus, making the growth conditions more critical. At higher concentrations (125 μg/ml and 250 μg/ml) the compound was toxic to the cells. It is interesting to note that the actual concentration of 3',4-0-dimethylcedrusin in Dragon's blood (0.0014% or about 14 μg/ml) is of the same magnitude as the biologically active concentration range. Fig. 4 shows the incorporation of tritiated thymidine, expressed as cpm, also as a function of the concentration of DMC. At first sight these results appear contradictory. Normally an increase of thymidine incorporation or DNA synthesis is expected in the concentration range where the number of cells increases. However, in this experiment, an inhibition of thymidine incorporation was observed. This indicated that 3',4-0-dimethylcedrusin did not stimulate cell proliferation, but rather that it had a protective effect against degradation of the cells in a starvation medium with only 2% HAS, as in the negative control experiment.

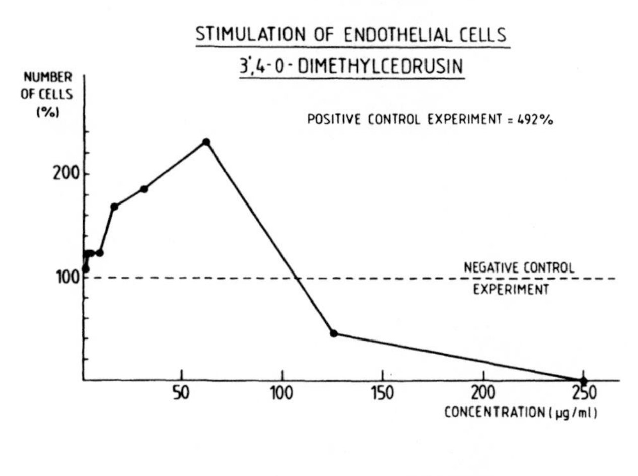

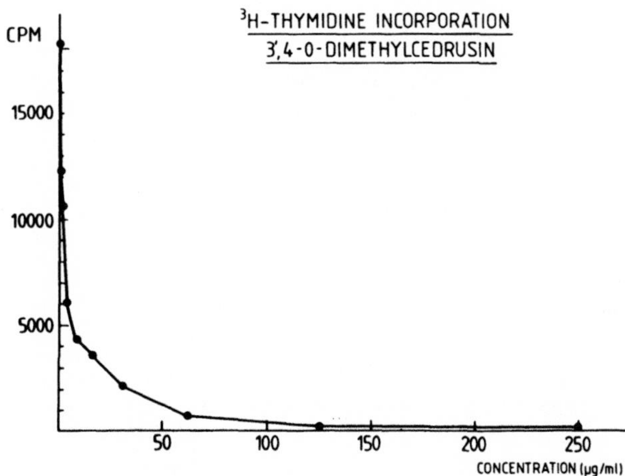

Figure 4. Thymidine incorporation assay for 3',4-0-dimethylcedrusin.

A similar experiment was carried out with taspine. Both graphs (number of cells and thymidine incorporation vs. concentration) are shown in Fig. 5. As mentioned before, this alkaloid is highly cytotoxic; at all concentrations tested, the number of cells was lower than in the negative control experiment, and it decreased as the concentration increased. In this experiment, both graphs have the same appearance. The decrease of thymidine incorporation was due only to the cytotoxicity.

The inhibition of thymidine incorporation by 3',4-0-dimethylcedrusin has been calculated and at non-toxic concentrations such as 62.5 µg/ml, about 98% of the thymidine incorporation is inhibited. At only 2 µg/ml, the inhibition is about

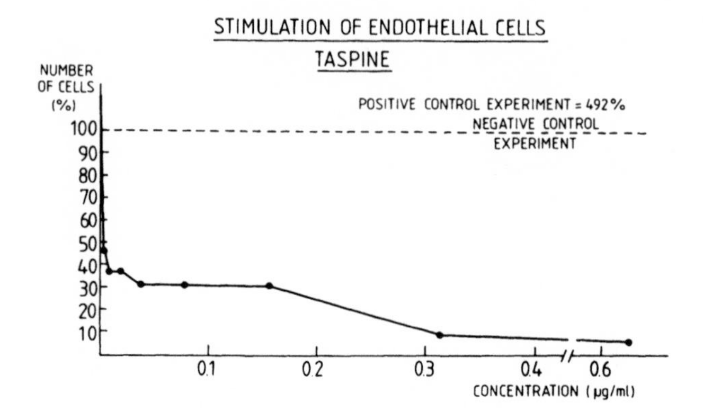

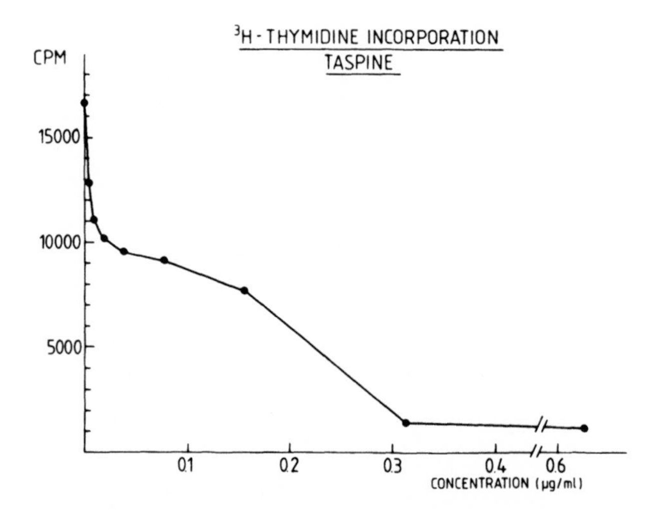

Figure 5. Thymidine incorporation assay for taspine.

50%. Compounds inhibiting thymidine incorporation (or DNA synthesis), a sensitive index of cell proliferation, without being toxic may be useful as antitumor agents. It is interesting to note that in the traditional South American medicine, Dragon's blood is not only used for wound healing but also for treatment of cancer.[2]

From our *in vitro* experiments we concluded that 3',4-0-dimethylcedrusin showed a protective effect on endothelial cells in a starvation medium. The alkaloid taspine, which previously was claimed to be the active principle of Dragon's blood acting by increasing the migration of human foreskin fibroblasts,[14] showed no activity in our assays.

In Vivo Evaluation

We also evaluated the wound healing activity of Dragon's blood and its constituents by *in vivo* experiments on rats. Because of the low yield of 3',4-0-dimethylcedrusin in *Sangre de Drago*, it was decided to synthetize this compound using a biomimetic procedure.[5] As shown in Fig. 6 oxydative dimerisation of methylferulate yielded racemic (E)-methyl-3-[2,3-dihydro-2-(4-hydroxy-3-methoxyphenyl)-7-methoxy-3-methoxycarbonyl-1-benzofuran-5-yl]-propenoate. Subsequent reactions leading to racemic 3',4-0-dimethylcedrusin included methylation of the phenolic hydroxyl group, saturation of the double bond of the C_2 side chain and reduction of both ester functions to primary alcohols. Since Cai et al.[15] found that the blood-red sap of *Croton lechleri* contains proanthocyanidins, varying from monomers to heptamers, as major constituents, we also prepared a mixture of oligomeric proanthocyanidins by a condensation reaction of racemic taxifolin and (+)-catechin under nitrogen.[5]

The *in vivo* wound healing activity of *Sangre de Drago* was compared to that of 3',4-0-dimethylcedrusin, taspine, the synthetic polyphenolic mixture, and the polyphenolic fraction isolated from *Sangre de Drago*. The pure compounds

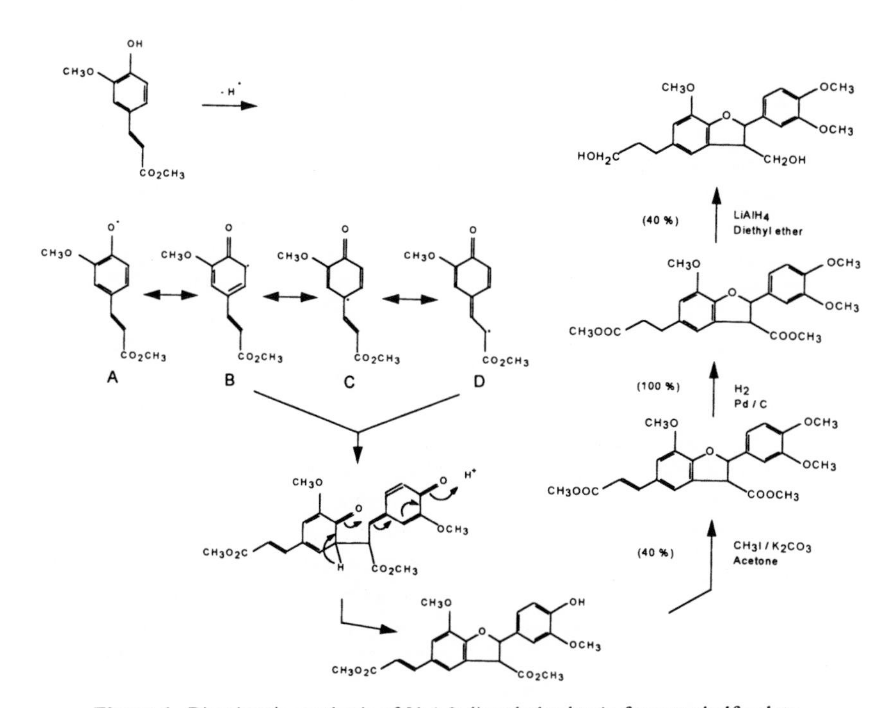

Figure 6. Biomimetic synthesis of 3',4-0-dimethylcedrusin from methylferulate.

3',4-0-dimethylcedrusin and taspine were applied as 0.0014% and 0.09% ointments respectively, which consisted of PEG 4000 (29.5%), PEG 400 (53%), cetylalcohol (4.2%) and water (13.3%) and which correspond to their actual concentrations in *Sangre de Drago*. Both synthetic and isolated polyphenolic fractions were applied as a 7% solution in distilled water, which also corresponded to their concentration in *Sangre de Drago*.

Several experiments were carried out on female rats, which were anesthetized, shaved, disinfected and wounded. Circular excised wounds with a diameter of about 3 cm were made by cutting of the epidermis of the rat's back, and then applying 2 ml of boiling distilled water to the wounds. Treatment started 1 hour after the injury was made, and consisted of applying about 0.5 ml of an ointment or solution containing the product or fraction to be tested, twice daily for 18 days, then once a day until the animals were sacrified. The result of each treatment was compared to untreated rats and rats treated with a placebo ointment. Each treatment was done on a pair of rats. Wound healing was examined daily during the experiment, and pictures were taken after 7, 14, 21 and 28 days. In particular, the area and depth of the lesions and the evolution of the healing process, in terms of tissue contraction and crust formation, were accurately followed up. After 1 month, all rats were sacrificed. The wound area of each rat was carefully removed and maintained in 4% formaldehyde at 4°C. Slices were prepared for microscopic evaluation. Macroscopic evaluation revealed that those wounds treated with crude *Sangre de Drago* and both the synthetic and isolated polyphenolic fraction were almost immediately (i.e. on day 1) covered with a thick, dark crust.

The results of the tissue contraction are shown in Fig. 7. After only one day, tissue contraction occurred in the rats treated with *Sangre de Drago* (D) and the aqueous solutions of the polyphenolic fraction, isolated from *Sangre de Drago* (E) and the synthetic procyanidins (F), whereas treatment with an ointment containing DMC (A), taspine (B) or PEG (C) delayed the initial tissue contraction. After about 11 days of treatment, a second stage of tissue contraction was observed, and all curves converged. For both wounds the ointment containing 3',4-0-dimethylcedrusin appeared to produce the largest tissue contractions, although the differences observed were rather small. Evaluation of the microscopic preparations of those rats treated with crude *Sangre de Drago* revealed that almost no difference could be observed between old, undamaged and newly formed tissue, indicating an effective healing process. Formation of a new epidermal layer was nearly complete, and better than in untreated rats or rats treated with taspine or placebo ointment. New hair follicles were present. Microscopic preparations from rats treated with 3',4-0-dimethylcedrusin also showed the presence of a new epidermal layer, less pronounced than after treatment with *Sangre de Drago*, but better than with other treatments or for the untreated rats. The formation of new hair follicles, although less pronounced was also observed.

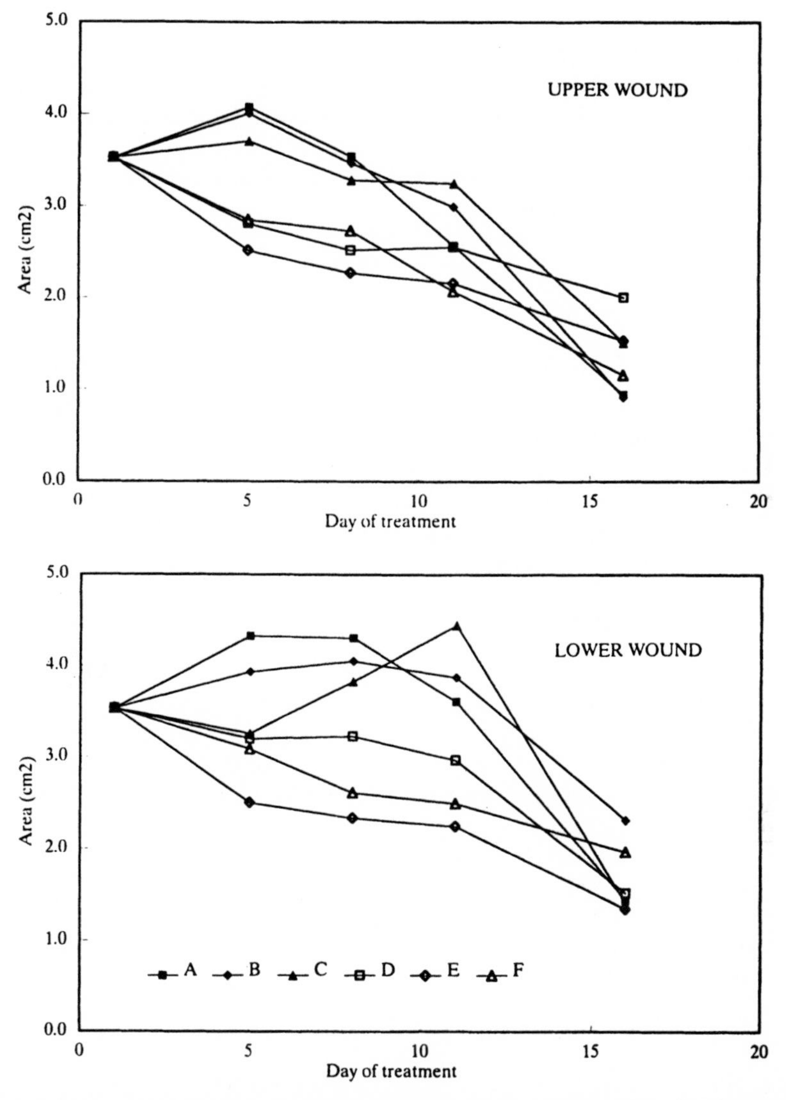

Figure 7. 3',4-0-dimethylcedrusin 0.0014% in PEG; (B) taspine 0.09% in PEG; (C) PEG control experiment; (D) Sangre de Drago; (E) polyphenolic fraction from Sangre de Drago; (F) synthetic procyanidins.In conclusion, the *in vivo* experiments on rats confirmed the wound healing effect of *Sangre de Drago*, as it is used for this purpose in traditional South American medicine. 3,4-0-Dimethylcedrusin, identified as the biologically active principle by *in vitro* bioassay-guided isolation, also improved wound healing *in vivo*. The wound healing effect of crude *Sangre de Drago* was better than that observed for ointments containing 3',4-0-dimethylcedrusin. This possibly was due to the physical effect of the polyphenols which precipitate proteins and form a dark crust covering the wound.

In conclusion, the *in vivo* experiments on rats confirmed the wound healing effect of *Sangre de Drago*, as it is used for this purpose in traditional South American medicine. 3',4-0-Dimethylcedrusin, identified as the biologically active principle by *in vitro* bioassay-guided isolation, also improved wound healing *in vivo*. The wound healing effect of crude *Sangre de Drago* was better than that observed for ointments containing 3',4-0-dimethylcedrusin. This possibly was due to the physical effect of the polyphenols which precipitate proteins and form a dark crust covering the wound.

MEDICINAL PLANTS WITH ANTIVIRAL PROPERTIES

The results of our continuous screening program of medicinal plants for antiviral properties showed that the 80% ethanolic extracts of the stem bark of *Pavetta owariensis* Beauv. (Rubiaceae) and of the leaves of *Euphorbia grantii* Oliv. (Euphorbiaceae) exhibited pronounced antiviral properties. The former plant, which is named worm's tree, is used throughout Guinea Conakry as a specific anthelminthic against *Ascaris lumbricoides* and *Schistosomas*,[16] whereas the latter is used in Rwanda against childhood diseases including poliomyelitis.[17]

Antiviral Test Methodology

Since both the methodology used for determination of antivíral activity and the interpretation of results vary greatly between laboratories, and thus are not comparable, simple proocedures and guidelines for evaluating antiviral and/or virucidal activities are urgently needed. Various cell culture-based assays currently are available and can be succesfully applied to the antiviral or virucidal determination of single substances or mixtures of compounds e.g. plant extracts. Antiviral agents interfere with one or more dynamic processes during virus biosynthesis and, consequently, are candidates for clinically useful antiviral drugs. Virucidal substances, in contrast, inactivate virus infectivity extracellularly, and rather are candidates for antiseptics which exhibit a broad spectrum of germicidal activities.[18]

Extracellular Virucidal Evalation

Preincubated (25°C) plant extracts or their twofold dilutions (e.g., 1/2 to 1/16) (1 ml), dissolved in physiological buffer, are mixed thoroughly with the same volume (1 ml) of a preincubated (25°C) virus suspension (e.g., 10^6 PFU ml^{-1} or TCD_{50} ml^{-1}) in physiological buffer. The mixture is incubated at 25°C for 5 min. The incubation is stopped by addition of a tenfold volume (i.e., 20 ml) of ice-cold maintenance medium, and the mixture is immediately filtered through a 0.22-μm filter to eliminate all possible precipitate. The ice-cold filtrate is filtered through a 0.01-μm filter for enveloped viruses, or a 10,000-MW membrane filter (Amicon, ultrafiltration system) for nonenveloped viruses, to concentrate residual virus on the filter and separate it from possibly cytotoxic plant components that pass through the filter. A small volume (0.2 ml) of the original sample is left on the filter so that the filter never becomes dry. The residual virus is removed. The filter is washed with maintenance medium supplemented with 5% serum (1 ml), sonicated in an ice-bath for 30 s, mixed with the residual virus suspension (0.2 ml), and titrated in tenfold dilutions at 37°C by plaque formation (plaque test, PT) or in microtiter plates, according to the end point titration technique (EPTT). A

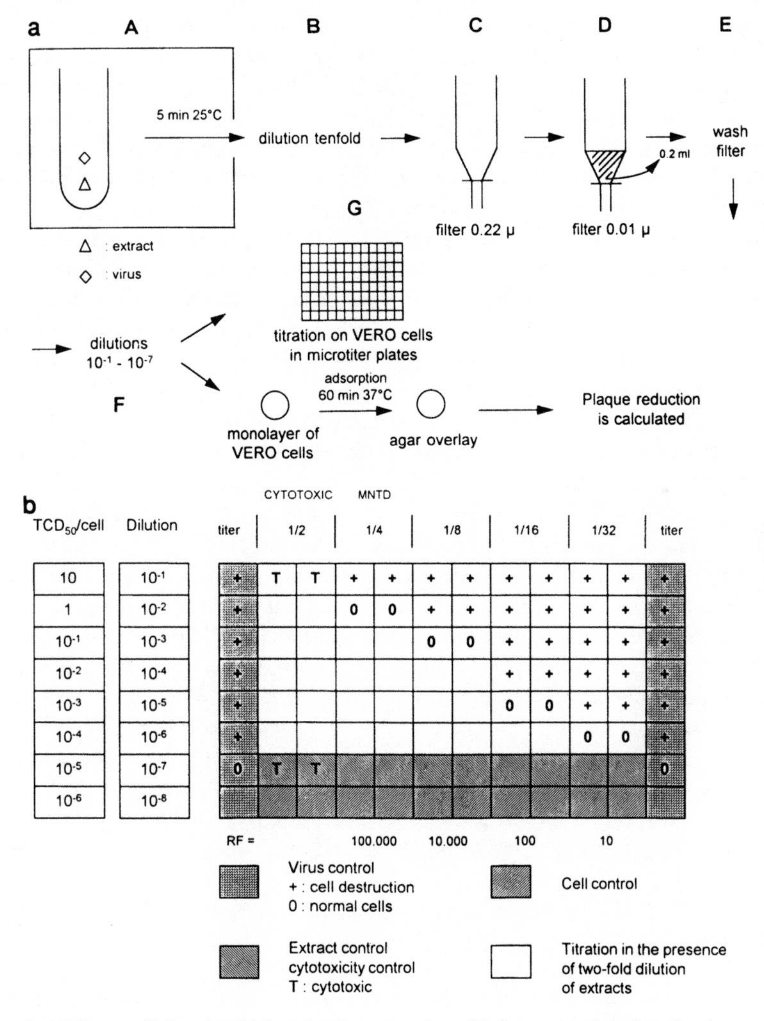

Figure 8. (a) Extracellular virucidal evaluation procedure.(b) *In vitro* antiviral evaluation procedure.

virus control in physiological buffer, containing no plant extract is carried out simultaneously.[19,20] Fig. 8a shows the scheme of this virucidal test methodology.

In Vitro Antiviral Evaluation

The EPTT technique (Fig. 8b) is performed on preemptied confluent monolayers of Vero or other cells, grown in the holes (e.g., 96) of microtiter plates, which are infected with serial tenfold dilutions of a virus suspension (100 μl). Starting with monolayers containing 10^4 cells per hole and a virus suspension of, for example, 10^7 TCD_{50} ml^{-1} or PFU ml^{-1}, the first monolayers of cells are infected with a multiplicity of infection (MOI) 10. By further serial tenfold dilutions of the virus suspension, the MOI decreases from 10 to 10^{-4}. The virus is allowed to absorb for 60 min at 37°C, after which serial twofold dilutions of plant extracts or test compounds in maintenance medium, supplemented with 2% serum and antibiotics, are added. The plates are incubated at 37°C, and the viral cytopathogenic effect (CPE) is recorded daily by light microscopy during at least 1 week. Cytotoxicity controls (uninfected, but treated cells) and cell controls (uninfected, untreated cells) are run at each treatment concentration, and virus controls (infected, but untreated cells) at each viral dilution. Toxic doses of the extracts (T) are considered to be dilutions that cause destruction and degeneration of the monolayer, so that no virus titer can be determined. The antiviral activity is expressed as the virus titer reduction at the maximal nontoxic doses (MNTD) of the test substance, i.e., the highest concentration (μg ml^{-1}) or lowest dilution (1/n) that does not affect the monolayers under the conditions of the antiviral test procedure.

Viral titer reduction factors (RF, i.e., the ratio of the viral titer reduction in the absence [virus control]and presence of the MNTD of the test sample) of 1 x 10^3 to 1 x 10^4 indicate a pronounced antiviral activity and are suitable as selection criteria for further investigation of plant extracts. This system allows the correlation of all possible multiplicities of infection (MOI) values in the same microtiter plate, with decreasing amounts of plant extracts, so that the nontoxic concentration of plant extracts can be determined. It can be stated as a general rule that the detected antiviral activity should be stable in at least two subsequent dilutions of nontoxic concentrations of the extract; otherwise the activity is directly correlated to the toxicity of the extract or is only virucidal. Moreover, a true antiviral product has to protect the cells that have been infected with low virus dilutions (starting from 0.1 PFU per cell onwards).[19,20]

Antiherpes (HSV) and/or Antihuman Immunodeficiency Viruse (HIV) Agents

It was found that ethanolic and acetone extracts prepared from the stembark of both "white" and "red" bark varieties of *Pavetta owariensis* exhibit antiviral properties against *Herpes simplex* and *Coxsackie* viruses,

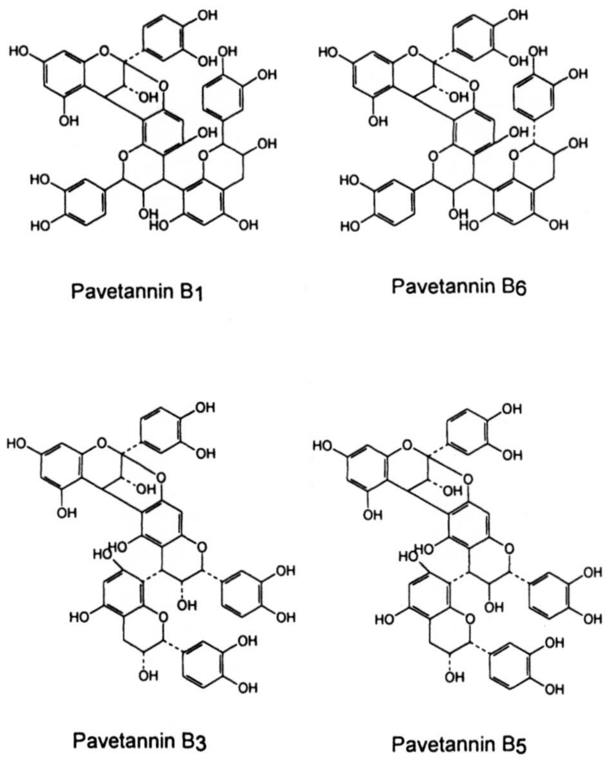

Pavetannin B1

Pavetannin B6

Pavetannin B3

Pavetannin B5

Figure 9. Chemical structures of pavetannins B. Pavetannin B_1, pavetannin B_3, pavetannin B_5 and pavetannin B_6.

two members of our normal screening battery, and also against human immunodeficiency virus (HIV), responsible for the pandemic immunosuppressive disease, acquired immunodeficiency syndrome (AIDS).[21] A detailed bioassay-guided isolation of the different active fractions showed doubly-linked proanthocyanidines to be responsible for the antiviral effects of the plant.[16,21]

Besides the monomeric flavan-3-ols, (+)catechin, (-)epicatechin and (+)-ent-epicatechin, four dimeric (pavetannins A), six trimeric (pavetannins B), four tetrameric proanthocyanidins (pavetannins C) and one pentameric proanthocyanidin (pavetannin D) were isolated, purified and characterized.[22,23,24] Structures of four different trimers are shown in Fig. 9. Pavettanin B_1 was identified as epicatechin-(4β → 8, 2β → 0 → 7)-epicatechin-(4α → 8)-ent-epicatechin, pavetannin B_3 is epicatechin-(4β → 6, 2β → 0 → 7)-epicatechin-(4α → 8)-epicatechin, pavetannin B_5 is epicatechin (4β → 6, 2β → 0 → 7)-catechin, (4α →

Table 2. Antiviral potency of proanthocyanidins against *Herpes simplex* and Coxsackie B_2

Compound	Concentration µg/ml	*Herpes simplex*	*Coxsackie B_2*
procyanidin A-2	250	10^4(T/4)	10^3(T/4)
pavetannin A-1	125	10^3	10^2
	62.50	10^2	10
	31.25	10	1
cinnamtannin B-1	125	T	10^3(T/4)
	62.50	10^4(T/4)	10^2
	31.25	10^2	10
pavetannin B-1	125	T	10^3(T/2)
	62.50	10^4(T/4)	10^3(T/4)
	31.25	10^2	10^2
	15.62	10	10
pavetannin B-2	125	T	nt
	62.50	10^4(T/2)	
	31.25	10^3(T/4)	
	15.62	10	
cinnamtannin B-2	62.50	T	nt
	31.25	10^2(T/2)	
	15.62	10(T/4)	
pavetannin C-1	62.50	T	10^2(T/4)
	31.25	10^2(T/2)	10
	15.62	10(T/4)	1
pavetannin D-1	31.25	T	10^2(T)
	15.62	10^2(T/2)	10(T/4)

- The antiviral activity is expressed as the reduction factor of the viral titer
- T, T/2, T/4 : cytotoxicity scale
- nt = not tested

8)-epicatechin and pavetannin B_6 is epicatechin-(4β → 8, 2β → 0 → 7)-epicatechin-(4α → 8)-catechin.

These compounds not only were shown to possess virucidal properties, but also were active in the antiviral EPTT in concentrations varying from 31.25 to 125 µg/ml[21](Table 2). The antiviral activity and the cytotoxicity seemed to increase with the degree of condensation and consequently the molecular weight. This parallels the capacity of tannins to bind to proteins. In general it is believed that polyphenols act by associating with proteins of viral particles and/or host cell surfaces, resulting in a reduction or prevention of viral adsorption. It is questionable, whether such compounds will have therapeutic efficacy against viral infections in animal models. They might, however, be considered as suitable candidates for investigating their potential in counter-acting sexual transmission of herpes and HIV-infections.

Antirhinovirus Agents

Human rhinoviruses are one of the major causes of the common cold (30-50%), compared to corona viruses (10 to 20%) and adeno, parainfluenza, respiratory syncytial, influenza and enteroviruses (< 5%). Viral infection of nasal cells does not lead to cell necrosis and mucosal damage, but rather induces host responses including elaboration of inflammatory mediators such as kinins, influxes of inflammatory cells including polymorphonuclear leucocytes, and probably neuroreflexes with associated cholinergic stimulation and neuropeptide release. This leads to the manifestations of illness including rhinorrhea, cough, sneezing and sore throat and to nasal obstruction and increased mucus production.[25] Studies indicate that the average preschool child experiences 6 to 10 colds per year and the average adult has 2 to 4 colds per year.[26] The microbial goals in treating common colds are to reduce their symptom burden, reduce the risk of complications, and decrease the likehood of spreading infections to contacts. The latter could be achieved by reducing the concentration of virus in respiratory secretions by an antiviral agent.[27]

A few years ago, in their antiviral screening program of pure microbial and plant products, Ishitsuka et al.[28] found 5,4'-dihydroxy,3,7,3'-trimethoxyflavone (3,7,3'-trimethylether of quercetin or 3,7,3'-TMQ), which was originally isolated from the Chinese medicinal herb *Agastache rugosa* Kuntze, to be highly active in tissue cultures against all picornaviruses except Mengovirus. Independently, Van Hoof et al.[18,29] found several 3-methoxyflavones, which were identified as derivatives of the 3-methylethers of quercetin (3-MQ) and kaempferol (3-MK), to be responsible for the pronounced antiviral properties of the alcoholic extracts prepared from different African *Euphorbia* spp. (Fig. 10). Among the many derivatives of 3,7,3'-TMQ that were synthesized and tested one chalcone, 2'-hydroxy-4'-ethoxy-4,6'-dimethoxychalcone, emerged as a new type of antiviral agent exclusively active against many human rhinovirus serotypes and, consequently, a candidate drug for the treatment of the common cold[30] (Fig. 11; 2). The antirhinovirus activity of flavan was discovered serependitously during an *in vitro* screening program utilizing the plaque inhibition test. Struc-

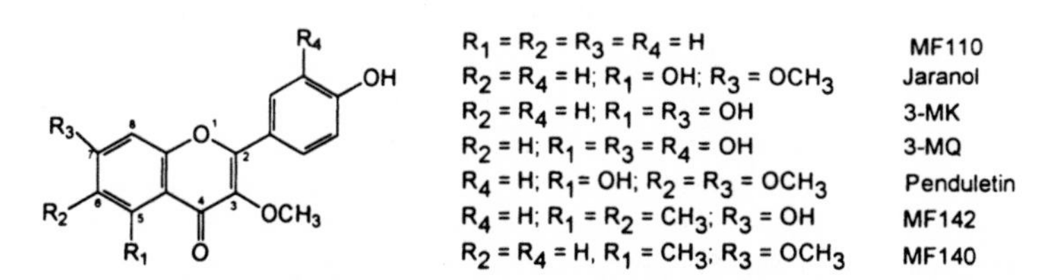

Figure 10. Chemical structures of antirhinovirus 2-methoxyflavones. 4'-Hydroxy-3methoxyflavone (MF 110), 4'-5-dihydroxy-3,7-dimethoxyflavone (jaranol) ; 3-methylether of kaempferol (3-MK) ; 3-methylether of quercetin (3-MQ) ; 4'-7-dihydroxy-3-methoxy-5,6-dimethylflavone (MF 142) , 4'-hydroxy-3,7-dimethoxy-5-methylfavone (MF 140).

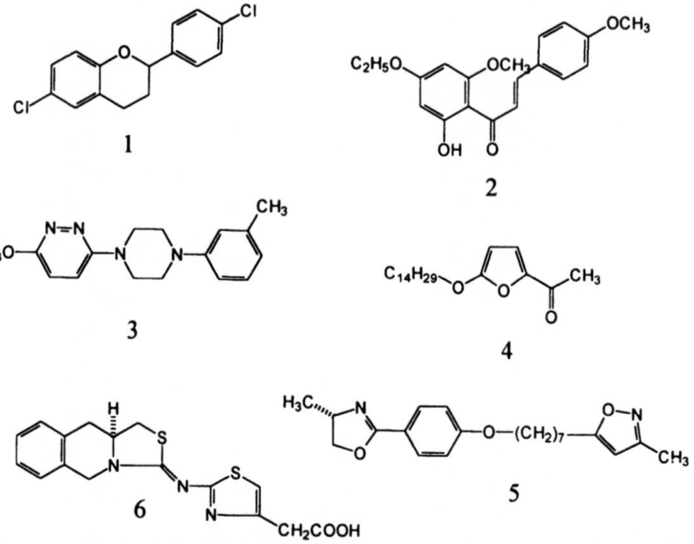

Figure 11. Chemical structures of antirhinovirus agents recently evaluated in human volunteers 1. 4'-6-dichloroflavan 2. 4'-ethoxy-2'-hydroxy-4,6'-dimethoxychalcone (R0-09-0410) 3. 3-methoxy-6-[4-(3-methylphenyl)-1-piperazinyl]pyrazidine (R 61837) 4. 1-(5-tetradecycloxy-2-furanyl)ethanone (RMI 15.731) 5. (S)-(-)-5-[7-[4-(4,5-dihydro-4-methyl-2-oxazolyl)-phenoxy]heptyl]-3-methylisoxazole (WIN 52084) 6. 2-[(1,5-10a-tetrahydro-3H-thiazolo[3,4-b]-isoquinilin-3-ylindene)amino]-4-thiazole acetic acid (44-081 R.P.)

ture-activity relationship studies led to the 4',6-dichloroflavon derivative, being the most potent agent against several rhinovirus serotypes[31,32] (Fig. 11; 1).

Subsequently, it was shown that flavans and chalcones inactivate rhinoviruses directly by binding to or interacting with specific sites on the viral capsid proteins. While little or no interference with viral attachment or penetration of the host cell membrane was observed, the uncoating process in the host cell was inhibited through stabilisation of the protein capsid of the virus and prevention of the conformational changes required for release of viral RNA.[33,34] At the same time several pharmaceutical companies developed a number of synthetic capsid-binding drugs with prominent antirhinovirus properties. Most were rhinovirus specific, but all of these agents had substantial serotype related variability in antiviral activity. The concentrations inhibiting rhinovirus replication *in vitro* varied up to 100-fold for different serotypes, indicating that the binding sites on the viral capsid proteins were highly specific.[35] X-Ray crystallographic structural analysis has determined that the precise binding site of one of the synthetic capsid-binding drugs *viz* (S)-(-)-5-[7-[4-(4,5-dihydro-4-methyl-2-oxazolyl)phenoxy]heptyl]—3-methylisoxazole, which has an isoxazole at one end and an oxazolyl-phenoxy group at the other end of an alkyl chain (Fig. 11; 5), to rhinovirus type 14 is the interior of viral protein 1 (VP1), one of the three external polypeptides of the protomers of the protein shell of the virus.[36] Changes in the amino acids of this binding pocket may affect the ability of a specific agent to bind to the capsid, and thus explain the different susceptibilities of different rhinovirus serotypes. The binding sites for some of these agents may be the same or lie very close to one another.[37] A potential limitation of the use of these compounds is that drug-resistant mutants can be selected readily under *in vitro* conditions.[38]

Clinical trials have found discrepancies between the *in vivo* and *in vitro* antiviral activities of these compounds. Orally administered dichloroflavan and phosphorylated chalcone were ineffective in the prophylaxis of experimental rhinovirus colds.[39,40] Intranasal preparations of both compounds failed to reduce infection rates or protect against illness, probably because no adequate levels of drugs were achieved in nasal mucosal cells.[41,42] Similarly, *in vivo* test results of the synthetic antiviral agents 4 (Fig. 11; 4) and 6 (Fig. 11; 6) did not show significant prophylactic activity as compared to placebo.[35,43] The antirhinovirus compound 5 (Figure 11; 5), which is a 6-(1-piperazinyl)pyridazine derivative, however, caused marked reductions in nasal symptoms and mucus weights, when it was administered intranasally in frequent doses beginning 1 h before and continuing for 6 days after experimental rhinovirus challenge with a very susceptible serotype.[35] The relative success of the latter experiment might probably be ascribed to the efficient absorption through the nasal mucosa of the hydrophobic antiviral agent from a pharmaceutical composition containging cyclodextrines (European Patent Application N°. 88201288.3, 1988).

The 3-methoxyflavones, 3-MQ and 3-MK, however, have been shown not to interact with the capsid proteins of picornaviruses, but rather to interfere with an early stage in the viral RNA-synthesis. Although their exact mode of action is not yet completely understood, the most recent results suggest that the target for the compounds is the membrane-bound virus replication complex in which + strand viral RNA is normally produced. It is possible that the compounds have a specific affinity for a protein compound in this complex.[44,45,46] In contrast to the capsid-binding antivirals no drug-resistant mutants have been detected in the presence of 3-methoxyflavones.[37]

The attractive mechanism of action, the pronounced and broad-spectrum antiviral activity, and the lack of resistance-induction by these flavones prompted us to explore this class of flavonoids. From a large screening program of naturally occurring 3-methoxyflavones jaranol and penduletin (Fig. 10) emerged as the most *in vitro* active substances against polio- and rhinoviruses. In order to establish a structure-activity relationship, a series of A-ring substituted analogues of 4'-hydroxy-3-methoxyflavone (Fig. 10, MF 110) were synthesized and tested for antiviral activity. The most interesting compound was 4',7-dihydroxy-3-methoxy-5,6-dimethylflavone (Fig. 10, MF 142) possessing *in vitro* TI_{99}-values of > 1000 and > 200 against poliovirus type 1 and rhinovirus type 15 respectively

(EPTT). This compound was then tested against a panel of 17 rhinoviruses. The median antiviral inhibitory value against them will accurately predict the median value against another 100 rhinovirus serotypes.[47,48] The substance inhibited all 17 different rhinovirus serotypes of the panel having 50% minimal inhbitory concentrations (MIC_{50}) ranging from 0.016 to 0.5 μg/ml. The corresponding values of a moderately active analogue such as 4'-hydroxy-3,7-dimethoxy-5-methylflavone (Fig. 10, MF 140) were 10 to 25 times higher.[49,50]

It was also found that in contrast to quercetin, MF 142 was not mutagenic in concentrations up to 2.5 mg in the Ames-test.[51] Since some 3-methoxyflavones, when administered intraperitoneally, have been shown to protect mice from potentially lethal infections of *Coxsackie* B_4[18], the most antivirally active substance of this study *viz.* MF 142 should be considered as a promising candidate for clinical studies in human volunteers.

CONCLUSION

The question of whether ethnopharmacology can contribute to the development of antiviral. drugs can be answered positively without too much premature optimism.[52] Indeed, the screening of a relatively low number of randomly collected plant substances has afforded a remarkably high ratio of active leads in comparison with the screening programs of synthetic compounds. As well, the testing of plants, selected on the basis of ethnopharmacological data, has been among the most succesful programs of screening plants for antiviral activity. Moreover, contrary to antibacterial and antifungal plant substances, several antiviral plant compounds have exhibited *in vitro* and *in vivo* antiviral activities competitive with those found for synthetic antiviral drugs, which are currently in various stages of development.

Finally, natural products have been shown to interfere with many viral targets ranging from adsorption of the virus to the host cell to release from it. These may result in mechanisms of action complementary to those of existing antiviral drugs. Consequently, mass screening of plant extracts should be started and/or continued, and naturally occurring leads should be improved by structure-activity studies until optimal antiviral activity with an acceptable therapeutic index is obtained. Development of effective clinically useful antiviral agents will, however, only be made possible by the willing collaboration of governments, academics and pharamceutical industries.

ACKNOWLEDGMENTS

These studies were supported by grants of the Belgian Fund for Scientific Research (NFWO) and the Flemish Government (Concerted Action). The authors are very grateful to all colleagues and collaborators, whose names are mentioned in the reference list.

REFERENCES

1. MARINO-BETTOLO, R., SCARPATI, M.L. 1979. Alkaloids of *Croton draconoides*. Phytochemistry. 18: 520-524.
2. HARTWELL, J.L. 1969. Plants used against cancer. A survey. Lloydia. 32: 153-194.
3. HECKER, E. J. 1981. Cocarcinogenesis and tumor promoters of the diterpene ester type as possible carcinogenic risk factors. Cancer Res. Clin. Oncol. 99: 103-124.
4. EVANS, F.J., TAYLOR, S.E. 1983. Pro-inflammatory, tumour-promoting and antitumour diterpenes of the plant families Euphorbiaceae and Thymelaceae. Prog. Chem. Org. Nat. Prod. 44: 1-99.
5. PIETERS, L. 1992. The biologically active constituents of *Sangre de Drago*, a traditional South American drug, Ph.D. Thesis, University of Antwerp (UIA), Belgium. 45-211.
6. CHENG, C.Y., MARTIN, D.E., LIGGETT, C.G., REECE, M.C., REECE, A.C. 1988. Fibronectin enhances healing of excised wounds in rats. Arch. Dermatol. 124: 221-225.
7. CLARK, R.A.F. 1988. Potential role of fibronectin in cutaneous wound repair. Arch. Dermatol. 124: 201-206.
8. GOUSTON, A.S., LEOF, E.B., SCHIPLEY, C.D., MOSES, H.L. 1982. Growth factors and cancer. Cancer Res. 46: 1015-1029.
9. TEN DIJKE, P., IWATA, K.K. 1989. Growth factors for wound healing. Biotechnology. 7: 793-798.
10. VANDEN BERGHE, D.A., YANG, Q.H., TOTTÉ, J., VLIETINCK, A.J. 1993. Specific stimulation of human endothelial cells by *Triticum vulgare* extract and its biologically active fraction. Phytother. Res. 7: 172-179.
11. JAFFE, E.A., NACHMAN, K.L., BECKER, C.C., MINICH, C.P. 1973. Culture of human endothelial cells derived from umbilical veins. Identification morphological immunological criteria. J. Clin. Invest. 52: 2745-2749.
12. PIETERS, L., DE BRUYNE, T., CLAEYS, M., VLIETINCK, A.J., CALOMME, M., VANDEN BERGHE, D. 1993. Isolation of a dihydrobenzofuran lignan from South American dragon's blood (*Croton* spp.) as an inhibitor of cell proliferation. J. Nat. Prod. 56: 899-906.
13. PIETERS, L.A.C., VANDEN BERGHE, D.A., VLIETINCK, A.J. 1990. A new dihydrobenzofuran lignan from *Croton erythrochilus* (Muell.) Arg. (Euphorbiaceae). Phytochemistry 29: 348-349.
14. VAISBERG, A.J., MILLER, M., DEL CARMEN PLANCS, M., CORDOVA, J.L., ROSAS DE AGUSTI, E., FERREYA, R., DEL CARMEN MUSTIGA, M., CARBIN, L., HAMMOND, C.B. 1989. Taspine is the cicatrizant principle in Sangre de Drago extracted from *Croton lechleri*. Planta Med. 55: 140-143.
15. CAI, Y., EVANS, F.J., ROBERTS, M.F., PHILLIPSON, J.D., ZENK, M.H., GLEBA, Y.Y. 1991. Polypenolic compounds from *Croton lechleri*. Phytochemistry. 30: 2033-2040.
16. BALDÉ, A.M., VAN HOOF, L., PIETERS, L.A., VANDEN BERGHE, D.A., VLIETINCK, A.J. 1990. Plant antiviral agents. VII. Antiviral and antibacterial proanthocyanidins from the bark of *Pavetta owariensis*. Phytother. Res. 4: 182-188.
17. VAN HOOF, L., VANDEN BERGHE, D.A., HATFIELD, G.M., VLIETINCK, A.J. 1984. Plant antiviral agents. V. 3-Methoxyflavones as potent inhibitors of viral induced block of cell synthesis. Planta Med. 50: 513-517.
18. VANDEN BERGHE, D.A., VLIETINCK, A.J., VAN HOOF, L. 1986. Plant products as potential antiviral agents. Bull. Inst. Pasteur. 84: 101-147.
19. VANDEN BERGHE, D.A., VLIETINCK, A.J. 1991. Screening methods for antibacterial and antiviral agents from higher plants. In: Methods in Plant Biochemistry (K. Hostettmann, ed.), Academic Press London. pp. 47-69.

20. VANDEN BERGHE, D.A., HAEMERS, A., VLIETINCK, A.J. 1993. Antiviral agents from higher plants and an example of structure-activity relationship of 3-methoxyflavones. In: Bioactive Natural Products : Detection, Isolation and Structural Determination (S.M. Colegate, R.J. Molyneux eds.), CRC Press, Boca Raton, Florida. pp.405-440.
21. BALDE, A.M., CALOMME, M., PIETERS, L., CLAEYS, M., VANDEN BERGHE, D.A., VLIETINCK, A.J. 1991. Structure and antimicrobial activity relationship of doubly-linked procyanidins. Planta Med. 57: Suppl. 2, A42-A43.
22. BALDÉ, A.M., PIETERS, L.A., GERGELY, A., KOLODZIEJ, H., CLAEYS, M., VLIETINCK, A.J. 1991. A-type proanthocyanidins from stem-bark of *Pavetta owariensis*. Phytochemistry. 30: 337-342.
23. BALDÉ, A.M., PIETERS, L.A.C., WRAY, V., KOLODZIEJ, H., VANDEN BERGHE, D.A., CLAEYS, M., VLIETINCK, A.J. 1991. Dimeric and trimeric proanthocyanidins possessing a doubly-linked structure from *Pavetta owariensis*. Phytochemistry 30: 4129-4135.
24. BALDÉ, A.M., DE BRUYNE, T., PIETERS, L., CLAEYS, M., VANDEN BERGHE, D., VLIETINCK, A.J., WRAY, V., KOLODZIEJ, H. 1993. Proanthocyanidins from stem bark of *Pavetta owariensis*. 3. NMR study of acetylated trimeric proanthocyanidins possessing a doubly-linked structure. J. Nat. Prod. 56: 1078-1088.
25. GREGG, I. 1983. Provocation of airflow limitation by viral infection: implication for treatment. Eur. J. Respir. Dis. 64: 369-379.
26. GWALTNEY, J.M.JR. 1985. The common cold. In: Principles and practices of infectious diseases, 2nd ed., (G.L. Mandell, R.G. Douglas, J.E. Bennett, J.E., eds.), John Wiley and Sons, Inc., New York, pp. 351-355.
27. LOWENSTEIN, S.R., PARRINO, T.A. 1987. Management of the common cold. Adv. Intern. Med. 32: 207-234.
28. ISHITSUKA, H., OHSAWA, C., OHIWA, T., UMEDA, I., SUHARA, Y. 1982. Antipicornavirus flavone R0-09-0179. Antimicr. Agents Chemother. 22: 611-616.
29. VAN HOOF, L., VANDEN BERGHE, D.A., VLIETINCK, A.J. 1982. Antiviral compounds of African Euphorbia species. Abstracts 4th Int. Conf. Comparative Virology., Alberta, Canada, 232.
30. ISHITSUKA, H., NINOMIYA, Y.T., OHSAWA, C., FUJU, M., SUHARA, Y. 1982. Direct and specific inactivation of rhinovirus by chalcone R0-09-0410. Antimicr. Agents Chemother. 22: 617-621.
31. BAUER, D.J., SELWAY, J.W.T., BACHEDOR, J.F., TISDALE, M., CADWELL, I.C., YOUNG, D.A.B. 1981. 4',6-Dichloroflavan (BW 683C), a new antirhinovirus compound. Nature 292: 369-370.
32. BAUER, D.J., SELWAY, J.W.T. 1983. A novel method for detecting the antiviral activity of flavans in their vapour phase. Antiviral Res. 3: 235-239.
33. NINOMYA, Y., UHSAWA, C., AYOAMA, M., UMEDA, I., SUHARA, Y., ISHITSUKA, H. 1984. Antiviral agents. RO-09-0410 binds to rhinovirus specifically and stabilizes the virus conformation. Virology 134: 269-276 .
34. TISDALE, M., SELWAY, J.W.T. 1984. Effect of dichloroflavan (BW 683C) on the stability and uncoating of rhinovirus type 1B. Antimcirob. Agents Chemother. 14: 97-105.
35. SPERBER, S.J., HAYDEN, F.G. 1988. Chemotherapy of rhinovirus colds. Antimicrob. Agents Chemother. 32: 409-419.
36. SMITH, T.J., KREMER, M.J., LUO, M., VRIEND, E., ARNOLD, G., KAMER, M.G., ROSSMANN, M.A., MCKINLEY, A., DIANA, G.D., OTTO, M.J. 1986. The site of attachment in human rhinovirus 14 for antiviral agents that inhibit uncoating. Science 233: 1286-1293.
37. NINOMIYA, Y., AOYAMA, M., UMEDA, I., SUHARA, Y., ISHITSUKA, H. 1985. Comparative studies on the mode of action of the antirhinovirus agents R0-09-0410, R09-0179,

RMI-15.731, 4',6-dichloroflavan and enviroxime. Antimicrob. Agents Chemother. 27: 595-599.

38. SELWAY, J.W.T. 1986. Antiviral activity of flavones and flavans. Prog. Clin. Biol. Res. 213: 521-526.
39. PHILLPOTTS, R.J., WALLACE, J., TYRRELL, D.A.J., FREESTONE, D.S., SHEPHERD, W.M.J. 1983. Failure of oral 4',6-dichloroflavan to protect against rhinovirus infection in man. Arch. Vir. 75: 115-121.
40. PHILLPOTTS, R.J., HIGGINS, P.G., WILLMAN, J.S., TYRRELL, D.A.J., LENOX-SMITH, I.J. 1984. Evaluation of the antirhinovirus chalcone R0-09-0415 given orally to volunteers. Antimicrob. Agents Chemother. 14: 403-419.
41. AL-NAKIB, W., WILLMAN, J., HIGGINS, P.G., TYRRELL, D.A.J., SHEPHERD, W.M., FREESTONE, D.S. 1987. Failure of intranasally administered 4',6-dichloroflavan to protect against rhinovirus in man. Archiv. Virol. 92: 255-260.
42. AL-NAKIB, W., HIGGINS, P.G., BASROU, I., TYRRELL, D.A.J., LENOX-SMITH, I., ISHITSUKA, H.J. 1987. Intranasal chalcone R0-09-0410 as prophylaxis against rhinovirus infection in human volunteers. Antimicrob. Agents Chemother. 20: 887-892.
43. ZERIAL, A., WERNER, G.H., PHILLPOTTS, P.J., WILLMAN, J.S., HIGGINS, V.G., TYRRELL, D.A.J. 1985. Studies on 44081 RP, a new antirhinovirus compound in cell cultures and in volunteers. Antimicrob. Agents Chemother. 27: 846-850.
44. CASTRILLO, J., VANDEN BERGHE, D., CARRASCO, V. 1986. 3-Methylquercetin is a potent and selective inhibitor of poliovirus RNA synthesis. Virology 152: 219-227.
45. VRIJSEN, R., EVERAERT, L., VAN HOOF, L.M., VLIETINCK, A.J., VANDEN BERGHE, D.A., BOEYÉ, A. 1987. The poliovirus induced shut-off of cellular protein synthesis persists in the presence of 3-methylquercetin, a flavonoid which blocks viral protein and RNA synthesis. Antiviral Res. 7: 35-42.
46. LOPEZ-PILA, J.M., KOPECKA, R., VANDEN BERGHE, D. 1989. Lack of evidence for strand-specific inhibition of poliovirus RNA-synthesis by 3-methylquercetin. Antiviral Res. 11: 47-54.
47. ANDRIES, K., DEWINDT, B., SNOECKS, J., WOUTERS, L., MOREELS, H., LEWI, P.J., JANSSEN, P.A.J. 1990. Two groups of rhinoviruses revealed by a panel of antiviral compounds present sequence divergence and differential pathogenicity. J. Virol. 64: 1117-1123.
48. ANDRIES, K., DEWINDT, B., SNOECKS, J., WILLEBORDS, R., STOKBROECK, X., LEWI, P.J. 1991. A comparative test of fifteen compounds against all known human rhinivorus serotypes as a basis for a more rational screening program. Antiviral Res. 16: 213-225.
49. VLIETINCK, A.J., VANDEN BERGHE, D.A., HAEMERS, A. 1988. Present status and prospects of flavonoids as antiviral agents. Prog. Clin. Biol. Res. 215: 283-299.
50. DE MEYER, N., VLIETINCK, A.J., PANDEY, H.K., MISHRA, L., PIETERS, L.A.C., VANDEN BERGHE, D.A., HAEMERS, A. 1990. Synthesis and antiviral properties of 3-methoxyflavones. In: Flavonoids in Biology and Medicine III. Current Issues in Flavonoid Research (N.P. Das, ed.,. National University of Singapore, Singapore, pp. 403-414.
51. DE MEYER, N., HAEMERS, A., MISHRA, L., PANDEY, H.K., PIETERS, L.A.C., VAN HOOF, L., VANDEN BERGHE, D.A., VLIETINCK, A.J. 1991. 4'-Hydroxy-3-methoxyflavones with potent antipicornavirus activity. J. Med. Chem. 34: 736-746.
52. VLIETINCK, A.J., VANDEN BERGHE, D.A. 1991. Can ethnopharmacology contribute to the development of antiviral drugs? J. Ethnopharmacol. 32: 141-153.

Chapter Seven

BIOLOGICALLY ACTIVE COMPOUNDS FROM CHILEAN MEDICINAL PLANTS

Hermann M. Niemeyer

Departamento de Ciencias Ecológicas
Facultad de Ciencias
Universidad de Chile
Casilla 653, Santiago, Chile

INTRODUCTION

The geographical isolation of Chile provided by the Andean range, the Pacific Ocean, and the Atacama desert, has produced a unique flora of *ca.* 5000 species, with a high degree of endemism (*ca.* 50%). The land has been inhabited for over 20,000 years by peoples who have developed rich traditions in the medicinal use of native plants, continuing to a large extent up to the present.

The extant traditional pharmacopoeia consists of nearly 300 native species, of which only some 90 have been studied chemically. These chemical studies have been, for the most part, limited in scope and have neither been guided by

Phytochemistry of Medicinal Plants
Edited by John T. Arnason et al., Plenum Press, New York, 1995

pharmacological bioassays, nor have isolated compounds been tested in pharmacologically relevant systems. As a consequence, in most cases the traditional use of the plant cannot be associated with the presence of any given compound.

This paper presents the diversity of Chilean medicinal plants and reviews the present status of their chemistry and bioactivity.

THE MEDICINAL FLORA OF CHILE

The Flora of Chile

The continental territory of the Republic of Chile is a long and thin strip of land set between the Andes and the Pacific Ocean stretching over 38° of latitude (18° to 56°S). Within this area, ten different vegetational zones may be distinguished,[1] ranging from one of the driest deserts on earth, the Atacama desert, to the cool temperate rain forests in the South of the country with average annual rainfalls in excess of 4.5 m. The geographical isolation of the country as well as the variety of vegetational zones it contains has allowed the development of a land flora comprising 184 families (18 Pteridophyta, 4 Gymnospermae, 132 Dicotyledoneae and 30 Monocotyledoneae) with a total of 5082 species, including 5012 native species of which 2561 are endemic.[2] The degree of endemism (51%) is higher than that expected from surface area.[3] The diversity of the Chilean flora is also exceptional, with 496 monotypic genera (Fig. 1). Major families and genera represented are shown in Fig. 1, together with their respective degrees of endemism.

The Peoples of Chile

When the Spanish "Adelantado" Diego de Almagro set foot on Chile in 1536 to start the conquest of the land in the name of the King of Spain, he found a mosaic of peoples inhabiting the territory, from the high Andean plateau, through the narrow valleys of the semi-arid northern region to the long central valley of Central Chile. Many of these peoples had originally come into the area more than 20,000 years before.[4] The region north of parallel 24°S was inhabited by Aymara who adopted the traditions of the late Tiwanaku empire centered around Lake Titicaca, and subsequently incorporated the culture of the Inca invaders of the late fifteenth century.[5] They occupied areas mainly in the high plateau in the north-eastern corner of Chile contiguous to the Titicaca periphery and short valleys descending from it to the Pacific Ocean. They maintained active communications with other, largely Aymara people, inhabiting areas in the high plateau which are now part of Argentina, Bolivia and Perú.

South of this area, from the Salado river (27°S) to the Choapa valley (32°S), the territory, consisting of rich valleys separated by arid mountain ranges, was occupied by Diaguita. Their culture was derived from the Animas complex,[6] with elements common to Diaguita and other peoples from the eastern side of the

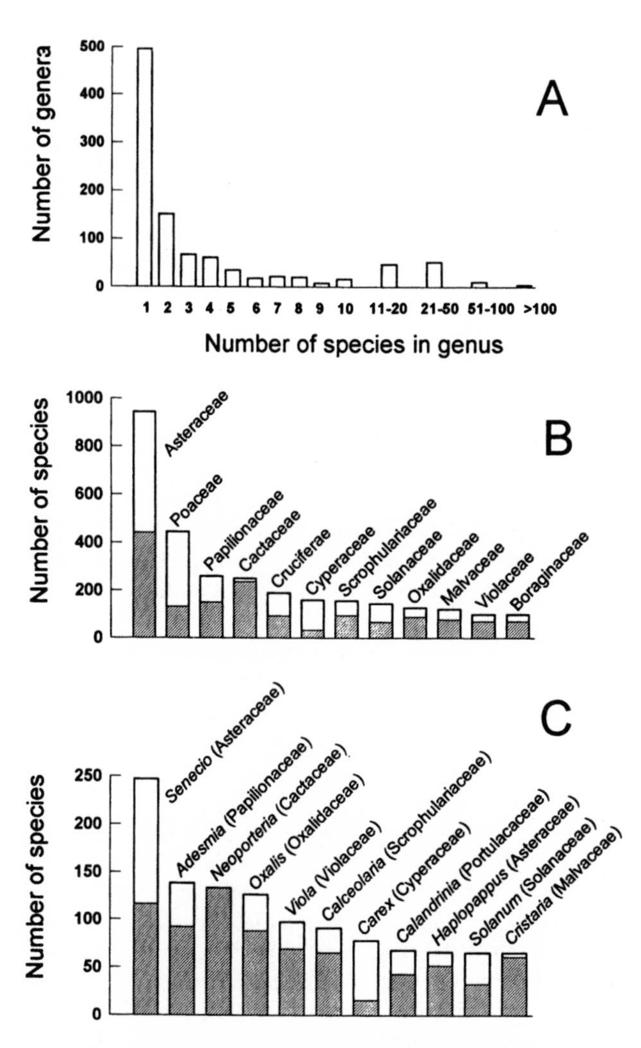

Figure 1. Characteristics of the flora of continental Chile: diversity (A), major families (B), and major genera (C). Shaded areas represent the number of endemic species.

Andes.[7] By the time of the Spanish conquest, they had suffered a high degree of acculturation originating in the Inca invasion.

The area around the valley of Santiago (33°S to 34°S) was chosen by Captain General Don Pedro de Valdivia in 1541 to found a city, which was to become the capital of the territory. It was a transition area inhabited by peoples of Mapuche origin receiving the influence of the Diaguita culture from the North and the Mapuche culture from the South, with an element of acculturation by the Incas.[8,9]

South of the Santiago basin, the country was sparsely inhabited up to the Itata river. South of this river (37° to 43°S) the lowlands were occupied by the Mapuche, a warring people with rich cultural traditions who managed to stop the Inca invasion from the North. The highlands were occupied by the Pehuenches, a nomadic, hunting and gathering people.[10]

The southernmost region of Chile (South of 43°S) was left uncolonized by newcomers until the end of last century. Its original inhabitants were largely exterminated through war and disease introduced by sporadic contacts with European navigators,[11] and by the systematic genocide by "white" farmers around 1900.

Native Medicinal Plants from Chile

At the onset of the European conquest and colonization, the native peoples of Chile had developed a rich tradition of healing diseases through the use of plants. These traditions were enriched further with the arrival of European settlers through the incorporation of part of the European traditional pharmacopoeia, and by the use of native plants with external characteristics (morphology, taste, aroma, etc.) similar to those known to Europeans. Examples of the former are *Chamomilla matricaria* (Asteraceae) and *Cichorium intybus* (Asteraceae) which are species from Europe incorporated into the traditional Chilean pharmacopoeia (these species are excluded from Table 1, which only refers to native species). A likely example of the latter is *Artemisia copa*, which is related to *Artemisia absinthium*, a common medicinal plant in Europe.

Table 1 shows the diversity of native species used in traditional medicine in Chile, along with distribution, parts of the plant used in traditional practices, and data on phytochemical studies with reference to the recent literature. Two general features of native Chilean medicinal plant species are worth noting: i) the degree of endemism (*ca.* 30%) is substantially lower than that of the Chilean flora in general (51%). This may reflect the permanent communication the people of what is now Chile maintained with their neighbors which led to the sharing of knowledge about medicinal plants growing in the whole range of the culture extending beyond present-day political boundaries; ii) the proportion of species with medicinal use in each of the richest plant families is fairly constant, as demonstrated by a significant linear correlation (Fig. 2). This may be taken as support for the idea that the search for healing properties of plants was a trial-and-error activity having to do with natural frequency of occurrence.

APPROACHES TO THE DISCOVERY OF BIOACTIVE COMPOUNDS

The use of natural compounds as leads for the development of commercial drugs is a well established strategy among pharmaceutical companies. This

Table 1. Medicinal species of the native flora of Chile

Family	Species[a]	Dist[b]	Pt[c,d]	Pt[c,e]	Chem[f]	Ref
PTERIDOPHYTA						
Adiantaceae	Adiantum chilense	NE	All			
Adiantaceae	Notholaena tomentosa	NE				
Adiantaceae	Pellaea myrtillifolia	E				
Blechnaceae	Blechnum chilense	NE		Ap	Et	50
Equisetaceae	Equisetum bogotense	NE	All			
Lophosoriaceae	Lophosoria quadripinnata	NE	Nf	Ap	Et	50
Lycopodiaceae	Lycopodium gayanum	NE				
Polypodiaceae	Polypodium feuillei	NE	R	R	Et	51
Salviniaceae	Azolla filiculoides	NE			PE-Bz	52
GYMNOSPERMAE						
Araucariaceae	Araucaria araucana	NE	r,S	S	Hx	53
Cupressaceae	Austrocedrus chilensis	NE	L,F			54
Cupressaceae	Fitzroya cupressoides	NE	r	L	Pt	55
Ephedraceae	Ephedra chilensis	NE	R.St			56
ANGIOSPERMAE: DICOTYLEDONEAE						
Aizoaceae	Carpobrotus aequilaterus	NE	F			
Anacardiaceae	Lithrea caustica	E	L	B	PE	57
Anacardiaceae	Schinus latifolius	E	r,B,l	L	MeW	58
Anacardiaceae	Schinus molle	NE	L,F			
Anacardiaceae	Schinus polygamus	NE	L,B,l	L	MeW	58
Apiaceae	Apium australe	NE	L,R			
Apiaceae	Apium panul	NE	L,R			
Apiaceae	Azorella compacta	NE	All			
Apiaceae	Azorella lycopodioides	NE				
Apiaceae	Azorella madreporica	NE	r,All			
Apiaceae	Bolax gummifera	NE	r			
Apiaceae	Eryngium depressum	E				
Apiaceae	Eryngium paniculatum	NE				
Apiaceae	Eryngium rostratum	NE	All			
Apiaceae	Laretia acaulis	NE	r			
Apiaceae	Mulinum crassifolium	NE	All	Ap	Et,PE	59
Apiaceae	Mulinum spinosum	E	R			
Apiaceae	Sanicula graveolens	NE	F			
Apocynaceae	Elytropus chilensis	NE	R,L,St	St	MeW	60
Araliaceae	Pseudopanax laetevirens	NE	L			

Table 1. *Continued*

Aristolochiaceae	Aristolochia chilensis	E	R	R,L	ml	37,61-63
Asclepiadaceae	Asthepanus geminiflorus	E				
Asclepiadaceae	Cynanchum pachyphyllum	NE				
Asteraceae	Aristeguietia salvia	E		Ap	Et	64
Asteraceae	Artemisia copa	E				
Asteraceae	Baccharis boliviensis	NE		Ap	PE&Et	65
Asteraceae	Baccharis confertifolia	NE		Ap	DCM	66
Asteraceae	Baccharis linearis	NE	L,Fl			39,41
Asteraceae	Baccharis magellanica	NE		Ap	Me	67,68
Asteraceae	Baccharis nivalis	NE	All			
Asteraceae	Baccharis paniculata	E	All	Ap	Me	66,69
Asteraceae	Baccharis patagonica	NE	All	Ap	Me	68
Asteraceae	Baccharis pedicellata	NE	All	Ap	Et	70
Asteraceae	Baccharis racemosa	NE	All			
Asteraceae	Baccharis rhomboidalis NE	All	Ap	PE&Et	71	
Asteraceae	Centaurea chilensis	NE	L			72
Asteraceae	Chaetanthera sphaeroidalis	NE	All			
Asteraceae	Dasyphyllum diacanthoides	NE	B	B	PE	73
Asteraceae	Flaveria bidentis	NE		Ap	Me	74
Asteraceae	Flourensia thurifera	E				
Asteraceae	Gnaphalium viravira	E	Fl,Ap			
Asteraceae	Haplopappus angustifolius	E	Ap	Ap	PE	75
Asteraceae	Haplopappus baylahuen	NE	Ap			76
Asteraceae	Haplopappus linifolius	E	Ap			
Asteraceae	Haplopappus multifolius	E	Ap	L,St	EtW	29
Asteraceae	Haplopappus rigidus	NE				
Asteraceae	Haplopappus villanuevae	E				
Asteraceae	Helenium aromaticum	E				
Asteraceae	Helenium glaucum	E				
Asteraceae	Hypochaeris scorzonerae	NE	R			
Asteraceae	Hypochaeris tenuifolia	NE	R			
Asteraceae	Leptocarpha rivularis	E	L	L	Chl	77
Asteraceae	Madia sativa	NE				
Asteraceae	Moscharia pinnatifida	E				
Asteraceae	Nassauvia revoluta	NE	Ap		Me/EE/PE	78
Asteraceae	Parastrephia quadrangularis	NE	Ap	Ap	Et	79
Asteraceae	Perezia atacamensis	NE				
Asteraceae	Podanthus mitiqui	E	Ap	Ap	Et	80
Asteraceae	Podanthus ovatifolius	E	Ap	Ap	Et	25
Asteraceae	Proustia cuneifolia	NE	R,L	Ap	Me/EE/PE	81
Asteraceae	Proustia ilicifolia	NE		Ap	Me/EE/PE	81
Asteraceae	Proustia pyrifolia	NE				
Asteraceae	Senecio buglossus	E				
Asteraceae	Senecio eriophyton	NE	L			
Asteraceae	Senecio eruciformis	E				
Asteraceae	Senecio fistulosus	NE	L	Ap	PE	82
Asteraceae	Senecio glaber	E		Ap	Me/EE/PE	83

Table 1. *Continued*

Asteraceae	Senecio nutans	NE		Ap	Et	35
Asteraceae	Senecio oreophyton	NE				
Asteraceae	Senecio viridis	NE	L			
Asteraceae	Solidago chilensis	NE	L			
Asteraceae	Tagetes gracilis	NE				
Asteraceae	Tagetes minuta	NE				
Asteraceae	Tagetes multiflora	NE				
Asteraceae	Tessaria absinthioides	NE	r,Ap	Ap	EE/PE;DCM	84,85
Asteraceae	Trichocline aurea	E				
Asteraceae	Triptilion spinosum	NE	L,Fl	Ap	Me/EE/PE	86
Asteraceae	Werneria poposa	NE				
Berberidaceae	Berberis buxifolia	NE		B,R	PE	87
Berberidaceae	Berberis darwinii	NE	L,F			88
Berberidaceae	Berberis empetrifolia	NE	R	S,W	Et	89
Berberidaceae	Berberis linearifolia	E	R			
Bignoniaceae	Argylia adscendens	NE	R			
Bignoniaceae	Argylia radiata	NE	R	R,Ap	Et	90,91
Boraginaceae	Cryptantha gnaphalioides	E				
Boraginaceae	Tiquilia paronychioides	NE				
Brassicaceae	Cardamine flaccida	NE				
Buddlejaceae	Buddleja globosa	NE	L	Fl	Me	92
Cactaceae	Eulychnia acida	E	F			
Cactaceae	Opuntia soehrensii	NE	S			
Caesalpiniaceae	Balsamocarpon brevifolium	E	F			
Caesalpiniaceae	Caesalpinia angulata	E	Fl			
Caesalpiniaceae	Cassia hirsuta	NE	L			
Caesalpiniaceae	Senna stipulacea	NE	B,Br			
Callitrichaceae	Callitriche terrestris	NE	All			
Callitrichaceae	Callitriche verna	NE	All			
Campanulaceae	Lobelia tupa	NE	l			
Campanulaceae	Wahlenbergia linarioides	NE	L,St			
Caryophyllaceae	Corrigiola propinqua	E				
Caryophyllaceae	Corrigiola squamosa	NE				
Caryophyllaceae	Spergularia fasciculata	NE				
Celastraceae	Maytenus boaria	NE	L	R,S	m2	93
Convolvulaceae	Convolvulus hermanniae	NE	R			
Cunoniaceae	Weinmannia trichosperma	NE	L,B			
Chenopodiaceae	Atriplex deserticola	NE	All			
Chenopodiaceae	Chenopodium quinoa	NE	Fr			
Droseraceae	Drosera uniflora	NE	All			
Elaeocarpaceae	Aristotelia chilensis	NE	L,F	L,St	Et,Me	94
Elaeocarpaceae	Crinodendron hookerianum	E	L,B	L,St	EtW	27
Eucryphiaceae	Eucryphia glutinosa	E	L,Br			95
Euphorbiaceae	Argythamnia berterana	NE				
Euphorbiaceae	Colliguaja odorifera	E	l			96
Euphorbiaceae	Colliguaja salicifolia	E	l			96
Euphorbiaceae	Euphorbia lactiflua	E	S			96

Table 1. *Continued*

Euphorbiaceae	Euphorbia portulacoides	NE	R,l,r			
Fagaceae	Nothofagus obliqua	E				
Flacourtiaceae	Azara microphylla	NE	L,Br			
Frankeniaceae	Frankenia triandra	NE				
Gentianaceae	Centaurium cachanlahuen	NE	Ap	Ap	Me	97
Gentianaceae	Cicendia quadrangularis	NE				
Geraniaceae	Geranium berterianum	E	R			
Geraniaceae	Geranium core-core	NE	R,L			
Gesneriaceae	Mitraria coccinea	NE	L,B			
Gesneriaceae	Sarmienta repens	E				
Gunneraceae	Gunnera tinctoria	NE	R,L,St	Ap		98
Hydrangeaceae	Hydrangea serratifolia	NE	L,B			
Icacinaceae	Citronella mucronata	E	L			
Krameriaceae	Krameria cistoidea	E	R			
Krameriaceae	Krameria lappacea	NE	R			
Lamiaceae	Satureja gilliesii	E			DCM	99
Lamiaceae	Satureja multiflora	E	L,St			
Lamiaceae	Sphacele chamaedryoides	E	L			
Lamiaceae	Sphacele salviae	E	Lo			
Lamiaceae	Stachys albicaulis	E	L			
Lamiaceae	Stachys bridgesii	E				
Lamiaceae	Stachys grandidentata	E				
Lamiaceae	Stachys macraei	E				
Lauraceae	Cryptocarya alba	E	L,B	L	PE+Me	100
Lauraceae	Persea lingue	NE	B	B		101
Linaceae	Linum chamissonis	E	All			
Linaceae	Cliococca selaginoides	NE	All			
Malvaceae	Corynabutilon vitifolium	NE	L	L,St	Et	102
Malvaceae	Cristaria andicola	NE				
Malvaceae	Sphaeralcea obtusiloba	E				
Mimosaceae	Acacia caven	NE	B,Fr			
Mimosaceae	Prosopis alba	NE	Br,L			
Mimosaceae	Prosopis chilensis	NE	S,r,F			
Mimosaceae	Prosopis strombulifera	NE	F			
Monimiaceae	Laurelia sempervirens	E	L	W,B	Me	103,104
Monimiaceae	Peumus boldus	E	L	B	Et	105
Myrtaceae	Amomyrtus luma	NE				
Myrtaceae	Luma apiculata	NE	L,B,R			
Myrtaceae	Luma chequen	E	j,L			106
Myrtaceae	Myrceugenia exsucca	NE	L			
Myrtaceae	Myrceugenia obtusa	E				
Myrtaceae	Myrceugenia planipes	NE				
Myrtaceae	Ugni molinae	NE				
Onagraceae	Epilobium denticulatum	NE	L,Fl,B			107
Onagraceae	Oenothera acaulis	NE	All			
Onagraceae	Oenothera affinis	NE	All			
Oxalidaceae	Oxalis articulata	NE	L,All			

Table 1. *Continued*

Papaveraceae	Argemone subfusiformis	NE	S			
Papilionaceae	Adesmia emarginata	NE	Ap			
Papilionaceae	Anarthrophyllum andicola	E	L			
Papilionaceae	Geoffroea decorticans	NE	F			
Papilionaceae	Otholobium glandulosum	E	All L	PE		108
Papilionaceae	Sophora microphylla	NE	St,B	Ap,S	MeW	109
Phytolaccaceae	Asomeria coriacea	E	R	R	Me	110
Plumbaginaceae	Limonium guaicuru	E	R			
Polygalaceae	Monnina linearifolia	E	R			
Polygalaceae	Polygala gayi	E	R			
Polygalaceae	Polygala gnidioides	E	R			
Polygalaceae	Polygala thesioides	E	R			
Polygonaceae	Muehlenbeckia hastulata	NE	L,R			
Polygonaceae	Polygonum sanguinaria	E	All			
Polygonaceae	Rumex romassa	NE	L			
Portulacaceae	Calandrinia discolor	E	L			
Portulacaceae	Calandrinia longiscapa	E	L			
Primulaceae	Anagallis alternifolia	E	L,Fl			
Proteaceae	Embothrium coccineum	NE	L			
Proteaceae	Gevuina avellana	NE				
Proteaceae	Lomatia ferruginea	NE	W,L,B			
Proteaceae	Lomatia hirsuta	NE	L,B			111
Ranunculaceae	Anemone decapetala	NE	L,S			
Ranunculaceae	Anemone moorei	E				
Ranunculaceae	Caltha sagittata	NE	R			
Ranunculaceae	Ranunculus chilensis	NE	L,St			
Ranunculaceae	Ranunculus peduncularis	NE	L,St			
Rhamnaceae	Colletia hystrix	NE	B	St	Bz+Et	112
Rhamnaceae	Discaria chacaye	NE		L,St	Et	113
Rhamnaceae	Discaria trinervis	NE				
Rhamnaceae	Retanilla ephedra	NE	R	St	PE+Et	114
Rhamnaceae	Rhamnus difussus	E	L,Fr			
Rhamnaceae	Trevoa trinervis	E	B	Ap	Et	115
Rosaceae	Acaena argentea	NE	L			
Rosaceae	Acaena magellanica	NE	R,St			
Rosaceae	Acaena pinnatifida	NE	R			
Rosaceae	Acaena splendens	NE	All			
Rosaceae	Acaena trifida	NE	All			
Rosaceae	Fragaria chiloensis	NE	R,Fl			
Rosaceae	Geum quellyon	NE	R			
Rosaceae	Kageneckia oblonga	E	R,L	Ap	Chl&EA	116
Rosaceae	Margyricarpus pinnatus	NE	L,R			117
Rosaceae	Polylepis tarapacana	NE	B			
Rosaceae	Quillaja saponaria	E	B			118
Rosaceae	Tetraglochin alatum	NE	L,Br			
Rubiaceae	Nertera granadensis	NE		Ap	Me	119
Rubiaceae	Relbunium hypocarpium	NE	R			

Table 1. *Continued*

Rutaceae	Pitavia punctata	E	L	L,St	Bz	120
Salicaceae	Salix humboldtiana	NE	B			
Santalaceae	Myoschilos oblonga	NE	L,R			
Santalaceae	Quinchamalium chilense	NE	All			106
Santalaceae	Quinchamalium majus	E	All			
Sapindaceae	Bridgesia incisifolia	NE	L			
Saxifragaceae	Escallonia illinita	NE	L	L	PE,Chl,EA	121
Saxifragaceae	Escallonia pulverulenta	NE	L			
Saxifragaceae	Escallonia revoluta	E	L			
Saxifragaceae	Escallonia rubra	E	Br			
Saxifragaceae	Francoa appendiculata	NE	R			
Scrophulariaceae	Calceolaria arachnoidea	E	R			
Scrophulariaceae	Calceolaria stellariifolia	E				
Scrophulariaceae	Calceolaria thyrsiflora	NE	Lj	Ap	DCM	122
Scrophulariaceae	Euphrasia aurea	NE	All			
Scrophulariaceae	Gratiola peruviana	NE				
Scrophulariaceae	Ourisia coccinea	NE				
Solanaceae	Cestrum parqui	NE	L,St	Ap	Et	123
Solanaceae	Fabiana denudata	NE	All			
Solanaceae	Fabiana imbricata	NE	B,Br	Ap	Et	106,124
Solanaceae	Latua pubiflora	E	All	L	PE	125
Solanaceae	Solanum americanum	NE	j			
Solanaceae	Solanum crispum	NE	L,St			
Solanaceae	Solanum gayanum	NE	L	Ap	Me	126
Solanaceae	Solanum ligustrinum	NE	L			
Solanaceae	Solanum nigrum	E	L			
Solanaceae	Solanum tuberosum	E	T			
Solanaceae	Vestia foetida	E	B,L,F	Ap	Me	127
Thymelaeaceae	Ovidia pillopillo	E	B	Ap	Et	128
Valerianaceae	Valeriana lapathifolia	NE	R	Ap	Me	129
Valerianaceae	Valeriana papilla	E	R			
Verbenaceae	Glandularia laciniata	NE	Ap			
Verbenaceae	Lampaya medicinalis	NE	L			
Verbenaceae	Verbena erinoides	NE	Ap			
Verbenaceae	Verbena litoralis	NE	Lj			
Violaceae	Hybanthus parviflorus	E	R,L			
Violaceae	Viola capillaris	NE	Ap			
Violaceae	Viola corralensis	E				
Violaceae	Viola maculata	E	Ap			
Vitaceae	Cissus striata	NE	L			
Vivianiaceae	Viviania marifolia	NE	Ap			
Winteraceae	Drimys winteri	NE	L,B	L	Et	130,131
Zygophyllaceae	Larrea nitida	NE	L,Br	Ap	steam	132
Zygophyllaceae	Porlieria chilensis	E	St	St	Bu,PE	30

Table 1. *Continued*

ANGIOSPERMAE: MONOCOTYLEDONEAE

Alismataceae	Alisma plantago-acuatica	NE	
Amaryllidaceae	Alstroemeria aurea	NE	
Amaryllidaceae	Alstroemeria ligtu	NE	R
Amaryllidaceae	Bomarea salsilla	E	T
Bromeliaceae	Greigia sphacelata	E	S
Bromeliaceae	Puya chilensis	E	r
Iridaceae	Libertia chilensis	NE	L,R
Iridaceae	Libertia sessiliflora	E	R
Iridaceae	Sisyrinchium junceum	E	R
Liliaceae	Herreria stellata	E	R
Orchidaceae	Spiranthes diuretica	NE	
Philesiaceae	Lapageria rosea	E	R
Poaceae	Anthoxanthum utriculatum	NE	R
Poaceae	Bromus stamineus	NE	All
Poaceae	Chusquea culeu	NE	j
Poaceae	Chusquea quila	NE	St
Poaceae	Cortaderia atacamensis	NE	
Poaceae	Cortaderia rudiuscula	NE	
Poaceae	Distichlis spicata	NE	R
Poaceae	Paspalum vaginatum	NE	R
Typhaceae	Typha angustifolia	E	R

a: refs. 43-49.
b: E= endemic; NE= non-endemic
c: Ap= aerial parts; B= bark; W= wood; Br= branches; St= stems; L= leaves; Lo= leaf oils; Fl= flowers; F= fruit; S= seeds; R= roots; T= tubers; r= resin; l= latex; j= juice; Nf= new fronds
d: part of the plant used in traditional medicine
e: part of the plant used in chemical studies
f: W= water; Me= methanol; Et= ethanol; Bu= butanol; EE= ethylether; EA= ethyl acetate; Chl= chloroform; DCM= dichloromethane; Pt= pentane; Hx= hexane; Bz= benzene; PE= petroleum ether; m1= R: PE, L & St: Me or PE+Et; m2= R: MeW or Hx+EE, S: Me

approach normally begins with a plant species of ethnopharmacological importance and a bioassay germane to the activity of interest.[12,13] Such a strategy is not devoid of risks and ambiguities. These include: the transformation and loss that traditional knowledge has suffered through the impact of European civilization; the important variations in chemical composition of a plant, depending on the conditions under which it is was collected and stored;[14,15] and the variable and sometimes nonspecific use of a given plant species,[1 6] or the relation of some medicinal uses of plants to general properties of broad classes of compounds from which useful leads for industry can not be expected, e.g. tannins, whose astringent properties are useful in the healing of wounds. Additionally, it is not uncommon that little attention is paid either to the part of the plant which is used traditionally, or to the way medicines are prepared from them. Due to compartmentation of

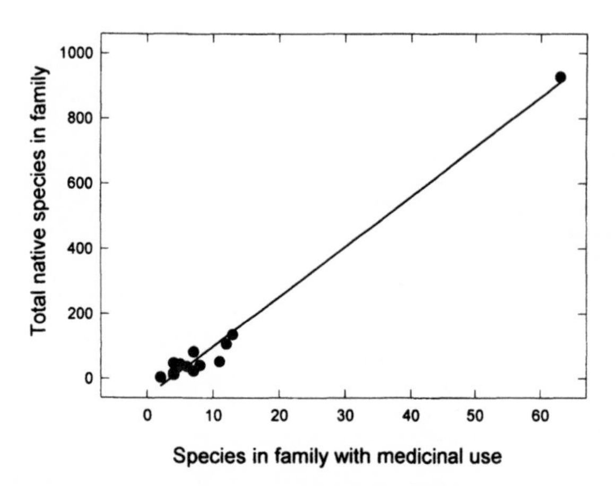

Figure 2. Medicinal use of species in the most prolific families of the native flora of continental Chile. Correlation coefficient for data shown is 0.99, and drops to 0.85 when last point is excluded.

metabolites within a plant and to their physical properties, modern extraction practices may yield compounds which bear little relation to those present in the preparations made in the traditional way.

An alternative to the ethnopharmacological approach is to use randomly chosen species and isolate bioactive compounds with the use of general bioassays. These may be correlated with pharmacologically or otherwise interesting activities.[17-20] In many cases, the activity niche may need to be discovered through specific tests. Bioassays available for testing pharmacological activities have been reviewed recently.[21] The traditional phytochemical practice of randomly isolating natural compounds may be of considerable usefulness if the compounds are subjected to bioassays. It may be claimed that every compound is in principle bioactive. The question is finding the proper niche where this bioactivity is expressed.

RESEARCH ON MEDICINAL PLANTS IN CHILE

Studies of Chilean medicinal plants have followed, for the most part, the traditional phytochemical approach. Studies have seldom analyzed the parts of the plant or the infusions or decoctions used in traditional medicine (Table 1). Most chemical studies have not been followed by relevant bioassays. As a consequence, very few proven bioactive compounds have been isolated from the Chilean native flora.

Bioassay-Guided Isolation of Bioactive Compounds

The fractionation of plant extracts using a bioassay related to the traditional use of the plant has been successful in some species. Studies on *Otholobium glandulosum* (formerly *Psoralea glandulosa*) (Papilionaceae) led to the isolation of angelicin (**1**) and the cyclobakuchiols (**2,3**).[22] An anti-inflammatory effect of the aqueous extract of the plant was suspected on the basis of its traditional use for healing wounds; hence the carrageenin-induced guinea pig paw edema test was used as a standard bioassay to test increasingly pure fractions of both aqueous and methanolic extracts of the plant. Compounds **1-3** showed anti-inflammatory activities comparable to those of non-steroidal anti-inflammatory drugs.

Solanum ligustrinum (Solanaceae) is a native shrub used in the treatment of fever. Bioassay guided fractionation of a methanolic extract of the plant using the bacterial endotoxin-induced fever test in rabbits, led to the isolation of scopoletin (**4**) and partially characterized steroidal glycoalkaloids, which exhibit strong antipyretic properties.[23]

Phytochemical Studies Coupled to Random Bioassays

Extracts or pure compounds derived from medicinal plants also have been studied in the search of bioactivity not related to the traditional medicinal use of the plant. The most extensive screening to date involved the antileukemic (PS: P-388 murine lymphocytic leukemia test) and cytotoxic (KB: human nasopharyngeal carcinoma test) activities of 519 extracts of plants belonging to the native and adventitious flora of Chile.[24] Three medicinal species were studied chemically: *Podanthus ovatifolius* and *Podanthus mitiqui* (Asteraceae), showing balsamic properties and used in the treatment of urinary infections, and *Crinodendron hookerianum* (Elaeocarpaceae), balsamic and used as an abortive. Eriflorin acetate (**5**) and ovatifolin (**6**), active in the PS and KB tests and eriflorin methacrylate (**7**), active in the PS test, were isolated from *P. ovatifolius*,[25] and ovatifolin and arturin (**8**), active in the KB test, were isolated from *P. mitiqui*.[26] Additionally, three cucurbitacins D, F and H (**9-11**), isolated from *Crinodendron hookerianum* (Elaeocarpaceae), were active in the KB test.[27]

Phytochemical Studies Coupled to Bioassays Related to the Activity of the Plant

The traditional phytochemical approach coupled to relevant bioassays also has been used to characterize natural compounds with various types of activities. The screening of 276 extracts of Chilean plants demonstrated antimicrobial activity for some medicinal species.[28] In the case of *Haplopappus multifolius* (Asteraceae), used as an antiseptic in diarrhea and in liver and bladder diseases, three coumarins, aesculetin (**12**), prenyletin (**13**), and haplopinol (**14**), inhibiting human disease-causing bacteria, were isolated.[29] Lignans isolated from *Porliera chilensis* (Zygophyllaceae), used against gout, syphilis and herpes, also have shown antimicrobial activity. *Meso*-dihydroguaiaretic acid (**15**), isopregomisin (**16**), and guayacasin (**17**), were active against a series of human disease-causing bacteria and fungi.[30] The same lignans also showed antioxidant activity comparable to the reference compound, n-propylgallate.[31]

The structure of boldine (**18**) suggested that it might possess antioxidant properties. Several different assays demonstrated its strong protective effect in systems undergoing lipid peroxidation or free radical-induced enzyme inactivation,[32,33] and it was therefore thought that these properties might be related to the long-forgotten use of its source plant, *Peumus boldus* (Monimiaceae), in the treatment of rheumatism. This hypothesis was supported by a more recent study which showed that boldine, administered orally, is a potent antiinflammatory-antipyretic drug which presumably owes its activity to the inhibition of cyclooxygenase, demonstrated *in vitro*.[34]

Senecio nutans (formerly *S. graveolens*) (Asteraceae) is a plant whose infusions are used in the high Andean plateau as a remedy for high altitude sickness. Since this use is presumably linked to circulatory physiology, one of the major compounds isolated, dihydroeuparin (**19**),[35] was subjected to a relevant bioassay and found to be a potent anti-hypertensive agent in rats at physiological concentrations.[36]

In investigations attempting to understand the lack of natural vegetation around plants of *Aristolochia chilensis* (Aristolochiaceae), the petrol extract of its roots was shown to contain terpenes which inhibit the germination of seeds of common food plants such as onion, carrot, lettuce, beans, radish and barley.[37] Although (-)-ß-bisabolene (**20**) and (+)-ß-sesquiphellandrene (**21**), the major components of the terpene mixture, were active, synergism by minor constituents of the mixture also was observed.[37]

Baccharis linearis (Asteraceae), a medicinal shrub which is seldom attacked by insects,[38] contains substantial amounts of oleanolic acid (**22**),[39,40] a widespread compound in plants. Oleanolic acid was shown to increase mortality and decrease weight and pupation rate of larvae of the corn ear worm, *Helicoverpa zeae*, when added to artificial diets on which the larvae fed.[41]

BIOINSECTICIDES FROM NATIVE CHILEAN PLANTS

An extensive bioassay guided screen of the native flora of continental Chile was undertaken in search of biopesticides. Petroleum ether and methanol extracts were sequentially prepared from each of 251 plant species. Three groups were studied: the first (73 species) consisted of medicinal plants; the second (90 species) consisted of species belonging to genera containing medicinal species; the third group (88 species) consisted of randomly chosen species not belonging

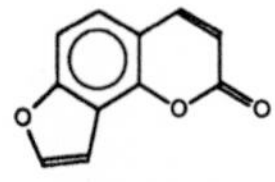

1: angelicin

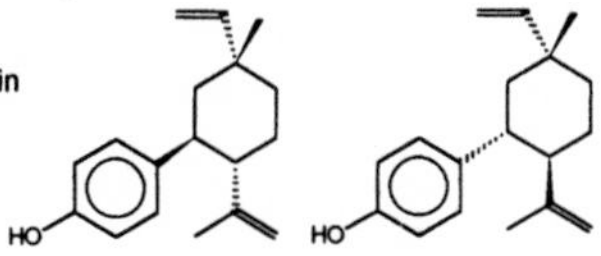

2: cyclobakuchiols : **3**

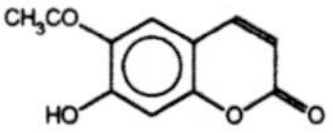

4: scopoletin

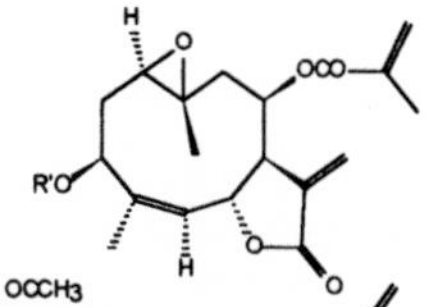

R'= OCCH3
5: eriflorin acetate
R'= OC
7: eriflorin methacrylate

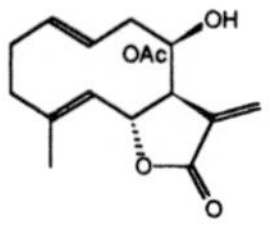

6: ovatifolin

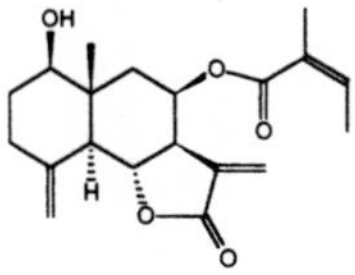

8: arturin

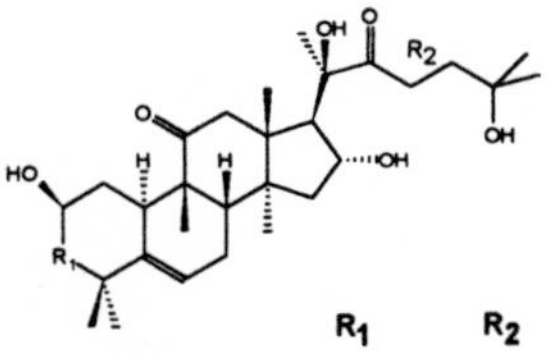

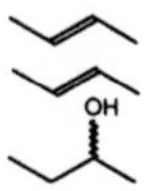

	R_1	R_2
9: cucurbitacin D	O=C	
10: cucurbitacin F	HO''' H	
11: cucurbitacin H	O=C	

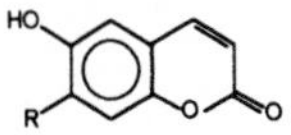

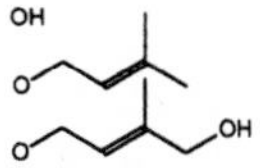

	R
12: aesculetin	OH
13: prenyletin	
14: haplopinol	

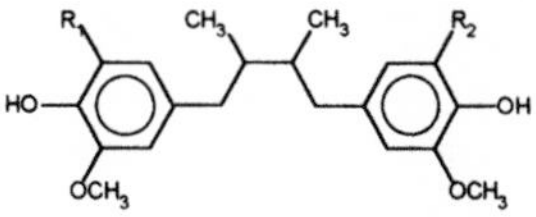

	R_1	R_2
15: *meso*-dihydroguaiaretic acid	H	H
16: isopregomisin	OCH_3	OCH_3
17: guayacasin	H	OCH_3

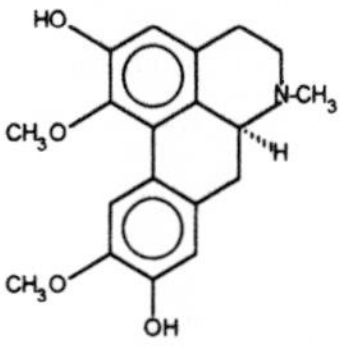

18: boldine

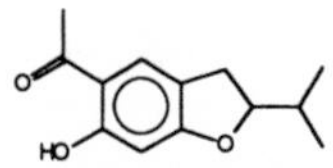

19 : dihydroeuparin

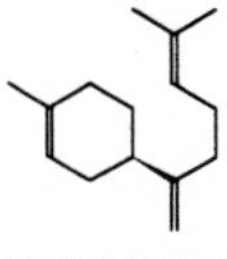

20: (-)-ß-bisabolene

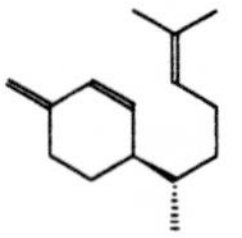

21: (+)-ß-sesquiphellandrene

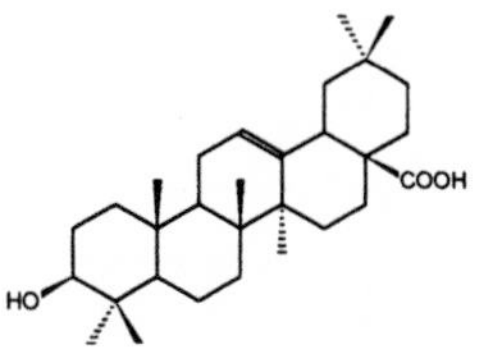

22 : oleanolic acid

to either of the two previous groups. The experimental design offered the oportunity of testing the hypothesis that the pesticide hit rate can be improved by choosing medicinal species or closely related ones, and that toxicity to humans can be thus decreased.

Contact mortality bioassays were designed for five organisms, which included the two-spotted mite, *Tetranychus urticae* (Acariformis), and four insects belonging to different orders: the pea aphid, *Acyrtosiphon pisum* (Hemiptera), the house fly, *Musca domestica* (Diptera), the cotton leafworm, *Spodoptera*

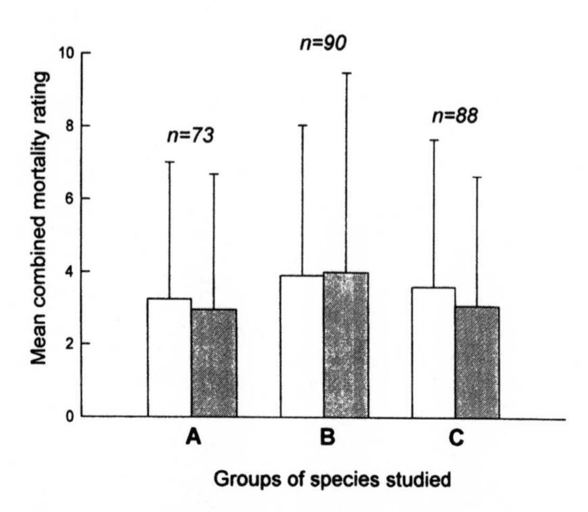

Figure 3. Insecticidal activity of petroleum ether (open) and methanol (shaded) extracts of species of the native flora of continental Chile (see text for definition of rating scale). Bars represent standard deviations of the mean of combined ratings (i.e. sum of ratings against five organisms) for each extract. Group A: medicinal species (i.e. listed in Table 1); Group B: non-medicinal species belonging to medicinal genera in Table 1; Group C: non-medicinal species at the species or genus levels (i.e. both species and genera absent from Table 1).

littoralis (Lepidoptera), and the Southern corn rootworm, *Diabrotica undecimpunctata* (Coleoptera). Extracts at a concentration of 2% were rated for mortality against each species on a 0 to 10 scale, equivalent to 0 to 100% mortality.

The extracts ranged from 0 (no mortality against any species) to 28 out of a possible maximum score of 50 (e.g. sample HMN-235p, with ratings of 10+10+0+0+8 for mite, aphid, fly, moth and beetle, respectively). Some extracts proved to be very specific, e.g. HMN-212m with 10+6+0+0+0 or HMN-117p with 0+10+0+4+0. Petroleum ether extracts were on the average equally as active as methanol extracts (Fig. 3). Figure 3 also shows that mortality scores for the three groups of plants studied did not differ significantly, suggesting that finding a bioinsecticide in a medicinal plant is as likely as in a non-medicinal one. The families where most activity was concentrated were the Asteraceae, in particular members of the tribes Helenieae, Mutisieae and Senecioneae, and the Santalaceae and Solanaceae.

The study included extracts of 16 species of the genus *Senecio* (Asteraceae), with pesticide bioactivity ranging from nil (0) to some of the highest determined (4+9+2+0+4) and (0+10+2+0+8) for HMN-87m and HMN-87p respectively). The average of all *Senecio* was 4.7 ± 5.3. This heterogeneity probably reflects the chemical complexity of the genus.[42] Work is in progress to isolate the active compound(s) from the most active extracts.

CONCLUSION

The vast assemblage of native medicinal plants of Chile has received little rational, multidisciplinary attention focusing on the isolation of active compounds that may serve as leads for the drug or agrochemical industries. A vast potential lies ahead for carrying out attractive science related to the synthesis and chemistry of analogs and understanding the mechanisms of action of active compounds, once these have been isolated and characterized.

ACKNOWLEDGMENTS

The author wishes to express his gratitude to the International Program in the Chemical Sciences at Uppsala University for continued support, in particular to Dr. Rune Liminga, its Director, for inspiring discussions, and to the British Technology Group for financial assistance. The work on the bioassay guided search of pesticides was carried out jointly with the Agrochemical Evaluation Unit at the University of Southampton, U.K.. Members of the unit, P. Jepson, M. Mead-Briggs, S. Vinall and S.D. Wratten, have greatly contributed to the generation of some of the data cited in the manuscript. The following have contributed their knowledge and criticism to the final form of this manuscript: Hans Niemeyer F., A.M. Humaña, M. Quezada and B.K. Cassels. These contributions are gratefully acknowledged.

REFERENCES

1. SCHMITHÜSEN, J. 1956. Die räumliche Ordnung der chilenischen Vegetation. Bonner Geogr. Abh. 17: 1-86.
2. ARTICORENA, C. 1992. Composición de la flora vascular de Chile. In: Flora Silvestre de Chile. (J. Grau, G. Zizka, eds.), K.G. Henssler, Frankfurt am Main, pp. 71-79.
3. MAJOR, J. 1988. Endemism: a botanical perspective. In: Analytical Biogeography. (A.A. Myers, P.S. Giller, eds.), Chapman and Hall, London, New York, pp. 117-146.
4. NUÑEZ, L. 1989. Los primeros pobladores (20.000? a 9.000 a.C.). In: Culturas de Chile. Prehistoria Desde Sus Orígenes Hasta los Albores de la Conquista. (J. Hidalgo, V. Schiappacasse, H. Niemeyer, C. Aldunate, I. Solimano, eds.), Editorial Andrés Bello. Santiago, Chile. pp. 13-32.
5. BERENGUER, J., DAUELSBERG, P. 1989. El norte grande en la órbita de Tiwanaku (400 a 1.200 d.C.). In: Culturas de Chile. Prehistoria Desde sus Orígenes Hasta los Albores de la Conquista. (J. Hidalgo, V. Schiappacasse, H. Niemeyer, C. Aldunate, I. Solimano, eds.), Editorial Andrés Bello. Santiago, Chile. pp. 129-180.
6. CASTILLO, G. 1989. Agricultores y pescadores del Norte Chico: el complejo Las Animas (800 a 1200 d.C.). In: Culturas de Chile. Prehistoria Desde sus Orígenes hasta los Albores de la Conquista. (J.Hidalgo, V. Schiappacasse, H. Niemeyer, C. Aldunate, I. Solimano, eds.) Editorial Andrés Bello. Santiago, Chile. pp. 227-263.
7. AMPUERO, G. 1989. La cultura diaguita chilena (1.200 a 1.470 d. C.). In Culturas de Chile. Prehistoria Desde sus Origenes Hasta los Albores de la Conquista. (J. Hidalgo, V. Schiap-

pacasse, H. Niemeyer, C. Alduante, I. Solimano. eds.), Editorial Andres Bello. Santiago, Chile, pp. 277-287.

8. FALABELLA, F., STEHBERG, R. 1989. Los inicios del desarrollo agrícola y alfarero: zona central (300 a. C. a 900 d. C.). In: Culturas de Chile. Prehistoria Desde sus Orígenes Hasta los Albores de la Conquista. (J. Hidalgo, V. Schiappacasse, H. Niemeyer, C. Aldunate, I. Solimano, eds.) Editorial Andrés Bello. Santiago, Chile. pp. 295-311.
9. DURAN, E., PLANELLA, M. T. 1989. Consolidación agroalfarera: zona central (900 a 1470 d. C.). In: Culturas de Chile. Prehistoria Desde sus Orígenes Hasta los Albores de la Conquista. (J. Hidalgo, V. Schiappacasse, H. Niemeyer, C. Aldunate, I. Solimano, eds.) Editorial Andrés Bello. Santiago, Chile. pp. 313-327.
10. VILLALOBOS, S. 1989. Los pehuenches en la vida fronteriza. Ediciones Universidad Católica de Chile. Santiago, Chile. 269 p.
11. MASSONE, M. 1989. Los cazadores de Tierra del Fuego (8.000 a.C. al presente). In: Culturas de Chile. Prehistoria Desde sus Orígenes Hasta los Albores de la Conquista. (J. Hidalgo, V. Schiappacasse, H. Niemeyer, C. Aldunate, I. Solimano, I., eds.) Editorial Andrés Bello. Santiago, Chile. pp. 349-366.
12. FARNSWORTH, N.R. 1990. The role of ethnopharmacology in drug development. In: Bioactive Compounds From Plants. (D.J. Chadwick, J.Marsh, eds.), CIBA Foundation Symposium 154. Wiley, Chichester, pp. 2-21.
13. COX, P.A. 1990. Ethnopharmacology and the search for newdrugs. In: Bioactive Compounds From Plants. (D.J. Chadwick, J. Marsh, eds.), CIBA Foundation Symposium 154. Wiley, Chichester, pp. 40-55.
14. WATERMAN, P.G., MOLE, S. 1989. Extrinsic factors influencing production of secondary metabolites in plants. In: Insect-Plant Interactions, Vol. 1. (E.A. Bernays, ed.), CRC Press, Inc., Boca Raton, Florida, pp. 107-134.
15. HARBORNE, J.B. 1990. Role of secondary metabolites in chemical defence mechanisms in plants. In: Bioactive Compounds From Plants. (Chadwick, D.J., J. Marsh, eds.), CIBA Foundation Symposium 154. Wiley, Chichester, pp. 128-139.
16. ALDUNATE, C., ARMESTO, J.J., CASTRO, V., VILLAGRAN, C. 1983. Ethnobotany of a pre-altiplanic community in the Andes of northern Chile. Econ. Bot. 37: 120-135.
17. CUTLER, H.G. 1984. A fresh look at the wheat coleoptile bioassay. Proceedings of the 11th Annual Meeting of the Plant Growth Regulator Society of America. pp. 1-9.
18. CUTLER, H.G. 1986. Isolating, characterizing, and screening mycotoxins for herbicidal activity. In: Advances in Allelopathy (A. Putnam, C.S. Tang, eds.), John Wiley & Sons, Inc. pp. 147-170.
19. MEYER, B.N., FERRIGNI, N.R., PUTNAM, J.E., JACOBSEN, L.B., NICHOLS, D.E., McLAUGHLIN, J.L. 1982. Brine shrimp: a convenient bioassay for active plant constituents. Planta med. 45: 31-34.
20. EINHELLIG, F.A., LEATHER, G.R., HOBBS, L.L. 1985. Use of *Lemna minor* L. as a bioassay in allelopathy. J. Chem. Ecol. 11: 65-72.
21. HAMBURGER, M., HOSTETTMANN, K. 1991. Bioactivity in plants: the link between phytochemistry and medicine. Phytochemistry 30: 3864-3874.
22. BACKHOUSE, C.N. 1993. Estudio químico y biológico de plantas chilenas con potencial acción antiinflamatoria. Estudio de la actividad antipirética, antiinflamatoria y toxicidad aguda de *Acaena splendens* Hook. et Arn. Pharm. D. thesis, Universidad de Chile, 242 p.
23. DELPORTE, C.L. 1993. Estudio de la actividad antipirética y toxicidad aguda de seis especies de la zona central de Chile. Aislamiento e identificación de los principios activos de *Solanum ligustrinum* Lood., Solanaceae, responsables de las actividades antipirética, hipotérmica y antiinflamatoria. Pharm. D. thesis, 183 p.

24. BHAKUNI, D.S., BITTNER, M., MARTICORENA, C., SILVA, M., WELDT, E., HOENEISEN, M., HARTWELL, J.L. 1976. Screening of Chilean plants for anticancer activity. J. Nat. Prod. 39: 225-243.
25. GNECCO, S., POYSER, J.P., SILVA, M., SAMMES, P.G., TYLER, T.W. 1973. Sesquiterpene lactones from *Podanthus ovatifolius*. Phytochemistry 12: 2469-2477.
26. HOENEISEN, M., SILVA, M., BOHLMANN, F. 1980. Sesquiterpene lactones of *Podanthus mitiqui*. Phytochemistry 19: 2765-2766.
27. BITTNER, M., POYSER, K.A., POYSER, J.P., SILVA, M., WELDT, E. 1973. Cucurbitacins and aromatic compounds from *Crinodendron hookerianum*. Phytochemistry 12: 1427-1431.
28. BHAKUNI, D.S., BITTNER, M., MARTICORENA, C., SILVA, M., WELDT, E., MELO, M.E., ZEMELMAN, R. 1974. Screening of Chilean plants for antimicrobial activity. Lloydia 37: 621-630.
29. CHIANG, M.T., BITTNER, T., SILVA, M., MONDACA, A., ZEMELMAN, R., SAMMES, P.G. 1982. A prenylated coumarin with antimicrobial activity from *Haplopappus multifolius*. Phytochemistry 21: 2753-2755.
30. TORRES, R., MODAK, B., URZUA, A., VILLARROEL, L., PALACIOS, Y. 1991. Guayacasina, un nuevo lignano de *Porlieria chilensis* Johnst.: determinación estructural y actividad antimicrobiana. Bol. Soc. Chil. Quím. 36: 249-252.
31. FAURE, M., LISSI, E., TORRES, R., VIDELA, L.A. 1990. Antioxidant activities of lignans and flavonoids. Phytochemistry 12: 3773-3775.
32. SPEISKY, H., CASSELS, B.K., LISSI, E.A., VIDELA, L.A. 1991. Antioxidant properties of the alkaloid boldine in systems undergoing lipid peroxidation and enzyme inactivation. Biochem. Pharmacol. 41: 1575-1581.
33. CEDERBAUM, A.I., KUKIELKA, E., SPEISKY, H. 1992. Inhibiton of rat liver microsomal lipid peroxidation by boldine. Biochem. Pharmacol. 44: 1765-1772.
34. BACKHOUSE, N., DELPORTE, C., GUIVERNAU, M., CASSELS, B.K. 1994. Antiinflammatory and antipyretic effects of boldine. Agents Actions, in press.
35. LOYOLA, L., PEDREROS, S., MORALES, G. 1985. p-Hydroxyacetophenone derivatives from *Senecio graveolens*. Phytochemistry 24: 1600-1602.
36. GALLARDO, R., ARAYA, B. 1982. Evaluación de las propiedades hipotensoras de un derivado de p-hidroxiacetofenona (SG_1), dihidroeuparina, en ratas normotensas. Proceedings of the Ninth Latin American Meeting of Pharmacology, CL-078, 86 p.
37. URZUA, A., RODRIGUEZ, R. 1992. Terpenos de raíz de *Aristolochia chilensis* inhibidores de germinación. Bol. Soc. Chil. Quím. 37: 183-187.
38. MONTENEGRO, G., JORDAN, M., ALJARO, M.E. 1980. Interactions between Chilean matorral shrubs and phytophagous insects. Oecologia 45: 346-349.
39. LABBE, C., ROVIROSA, J., FAINI, F., MAHU, M., SAN MARTIN, A. CASTILLO, M. 1986. Secondary metabolites from Chilean *Baccharis* species. J. Nat. Prod. 49: 517-518.
40. FAINI, F., HELLWIG, F., LABBE, C., CASTILLO, M. 1991. Hybridization in the genus *Baccharis :Baccharis linearis x B. macraei*. Biochem. Syst. Ecol. 19: 53-57.
41. ARGANDOÑA, V.H., FAINI, F.A. 1993. Oleanolic acid content in *Baccharis linearis* and its effects on *Heliothis* zeae larvae. Phytochemistry 33: 1377-1379.
42. DUPRE, S., GRENZ, M., JAKUPOVIC, J., BOHLMANN, F., NIEMEYER, H.M. 1991. Eremophilane, germacrane and shikimic acid derivatives from Chilean *Senecio* species. Phytochemistry 30: 1211-1220.
43. ZIN, J., WEISS, C. 1980. La salud por medio de las plantas medicinales. Editorial Salesiana, Santiago, Chile. 387 p.4. FARGA, C., LASTRA, J. 1988. Plantas medicinales de uso común en Chile. Tomo I. Paesmi, Santiago, Chile, 119 p.
44. FARGA, C., LASTRA, J. 1988. Plantas medicinales de uso común en Chile. Tomo I Paesmi, Santiago, Chile, 119 p.

45. FARGA, C., LASTRA, J., HOFFMANN, A. 1988. Plantas medicinales de uso común en Chile. Tomo II. Paesmi. Santiago, Chile, 112 p.
46. FARGA, C., LASTRA, J., HOFFMANN, A. 1988. Plantas medicianales de uso común en Chile. Tomo III. Paesmi, Santiago, Chile, 130 p.
47. MONTES, M., WILKOMIRSKY, T. 1985. Medicina tradicional chilena. Editorial de la Universidad de Concepción. Concepción, Chile. 205 p.
48. MUÑOZ, M., BARRERA, E., MEZA, I. 1981. El uso medicinal y alimentario de plantas nativas y naturalizadas en Chile. Publ. Ocas. Mus. Hist. Nat. 33: 3-91.
49. PACHECO, P., CHIANG, M.T., MARTICORENA, C., SILVA, M. 1977. Química de las plantas chilenas usadas en medicina popular. Concepción. 287 p.
50. NUÑEZ-ALARCON, J., QUIÑONES, M., CARMONA, M.T. 1987. O-Glicosil flavonoides y ácido clorogénico en algunos helechos de Chile: *Blechnum chilense* (Kaulf) C. Christ. *Blechnum penna marina* (Poir) Kuhn y *Lophosoria cuadripinnata* (Gmel) Mett. Bol. Soc. Chil. Quím. 32: 155-158.
51. BHAKUNI, D.S., GNECCO, S., SAMMES, P.G., SILVA, M. 1974. A new triterpene and catechin from *Polypodium feullei* Bertero. Rev. Latinoamer. Quím. 5: 109-115.
52. BECERRA, J., SAMMES, P.G., SILVA, M. 1978. Anticancer agents. Some constituents of *Azolla filiculoides* Lam. Rev. Latinoamer. Quím. 9: 203.
53. GARBARINO, J.A., OYARZUN, M.L., GAMBARO, V. 1987. Labdane diterpenes from *Araucaria araucana*. J. Nat. Prod. 50: 935-936.
54. CAIRNES, D.A., KINGSTON, D.G.I., MADHUSUDANA RAO, M. 1981. High performance liquid chromatography of podophyllotoxins and related lignans. J. Nat. Prod. 44: 34-37.
55. COOL, L.G., POWER, A.B., ZAVARINI, E. 1991 . Variability of foliage terpenes of *Fitzroya cupressoides*. Biochem. Syst. Ecol. 19: 421-432.
56. CASTLEDINE, R.M., HARBORNE, J.B., 1976. Identification of the glycosylflavones of *Ephedra* and *Briza* by mass spectrometry of their permethyl ethers. Phytochemistry 15: 803-804.
57. GAMBARO, V., CHAMY, M.C., VON BRAND, E., M.C., GARBARINO, J.A . 1986. 3-(Pentadec-10-enyl)-catechol, a new allergenic compound from *Lithraea caustica* (Anacardiaceae). Planta Med. 51: 20-22.
58. MANDICH, L.M., BITTNER, M., SILVA, M., BARROS, C. 1984. Phytochemical screening of medicinal plants. Studies of flavonoids. Rev. Latinoamer. Quím. 15: 80-82.
59. LOYOLA, L.A., MORALES, G. 1991. Mulinenic acid, a rearranged diterpenoid from *Mulinum crassifolium*. J. Nat. Prod. 54: 1404-1408.
60. CASTRO, L., SANCHEZ, F., CORTES, M., NARANJO, J. 1980. Agliconas esteroidales de *Elytropus chilensis* Muell. Arg. Anal. Quím. (Esp.) 76: 180-182.
61. URZUA, A., FREYER, A.J., SHAMMA, M. 1987. Aristolodione, a 4,5-Dioxoaporphine from Aristolochia chilensis. J. Nat. Prod. 50: 305-306.
62. URZUA, A., RODRIGUEZ, R. 1988. Sesquiterpenos de raíz de *Aristolochia chilensis* Miers. Bol. Soc. Chil. Quím. 33: 147-150.
63. URZUA, A., PRESLE, L. 1993. 3-epi-Austrobailignan-7: a 2,5-diaryl-3,4-dimethyltetrahydrofuranoid lignan from *Aristolochia chilensis*. Phytochemistry 34: 874-875.
64. GONZALEZ, A.G., BERMEJO, J., DIAZ, J., RODRIGUEZ, E., YAÑEZ, A., RAUTER, P., POZO, J. 1990. Diterpenes and other constituents of *Eupatorium salvia*. Phytochemistry 29: 321-323.
65. ORALES, G., MANCILLA, A., GALLARDO, O., TRUJILLO, R , LOYOLA, L., BRETON, J. 1990. Diterpenoids and flavonoids from *Baccharis boliviensis*. Bol. Soc. Chil. Quím. 35:257-263.
66. LABBE, C., FAINI, F., CASTILLO, M. 1990. Deterpenoids from Chilean *Baccharis* species. Phytochemistry 29: 324-329.

67. REYES, A., REYES, M., LEMAIRE, M., ROMERO, M. 1990. Compuestos fenólicos en dos especies de *Baccharis* chilenas. Rev. Latinoamer. Quím. 21: 1-2.
68. RIVERA, A.P., FAINI, F., CASTILLO, M. 1988. 15α-Hydroxy-ß-amyrin and patagonic acid from *Baccharis magellanica* and *Baccharis patagonica*. 51: 155-157.
69. RIVERA, A.P., ARANCIBIA, L., CASTILLO, M. 1989. Clerodane diterpenoids and acetylenic lactones from *Baccharis paniculata*. J. Nat. Prod. 52: 433-435.
70. FAINI, F., RIVERA, P., MAHU, M., CASTILLO, M. 1987. Neo-clerodane diterpenoids and other constituents from *Baccharis* species. Phytochemistry 26: 3281-3283.
71. SAN MARTIN, A., ROVIROSA, J., LABBE, C., GIVOVICH, A., MAHU, M., CASTILLO, M. 1986. Neo-clerodane diterpenoids from *Baccharis rhomboidalis*. Phytochemistry 25: 1393-1395.
72. NEGRETE, R.E., BACKHOUSE, N., CAJIGAL, I., DELPORTE, C., CASSELS, B.K., BREITMAIER, E., ECKHARDT, G. 1993. Two new antiinflamatory elemanolides from *Centaurea chilensis*. J. Ethnopharmacol. 40: 148-153.
73. REYES, A., VICUÑA, P., ZARRAGA, A. 1977. Algunos triterpenos presentes en *Flotowia diacanthoides* Less. Rev. Latinoamer. Quím. 8: 46-47.
74. CABRERA, J.L., JULIANI, H. 1979. wo new quercetin sulphates from leaves of *Flaveria bidentis*. Phytochemistry 18: 510-511.
75. SILVA, M., SAMMES, P. 1973. A new diterpenic acid and other constituents of *Haplopappus foliosus* and *H. angustifolius*. Phytochemistry 12: 1755-1758.
76. NUÑEZ-ALARCON, J., DOLZ, H., QUIÑONES, M., CARMONA, M. 1993. Epicuticular flavonoids from *Haplopappus baylahuen* and the hepatoprotective effect of the isolated 7-methyl aromadendrin. Bol. Soc. Chil. Quím. 38: 15-22.
77. MARTINEZ, R., AYAMANTE, I.S., NUÑEZ-ALARCON, J.A., ROMO DE VIVAR, A. 1979. Leptocarpin and 17,18-dihydroleptocarpin, two new heliangolides from *Leptocarpha rivularis*. Phytochemistry 18: 1527-1528.
78. BITTNER, M., JAKUPOVIC, J., BOHLMANN, F., SILVA, M. 1988. 5-Methylcoumarins from *Nassauvia* species. Phytochemistry 27: 3845-3847.
79. LOYOLA, L.A., NARANJO, J., MORALES, G. 1985. 5,7-Dihydroxy-3,8,3',4'-tetramethoxyflavone from *Parastrephia quadrangularis*. Phytochemistry 24: 1871-1872.
80. HOENEISEN, M., KODAMA, M., ITO, S. 1981. Sesquiterpene lactones of *Podanthus mitiqui*. Phytochemistry 20: 1743.
81. BITTNER, M., JAKUPOVIC. J., BOHLMANN, F., SILVA, M. 1989. Coumarins and guaianolides from further Chilean representatives of the subtribe Nassauviinae. Phytochemistry 28: 2867-2868.
82. VILLARROEL, L., TORRES, R., FAJARDO, V. 1987. Metabolitos secundarios de algunas especiesdel género *Senecio* de Chile. Bol. Soc. Chil. Quím. 32: 5-9.
83. DUPRE, S., GRENZ, M., JAKUPOVIC, J., BOHLMANN, F., NIEMEYER, H. M. 1991. Eremophilane, germacrane and shikimic acid derivatives from Chilean *Senecio* species. Phytochemistry 30: 1211-1220.
84. SILVA, M., BOHLMANN, F., ZDERO, C. 1977. Two further eremophilane derivatives from *Tessaria absynthioides*. Phytochemistry 16: 1302-1303.
85. GNECCO, S., BARTULIN, J., MONTECINOS, M., BECERRA, J. 1992. Chemical evaluation of Chilean species of plants as hydrocarbon-producing crops. Bol. Soc. Chil. Quím. 37: 335-340.
86. BITTNER, M., JAKUPOVIC, J., BOHLMANN, F., GRENZ, M., SILVA, M. 1988. 5-Methyl coumarins and chromones from *Triptilion* species. Phytochemistry 27: 3263-3266.
87. FAJARDO, V., URZUA, A., TORRES, R., CASSELS, B.K. 1979. Metabolitos secundarios de *Berberis buxifolia*. Contr. Cient. Tec. 35: 21-26.
88. VALENCIA, E., FAJARDO, V., FREYER, A.J., SHAMMA, M. 1985. Magallanesine: an isoindolobenzazocine alkaloid. Tet. Lett. 26: 993-996.

89. HUSSAIN, F., FAJARDO, V., SHAMMA, M. 1989. (-)-Natalamine, an aporphinoid alkaloid from *Berberis empetrifolia*. J. Nat. Prod. 52: 644-645.
90. BIANCO, A., PASSACANTILLI, P., GARBARINO, J.A., GAMBARO, V., SERAFINI, M., NICOLETTI, M., RISPOLI, C., RIGHI, G. 1991. A new non-glycosidic iridoid and a new bisiridoid from *Argylia radiata*. Planta Med. 57: 286-287.
91. BIANCO, A., MARINI, E., NICOLETTI, M., FODDAI, S., GARBARINO, J.A., PIOVANO, M., CHAMY, M.T. 1992. Bis-iridoid glucosides from the roots of *Argylia radiata*. Phytochemistry 31: 4203-4206.
92. MARIN, G., JIMENEZ, B., CORTES, M., PARDO, F., NUÑEZ-ALARCON, J., NARANJO, J., 1979. Estudio fitoquímico de *Buddleja globosa* Lam. (Buddlejaceae). Rev. Latinoamer. Quím. 10: 19-21.
93. GONZALEZ, A.G., MUÑOZ, O.M., RAVELO, A.G., CRESPO, A., BAZZOCCHI, I.L., SOLANS, X., RUIZ-PEREZ, C., RODRIGUEZ-ROMERO, V. 1992. A new sesquiterpene from *Maytenus boaria* (Celastraceae). Crystal structure and absolute configuration. Tet. Lett. 33: 1921-1924.
94. CESPEDES, C., JAKUPOVIC, J., SILVA, M., WATSON, W.H. 1990. Indole alkaloids from *Aristotelia chilensis*. Phytochemistry 29: 1354-1356.
95. SEPULVEDA-BOZA, S., DELHVI, S., CASSELS, B.K. 1993. Flavonoids from the twigs of *Eucryphia glutinosa*. Phytochemistry, 32: 1301-1303.
96. GNECCO, S., BARTULIN, J., BECERRA, J., MARTICORENA, C. 1989. N-Alkanes from Chilean Euphorbiaceae and Compositae species. Phytochemistry 28: 1254-1256.
97. VERSLUYS, C., CORTES, M., LOPEZ, J., SIERRA, J., RAZMILIC, I. 1982. A novel xanthone as secondary metabolite from *Centaurium cachanlahuen*. Experientia 38: 771-772.
98. BITTNER, M., SILVA, M., ROZAS, Z., JAKUPOVIC, J., STUESSY, T. 1994. Estudio químico del género *Gunnera* en Chile. Parte II. Metabolitos secundarios de dos especies continentales y dos especies de las Islas de Juan Fernández. Bol. Soc. Chil. Quím. 39: 79-83.
99. LABBE, C., CASTILLO, M., CONNOLLY, J. 1993. Mono and sesquiterpenoids from *Satureja gilliesii*. Phytochemistry 34: 441-444.
100. URZUA, A., TORRES, R., CASSELS, B.K., 1975. (+)-Reticulina en *Cryptocarya alba*. Rev. Latinoamer. Quím. 6: 102-103.
101. SEPULVEDA-BOZA, S., DELHVI, S., CASSELS, B. K. 1990. An aryltetralin lignan from *Persea lingue*. Phytochemistry 29: 2357-2358.
102. REYES, A., VEGA,L. 1983. Constitutents of *Corynabutilon vitifolium* (Cav.) Kearney. Rev. Latinoamer. Quím. 14: 138-139.
103. CASSELS, B.K., URZUA, A. 1985. Bisbenzylisoquinoline alkaloids of *Laurelia* sempervirens. J. Nat. Prod. 48: 671.
104. URZUA, A., VASQUEZ, L. 1993. Alcaloides de madera de *Laurelia sempervirens*. Análisis de trifluoroacetil derivados por CGL-EM. Bol. Soc. Chil. Quím. 38: 35-41.
105. ASENCIO, M., CASSELS, B.K., SPEISKY, H., VALENZUELA, A. 1993. (R)- and (S)-coclaurine from the bark of *Peumus boldus*. Fitoterapia 64: 455-458.
106. HORHAMMER, L., WAGNER, H., WILKOMIRSKY, M.T., IYENGAR, M.A. 1973. Flavonoide in einigen Chilenische Heilpflanzen. Phytochemistry 12: 2068-2069.
107. CROWDEN, R.K., WRIGHT, J., HARBORNE, J.B. 1977. Anthocyanins of *Fuchsia* (Onagraceae). Phytochemistry 16: 400-402.
108. ERAZO, S., GARCIA, R., DELLE MONACHE, F. 1990. Bakuchiol and other compounds from Psoralea glandulosa. Rev. Latinoamer. Quím. 21: 62.
109. HOENEISEN, M., SILVA, M., WINK, M., CRAWFORD, D., STUESSY, T. 1993. Alkaloids of *Sophora* of Juan Fernández islands and related taxa. Bol. Soc. Chil. Quím. 38: 167-171.
110. APPEL, H.H., PIOVANO, M., GARBARINO, J.A., GAMBARO, V. 1987. Phytolaccagenic acid and phytolaccagenin from *Anisomeria coriacea*. Planta Med. 52: 115.

111. MOIR, M., THOMSON, R. H. 1973. Naphthaquinones in *Lomatia* species. Phytochemistry 12: 1351-1353.
112. PACHECO, P., SILVA, M., SAMMES, P.G., TYLER, T.W. 1973. Triterpenoids of *Colletia spinosissima*. Phytochemistry 12: 893-897.
113. CORREA, C., URZUA, A., TORRES, R. 1987. 1,2,11-Trimethoxynoraporphine from *Discaria* chacaye (G. Don) Tort. Bol. Soc. Chil. Quím. 32: 105-106.
114. SILVA, M. 1967. Ursolic acid in *Retanilla ephedra*. J. Pharm. Sci. 56: 908-909.
115. BETANCOR, C., FREIRE, R., HERNANDEZ, R., SUAREZ, E., CORTES, M., PRANGE, T. H., PASCARD, C. 1983. Dammarane triterpenes of *Trevoa trinervis:* structure and absolute stereochemistry of trevoagenins A, B, and C1. J. Chem. Soc. Perkin Trans. 6:119-126.
116. CASSELS, B.K., URZUA, A., CORTES, M., GARBARINO,J.A. 1973. Triterpenoid constituents of *Kageneckia oblonga*. Phytochemistry 12: 3009.
117. FERRARI, F., DELLE MONACHE, F., JUAREZ, B.E., COUSSIO, J.D. 1972. Rosaceae flavonoids of *Margyricapus pinnatus*. Phytochemistry 11: 2647.
118. HIGUSHI, R., TOKIMITSU, Y., KOMORI, T. 1988. An acylated triterpenoid saponin from *Quillaja saponaria*. Phytochemistry 27: 1165-1168.
119. LUDWIG, H., NARANJO, J., CORTES, M. 1978. Ursolic acid and asperuloside from *Nertera granadensis*. Rev. Latinoam. Quím. 9: 21.
120. SILVA, M., CRUZ, A.M., SAMMES, P.G. 1971. Some constituents of *Pitavia punctata*. Phytochemistry 10: 3255-3258.
121. GARCIA, R., ERAZO, S., CANEPA, A., LEMUS, I. 1990. Metabolitos secundarios de *Escallonia illinita* Presl. An. Real. Acad. Farm. 56: 539-542.
122. CHAMY, M.C., PIOVANO, M., GARBARINO, J.A., MIRANDA, C., GAMBARO, V., RODRIGUEZ, M.L., RUIZ-PEREZ, C., BRITO, I. 1991. Diterpenoids from *Calceolaria thyrsiflora*. Phytochemistry 30: 589-592.
123. PEARCE, C.M., SKELTON, N.J., NAYLOR, S., RAJAMOORTHI, K., KELLAND, J., OELRICHS, P.B., SANDERS, J.K.M., WILLIAMS, D.H. 1992. Parquin and carboxyparquin, toxic kaurene glycosides from the shrub *Cestrum parqui*. J. Chem. Soc. Perkin Trans. 1: 593-600.
124. KNAPP, J.E., FARNSWORTH, N.R., THEINER, M., SCHIFF, P.L. 1972. Anthraquinones and other constituents of *Fabiana* imbricata. Phytochemistry 11: 3091-3092.
125. SILVA, M., MANCINELLI, P. 1959. Atropina en *Latua pubiflora* (Griseb.) Phil. Bol. Soc. Chil. Quím. 9: 49-50.
126. REYES, A., CUESTA, F., GOMEZ, M. 1988. Escopoletina y escopoletin-7-0-glucósido en *Solanum valdiviense* y *Solanum gayanum*. Rev. Latinoamer. Quím. 19: 33.
127. TORRES, R., MODAK, B., FAINI, F. 1988. (25R)- Isonuatigenina, una sapogenina esteroidal novedosa como marcador taxonómico entre *Cestrum parqui* y *Vestia lycioides*. Bol. Soc. Chil. Quím. 33: 239-241.
128. NUÑEZ-ALARCON, J., RODRIGUEZ, E., SCHMID, R.D., MABRY, T.J. 1973. 5-O-Xylosylglucosides of apigenin and luteolin 7- and 7,4'-methyl ethers from *Ovidia* pillopillo. Phytochemistry 12: 1451-1454.
129. HOENEISEN, M., SILVA, M., CHAVADEJ, S., BECKER, H. 1992. Valepotriatos de *Valeriana lapathifolia* Vahl. y *V. macrorhiza* Poepp. ex DC. . Bol. Soc. Chil. Quím. 37: 345-347.
130. SIERRA, J., LOPEZ, J., CORTES, M. 1986. (-)-3ß-Acetoxydrimenin from the leaves of *Drimys winteri*. Phytochemistry 25: 253-254.
131. REYES, A., REYES, M., ALVEAL, A., GARCIA, H. 1990. Compuestos fenólicos de *Drimys winteri* Forst variedad andina. Rev. Latinoamer. Quím. 21: 42.
132. BOHNSTEDT, C.F., MABRY, T.J., 1979. The volatile constituents of the genus *Larrea* (Zygophyllaceae). Rev. Latinoamer. Quím. 10: 128-131.

Chapter Eight

PHYTOCHEMICALS INGESTED IN TRADITIONAL DIETS AND MEDICINES AS MODULATORS OF ENERGY METABOLISM

Timothy Johns and Laurie Chapman

Centre for Nutrition and the Environment of Indigenous Peoples
Macdonald Campus of McGill University,
Ste. Anne de Bellevue, Quebec, H9X 3V9, Canada

INTRODUCTION

While the importance of phytochemicals as the active principles of herbal medicines is well-known, the role of non-nutritive plant constituents consumed on a routine basis as normal mediators of health is more recently being recognized. For human populations engaged in traditional forms of subsistence the ingestion of a range of chemicals from plant foods, condiments, beverages and medicines may be part of their normal ecology. The refined diets of industrial societies differ from those of pre-industrial humans in composition, digestibility and palatability. They may not, however, be equatable with health. Indeed, diseases that are widespread in industrial societies such as diabetes, cancer, cardiovascular disease, and dental caries are recognized as

Phytochemistry of Medicinal Plants, edited by John T. Arnason et al.
Plenum Press, New York, 1995.

having a dietary basis attributable in large part to consumption of lipids and carbohydrates. Increasingly the role of phytochemicals as factors in these diseases is the subject of investigation.

This paper considers recent data on the health-promoting role of natural products in relation to lipid and carbohydrate metabolism from a human chemical ecological perspective. Concurrently it considers evolutionary insights into phytochemical ingestion from the context of dietary and medicinal plant use by peoples with a traditional life-style.

CONVERGING PERSPECTIVES OF NUTRITION AND PHYTOCHEMISTRY

Research activities of nutritionists and food scientists have historically run parallel to those of phytochemists, pharmacognocists and natural products chemists. While the methodologies of analysis for determining constituents of plant materials are the same across these fields, the orienting paradigms, problems addressed and target applications tend to be different. Compounds of primary interest to food chemists such as vitamins, minerals, amino acids, carbohydrates and lipids are usually of less interest to phytochemists than the secondary metabolites of plants. Natural products chemistry focuses on compounds as the potential sources of new pharmaceuticals and other industrial chemicals. Interests of nutritionists and phytochemists have typically overlapped, however, in the negative aspect of allelochemicals, that is in their role as toxins or antinutrients. The recognition that alkaloids, saponins, polyphenolics and other compounds have positive and complex roles in physiological processes affecting human health presents a new point of convergence with potential opportunities for multi-disciplinary research.

Impetus for this convergence comes from at least two sources. The first is the growing body of epidemiological data that links diets high in plant foods with reduced incidence of chronic degenerative diseases[1] such as non-insulin dependent diabetes mellitus (NIDDM), cancer[2] and coronary heart disease (CHD).[3] Secondly, the development of the chemical ecological perspective in basic biology and chemistry provides an organizing framework from which the importance of allelochemicals in the complex balance of physiological processes affecting health can make some sense. The recognition of allelochemicals as determinants of well-being of insects, mammals and other animals is a logical entry point to a field of human chemical ecology.[4] The scrutiny to which matters affecting human health are subjected has stimulated considerable research with laboratory animals on the mechanisms by which constituents of plants mediate the etiology of major societal diseases. This growing body of supporting data in turn provides a boost to studies of the interactions of animals, including humans, and plants.

CROSS-CULTURAL EPIDEMIOLOGY AND HUMAN CHEMICAL ECOLOGY

Cross-cultural comparisons form the basis for the understanding of the role of dietary factors in disease etiology. The so-called "French Paradox" posed by the different rates of atherosclerosis and CHD among some populations of industrialized European countries with similarly high intake of saturated fat has been attributed to a lower consumption of whole milk and greater consumption of plant foods and their associated antioxidants, saponins and fiber.[3] Epidemiological studies show differences in rates of cancer incidence among countries. For example, colon and rectal cancer is common in Europe and North America but is rare in Africa and Asia.[5,6] In Japan such cancers have increased over the last three decades. Dietary factors such as the portion of total calories from fat, dietary fibre, and the content of fruits and vegetables[7] including specific protective vegetables such as crucifers and alliums[8,6,9] are among the many environmental differences that distinguish countries with high and low rates of these and other cancers. NIDDM can reach epidemic proportions in societies undergoing cultural and dietary change. This is the most common form of diabetes in North America and its greater occurrence is the major reason for the increasing levels of diabetes in North America from 1866 into the twentieth century. Populations adopting a western life-style such as Polynesians and Amerindians have suffered sharp increases in NIDDM rates in the last few decades.[10,11] Coronary heart disease, cancer, and NIDDM are so-called diseases of affluence and, as demonstrated in a recent study comparing diet and diseases patterns between rural China and the United States1,[1] their incidence is associated with various factors that relate to the proportion of foods of plant origin in the diet.

Studies that consider human populations in a traditional ecological context offer specific insights into the role of dietary factors in disease incidence. For example, the classic work of Burkitt[12] in associating low incidence of colon cancer among Ugandans with the high intake of vegetable matter stimulated extensive research on the role of dietary fibre in intestinal function and cancer incidence.[13] Furthermore, consider diabetes among populations of indigenous peoples who have given up traditional diets in favour of diets that are typically more processed and higher in fats and calories. [11] Among the Pima Indians of the American Southwest and Northern Mexico, of whom approximately 50% of adults are afflicted by diabetes, traditional diets came from subsistence farming, hunting and gathering. Plant foods consumed in the last century included crops such as maize, wheat, tepary beans (*Phaseolus acutifolius* A. Gray) and squash and gathered species such as cactus buds and fruit, mesquite beans (*Prosopis* spp.), wild berries and wild greens. Boyce and Swinburn[14] estimated that 70-80% of the calories in this traditional diet came from carbohydrate; many of these carbohydrates were high in fibre, with low glycemic indices[15] (see below). The actual benefits of adopting traditional diets have been demonstrated among Australian Aborigines[16] and Hawaiians.[17] The latter study emphasized the consumption of traditional Hawaiian sources of complex carbohydrate. These diets led to improvements in both measures of blood sugars and blood lipids.[17]

In relation to the understanding of dietary factors contributing to coronary heart disease, the high-fat diets of Inuit present a classical case by which a dietary component with a protective effect, specifically ω-3 fatty acids in fish, was identified.[18,19] Another example of insights from a traditional case is that of the Maasai pastoralists in East Africa. The low incidence of cardiovascular disease among the Maasai was studied extensively during the 1960s and 1970s in relation to the high cholesterol and fat containing diet of this population, but it was not satisfactorily explained.[20] The Maasai provide a paradox between diet and coronary heart disease comparable to that in Europe mentioned above, but in a traditional ecological context.

Until recently few of the studies linking diet and disease considered ingested phytochemicals as positive agents in disease prevention. The emphasis has been instead on fat, carbohydrate, protein and fiber. Fat is the single dietary constituent that is most consistently associated with chronic degenerative disease. However, its relative contribution to cancer, diabetes and CHD is confounded by the fact that diets low in fat are typically high in foods of plant origin. Considering that traditional diets are typically higher in unpalatable and potentially-toxic substances from plants than modern diets,[4,21] such ingestion patterns would seem likely to provide insight into the relationship between allelochemicals and health. Moreover, the importance of herbal medicine as a component of traditional ingestion patterns is often overlooked. For many cultures distinctions between food and medicine are weak or even non-existent. Without quantitative data on intake of medicinal plants and herbal beverages, our understanding of the importance of allelochemicals in the physiology of peoples engaged in traditional life-styles will remain underestimated.

PHYTOCHEMICAL MEDIATORS OF CHRONIC DEGENERATIVE DISEASES

Mediators of Diabetes

Non-insulin dependent diabetes mellitus typically affects adults, and although the disease is determined by both genetic and environmental factors, it is believed to have, in part, a dietary etiology.[22] In NIDDM, insulin secretion of pancreatic β-cells in response to blood glucose is impaired and target cells in peripheral tissues may be resistant to the effects of insulin. Patients suffering from NIDDM do not require insulin replacement for survival, although insulin may improve their glycemic control.[23] NIDDM is primarily a disease of energy balance and energy metabolism and is strongly associated with obesity.[24] Also associated with this form of diabetes are hyperlipidemia and a risk of cardiovascular disease.

Treatment of NIDDM involves changes in diet and lifestyle particularly directed at reducing obesity and oral hypoglycemic agents. Dietary recommendations include reduction in fat intake and greater intake of dietary fibre and complex carbohydrates. An important goal of treatment is to reduce the characteristically elevated blood glucose levels of diabetics and to reduce the exaggerated postprandial glycemic responses. The extent to which other ingested constituents such as non-nutrients in plant foods mediate the glycemic response or have other effects on the development or severity of NIDDM is poorly known.

Marles and Farnsworth[25] have written a recent critical review of the potential of plants as sources of new antidiabetic agents, and this search continues to be an area of active investigation. Others[26,27] provide broad surveys of plant constituents with antidiabetic activities. Hypoglycemic and/or anti-hyperglycemic activities have been recorded in assays with numerous plants, many of which are used as traditional herbal treatments of diabetes. Compounds with antidiabetic activities come from a range of chemical classes including polysaccharides and proteins, flavonoids, steroids and terpenoids and alkaloids.[27] Potential mechanisms by which compounds can reduce blood glucose include: 1) increasing insulin secretion; 2) decreasing hepatic glucose production; 3) increasing insulin action; 4) increasing peripheral glucose metabolism independently of insulin; 5) and decreasing nutrient ingestion.[28]

Oral agents currently employed in the treatment of diabetes include sulphonylureas, metformin and acarbose[28] (Fig. 1). The former class of compounds acts primarily by stimulating β-cells of the pancreas to release insulin. Metformin, a biguanide, appears to lower fasting and postprandial hyperglycemia by enhancing peripheral glucose uptake and reducing hepatic glucose production. Hypoglycemic agents that decrease nutrient ingestion may act by reducing intestinal absorption, by delaying gastric emptying (e.g. the insoluble fiber of guar gum[29]) or by delaying carbohydrate digestion through inhibition of glycosidase

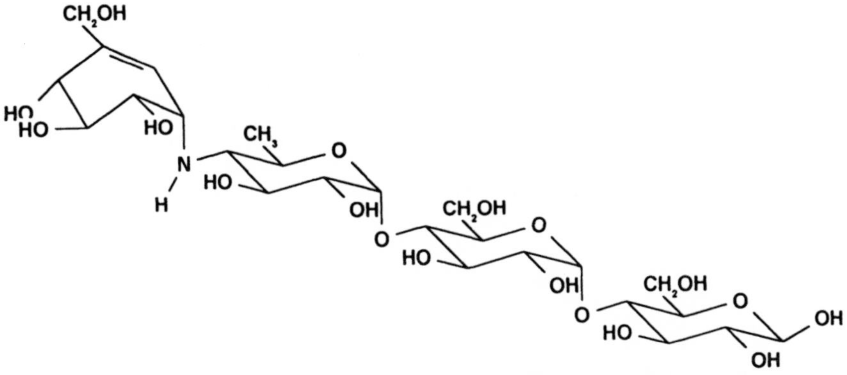

Figure 1. Structure of the oligosaccharide Acarbose from Actinomycetes.

enzymes or some other mechanism. The α-glucosidase inhibitor acarbose (Bay g5421; Bayer Pharmaceutical Co.) is an oligosaccharide that does not affect energy intake; rather instead of glucose from digested starch being absorbed in the top one third of the jejunum, it is absorbed over the full length of the small intestine.

Table 1 contains a summary of plants used as food, food additives or beverages from which hypoglycemic activity or antidiabetic agents have been identified whether acting on digestion or by some other possible mechanisms of action. The prevalent effect of naturally-occurring components of food and beverages with positive effects on diabetes and glycemic control, whether ingested by persons in industrial or non-industrial societies, is through reduction of the glycemic response from ingested carbohydrate. This reduction in glucose uptake occurs most likely from affects on nutrient availability and the rate of digestion. A glycemic index of foods is calculated from the area under the curve describing the change in blood glucose level that arises over a fixed time after consumption and is used as an indication of the beneficial effects of various sources of carbohydrates.[30,31] Specifically, it is a ratio of the glycemic response to a food portion, usually containing 50g available carbohydrate, versus the glycemic response to an equal portion of carbohydrate in white bread or glucose solution. Some compounds such as polyphenols, lectins, fiber and phytic acid are widespread in foods of plant origin. They are negatively correlated with the glycemic index and thus likely contribute to the differences in the glycemic index observed among many foods.[32] Thompson[33,34] has demonstrated the role of these compounds in legumes in reducing insulin response and post-meal glucose levels. α-Amylase inhibitors which occur widely in legumes[21] and other plants also are likely contributors to lower glycemic indices of certain foods. The mechanism of action in all of these cases is to slow down the digestion of carbohydrate. Through the binding of carbohydrate or of proteins associated with carbohydrate, by inhibiting the activity of digestive enzymes either directly or by chelating mineral cofactors, food constituents such as phytic acid can affect the rate at which glucose enters the blood (Fig.2).

α-Glucosidase inhibitors in food plants also may affect digestive processes, although compounds with this activity have not been systematically studied in diets. α-Glucosidase inhibitors isolated from natural sources include oligosaccharides[35] and aminocyclitols[36] from microorganisms, pentagalloylglucose from an alga,[37] and various polyhydroxy alkaloids[38] from higher plants (Fig.3). Such agents are competitive inhibitors with different specificities against various α-glucosidases. Acarbose, for example, reduces glycemic response by inhibiting glucoamylase and sucrase and to a lesser extent maltase.[28] An array of nitrogen-containing sugars, such as deoxynojirimycin (Fig. 3), from the roots of the mulberry (*Morus alba* L.) and other species of the same genus[38] show potent inhibition of intestinal α-glucosidases that is variable among the compounds from the same plant. These compounds have not reported from the edible fruit of the mulberry.

Table 1. Foods, food additives and beverages with reported hypoglycemic activity

Plant species	Part	Active constituent	Reference
Abelmoschus esculentus Moench (Malvaceae) okra	fruit, root	mucilage	98
Allium cepa L. (Liliaceae) onion	bulbs	allicin; allyl propyl disulphide	25
Allium sativum L. (Liliaceae) garlic	bulbs	allicin; allyl propyl disulphide	25
Arachis hypogaea L. (Fabaceae) peanut, ground-nut	nut		26
Balanites aegyptiaca Del. (Balanitaceae) desert date	fruit	extract	99
Bixa orellana L. (Bixaceae) annatto	seeds	extract	27
Blighia sapida Koenig (Sapindaceae) akee	fruit	hypoglycin A,B	25
Cajanus cajan (L.) Millsp. (Fabaceae) pigeon pea	seeds		26
Capsicum annuum L. (Solanaceae) chili pepper	fruits	capsaicin	27
Cinnamomum cassia Blume (Lauraceae) cassia bark tree		cinnamaldehyde	26
Coix lachryma-jobi L. var. *mayuen* Stapf. (Poaceae) Job's tears	seeds	coixans A,B & C	27
Dioscorea dumetorum Pax. (Dioscoreaceae) bitter yam	tubers	extracts	27
Dioscorea japonica Thunb. (Dioscoreaceae)	tubers	dioscorans A,B,C,D,E & F	27
Glycine max (L.) Merill (Fabaceae) soybean	seed		26
Ipomoea batata (L.) Lamk. (Convolvulaceae) sweet potato			26
Juglans regia L. (Juglandaceae) walnut		extract	26
Lablab purpureus (L.) Sweet (Fabaceae) lablab	pods		26
Lupinus albus L. (Fabaceae) white lupin	seeds	extracts	27
Lycopersicon esculentum Miller (Solanaceae) tomato			26
Momordica charantia L. (Cucurbitaceae) balsam-pear	fruits, leaf, stem	peptides, terpenoids	25

Table 1. *Continued*

Plant species	Part	Active constituent	Reference
Morus alba L. (Moraceae) white mulberry	leaves root bark	extract moran A alkaloids	26,27,28
Mucuna pruriens (L.) DC (Fabaceae) velvet bean	seed; fruit	infusion	26
Opuntia ficus-indica (Cactaceae) nopal	leaves	extract	100
Oryza sativa L. (Poaceae) rice	roots; external seed coats	oryzabrans A,B,C & D	26,27
Panax ginseng Mey. (Araliaceae) ginseng	roots	panaxans A,B,C,D & E; ginsenoside Rb_2	26,27
Phaseolus vulgaris L. (Fabaceae) common bean	seed	phytic acid; lectins; tannins	33,34
Piper nigrum L. (Piperaceae) black pepper			26
Psidium guajava L. (Myrtaceae) guava	leaves; fruits	extracts	26,27
Saccharum officinarum L. (Poaceae) sugarcane	aerial parts	saccharans A,B,C,D,E & F	26,27
Thea sinensis L. (Theaceae) tea	leaves	diphenylamine	27
Trigonella foenum-graecum L. (Fabaceae) fenugreek	seeds	trigonelline, coumarin	25

Studies on carbohydrate digestion of Aboriginal bushfoods suggests that, in addition to widespread compounds such as polyphenolics and phytates, traditional diets may have more specific constituents offering protection against diabetes.[39] Consider *Castanospermum australe* Cunn., *Acacia aneura* and *Dioscorea bulbifera* L. which were the three species among those tested by Thorburn et al.[39] which significantly reduced glucose and insulin responses in human subjects. *Castanospermum australe,* Moreton Bay chestnut, is the source of castanospermine and related indolizidine alkaloids (Fig.3) with strong α-glucosidase inhibiting activity.[40] Hypoglycemic alkaloids[41] and glycans[27] have been isolated from tubers of the genus *Dioscorea*. The slow digestibility of *Acacia* seeds may result from the viscous fibres that are characteristic of this genus.[42]

From an ecological perspective one would expect constituents that decrease nutrient digestion to be more widespread in plants than those affecting other aspects of glucose metabolism and insulin response. Inhibitors of α-glucosidases and α-amylases are more widely effective as agents of plant defense against organisms ranging from bacteria to mammals than agents affecting only

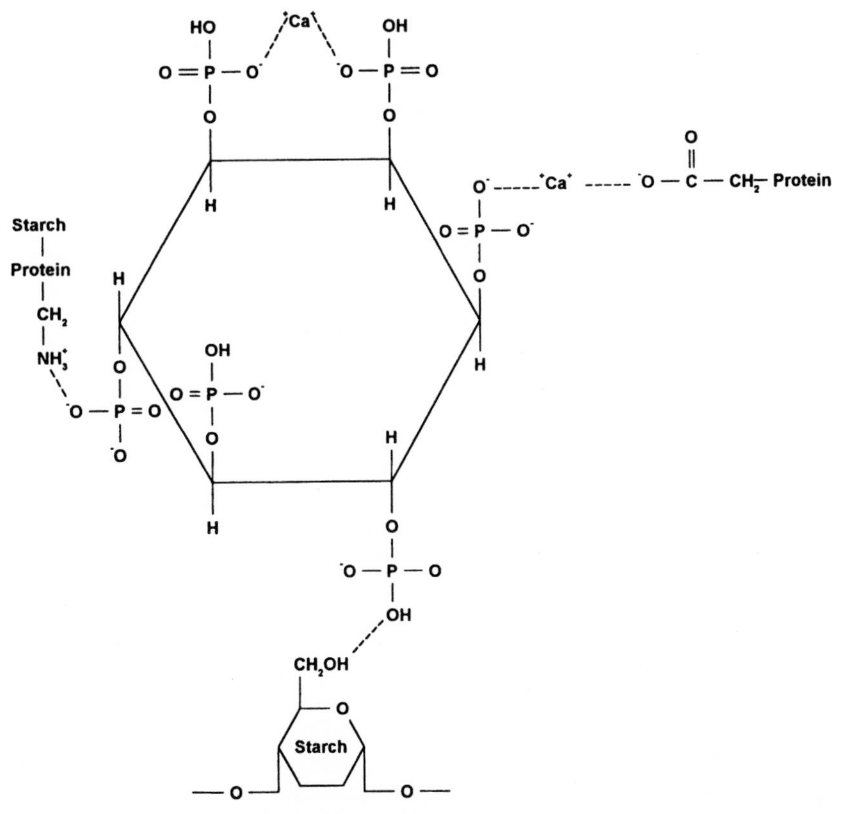

Figure 2. Possible interactions of phytic acid with proteins, minerals, and starch (after Thompson 1988).

physiological processes of higher animals such as insulin metabolism. This is not to say that medicinal plants or the numerous hypoglycemic principles isolated from plants[27] may not produce valuable agents affecting insulin-mediated glucose metabolism. However, such an event would seem more fortuitous than related to ecological function. On the contrary, Marles and Farnsworth[25] use an ecological argument based on the similarities in glucose metabolism between plants and animals to predict the likely occurrence of hypoglycemic agents in plants. Guanides are a good example of naturally-occurring substances from folk treatments for NIDDM which affectboth insulin and glucose metabolism.[43] Biguanides, such as metformin, are derived synthetically from two moieties of guanide separated by a chain of methylene groups.

Indigenous food and medicinal plants of other human communities may contain principles similar to the bushfoods consumed by the Australians, de-

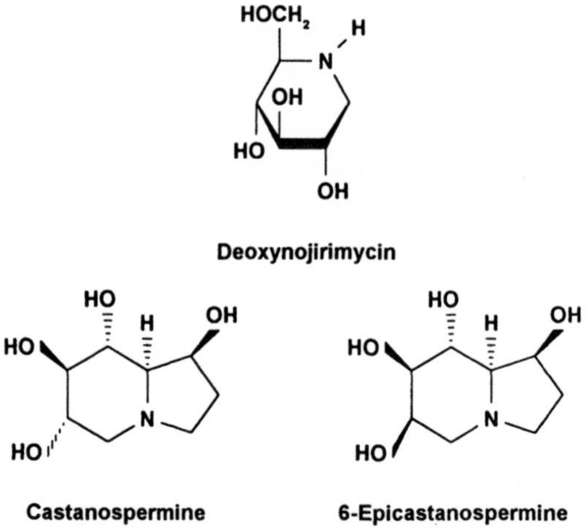

Figure 3. Examples of polyhydroxyalkaloids with α-glucosidase inhibiting activity.

scribed above,[39] and broader investigation of the hypoglycemic activity of ingested substances may help explain why traditional indigenous patterns of ingestion appear to confer protection against diabetes. Unfortunately, the data from studies on hypoglycemic activities of plants are of limited use in this regard. Apart from the paucity of specific plant studies, the majority of those conducted have used intraperitoneal rather than oral injection of rats, mice or other experimental animals. Usually the mechanism of the activity is unclear and in many cases likely non-specific. Thus, the literature offers little insight into the actions of substances consumed orally which may not be absorbed from the gastrointestinal tract and/or which exert their effects on digestive or absorptive processes.

Naturally-Occurring Antioxidants

Activated forms of oxygen such as free radicals and singlet oxygen are damaging to eukaryotic cells, and current understanding is that antioxidative characteristics are important in the preventive role that plant constituents play in relation to cancer, atherosclerosis and other diseases.[44] Important in the etiology of a number of diseases including cardiovascular disease is the attack by the hydroxyl radical on polyunsaturated fatty acids initiating a chain reaction of lipid peroxidation. Antioxidant nutrients (including the minerals selenium, copper, zinc and manganese) and vitamins E (α-tocopherol), C and A (including carotenoids with provitamin A activity) (Fig. 4) have been the subjects of extensive ongoing research.[44] Among their possible antioxidant functions, α-tocopherol and ascorbic acid are essential for the control of membrane damage caused by free radicals and for preventing lipid peroxidation. Retinoids, β-carotene and other carotenoids are important scavengers of singlet

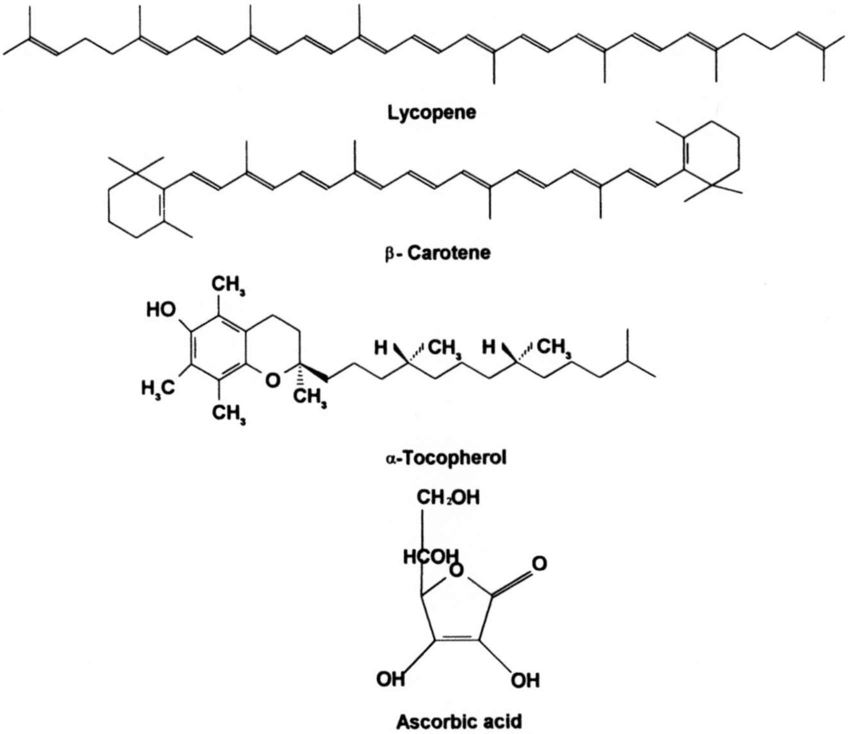

Lycopene

β- Carotene

α-Tocopherol

Ascorbic acid

Figure 4. Some lipophilic antioxidants of dietary inportance.

oxygen. This activity is independent of the nutritional role of vitamin A and, in fact, carotenes without vitamin A activity such as lycopene (Fig. 4) have greater singlet oxygen-scavenging ability than β-carotene.[45]

In addition to extensive research interest in carotenoids, the literature on the disease mediating activity of other natural antioxidants, particularly phenolics, is growing rapidly.[46] Interestingly, many of the phenolics with known antioxidant activities have been isolated from foods and beverages (Table 2). Typical of compounds under investigation are catechins and other flavonoids. *In vitro* studies suggest that water-soluble flavonoids protect against lipid peroxidation by localizing near the surface of phospholipid bilayers where they are able to scavenge aqueous oxygen radicals and thereby prevent the consumption of lipophilic α-tocopherol.[47]

Protectors against Cancer

High levels of dietary fat are believed to contribute to the development of certain types of cancer.[48] Although the mechanisms are unclear, peroxidation of

Table 2. Naturally-occurring antioxidants in food, beverages, and food additives

Chemical Compound(s)	Sources	References
ascorbic acid	fruits and vegetables	2,50
biphenyls	thyme	53
capsaicinopids, e.g. capsaicin	chili pepper	53
carotenoids, e.g. β-carotene, lycopene	green and yellow vegetables	50,55,101
catechins, e.g. (-)-epigallocatechin-gallate	green tea	2,52
carnosic acid	rosemary	50,53
diarylheptanoids, e.g. curcumin	ginger, tumeric	53
flavones, e.g. quercetin, robinetin, myricetin	widespread	50
flavonones, e.g. dihydroquercetin	widespread	50
gingerol	ginger	53
hydroxycinnamic acids, eg. chlorogenic acid	coffee, prunes, blueberries, apples, pears, sweet potato peels, grapes, oats, cereals, soybeans	2,50,52
isoflavones	soybeans	50
phenolic amides	pepper	53
rosmaric acid	rosemary	53
tannins	widespread	50,102
α-tocopherol	oils, green vegetables, whole grains	2,52,55

lipids and the formation of free radicals may be involved. Damage by free radicals to DNA is important in the etiology of cancer.

Naturally-occurring substances have been drawn on for decades as the sources of important cancer treatments,[49] although the recognition of phytochemicals as 'chemopreventives' has been more recent. The importance of diet in the incidence of cancer is a widely debated topic, although micronutrients present in fruit and vegetable appear to be important in reducing the incidence of cancer.[7]

The role of antimutagenic and anticarcinogenic compounds, particularly non-nutrients present in routinely consumed foods and beverages, is clouded by the complex nature of food constituents, including lipids, fibre and minerals, and their biological activities.

Stavric[2] has recently reviewed antimutagens and anticarcinogens from foods, and this paper simply summarizes the major compounds with chemopreventive activity. Although the mechanisms of action of such compounds are potentially varied and complex, many of the significant activities of chemopreventers are related to the antioxidant properties of compounds such as those listed in Table 2. Natural antioxidants occur in all higher plants,[50] and, among these, the antioxidant vitamins have received the most careful scrutiny as chemopreventers. Among its multiple and complex activities vitamin C acts in the stomach as a scavenger of nitrites and free radicals formed during the processing of food or in metabolic processes.[2] As the major antioxidant in the lipid phase of cells, vitamin E reduces lipid peroxidation and free radical formation, which in turn may result in a decrease in carcinogenesis. Nonetheless, studies in which supplements of antioxidant vitamins are administered to patients do not consistently lead to reductions in cancer incidence,[51] and the association of antioxidant vitamins and cancer is being questioned. Since the association of reduced cancer risk with a diet high in fruits and vegetables is strong, dietary constituents other than vitamins E and C or β-carotene may be making a larger contribution to this effect.

A variety of dietary plant flavonoids inhibit tumor development in experimental animals models.[52] Quercetin (Fig. 5) and rutin, which are antimutagenic, are among flavonoids widespread in fruits and vegetables. Similarly, isoflavones, which have been extensively studies in soybean, (Fig. 5) are potent antioxidants[50] and are associated with the antitumorigenic effect of this food.[2] Hydroxycinnamic acids, specifically chlorogenic acid, caffeic acid and ferulic acid, (Fig. 6) are

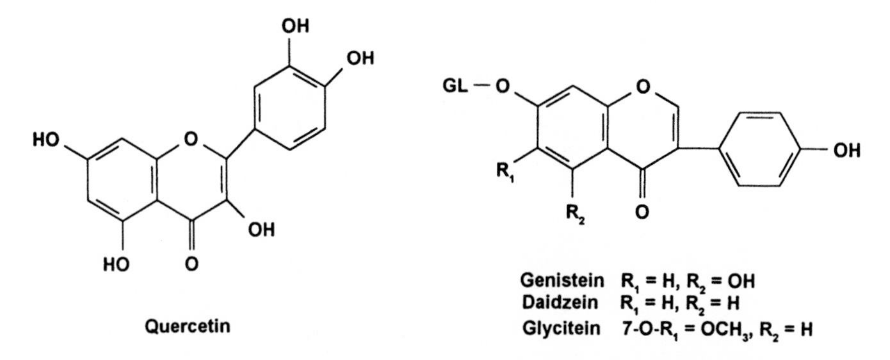

Figure 5. Examples of flavonoid and isoflavonoid antioxidants: Soybean isoflavonoids.

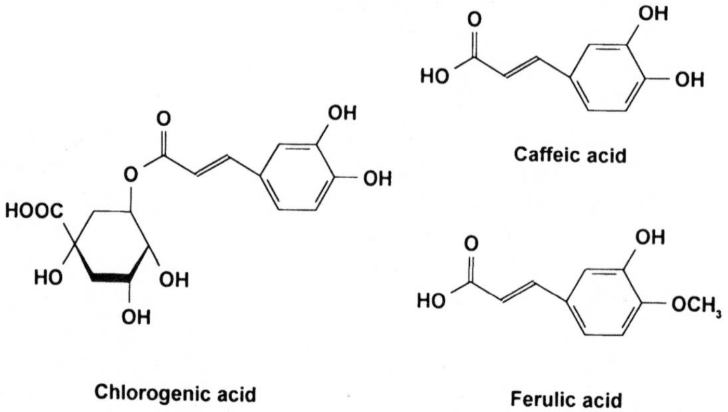

Caffeic acid

Chlorogenic acid

Ferulic acid

Figure 6. Structure of three common hydroxycinnamic acid derivatives.

similarly widespread in plant foods and beverages and have antioxidant[50] and tumor inhibiting activities.[52]

Polyphenolic compounds in green tea *(Camellia sinensis)*, most specifically (-) epigallocatechin gallate (Fig. 7), have been studied extensively as tumor inhibitors.[52] Carnosol from *Rosmarinus officinalis* L. (rosemary) and curcumin from *Curcuma longa* L. (tumeric) (Fig. 8) are examples of phenolics from common spices that are known to inhibit the initiation and development of tumours in animals.[52] *Rosmarinus officinalis* also contains the antioxidant rosmaric acid (Fig. 8).[50] Nakatani[53] summarizes data on the antioxidant activities of these and several other phenolics from spices. In addition, monoterpenoids from

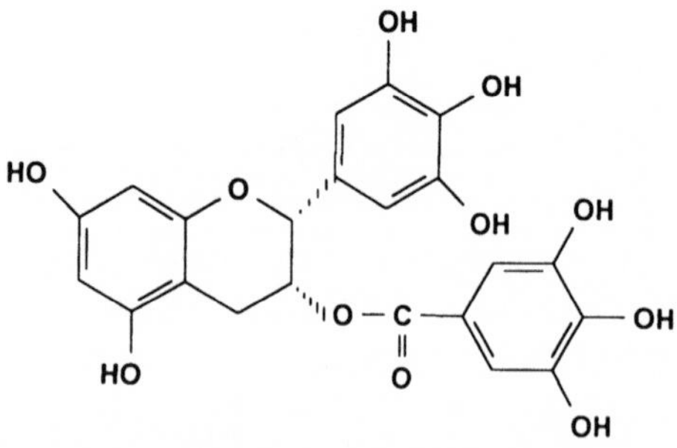

(-) Epigallocatechin-3-gallate

Figure 7. Structure of (-) Epigallocatechin-3-gallate.

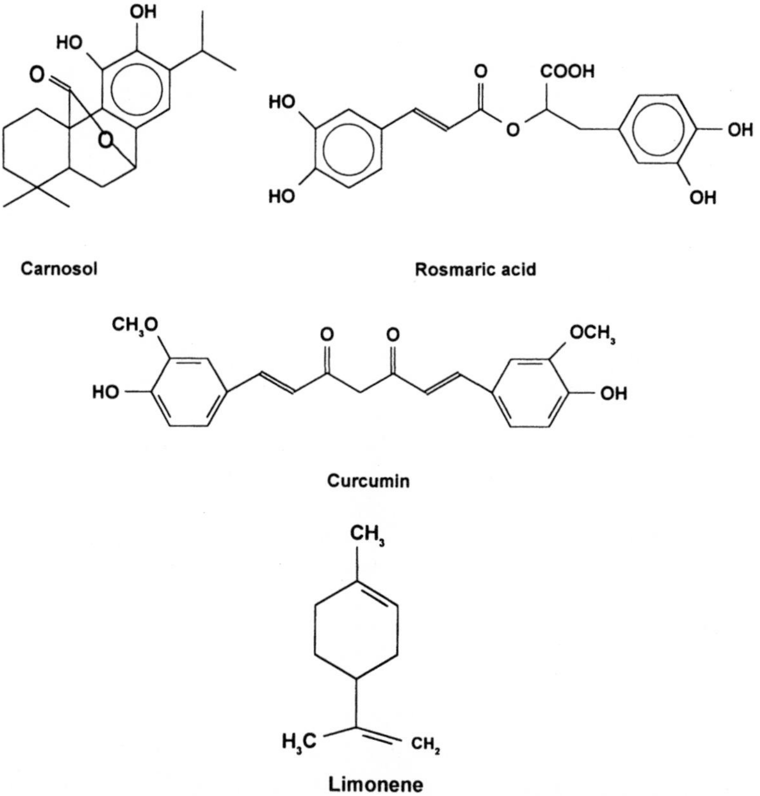

Carnosol

Rosmaric acid

Curcumin

Limonene

Figure 8. Examples of antioxidants from spices and citrus: Carnosol and Rosmaric acid from *Rosmarinus officinalis*; Curcumin form *Curcuma longa*; Linonene from *Citrus*.

spices, as well as *d*-limonene (Fig. 8) found in citrus peel and other sources, are antioxidants with tumor inhibiting properties.[2]

Protectors against Cardiovascular Disease

In light of evidence that elevated levels of plasma cholesterol and low-density-lipoprotein (LDL) cholesterol are risk factors, although not necessarily causal factors, for CHD, it is generally believed that treating these conditions through dietary, life-style and/or pharmacological interventions is advisable. Among dietary constituents, carotenoids and vitamin E receive the most attention for having a role in retarding the progression of atherosclerosis by protecting against lipoprotein peroxidation[54] particularly of polyunsaturated LDL cholesterol.[55]

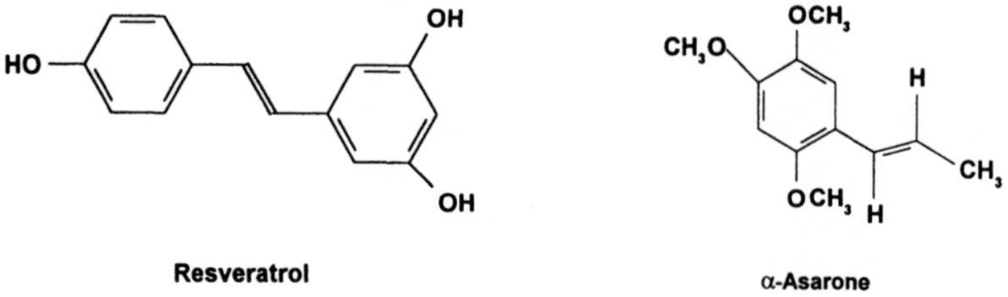

Resveratrol

α-Asarone

Figure 9. Structure of Resveratrol and α-Asarone.

Flavonoids[56] and other polyphenolics in food and beverages with antioxidants properties, as discussed above, may lower cholesterol through this mechanism. In addition, dietary tannins increase fecal bile acid excretion.[57] The apparent positive effects of red wine on platelet aggregation and lipid metabolism[58,59] may come from the stilbene, resveratrol,[60] (Fig. 9) polymeric tannins,[61] and other phenolics.[58] The hypocholesterolemic effects of "Mediterranean" diets and the "French paradox" of different rates in CHD among European countries have been attributed to the consumption of plant foods and their associated antioxidants, saponins and fiber.[3,62]

Saponins have been shown to have hypocholesterolemic effects in animals and humans.[34] They have an amphiphilic structure and may bind with dietary cholesterol and/or with bile acids. Saponins may prevent cholesterol absorption, interfere with its enterohepatic circulation, and increase its fecal excretion. Similarly phytosterols, such as β-sitosterol, inhibit the absorption of both endogenous and exogenous cholesterol and in moderate doses lower serum cholesterol.[63] Water-soluble dietary fibre such as pectin, β-glucan and certain gums also lower plasma cholesterol.[64,65]

Several drugs are currently available for the clinical treatment of hypercholesterolemia.[66] HMG CoA reductase inhibitors, for example, are fungal metabolites which reduce plasma cholesterol levels by inhibiting *de novo* cholesterol synthesis and increasing the receptor-mediated catabolism of low density lipoprotein (LDL). Chloride salts of anion-exchange resins such as cholestyramine bind bile acids in the intestine and lead to their fecal excretion. Bile acids suppress the enzymatic conversion of cholesterol to bile acids, and, as is suggested to occur with saponins, their removal increases the production of bile salts from cholesterol. None of these compounds is without side effects, and new methods of controlling hypercholesterolemia are sought. High plasma cholesterol is not a condition that can be recognized by traditional peoples lacking modern analytical methods, and it is unlikely that many new pharmacological agents for lowering cholesterol will be based on traditional remedies. Nonetheless, α-asarone (Fig. 9) has been recently isolated from bark of *Guatteria gaumeri* Greenman (Annonaceae), used in a traditional Mexican treatment of hypercholesterolemia and cholelithiasis[67]. Moreover, the variety of compounds in plants or foods of plant origin with similar beneficial effects that are discussed above suggests that naturally-occurring agents should not be dismissed from a role in controlling such conditions. Potentially useful in the treatment of the chronic risk-factors for CHD, which themselves are not debilitating, is the avoidance of powerful pharmaceutical agents and substitution of an increased intake of naturally-occurring constituents of normal diets. Further study of the potential benefits of these phytochemicals is therefore important.

CASE STUDY OF PHYTOCHEMICALS INGESTED IN A TRADITIONAL ECOLOGICAL CONTEXT

The Maasai and Batemi are tribes in East Africa which maintain a traditional subsistence lifestyle as pastoralists and agropastoralists respectively. Like most such populations their major health problems are related to undernutrition, sanitation and infectious disease, and they are relatively free of the diseases of industrial societies. Indeed, because of their low rates of atherosclerosis, cholelithiasis, coronary heart disease, and other diseases associated with saturated fat and cholesterol-rich diets the Maasai have been subjects of considerable biomedical research.[20] The Maasai obtain as much as 66% of their calories from fat[68] and despite subsistence on a diet high in milk, yogurt and meat they are not subject to hypercholesterolemia.[69]

Several theories, none of which is totally satisfactory, have been presented to account for this paradox. They include the following: i) physical fitness from a lifestyle of extensive walking is protective;[69] ii) the Maasai have a genetic trait that allows metabolic adaptation to high dietary cholesterol;[68] iii) fermented milk contains a hypocholesterolemic factor;[70] iv) calcium in milk prevents hypercholesterolemia;[71] v) the low caloric intake and variability in dietary adequacy is responsible for the low incidence of cardiovascular disease.[70]

Day et al.[72] suggested that low serum cholesterol of rural Maasai might be a result of herbs consumed, but published nothing subsequently in this regard. Many ancillary reports[70] point out limited contributions of plants to Maasai diets. Such accounts ignore data on Maasai ethnobotany[73] and are invalid. In fact, the Maasai, their neighbours the Batemi, and other East African ethnic groups routinely ingest wild plant materials as foods, as ingredients of milk and meat-based soups, as masticants, and as herbal medicines. In light of the numerous reports of roots and barks that are regularly added to fat-containing food by the Maasai and neighboring tribes,[74] we are reconsidering the epidemiological data on Maasai diet and cardiovascular status from the perspective of the more recent literature on the hypocholesterolemic and antioxidant activities of non-nutritive plant constituents in diet. Specifically, we are examining plant foods and food additives used by the Batemi and Maasai in relation to their possible role as hypocholesterolemic agents. The Batemi are a small tribe living in Northern

Tanzania who share numerous cultural features with their Maasai neighbours including use of local plant materials as normal dietary additives.

Based on a literature survey and an ethnobotanical study, we have tabulated a list of 25 species that the Maasai in both Kenya and Tanzania add to food.[73] Our ethnobotanical study among the Batemi identified the use of 64 wild species as part of the diet, 22 of which are specifically food additives.[75] Twelve of these correspond to the food additives of the Maasai.

We also have reported 41 medicinal plants ingested by the Batemi.[74] Herbal medicine is equally important among the Maasai. It is instructive that Maasai word for tree and medicine are one and the same, *ol-chani*. The common use of medicinal plants by the Batemi as dietary constituents is prominent. Over 60% of plant species and of remedies may be prepared in soup, milk or other food. Our Batemi and Maasai informants state consistently that they usually consume meat in the form of soup, and that soups almost always contain additives of roots or bark.

Although these additives have medicinal properties, most of the plant materials are not used necessarily in a curative manner; rather their purported purpose is preventative or to provide routine relief of chronic afflictions. Among the 12 most frequently reported food additives, *Pappea capensis* (Spreng)Eckl. & Zeyh., *Croton dichtygamous* Pax. and *Strychnos henningsii* Gilg. are general purpose tonics, while *Harrisonia abyssinica* Oliv., *Acacia nilotica* (L.) Del., *Myrsine africana* L., *Salvadora persica* L., *Warburgia salutaris* (Bertol.f.)Chiov., and *Acacia nubica* Benth. are used as agents to treat body and joint pain and along with tonics serve the general purpose of making one feel better. As a digestive aid *Acacia nilotica* also is used in a routine manner. It is as well employed as a flavoring of soup and milk without an intended medicinal role. *Albizia anthelmintica* Brongn. primarily is used by the Batemi to treat malaria and other fevers.[74] It is an emetic and is also used by the Maasai and other tribes as a purgative to treat tapeworm.[76]

We screened 16 of these species plus resin of *Commiphora africana* (A. Rich) Engl., a locally-important masticant, for the presence of saponins and phenolics.[73,77] Saponin presence was assessed based on congruity of a number of methods including froth test, thin layer chromatography (TLC), molluscicidal activity and hemolytic activity.[73] Tannin content was examined based on the precipitation of bovine serum albumin, and total phenolic content and water-soluble phenolic content was determined using the Folin-Ciocalteau reagent.[73] Sixty-three percent (63%) of the Batemi additives and 75% of these known to be also used by the Maasai contain potentially hypocholesterolemic saponins and/or phenolics.

Since hemolytic and membranolytic activity of some saponins may be related to potential hypocholesterolemic activity,[78,79] we characterized plant extracts based on their hemolytic activity in an assay with bovine erythrocytes.[77,80] In addition, in relation to the possible mechanism of action of saponins with cholesterol metabolism we examined the effects of cholesterol, cholesterol-analogues, conjugated bile salts and non-conjugated bile salts on hemolytic activity.[77,80] A further *in vitro* assay, related to the proposed mechanisms of saponin action, examined the binding capacity of plant extracts to radiolabelled cholesterol, either solubilized in ethanol or in micellar form.[77,80] Strongest hemolytic activity was detected in *Acacia goetzii* Harms, *Warburgia salutaris*, *Albizia anthelmintica* and fruit of *Myrsine africana*. All of these species were among those judged to contain saponins, with the latter two particularly high. In each case, micellar cholesterol, free cholesterol and β-sitosterol inhibited hemolysis suggesting an interaction of these compounds with the hemolyzing agent. In the study of binding by radiolabelled cholesterol, *A. anthelmintica* and *A. goetzii* were significantly active compared to controls. The fact that saponins have been identified from *A. anthelmintica*,[81] (Fig. 10) and that *A. anthelmintica* is the food additive with the highest levels of saponins among the plants we studied, supports the hypothesis that at least some of the saponins added to the diet of the Batemi and Maasai contribute to the low serum cholesterol levels of this population. *In vivo* and clinical studies with animals and humans are required to further test this hypothesis.

Like those employed by the Batemi, most of the plants the Maasai add to milk and soup are described as flavours, tonics or useful for prevention or alleviation of a range of common ailments. The effects of these rather non-specifically used plants and their allelochemical constituents on dietary ecology may

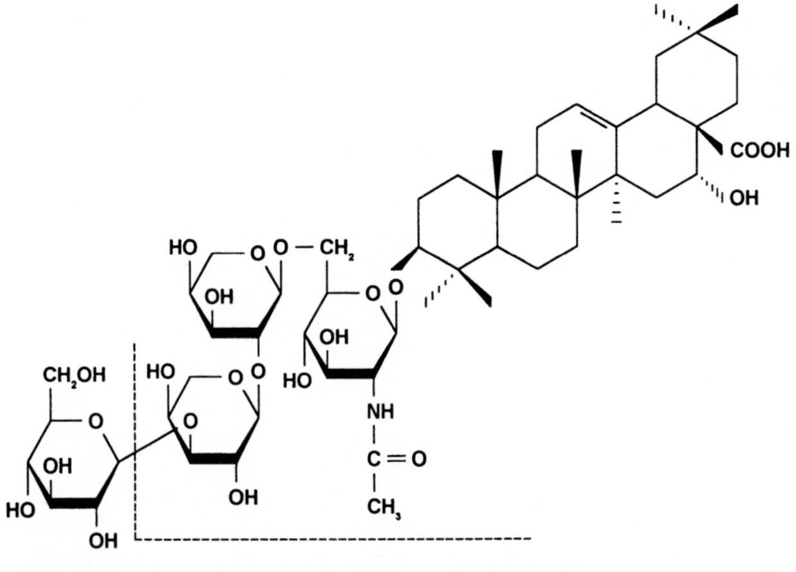

Figure 10. Glycosides of echinocystic acid from *Albizia anthelmintica* (Carpani et al., 1989).

extend beyond the immediately stated roles. Likewise, plants ingested in a number of other forms probably contribute considerable amounts of phytochemicals.

Those wild food plants which are known to have medicinal properties may have a broader physiological role than simply as nutrient sources. Of particular note are chewed gums and resins from such species as *Commiphora* spp., *Euphorbia cuneata* Vahl and *Ficus* spp., and barks and roots that are chewed as thirst quenchers such as *Lannea schimperi* (A. Rich.) Engl. and *Vigna* sp.. Several leafy vegetables consumed by the Batemi are reported as medicinal plants,[82] and contain known secondary metabolites. Worthy of specific mention are: glucosinolates[83] from *Gynandropsis gynandra* Briq.; glycoalkaloids[84,85] from *Solanum nigrum* L.; and saponins[85] from *Tribulus terrestris* L.. Honey beer is the most popular beverage in the area, and this is always made with the addition of roots of *Aloe volkensii* Engl., a member of a medicinally important genus characterized by anthroquinone derivatives and other physiologically active compounds.[86,87] *Aloe* spp. are used as antidiabetics in folk medicine and the leaves have hypoglycemic activity in animal and clinical studies.[26,27] Hypoglycemic activity of the bitter principle[88] and of glycans[27] of different species has been reported from intraperitoneal assays. While fructose, the predominant sugar in honey, produces a lower glycemic response than glucose,[89] the potential role of the hypoglycemic additives in further reducing the effects of this beverage on glucose and insulin levels is intriguing.

Resin of *Commiphora* spp. appears to be the most important masticant among the Maasai and Batemi. According to anecdotal reports, Maasai women chew *Commiphora* resin daily. Hypolipidemic activity has been extensively studied in resin of *Commiphora mukul*[90] and a similar effect from chewing African members of the genus is likely. Such activity has been attributed to sterols, and *Commiphora* species from East Africa and elsewhere contain triterpenes with anti-inflammatory activity.[91]

Cultivated crops grown and consumed by the Batemi, and much less so by the Maasai, may also contribute secondary compounds with anticholestremic, hypoglycemic and other physiological roles. The most important cereal staples are maize and sorghum, the latter being noteworthy for its high tannin content.[92] Legumes comprise a major component of the Batemi diet and contain an array of saponins, polyphenols and other secondary compounds.[21] It is generally recognized that soluble dietary fibre in grain legumes is important in low cholesterol levels associated with legume consumption.

GENERAL CONCLUSIONS

There are many ways in which ingested plant constituents contribute to health, many of them undoubtedly subtle. For example, vitamins and minerals contained in the roots, barks, fruits and gums consumed by the Batemi and Maasai are important nutritionally. Antibiotic compounds in these plants may prevent and control parasitic disease, and other compounds with immunostimulatory properties may have a range of potential effects (cf Wagner, this volume).

The chronic diseases of industrial societies are in part related to energy, and an effect of the reduction of levels of ingestion of phytochemicals on energy metabolism is a potential explanatory factor for changes in disease patterns since the industrial revolution. Inhibition of carbohydrate digestion leading to slowing of the glycemic response is important in controlling diabetes and may reduce concentrations of serum triglycerides and LDL cholesterol without significant reductions in high-density lipoprotein (HDL) cholesterol.[32] Based on our initial investigation we hypothesize that saponins, polyphenols and water-soluble dietary fibre in foods, medicines and masticants contributes to the phenomenon of low rate of atherosclerosis and CHD of the Maasai.

Plant secondary chemicals in the diet of industrial societies are usually viewed as undesirable (as toxins or antinutrients) or unimportant, and they are actively removed or simply ignored. However, the extensive consumption of these compounds by traditional agriculturalists, pastoralists and hunter-gatherers, as well as by hominoid apes,[93] supports the notion that these compounds are not only a normal part of dietary ecology, but may have become essential to physiological homeostasis.[4] Consider that humans, several primate species and guinea pigs are the only mammals without the capacity to synthesize vitamin C; this remains one of the most important antioxidants for humans but the enzyme necessary for its biosynthesis may have been lost during primate evolution because ample supplies of ascorbic acid were present in the diet.[94] Organic molecules in the diet, likewise, could be determinants of human gut morphology. Without the inhibitory effects of various allelochemicals on rates of carbohydrate digestion, the human small intestine might not need to be as long as it currently is. Growing evidence for the role of non-nutrients such as those with anticholestremic activity, antioxidants and those mediating carbohydrate metabolism in physiological processes should be assessed not simply from a clinical perspective, but in relation to disease prevention.

Recognition that the chronic diseases of industrial societies have a dietary basis has led to recommendations to reduce dietary fat and to increase consumption of fiber, antioxidant vitamins and complex carbohydrates. This is a diet higher in plant material and consequentially closer to the diet of our hominid ancestors and of contemporary populations engaged in traditional lifestyles. It is also undoubtedly a diet with higher levels of alkaloids, phenolics, saponins and other compounds.

Recommendations to reduce fat to less than 30% of calories and dietary cholesterol to < 300mg/day[95] are consistent with diets of many, but not all, contemporary subsistence and hunter-gatherer groups.[96] The high fat-containing diets of Inuit and Amerindians in the Arctic or of African pastoralists support the possibility that the diets of human ancestors could have been higher in fat than current wisdom generally accepts. Animal derived fat and protein became a component in the diet of our ancestors as scavenging and hunting took on importance at the beginning of evolution of the genus *Homo* some two million years ago.[97] Nonetheless, leaves, fruits and roots were essential components of hominid diets, and our dietary physiology and detoxication mechanisms, and our requirements for fat and carbohydrates, evolved in an ecological context where plant secondary compounds were difficult to avoid.[4] The actions of plant secondary compounds in mediating lipid and carbohydrate metabolism suggest that physiological homeostasis could include both plant non-nutrients and fats in higher levels than currently recommended, and that in ecological contexts that provide for significant consumption of plants a diet low in fats is not a necessary prerequisite to health.

Humans have engaged in food processing for at least 500,000 years, although the sophistication of the methods has evolved. A major impact of the development of simple techniques such as cooking, leaching, fermenting, or grinding of plant foodstuffs was to make more available concentrated carbohydrates from roots, tubers and seeds, that were otherwise toxic.[98] While the use of fire and other processing techniques during the course of human evolution may have reduced the ingestion of allelochemicals somewhat, this effect was minimal in comparison to the sophisticated food processing techniques developed in the industrial age. Human appetite, taste preferences and dietary consumption are regulated by the need for energy and energy consumption is essential to our survival. In traditional contexts, consumption of energy sources is coupled with the consumption of a variety of naturally-occurring substances including phytochemicals. In modern contexts, technological advances in food processing has given humankind the capacity to decouple the relationship between energy sources and phytochemicals. While a vast amount of research remains to fully elucidate the negative and positive effects of phytochemicals on human physiology, it is a reasonable expectation that in the future we will move to recouple the historical relationship. Recent interest of the food industry in chemopreventives and in producing foods with optimized health-related properties[31,99] suggests that this future may be not be so far off.

ACKNOWLEDGMENTS

We are grateful for financial support from the Natural Sciences and Engineering Research Council of Canada (NSERC). In addition we thank the Batemi people of Tanzania, R.L.A. Mahunnah and E.N. Mshiu of the Institute of Traditional Medicine in Dar es Salaam, M.L. Parkipuny, D. Luckert and T. Ticktin for assistance with this research.

REFERENCES

1. CAMPBELL, C., JUNSHI, C. 1994. Diet and chronic degenerative diseases: perspectives from China. Am. J. Clin. Nutr. 59: 1153-1161S.

2. STAVRIC, B. 1994. Antimutagens and anticarcinogens in foods. Food Chem. Toxicol. 32: 79-90.
3. ARTAUD-WILD, S.M., CONNOR, S.L., SEXTON, G., CONNOR, W.E. 1993. Differences in coronary mortality can be explained by differences in cholesterol and saturated fat intakes in 40 countries but not in France and Finland. Circulation 88: 2771-2779.
4. JOHNS, T. 1990. With Bitter Herbs They Shall Eat It: Chemical Ecology and the Origins of Human Diet and Medicine. University of Arizona Press, Tucson. p.356.
5. COLDITZ, G.A., WILLETT, W.C. 1991. Epidemiologic approaches to the study of diet and cancer. In: Cancer and Nutrition, (R.B. Alfin-Slater, D. Kritchevsky, eds.), Plenum Press, New York, pp. 51-67.
6. HIGGINSON, J., SHERIDAN, M.J. 1991. Nutrition and human cancer. In: Cancer and Nutrition, (R.B. Alfin-Slater, D. Kritchevsky, eds.), Plenum Press, New York, pp. 1-50.
7. WILLET, W.C. 1994. Micronutrients and cancer risk. Am. J. Clin. Nutr. 59: 1162S-1165S.
8. BEECHER, C.W.W. 1994. Cancer preventive properties of varieties of *Brassica oleracea*: a review. Am. J. Clin. Nutr. 59: 1166S-1170S.
9. BIRT, D.F., BRESNICK, E. 1991. Chemoprevention and nonnutrient components of vegetables and fruits. In: Cancer and Nutrition,(R.B. Alfin-Slater, D. Kritchevsky, eds.), Plenum Press, New York, pp. 221-260.
10. BENNETT, P. 1983. Diabetes in developing countries and unusual populations. In: Diabetes in Epidemiological Perspective, (J.I. Mann, K. Pyorala, A. Teuscher, eds.), Churchill Livingstone, Edinburgh, pp. 43-57.
11. DIAMOND, J.M. 1992. Human evolution - diabetes running wild. Nature 357: 362-363.
12. BURKITT, D.P. 1971. Epidemiology of cancer of the colon and rectum. Cancer 28: 3-13.
13. KRITCHEVSKY, D., KLURFELD, D.M. 1991. Dietary fiber and cancer, In: Cancer and Nutrition, (R.B. Alfin-Slater, D. Kritchevsky, eds.), Plenum Press, New York, pp. 127-140.
14. BOYCE, V.L., SWINBURN, B.A. 1993. The traditional Pima Indian diet - composition and adaptation for use in a dietary intervention study. Diabetes Care 16: 369-371.
15. BRAND, J.C., SNOW, B.J., NABHAN, G.P., TRUSWELL, A.S. 1990. Plasma glucose and insulin responses to traditional Pima Indian meals. Am. J. Clin. Nutr. 51: 416-420.
16. O'DEA, K. 1984. Marked improvement in carbohydrate and lipid metabolism in diabetic Australian Aborigines after temporary reversion to traditional lifestyle. Diabetes 33: 596-603.
17. SHINTANI, T.T., HUGHES, C.K., BECKHAM, S., OCONNOR, H.K. 1991. Obesity and cardiovascular risk intervention through the ad libitum feeding of traditional Hawaiian diet. Am. J. Clin. Nutr. 53: S1647-S1651.
18. BANG, H.O., DYERBERG, J., SINCLAIR, H.M. 1980. The composition of the Eskimo food in northwestern Greenland. Am. J. Clin. Nutr. 33: 2657-2661.
19. INNIS, S.M., KUHNLEIN, H.V., KINLOCH, D. 1988. The composition of red cell membrane phospholipids in Canadian Inuit consuming a diet high in marine mammals. Lipids 23: 1064-1068.
20. MCGILL, H.C. 1979. The relationship of dietary cholesterol to serum cholesterol concentration and to atherosclerosis in man. Am. J. Clin. Nutr. 32: 2664-2702.
21. JOHNS, T. 1994. Defense of nitrogen-rich seeds constrains selection of reduced toxicity during the domestication of the grain legumes. In: Advances in Legume Systematics 5, (J.I. Sprent, D. McKey, eds.), Royal Botanic Gardens, Kew, pp.151-167
22. MANN, J., HOUSTON, A. 1983. The aetiology of non-insulin dependent diabetes mellitus. In: Diabetes in Epidemiological Perspective, (J.I. Mann, K. Pyorala, A. Teuscher, eds.), Churchill Livingstone, Edinburgh, pp 122-164.
23. GILL, G.V. 1991. Non-insulin-dependent diabetes mellitus. In: Textbook of Diabetes, (J. Pickup, G. Williams, eds.), Blackwell Scientific Publications, Oxford, pp. 24-29.

24. LEAN, M.E.J., MANN, J.I. 1991. Obesity, body fat distribution and diet in the aetiology of non-insulin-dependent diabetes mellitus. In: Textbook of Diabetes, (J. Pickup, G. Williams, eds.), Blackwell Scientific Publications, Oxford, pp. 181-191.
25. MARLES, R.J., FARNSWORTH, N.R. 1994. Plants as sources of antidiabetic agents. Economic and Medicinal Plant Research. 6: 149-187.
26. ATTA-UR-RAHMAN, ZAMAN, K. 1989. Medicinal plants with hypoglycemic activity. J. Ethnopharmacol. 26: 1-55.
27. IVORRA, M.D., PAYA, M., VILLAR, A. 1989. A review of natural products and plants as potential antidiabetic drugs. J. Ethnopharmacol. 27: 243-275.
28. LEBOVITZ, H.E. 1992. Oral antidiabetic agents. Drugs 44(Suppl. 3): 21-28.
29. LECLERC, C.J., CHAMP, M., BOILLOT, J., GUILLE, G., LECANNU, G., MOLIS, C., BORNET, F., KREMPF, M., DELORTLAVAL, J., GALMICHE, J.P. 1994. Role of Viscous Guar Gums in Lowering the Glycemic Response After a Solid Meal. Am. J. Clin. Nutr. 59: 914-921.
30. TROUT, D.L., BEHALL, K.M., OSILESI, O. 1993. Prediction of glycemic index in starchy foods. Am. J. Clin. Nutr. 58: 873-878.
31. BJORCK, I., ASP, N.-G. 1994. Controlling the nutritional properties of starch in foods - a challenge to the food industry. Trends Food Sci. Technol. 5: 213-217.
32. JENKINS, D.J.A., JENKINS, A.L., WOLEVER, T.M.S., VUKSAN, V., RAO, V. 1994. Low glycemic index: lente carbohydrates and physiological effects of altered food frequency. Am. J. Clin. Nutr. 59(Suppl): 706S-709S.
33. THOMPSON, L.U. 1988. Antinutrients and blood glucose. Food Technol. 42: 123-132.
34. THOMPSON, L.U. 1993. Potential health benefits and problems associated with antinutrients in foods. Food Res. Int. 26: 131-149.
35. SCHMIDT, D.D., FROMMER, W., JUNG, B., MULLER, L., WINGENDER, W., TRUSCHEIT, E., SCHAFER, D. 1977. α-Glucosidase inhibitors. Naturwissenschaften 64: 535-536.
36. KAMEDA, Y., ASANO, N., YOSHIKAWA, M., TAKEUCHI, M., YAMAGUCHI, T., MATSU, K., HORII, S., FUKAASE, H. 1984. Valiolamine, a new α-glucosidase inhibiting aminocyclitol produced by *Streptomyces hygroscopicus*. J. Antibiotics 37: 1301-1307.
37. CANNELL, R.J.P., FARMER, P., WALKER, J.M. 1988. Purification and characterization of pentagalloylglucose, an α-glucosidase inhibitor/antibiotic from the freshwater green alga *Spirogyra varians*. Biochem. J. 255: 937-941.
38. ASANO, N., OSEKI, K., TOMIOKA, E., KIZU, H., MATSUI, K. 1994. N-containing sugars from *Morus alba* and their glucosidase inhibitory activities. Carbohydrate Res. 259: 243-255.
39. THORBURN, A.W., BRAND, J.C., TRUSWELL, A.S. 1987. Slowly digested and adsorbed carbohydrate in traditional bushfoods: a protective factor against diabetes? Am. J. Clin. Nutr. 45: 98-106.
40. SAUL, R., GHIDONI, J.J., MOLYNEUX, R.J., ELBEIN, A.D.. 1985. Castanospermine inhibits α-glucosidase activities and alters glycogen distribution in animals. Proc. Natl. Acad. Sci. USA 82: 93-97.
41. IWU, M.M., OKUNJI, C.O., OHIAERI, G.O., AKAH, P., CORLEY, D., TEMPESTA, M.S. 1990. Hypoglycemic activity of disocoretine from tubers of *Dioscorea dumetorum* in normal and alloxan diabetic rats. Planta Medica 56: 264-267.
42. THORBURN, A.W., BRAND, J.C., CHERIKOFF, V., TRUSWELL, A.S. 1987. Lower postprandial plasma glucose and insulin after addition of *Acacia coriacea* flour to wheat bread. Aust. New Zealand J. Med. 17: 24-26.
43. PERL, M. 1988. The biochemical basis of the hypoglycemic effects of some plant extracts. In: Herbs, Spices and Medicinal Plants: Recent Advances in Botany, Horticulture, and Pharmacology, Volume 3, (L.E. Craker, J.E. Simon, eds.), Oryx Press, Phoenix, pp. 49-70.

44. DIPLOCK, A.T. 1991. Antioxidant nutrients and disease prevention - an overview. Am. J. Clin. Nutr. 53: S189-S193.
45. DI MASCIO, P., KAISER, S., SIES, H. 1989. Lycopene as the most efficient biological carotenoid singlet oxygen quencher. Arch. Biochem. Biophys. 274: 532-538.
46. HUANG, M.-T., HO, C.-T., LEE, C.Y.,(eds.), 1992. Phenolic Compounds in Food and Their Effects on Health II: Antioxidants and Cancer Prevention. American Chemical Society, ACS Symposium Series 507, Washington, DC.. p.402.
47. TERAO, J., PISKULA, M., YAO, Q. 1994. Protective effect of epicatechin, epicatechin gallate and quercetin on lipid peroxidation in phospholipid bilayers. Arch. Biochem. Biophys. 308: 278-284.
48. CARROLL, K.K. 1991. Dietary fats and cancer. Am. J. Clin. Nutr. 53: 1064S-1067S.
49. CORDELL, G.A., BEECHER, C.W.W., PEZZUTO, J.M. 1991. Can ethnopharmacology contribute to the development of new anticancer drugs? J. Ethnopharmacol. 32: 117-133.
50. PRATT, D.E. 1992. Natural antioxidants from plant material. In: Phenolic Compounds in Food and Their Effects on Health II: Antioxidants and Cancer Prevention. (M.-T. Huang, C.-T. Ho, C.Y. Lee, eds.), American Chemical Society, ACS Symposium Series 507, Washington, DC, pp. 54-71.
51. GREENBERG, E.R., BARON, J.A., TOSTESON, T.D., FREEMAN, D.H., BECK, G.J., BOND, J.H., COLACCHIO, T.A., COLLER, J.A., FRANKL, H.D., HAILE, R.W., MANDEL, J.S., NIERENBERG, D.W., ROTHSTEIN, R., SNOVER, D.C., STEVENS, M.M., SUMMERS, R.W., VANSTOLK, R.U. 1994. Clinical trial of antioxidant vitamins to prevent colorectal adenoma. New England J. Med. 331: 141-147.
52. HUANG, M.T., FERRARO, T. 1992. Phenolic compounds in food and cancer prevention. In: Phenolic Compounds in Food and Their Effects on Health II: Antioxidants and Cancer Prevention. (M.-T. Huang, C.-T. Ho, C.Y. Lee, eds.), American Chemical Society, ACS Symposium Series 507, Washington, DC, pp. 8-34.
53. NAKATANI, N. 1992. Natural antioxidants from spices. In: Phenolic Compounds in Food and Their Effects on Health II: Antioxidants and Cancer Prevention. (M.-T. Huang, C.-T. Ho, C.Y. Lee, eds.), American Chemical Society, ACS Symposium Series 507, Washington, DC, pp. 72-86.
54. ESTERBAUER, H. 1989. Role of vitamin E and carotenoids in preventing oxidation of low density lipoproteins. Ann. N.Y. Acad. Sci. 570: 254-267.
55. ABBEY, M., NESTEL, P.J., BAGHURST, P.A. 1993. Antioxidant vitamins and low-density-lipoprotein oxidation. Am. J. Clin. Nutr. 58: 525-532.
56. STAVRIC, B., MATULA, T.I. 1992. Flavonoids in foods - their significance for nutrition and health. In: Lipid - Soluble Antioxidants: Biochemistry and Clinical Applications, (A.S.H. Ong, L. Packer, eds.), Birkhauser Verlag, Basel, Switzerland, pp. 274-294.
57. LEVRAT, M.-A., TEXIER, O., REGERAT, F., DEMIGNE, C., REMESY, C. 1993. Comparison of the effects of condensed tannin and pectin on cecal fermentations and lipid metabolism in the rat. Nutr. Res. 13: 427-433.
58. FRANKEL, E.N., KANNER, J., GERMAN, J.B., PARKS, E., KINSELLA, J.E. 1993. Inhibition of oxidation of human low-density lipoprotein by phenolic substances in red wine. Lancet 341: 454-457.
59. SEIGNEUR, M., BONNET, J., DORIAN, B., BENCHIMOL, D., DROUILLET, F. 1990. Effect of the consumption of alcohol, white wine and red wine on platlet function and serum lipids. J. Appl. Card. 5: 215-222.
60. SIEMANN, E.H., CREASY, L.L. 1992. Concentration of the phytoalexin resveratrol in wine. Am. J. Enol. Vitic. 43: 49-52.
61. TEBIB, K., BITRI, L., BESANCON, P., ROUANET, J.-M. 1994. Polymeric grape seed tannins prevent plasma cholesterol changes in high-cholesterol-fed rats. Food Chem. 49: 403-406.

62. JAMES, W.P.T., DUTHIE, G.G., WAHLE, K.W.J. 1989. The Mediterranean diet: protective or simply non-toxic? European J. Clin. Nutr. 43: 31-41.
63. FRASER, G.E. 1994. Diet and coronary heart disease: beyond dietary fats and low-density-lipoprotein cholesterol. Am. J. Clin. Nutr. 59: S1117-S1123.
64. HASKELL, W.L., SPILLER, G.A., JENSEN, C.D., ELLIS, B.K., GATES, J.E. 1992. Role of water-soluble dietary fiber in the management of elevated plasma cholesterol in healthy subjects. Am. J. Card. 69: 433-439.
65. JENSEN, C.D., SPILLER, G.A., GATES, J.E., MILLER, A.F., WHITTAM, J.H. 1993. The effect of acacia gum and water-soluble dietary fiber mixture on blood lipids in humans. J. Am. College Nutr. 12: 147-154.
66. BROWN, M.S., GOLDSTEIN, J.L. 1990. Drugs used in the treatment of hyperlipoproteinemias. In: Goodman and Gilman's The Pharmacological Basis of Therapeutics, 8th Edition, (A.G.Goodman, T.W. Rall, A.S. Nies, P. Taylor, eds.), Macmillan Publishing Company, New York, pp. 874-896.
67. CHAMORRO, G., SALAZAR. M., SALAZAR, S., MENDOZA, T. 1993. Farmacologia y toxicologia de *Guatteria gaumeri* y α-asarona. Revista de Investigacion Clinica 45: 597-604.
68. BISS, K., HO, K.-J., MIKKELSON, B., LEWIS, L., TAYLOR, C.B. 1971. Some unique biological characteristics of the Masai of East Africa. New England J. Med. 284: 694-699.
69. MANN G., SPOERRY, A., GRAY, M., JARASHOW, D. 1972. Atherosclerosis in the Masai. Am. J. Epidemiology 95: 26-37.
70. GIBNEY, M., BURSTYN, P. 1980. Milk, serum cholesterol and the Maasai. Atherosclerosis 35: 339-343.
71. HOWARD, A.D. 1977. The Masai, milk and the yogurt factor: an alternative explanation. Atherosclerosis 27: 383-385.
72. DAY, J., CARRUTHERS, M., BAILEY, A., ROBINSON, D. 1976. Anthropometric, physiological and biochemical differences between urban and rural Maasai. Atherosclerosis 23: 357-361.
73. JOHNS, T., MAHUNNAH, R.L.A., SANAYA, P., CHAPMAN, L., TICKTIN, T. 1994. Hypocholesterolemic constituents in plant dietary additives of a traditional subsistence community, the Batemi of Ngorongoro District, Tanzania. Submitted.
74. JOHNS, T., MHORO, E.B., SANAYA, P., KIMANANI, E.K. 1995. Herbal remedies of the Batemi of Ngorongoro District, Tanzania: a quantitative appraisal. Economic Botany 48: 90-95.
75. JOHNS, T., MHORO, E.B., SANAYA. 1995. Food plants and masticants of the Batemi of Ngorongoro District, Tanzania. Submitted.
76. KOKWARO, J.O. 1976. Medicinal plants of East Africa. East African Literature Bureau, Kampala, p.384.
77. CHAPMAN, L. 1994. *In vitro* Hypocholesterolemic Potential of Dietary Additives Used by the Batemi and Maasai People. M.Sc. Thesis, McGill University, Montreal., p.94
78. PRICE, K., JOHNSON, I., FENWICK, G. 1987. The chemistry and biological significance of saponins in foods and feedingstuffs. CRC Crit. Rev. Food Sci. Nutr. 26: 27-135.
79. STORY, J., LEPAGE, S., PETRO, M., WEST, L., CASSIDY, M., LIGHTFOOT, F., VAHOUNY, G. 1984. Interactions of alfalfa plant and sprout saponins with cholesterol *in vitro* and in choleterol-fed rats. Am. J. Clin. Nutr. 39: 917-929.
80. CHAPMAN, L., JOHNS, T., MAHUNNAH, R.L.A. 1995. Hypocholesterolemic potential of bark of *Albizia anthelmintica* and other non-nutrient wild plant food additives used by Maasai. Submitted.
81. CARPANI, G., ORSINI, F., SISTI, M., VEROTTA, L. 1989. Saponins from *Albizzia anthelmintica*. Phytochemistry 28: 63-866.

82. JOHNS, T., KOKWARO, J.O., KIMANANI, E.K. 1990. Herbal remedies of the Luo of Siaya District, Kenya: establishing quantitative criteria for consensus. Economic Botany 44: 369-381.
83. HASAPIS, X., MACLEOD, A.J., MOREAU, M. 1981. Glucosinolates of nine Cruciferae and two Capparaceae species. Phytochemistry 20: 2355-2358.
84. IWU, M.M. 1993. Handbook of African medicinal plants. CRC Press: Boca Raton, p.435.
85. SAMUELSSON, G., FARAH, M.H., CLAESON, P., HAGOS, M., HEDBERG, O., WARFA, A.M., HASSAN, A.O., ELMI, A.H., ABDURAHAM, A.D., ELMI, A.S., ABDI, Y.A., ALIN, M.H. 1993. Inventory of plants used in traditional medicine in Somalia. 4. Plants of the families Passifloraceae-Zygophyllaceae. J. Ethnopharmacol. 38: 1-29.
86. FARAH, M.H., ANDERSSON, R., SAMUELSSON, G. 1992. Microdontin-A and microdontin-B - two new aloin derivatives from *Aloe microdonta*. Planta Medica 58: 88-93.
87. NASH, R.J., BEAUMONT, J., VEITCH, N.C., REYNOLDS, T., BENNER, J., HUGHES, C.N.G., DRING, J.V., BENNETT, R.N., DELLAR, J.E. 1992. Phenylethylamine and piperidine alkaloids in *Aloe* species. Planta Medica 58: 84-87.
88. AJABNOOR, M.A. 1990. Effect of aloes on blood glucose levels in normal and alloxan diabetic mice. J. Ethnopharmacol. 28: 215-220.
89. DELARUE, J., NORMAND, S., PACHIAUDI, C., BEYLOT, M., LAMISSE, F., RIOU, J.P. 1993. The contribution of naturally labelled 13C fructose to glucose appearance in humans. Diabetologia 36: 338-345.
90. SATYAVATI, G.V. 1988. Gum guggul (*Commiphora mukul*) - the success story of an ancient insight leading to a modern discovery. Indian J. Med. Res. 87: 327-335.
91. DUWIEJUA, M., ZEITLIN, I.J., WATERMAN, P.G., CHAPMAN, J. 1993. Anti-inflammatory activity of resins from some species of the plant family Burseraceae. Planta Medica 59: 12-16.
92. HAHN, D.H., ROONEY, L.W., EARP, C.F. 1984. Tannins and phenols of sorghum. Cereal Foods World 29: 776-779.
93. OHIGASHI, H., HUFFMAN, M.A., IZUTSU, D., KOSHIMIZU, K., KAWANAKA, M., SUGIYAMA, H., KIRBY, G.C., WARHURST, D.C., ALLEN, D., WRIGHT, C.W., PHILLIPSON, J.D., TIMON-DAVID, P., DELMAS, F., ELIAS, R., BALANSARD, G. 1994. Toward the chemical ecology of medicinal plant use in chimpanzees: the case of *Vernonia amygdalina*, a plant used by wild chimpanzees possibly for parasite-related diseases. J. Chem. Ecol. 20: 541-553.
94. PAULING, L. 1970. Evolution and the need for ascorbic acid. Proc. Natl. Acad. Sci. USA 67: 163-1648.
95. NATIONAL RESEARCH COUNCIL. 1989. Recommended Dietary Allowances: 10th ed. National Academy Press, Washington, D.C., p. 284.
96. EATON, S.B., KONNER, M. 1985. Paleolithic nutrition: a consideration of its nature and current implications. New Engl. J. Med. 312: 283-289.
97. ISAAC, G.L., CRADER, D. 1981. To what extent were early hominids carnivorous? an archaeological perspective. In: Omnivorous primates: gathering and hunting in human evolution. (R.S.O. Harding, G. Teleki, eds.), Columbia University Press, New York, pp. 37-103.
98. JOHNS, T., KUBO, I. 1988. A survey of traditional methods employed for the detoxification of plant foods. J. Ethnobiology 8: 81-129.
99. CARAGAY, A.B. 1992. Cancer-preventative foods and ingredients. Food Technol., April: 65-68.
100. TOMODA, M., SHIMIZU, N., GONDA, R., KANARI, M. 1989. Anticomplementary and hypoglycemic activity of okra and hibiscus mucilages. Carbohydrate Res. 190: 323-328.

101. KAMEL, M.S., OHTANI, K., KUROKAWA, T., ASSAF, M.H., EL-SHANAWANY, M.A., ALI, A.A., KASAI, R., ISHIBASHI, S., TANAKA, O. 1991. Studies on *Balanites aegyptiaca* fruits, an antidiabetic Egyptian folk medicine. Chem. Pharm. Bull. 39: 1229-1233.
102. FRATI, A.C., JIMENEZ, E., ARIZA, C.R. 1990. Hypoglycemic effect of *Opuntia ficus-indica* in non-insulin-dependent diabetes-mellitus patients. Phytotherapy Res. 4: 195-197.
103. OLSON, J.A. 1993. Vitamin A and carotenoids as antioxidants in a physiological context. J. Nutr. Sci. Vitaminol. 39: S57-S65.
104. OKUDA, T., YOSHIDA, T., HATANO, T. 1992. Antioxidant effects of tannins and related polyphenols. In: Phenolic Compounds in Food and Their Effects on Health II: Antioxidants and Cancer Prevention. (M.-T. Huang, C.-T. Ho, C.Y. Lee, eds.), American Chemical Society, ACS Symposium Series 507, Washington, DC, pp 87-97.

Chapter Nine

APPLICATIONS OF LIQUID CHROMATOGRAPHY-MASS SPECTROMETRY TO THE INVESTIGATION OF MEDICINAL PLANTS

Jean-Luc Wolfender and Kurt Hostettmann

Institut de Pharmacognosie et Phytochimie
Université de Lausanne
BEP, CH-1015 Lausanne, Switzerland

INTRODUCTION

Efficient detection and rapid characterization of natural products play an important role as analytical support in the work of phytochemists. The identification of a metabolite at the earliest stage of separation is a strategic element for

Phytochemistry of Medicinal Plants, edited by John T. Arnason et al.
Plenum Press, New York, 1995.

guiding an efficient and selective isolation procedure. Chromatographic analyses are used to "pilot" the preparative isolation of natural products (optimization of the experimental conditions, checking the different fractions throughout the separation), and to control the final purity of the isolated compounds. For chemotaxonomic purposes, the botanical relationships between different species can be shown by chromatographic comparison of their chemical composition. Comparison of chromatograms, used as fingerprints between authentic samples and unknowns, permits identification of drugs and/or the search for adulterated products. The selective detection of a given product in a complex mixture allows good quantitative measurement as well as precise chemotaxonomic comparison. Furthermore, in many applications it may be necessary not only to detect but also to identify compounds in extracts. With conventional detection methods, the identity of peaks can be confirmed only from their retention time and by comparison with authentic samples. In order to get more information on the metabolites of interest and detect them with satisfactory sensitivity and selectivity, a powerful and universal detection tool is needed.

A crude plant extract is a complex mixture which may contain hundreds or thousands of different metabolites.[1] The chemical nature of these constituents differs considerably between extracts, and the variability of the physicochemical, as well as the spectroscopic parameters of the compounds, causes numerous detection problems. Although different types of liquid chromatography (LC) detectors such as UV, RI, fluorescence, electrochemical, evaporative light scattering exist, none permits, within the same analysis, the detection of all the secondary metabolites encountered in a plant extract. Each method has its own selectivity. For example, a product having no important chromophore cannot be detected by a UV monitor. However, as every natural compound possesses a given molecular weight, mass spectrometry (MS) can be considered a "universal" detection technique.[2] At present, MS is the most sensitive method of molecular analysis. It has the potential to yield information on molecular weight as well as on structure. Furthermore, due to its high power of mass separation, very good selectivities can be obtained, a factor of utmost importance in trace analysis.

For volatile non-polar compounds from essential oils, MS detection used in combination with gas chromatography (GC) separation has been applied successfully.[3] Indeed GC-MS has become essential for applications in this field and has been used routinely for two decades as the method of choice.[4,5] The extensive use of computerized EI databases permits good on-line identification of the volatile constituents. MS detection of non-volatile polar molecules (which generally represent the main group of metabolites in a crude plant extract) is not simple. These metabolites normally require a chromatographic separation step such as high performance liquid chromatography (HPLC) which usually provides good separation efficiency.[6,7] On-line analysis requires the coupling of an LC technique to the mass spectrometer. This coupling is not straightforward since the normal operating conditions of a mass spectrometer (high vacuum, high temperatures, gas-phase operation, and low flow rates) are diametrically opposed to those used in HPLC, namely liquid-phase operation, high pressures, high flow rates, and relatively low temperatures.[8] Because of their basic incompatibilities, on-line coupling of these instrumental techniques has been difficult to achieve.

In this chapter, the role of LC-MS in phytochemical analyses will be illustrated by recent applications performed in our laboratory. They cover different uses of this new technique. The detection capability of LC-MS, notably for compounds having weak chromophores, will be demonstrated. The analysis of polar labile metabolites such as glycosides will be shown, and results obtained with two LC-MS interfaces will be compared. Finally, the usefulness of the on-line information obtained both by LC-MS and by photodiode array LC-UV for screening extracts will be emphasized. The role of these hyphenated techniques for the early recognition of metabolites in plant extracts is of great importance.

CHARACTERISTICS AND RANGE OF APPLICATION OF LC-MS INTERFACES

In LC-MS three general problems must be considered: the amount of column effluent that has to be introduced in the MS vacuum system; the composition of the eluent; and the type of compounds to be analyzed. To address these problems, a number of interfaces have been constructed over twenty years of research.[2] LC-MS interfaces must accomplish nebulization and vaporization of the liquid, ionization of the sample, removal of the excess solvent vapor, and extraction of the ions into the mass analyzer. Different techniques involve variations in the methods by which these steps are accomplished.

The goal here is not to discuss mechanisms and modes of action of all these different interfaces, since several reviews and books have been devoted to this aspect.[2,9] No real universal interface has yet been constructed, and each LC-MS interface has characteristics that will be dependent on the nature of the compounds being analyzed. They all present a certain selectivity of response that is dependent on their individual modes of ionization (EI, CI, FAB, ESP, etc.). Electron impact (EI), for example, will give a satisfactory LC-MS spectrum for a small stable molecule, while it will not be suitable for the analysis of a large polar and thermolabile molecule. In LC-MS, the same rules that govern the ionization of pure compounds in the direct insertion mode basically are preserved. Furthermore, each interface is compatible with certain types of LC-conditions (composition of eluent, flow rate), and most of them work with reversed HPLC systems. For all, the use of non-volatile buffers is forbidden.[2]

Despite these restrictions a number of LC-MS interfaces are suitable for the analysis of plant secondary metabolites. In our approach, used for the HPLC screening of crude plant extracts, two interfaces, thermospray (TSP) and continuous flow fast atom bombardment (CF-FAB), have been employed. They cover the ionization of relatively small non-polar products (aglycones, 200 amu) to high polar molecules (glycosides, 2000 amu).

Thermospray (TSP)

The TSP interface is one of the most popular.[10] It is capable of introducing liquid aqueous phase into the MS at a flow-rate compatible with that usually employed for standard HPLC conditions (organic solvents and water mixtures either in isocratic or gradient mode). TSP provides mass spectra that are closely related to those from chemical ionization (CI), and thus it is well suited for the analysis of moderately polar molecules in a mass range from 200 to 1000 amu or more.

In the thermospray system, the LC effluent is partially vaporized and nebulized directly in the heated vaporizer probe to produce a supersonic jet of vapor containing a mist of fine droplets or particles. As the droplets travel at high velocity through the heated ion source, they continue to vaporize. A portion of the vapor and ions produced escapes into the vacuum system through a sampling cone, and the remainder is pumped away by a mechanical vacuum pump.[9] Three ionization modes are available with this interface: chemical ionization (CI) initiated by a conventional electron bombardment using a heated filament ('filament-on' mode); CI with a discharge electrode ('discharge-on' mode); CI with ammonium acetate or some other volatile buffer in the mobile phase ('thermospray buffer' ionization); and ion evaporation. The latter mechanism may be operative in TSP buffer ionization mode as well.[2]

The chemistry of ionization plays an important role and must be taken into account when analyzing samples with TSP LC-MS. Usually, as for CI, the values of the proton affinity of substances entering the eluent must be considered. For the analysis of plant metabolites (150-1000 amu), TSP usually is operated in the positive ion mode using 'thermospray buffer' ionization. This affords spectra similar to desorption/chemical ionization (NH_3, positive ion mode).

In order to perform satisfactory TSP LC-MS of plant metabolites, many interdependent experimental parameters must be optimized: the mobile phase composition (amount and type of buffer and organic modifier); the solvent flow-rate; the temperature of the vaporizer and the ion source; and the geometry, position and potential of the repeller electrode. The tuning of these parameters is of utmost importance in obtaining good sensitivity and selectivity for any given type of substance. Several articles have emphasized the problems of thermospray optimization.[11]

Continuous Flow Fast Atom Bombardment (CF-FAB)

For compounds too polar, too labile or too non-volatile to be amenable to TSP LC-MS, CF-FAB LC-MS is a good alternative. The continuous flow FAB interface is simple, and was devised to minimize the drawbacks of standard FAB technique while retaining its inherent advantages.[12] Basically, in CF-FAB a direct insertion probe containing a fused silica capillary transfer tube is introduced directly into the MS ion source. The capillary terminates at the target of the probe, which (as in static FAB) is bombarded by xenon atoms having a translational energy of about 8kV. The sample solution flows through the capillary onto the target where ions are produced as a result of the bombardment process.[13] The samples are injected into a solvent mixture containing a non-volatile matrix such as glycerol (2-10%) at a flow rate of 5-10 μl/min. Due to constant replenishment of the liquid surface on the target and the need for minimal quantities of glycerol, CF-FAB is more sensitive than standard FAB and the spectra recorded are much cleaner. Due to the slow flow rate used, no additional pumping is needed. With this technique, LC separations can be achieved on capillary (syringe pumps), microbore, or standard HPLC columns with the use of an efficient splitter to cope with the restriction of flow rate imposed.

Setup Used for LC-MS and LC-UV Analysis

In order to keep LC-MS conditions close to those employed in normal HPLC (1ml/min, gradient with aqueous solvent systems) and with the aim of using the same columns without changing chromatographic conditions, the LC-UV and TSP LC-MS or LC-UV and CF-FAB LC-MS configurations shown in Fig. 1 are employed. A pump equipped with a gradient controller provides the eluent for the HPLC separation. A reversed-phase column (i.d. about 4 mm) is most often used. At the column outlet the eluent passes through a photodiode array detector equipped with a high pressure cell. At the exit of the UV detector, two configurations,

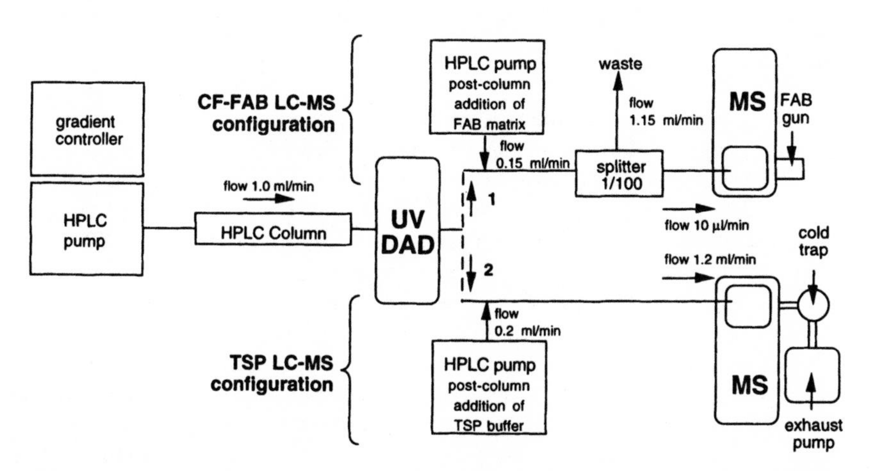

Figure 1. Schematic representation of the experimental setup used for LC-UV and TSP or CF-FAB LC-MS analysis. The mass spectrometer used is a quadrupole instrument (Finnigan MAT TSQ 700).

according to the LC MS mode chosen (TSP or CF-FAB), are possible. For TSP LC-MS operation, post-column addition of the buffer needed for 'TSP buffer' ionization is provided by an additional pump (usually 0.5M ammonium acetate at 0.2 ml/min). The total eluent (1.2 ml/min) containing the buffer passes through the TSP interface, and the exhaust eluent is pumped away by a mechanical pump and trapped in a cold trap. For CF-FAB operation, the additional pump allows the post-column addition of the glycerol matrix needed for FAB ionization (usually 50% glycerol, 0.2 ml/min). The viscous matrix is mixed with the eluent using a visco mixer. The mixture is then split with an accurate splitter (1/100), and 10 μl/min of the total eluent enters the CF-FAB interface through a fused silica capillary. Using such a setup, standard HPLC conditions for crude extract analysis (1ml/min, 4 mm i.d column) can be maintained without alteration in either TSP or CF-FAB LC-MS modes. DAD LC-UV detection is not affected by the buffer or the matrix used.

TSP LC-MS: AN EFFICIENT AND SELECTIVE DETECTION TOOL FOR PLANT METABOLITES

In many phytochemical analyses, detection is a problem. When compounds possess a strong chromophore, a direct LC-UV analysis is generally sufficient when a satisfactory LC resolution is obtained. When a compound of interest has only a weak chromophore and is hidden by other strongly UV-active compounds, its detection is not so straightforward. Often a long sample preparation procedure which includes prepurification and/or derivatization is needed. To avoid this problem, LC-MS may be the solution of choice due to its great selectivity and 'universality' of detection. Three different examples of TSP LC-MS use will emphasize this aspect.

Ginkgolides

Preparations containing *Ginkgo biloba* L. (Ginkgoaceae) leaf extracts have become a major market, with estimated annual sales of US$ 500 million worldwide.[14] The antagonistic activity of *G. biloba* to platelet aggregation induced by the platelet aggregation factor (PAF) is associated with the non-flavonoid fraction, composed of the diterpene ginkgolides (**1**, **2**, **4**, **5**) and of bilobalide (**3**).[15] The compounds are difficult to identify (low concentrations, weak chromophores), and often missed in standard HPLC analyses for phenolics. Hence methods for their determination in G. biloba preparations are urgently needed.

TSP mass spectra of ginkgolides revealed only a strong quasi-molecular ion $[M+NH_4^+]$ peak without other adduct species, as is the case with desorption/chemical ionization (D/CI, NH_3 positive-ion mode). No subsequent fragmentation was shown (**1**, Fig. 2). However, since ginkgolides are easily ionized by TSP, on-line TSP/LC-MS was demonstrated to be an efficient method to detect them (Fig. 2). Ginkgolides were not visible in the UV trace, but the total ion current (TIC) trace showed the presence of some of the compounds. The specific display of ion traces

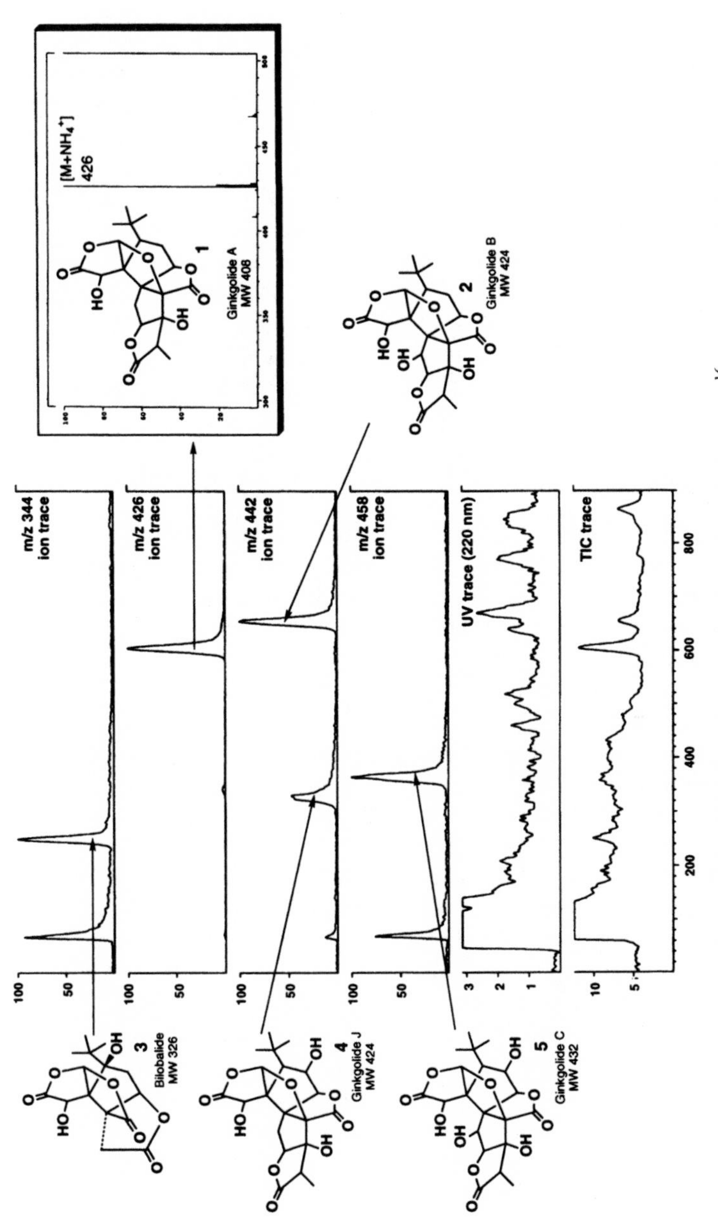

Figure 2. TSP LC-MS of Ginkgo biloba extract and TSP mass spectrum of gingkolide A (1)[16] . Reprinted from 40 with permission.

for masses corresponding to their respective pseudomolecular $[M+NH_4^+]$ ions allowed the identification and quantification of all ginkgolides in the extract.

Repetitive quantitative analyses showed good reproducibility (standard deviation of about 10%), as long as all ionization parameters (source block temperature, vaporizer temperature, aerosol temperature, pressure in the source) were well stabilized. Different temperatures for the source and aerosol were tried, and the best results were obtained with the block source temperature kept at 190°C and the vaporizer temperature fixed at 100°C. Ionization was induced by the use of ammonium acetate and the filament was operated at 600 V with a 0.200 mA emission current.[16] Using these parameters with isocratic elution (methanol-water, 40:60) it was possible to analyze different phytotherapeutic preparations containing *Ginkgo biloba* . Low levels of detection, in the range of 1 ng (*ca.* 0.3 pmol/l) were obtained. Even homeopathic preparations could be analyzed with this technique. The sensitivity of TSP LC-MS analysis of ginkgolides is comparable to that obtained with GC-MS.[17]

Artemisinin

The antimalarial compound artemisinin (6) is found in extracts of *Artemisia annua* L. (Asteraceae), a Chinese plant known locally as Qinghao. Artemisinin represents one of the remarkable success stories of antimalarial compounds from plants. This metabolite is a sesquiterpene lactone, too complex to be synthesized on a large scale. The only way to obtain the active principle is by isolation from dried plant material. To increase the yield of artemisinin, different genotypes of *A. annua* were cultured.

Unlike sesquiterpenes, sesquiterpene lactones such as artemisinin are not readily amenable to GC due to on-column thermal decomposition. Hydroxylated products have to be derivatized as trimethylsilyl derivatives to be analyzed by this method.[18] HPLC-UV determination in a crude plant extract is not straightforward because it lacks a suitable chromophore for detection with conventional UV detectors. As with ginkgolides, artemisinin also occurs together with phenolic components that interfere with its analysis. Consequently, TSP LC-MS represents a good alternative.

In contrast to ginkgolides (**1-5**), the TSP mass spectrum of artemisinin (**6**) shows different quasi-molecular peaks *i.e.*, at *m/z* 283 $[M+H]^+$, 300 $[M+NH_4^+]$ and 341 $[M+CH_3CN+NH_4^+]$ (Fig. 3). The base peak in this spectrum is the $[M+H]^+$ ion, as is the case for the D/CI spectrum (NH_3 or other reagent gas, positive-ion mode). Peaks resulting from the subsequent elimination of water (*m/z* 265) and HCOOH (*m/z* 237) are also visible in the TSP spectrum. Chromatographic analysis of the dichloromethane extract of the aerial parts of *A. annua* showed, even at lower wavelengths, a UV trace that was not suitable for artemisinin detection. Phenolic components were present, co-migrating with artemisinin, and interfering with its detection. However, observation of the TIC trace, together with selective display of the *m/z* 283 $[M+H]^+$ ion trace, permitted an efficient

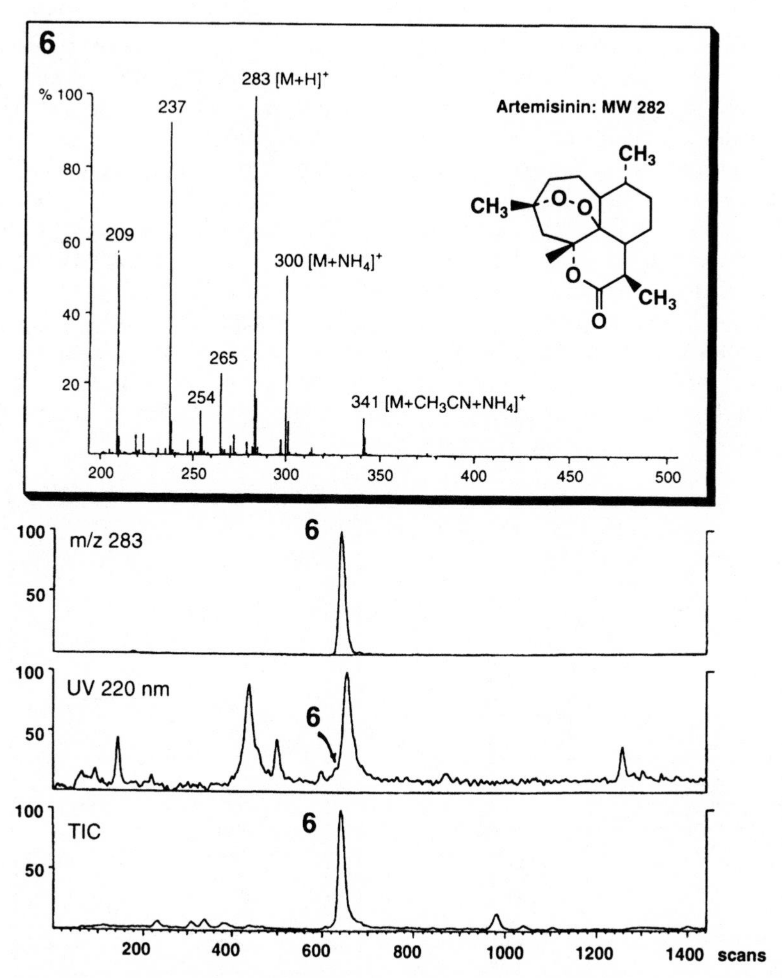

Figure 3. TSP LC-MS of the dichloromethane extract of the aerial parts of *Artemisia annua* (Asteraceae). Artemisinin (6) is shown by an arrow in the UV trace. Reprinted from 40 with permission.

detection (Fig. 3).[19] Using the same technique, arteether, an artemisinin derivative, was studied in rat liver microsome preparations.[20]

Chlorinated Sesquiterpene Lactones

In addition to being a powerful tool for the detection of plant metabolites having poor UV chromophores, TSP LC-MS provides qualitative information such as molecular weight and information on significant fragments. In this regard,

TSP LC-MS was used to demonstrate the artifactual formation of chlorinated sesquiterpene lactones in different lipophilic extracts of the neurotoxic thistle *Centaurea solstitialis* (Asteraceae).[21] This plant is known in central and northern California to cause a distinct neurotoxic disorder in horses. Sesquiterpene lactones have been suggested as possible toxins, and *C. solstitialis* is known to contain sesquiterpene lactone epoxides such as the highly neurotoxic repin (**7**). The presence of chlorinated derivatives also has been reported.[22] The possible artifactual formation of these latter compounds was investigated by analyzing $CHCl_3$ and toluene extracts of *C. solstitialis*.[21] TSP was performed in the positive ion mode using ammonium acetate buffer with the filament off mode. The presence of sesquiterpene lactone chlorohydrins was investigated by displaying the pseudomolecular ions of all possible chlorinated derivatives corresponding to the known sesquiterpene lactone epoxides. The TSP spectra of the compounds of interest (Fig. 4) were extremely easy to interpret under the conditions chosen. The

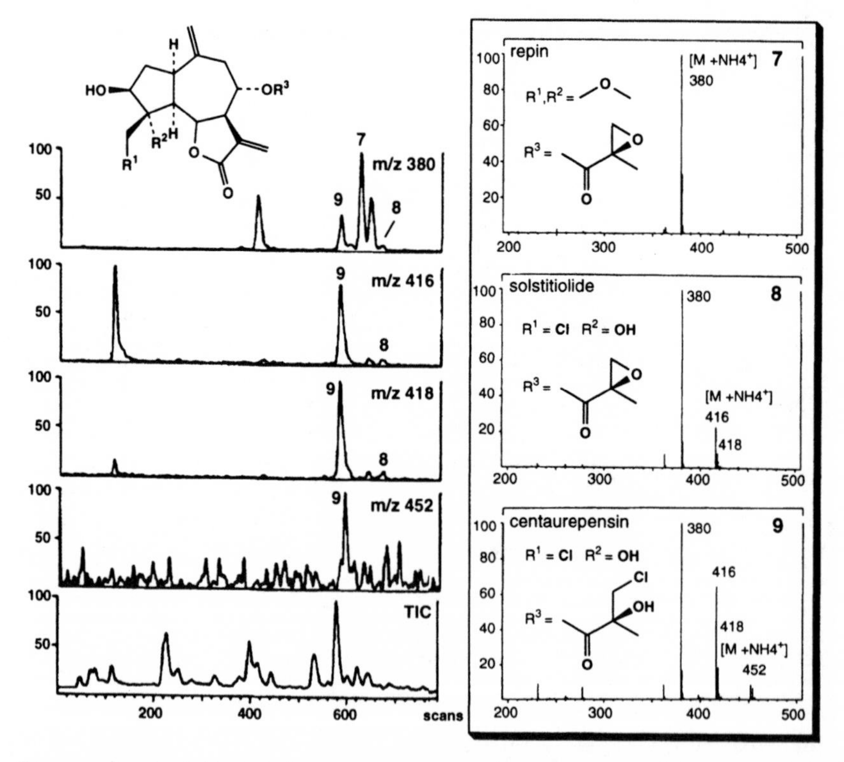

Figure 4. Search of chlorinated sesquiterpene lactones in crude extracts of *Centaurea solstitialis* (Asteraceae).[21] TSP LC-MS of an $CHCl_3$ extract standing in $CHCl_3$ for 6 days (r.t. daylight). Peak 7 corresponds to genuine repin, peak 8 to the monochlorohydrin solstitiolide and peak 9 to the dichlorohydrin centaurepensin. Reprinted from Ref. 40 with permission.

NH_4^+ adduct pseudomolecular ion was by far the most abundant signal. The $[M+H]^+$ ion was usually less than 10%, and fragment peaks were virtually absent. In the extract, monitoring of the $[M+NH_4]^+$ ions allowed rapid detection of the individual sesquiterpene lactones. In the case of chlorohydrin artifacts such as solstitiolide (**8**), the presence of a chlorine atom was detected (the appearance of two pseudomolecular ions separated by 2 amu and in an approximate ratio of 3:1, together with a base peak resulting from the loss of HCl). Dichloro-compounds such as centaurepensin (**9**) exhibited more complex spectra; ammonium adducts at M, M+2 and M+4 (100:71:14) corresponded well to the calculated abundance in dichloro-compounds (Fig. 4). The selective monitoring of the isotope peaks allowed a reliable localization and identification of the chlorine-containing molecules.[21] From this work, we deduced that chlorohydrins reported from yellow star thistle likely are artifacts formed through rapid reactions of the lactone epoxy ring with traces of HCl found in solvents such as $CHCl_3$ or CH_2Cl_2. This method was also applied successfully to the search for chlorinated valepotriates in *Valeriana* species.[23]

COMPLEMENTARITY OF TSP AND CF-FAB INTERFACES FOR THE LC-MS ANALYSIS OF GLYCOSIDES

Natural products often exist in the form of glycosides. These conjugates may occur together with their respective aglycones in the plants. Glycosides are thermally labile, polar and non-volatile compounds. Their direct mass spectral investigation requires soft ionization techniques such as desorption chemical ionization (D/CI) or fast atom bombardment (FAB), if information on molecular weights or sugar sequences is desired. Using LC-MS, comparable ionization is obtained with TSP or CF-FAB interfaces, respectively. Concerning D/CI (NH_3, positive ion mode) it can be stated that TSP (ammonium acetate, positive ion mode) allows the observation of the pseudomolecular ions of mono-, di- and, in certain cases, triglycosides.[24] CF-FAB, like FAB, permits the detection of larger glycosides and pseudomolecular ions of compounds bearing up to eight sugar units. Often TSP and CF-FAB provide complementary information. CF-FAB, for example, will produce important molecular weight information while fragments will be identified by TSP. For relatively small glycosides, the use of both techniques gives a confirmation of the nature of the pseudomolecular ions. TSP and CF-FAB (see setup in Fig. 1) have been used for the analysis of various glycosides. Examples of secoiridoid and triterpene glycosides will be discussed here.

Saponins

Saponins are glycosides that occur commonly in higher plants. They are biosynthesized by more than 500 species belonging to almost 80 different families. Their aglycone part is a triterpene group that usually has an oleanane, ursane or damarane skeleton or a steroid group. Monodesmosidic saponins (glycosylated in position 3 with a free α-carboxylic group in position 28) are known to exhibit important molluscidal activities.[25] Schistosomiasis or (bilharzia) affects millions of people living in African, Asian and South-American countries. The disease is linked with certain species of aquatic snails which serve the parasite as intermediate host. Natural products are of special importance for the control of schistosomiasis, since they are less expensive than synthetic compounds. In this respect a broad screening of snail-killing plants was undertaken. Several plants containing saponins have been studied.

A rapid assessment of the saponin content of a plant extract can be obtained by complementary TSP and CF-FAB LC-MS analyses. For example, this approach was used in the study of the molluscidal water extract of the fruit of *Swartzia madagascariensis* (Leguminosae), a common tree widespread in Africa.[26] The TSP LC-MS analysis was carried out in the positive ion mode. The extract showed the presence of triterpene glycosides (Fig. 5) probably derived from oleanolic acid (MW 456). Strong ionization of this aglycone moiety was observed for several peaks in the extract (**10, 11, 12**, Fig. 5a). The ions characteristic of this sapogenin were *m/z* 439 $[A+H-H_2O]^+$ and another adduct *m/z* 502 [A+ 46]+ (TSP spectra, Fig. 6a). Only very weak or no pseudomolecular ions at all of the corresponding saponins were recorded. For **10**, a distinctive ion at *m/z* 796 and a fragment at *m/z* 650 were characteristic for a saponin bearing a diglycosidic moiety consisting of a terminal deoxyhexose unit (-146 amu) and a glucuronic acid (-176 amu) moiety. As rhamnose is usually the only deoxyhexose known for saponins, it can be assumed from the on-line MS data that **10** is a saponin derived form oleanolic acid, substituted by a glucuronic unit and a rhamnose in the terminal position. The TSP spectra of saponins **11** and **12** were less clear. In both cases, characteristic signals for the oleanolic acid moiety were present, but only a very weak ion at *m/z* 941 was indicative of a tri- or higher glycosylation (TSP spectra, Fig. 6a). This ion also was barely detectable on the single ion current trace (Fig. 5a). For these metabolites, the TSP analysis alone could not provide adequate structural information. Thus, in order to confirm the information obtained by TSP and to get accurate MS data, the same extract was submitted to a second LC-MS analysis using the CF-FAB interface under the same LC conditions. CF-FAB analysis was carried out in the negative ion mode using glycerol as matrix (Fig. 5b). All the saponins found in the extract exhibited intense $[M-H]^-$ pseudomolecular ions and very weak ions characteristic for the aglycone moiety [A-H]-(*m/z* 455) and $[A-H-H_2O]^-$ (*m/z* 437). Furthermore, different characteristic cleavages were distinctive. For **10**, ions at *m/z* 777 $[M-H]^-$ *m/z* 631 $[M-H-146]^-$ and *m/z* 455 $[A-H]^-$ confirmed the results obtained with TSP. For **11**, the identity of the very weak *m/z* 941 recorded in the TSP spectrum was confirmed to be a protonated pseudomolecular ion $[M+H]^+$. An intense $[M-H]^-$ ion at *m/z* 939 was observed in the corresponding CF-FAB spectrum (Fig. 5b). The different fragments recorded in the CF-FAB spectrum of **11** (*m/z* 777 $[M-H-162]^-$ and *m/z* 793 $[M-H-146]^-$) confirmed that **11** is probably similar to **10** with one more hexose

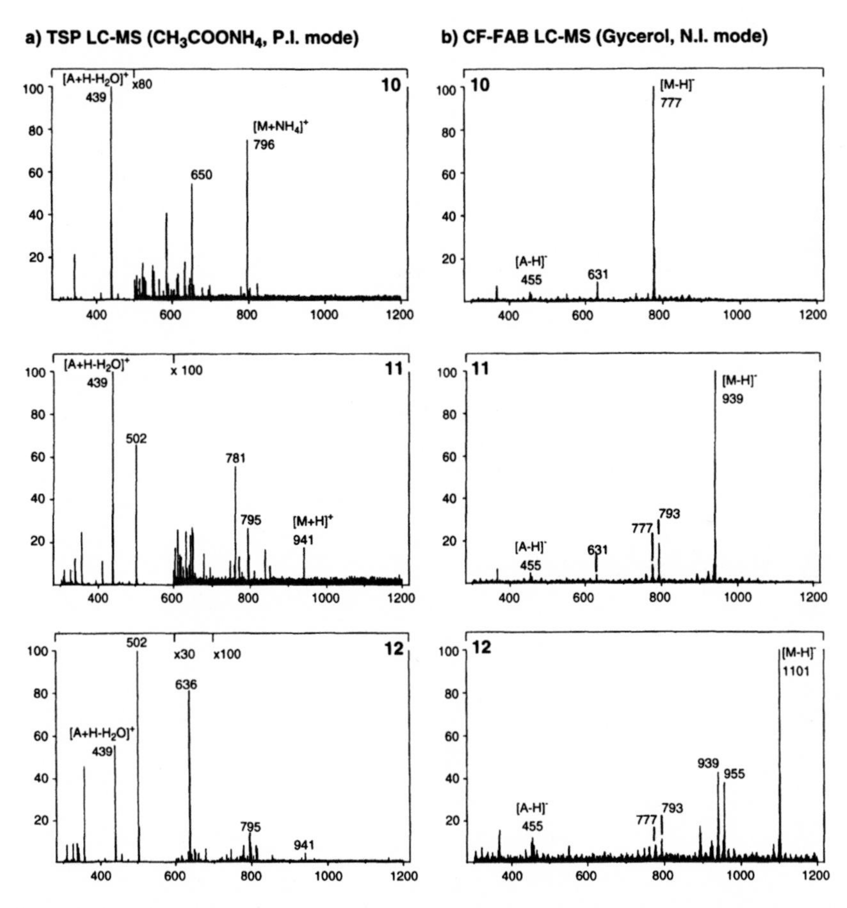

Figure 5. Combined TSP (a) (positive ion mode) and CF-FAB (b) (negative ion mode) LC-MS of the fruit water extract of *Swartzia madagacariensis* (Leguminosae).

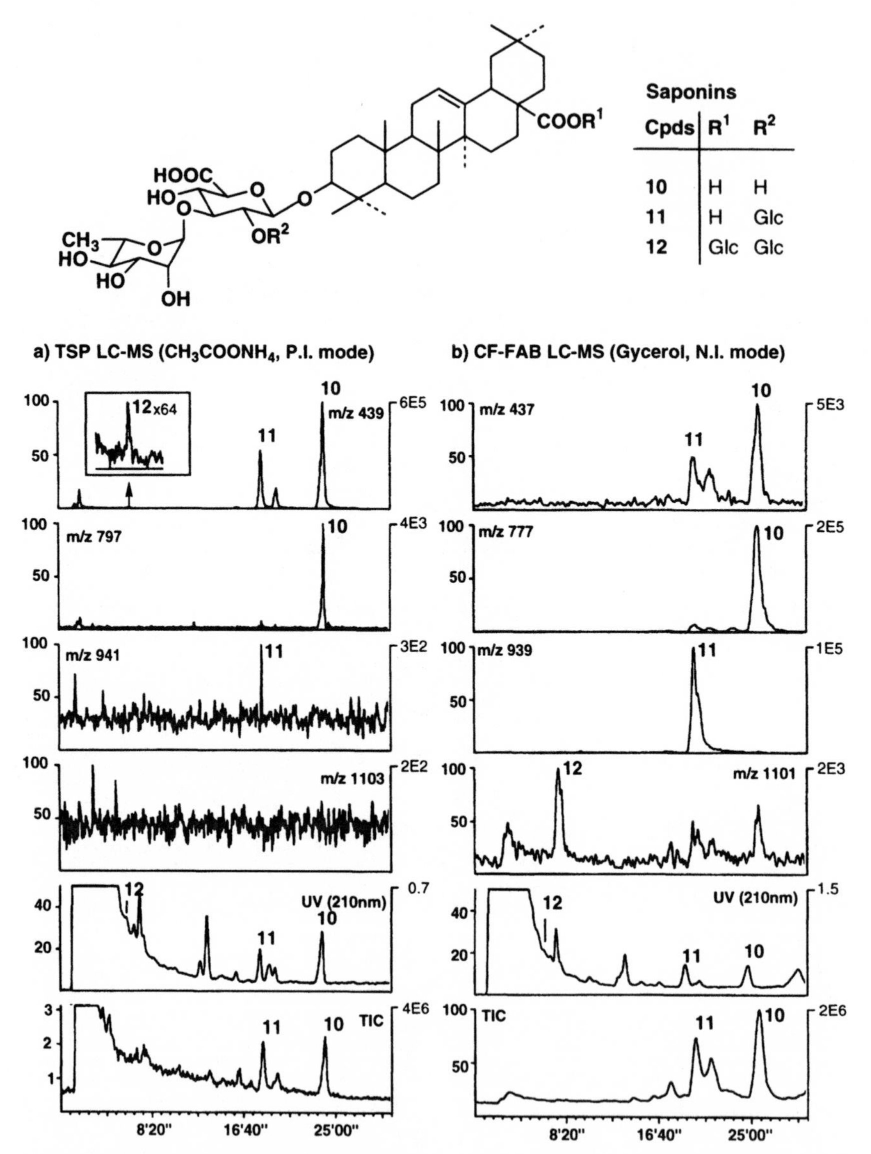

Figure 6. TSP and CF-FAB MS spectra of saponins 10, 11 and 12 from the fruit water extract of *Swartzia madagacariensis* (Leguminosae).

unit in position 28 or branched on the diglycoside moiety. The CF-FAB spectra of **12** exhibited an intense pseudomolecular ion at *m/z* 1101 $[M-H]^-$. This indicated that 12 has one more hexose unit (164 amu) than **11**. Saponin **12** was thus a tetraglycosylated triterpene. This was also confirmed by its high polarity (Fig. 5). The pseudomolecular ion $[M+H]^+$ of **12** was not detected during the TSP analysis.

This example of LC-MS shows well the complementarity of both TSP and CF-FAB ionizations. While TSP is sensitive and produces intense aglycone ions for the different saponins, permitting their assignment according to the nature of their sapogenin part, it fails to give the molecular weight information for tetraglycosylated triterpenes such as **12**. On the other hand, CF-FAB (soft ionization) produces intense pseudomolecular ions for saponins but in some cases the aglycone is more difficult to identify. This technique, due to the very low flow rate used (10 μl/min), fails to give sharp chromatographic peaks. Nevertheless, the use of CF-FAB is not necessary in all cases. For example, TSP LC-MS is a valuable method for the analysis of mono- and diglycosidic saponins, related to aridanin, in the fruit methanolic extract of another leguminous plant *Tetrapleura tetraptera*.[27]

Secoiridoid Glycosides

Secoiridoid glycosides are monoterpenes derived from secologanin. They represent the bitter principles of many plant families, and are especially common in the Cornales, Dipsacales and Gentianales orders. They may occur as esters with different acids, especially with biphenylcarbonic acids which enhance their bitter taste. Plants containing these compounds are used in many herbal preparations for treatment of stomach ailments.[28]

Monoterpenes, like triterpene glycosides, have been analyzed in LC-MS by both TSP and CF-FAB ionization. As secoiridoid glycosides occur mainly as monoglucosides, TSP is often the method of choice for their detection and identification. The bitter principles of various Gentianaceae species have been screened by this mode.[29,30] However, in certain cases CF-FAB LC-MS was found essential for the screening of unknown larger secoiridoids. For example, the methanolic extract of *Gentiana rhodentha*, a Chinese Gentianaceae species, was screened by TSP LC-MS and LC-UV using our routine procedure for the Gentianaceae (see xanthones below) in order to obtain rapid and precise information on its composition.[31]

HPLC-UV analysis showed four different peaks (**14, 15, 16, 17**) with retention times less then 10 min (Fig. 7). UV spectra showed only one maximum (around 240 nm) that was characteristic for a chromophore containing an α-β-unsaturated ketone function, and attributable to the secoiridoid glycoside, swertiamarin (**15**) (MW:374), kingiside (**16**) (MW:404), epi-kingiside (**17**) (MW:404) and sweroside (**14**) (MW:358). A predominant component (**18**) (rt. 9'40", MW:422) was identified as the xanthone C-glycoside mangiferin by comparison with an authentic sample. The slower running peaks (**13, 19, 20** and **21**) (rt. 20-25

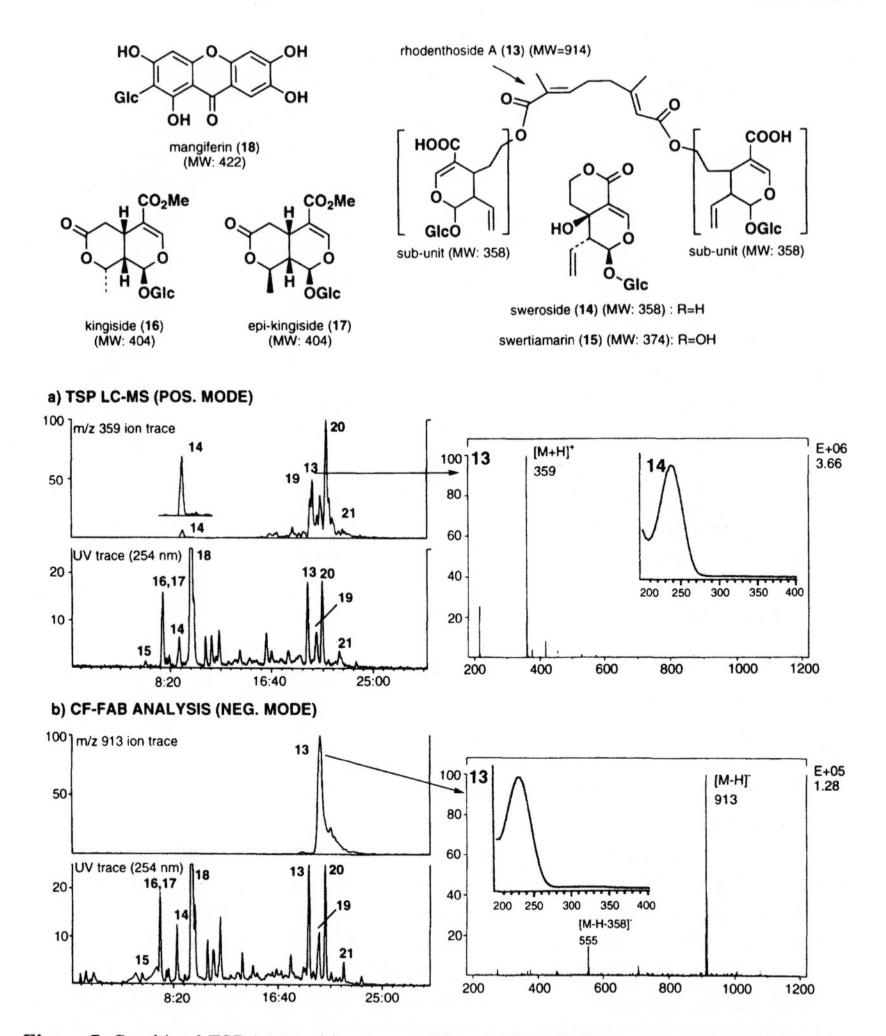

Figure 7. Combined TSP (a) (positive ion mode) and CF-FAB (b) (negative ion mode) LC-MS of the enriched BuOH fraction of the methanolic extract of *Gentiana rhodentha* (Gentianaceae).[32]

min) (Fig. 7b) also exhibited the same characteristic UV spectra of secoiridoids (**13** in Fig. 7b). These compounds, which were less polar than the common secoiridoids, were studied in more detail. A TSP LC-MS analysis with ammonium acetate as buffer (positive ion mode) was carried out (Fig. 7a). Under these conditions, the technique usually gives intense pseudomolecular $[M+H]^+$ or $[M+NH_4]^+$ ions.[29] The analysis revealed the presence of sweroside **14** (MW: 358) in trace amounts (tr. 9'50" min) (Fig. 7a). This compound exhibited a characteristic TSP-MS spectrum with an intense pseudomolecular ion $[M+H]^+$ at *m/z* 359. The display of the single ion trace *m/z* 359 allowed the specific assignment of peak **14**, but it also showed important signals corresponding to the less polar "secoiridoid-like" compounds, **13, 19, 20** and **21** (Fig. 7a). Indeed the TSP spectra recorded for all four were identical, and all exhibited an intense ion at 359 amu and no ion at higher masses. These first results from the crude extract of *G. rhodentha* showed that all compounds had the same UV and TSP-MS spectra as sweroside (**2**). Their chromatographic behavior, however, was quite different. In order to obtain complementary information on the constituents, a second LC-MS analysis with CF-FAB was done under the same HPLC conditions. The total ion current recorded for the whole chromatogram showed an important MS response for compounds **13, 19, 20** and **21**, while the more polar metabolites were only weakly ionized (Fig 7b). The CF-FAB spectrum of **13** recorded on-line exhibited an intense pseudomolecular ion $[M-H]^-$ at *m/z* 913 together with a weak ion at *m/z* 555 corresponding to the loss of a "sweroside like" unit $[M+H-358]^+$ (**13**, Fig. 7b). This complementary information indicated clearly that the molecular weight of **13** was 914 amu. For the same compound, only a fragment corresponding to a "sweroside like" unit *m/z* 359 was recorded during the TSP LC-MS analysis (**13,** Fig. 7a). According to the different results obtained in the HPLC screening, it was concluded that **13** was probably a type of moderately polar large secoiridoid containing at least one unit similar to sweroside (**14**). The CF-FAB spectra of compounds **19, 20** and **21** exhibited pseudomolecular ions $[M-H]^-$ at *m/z* 1271, 1629 and 1643 respectively. Thus, these compounds have even higher molecular weights than **13** and all should have at least a common sweroside-type unit.

Following the LC-MS screening results, a targeted isolation of **13, 19, 20** and **21** was undertaken, and full structure determination of **13** showed it to consist of two secoiridoid units linked together with a monoterpene unit through two ester groups[32] (Fig. 7). This compound is a new natural product. The structure determinations of **19, 20** and **21** are still in progress. This example demonstrates well the use of both LC-MS ionization techniques for targeting unknowns. TSP LC-MS, indicated that compounds **14** and **13, 19, 20** and **21** had common sub-units (identical fragments (MW 358), while CF-FAB allowed the on-line molecular weight determination of all these types of oligomers. The combined information thus allowed early recognition of this type of large secoiridoid glycosides.

COMBINED LC-UV AND LC-MS ANALYSIS AS A STRATEGIC SCREENING TOOL

When searching plant extracts for UV active compounds such as polyphenols, a multidimensional approach to their chromatographic analysis also is significant. By combining LC-MS[29] and LC-UV,[33] a large amount of preliminary information can be obtained about the constituents of an extract before isolation of the compounds. Indeed UV spectra recorded on-line give useful information (type of chromophore or pattern of substitution) complementary to that obtained with LC-MS. To illustrate this approach, examples of on-line LC-UV-MS analyses of xanthones and flavones in crude extracts are given.

Xanthones

Xanthones are dibenzo-g-pyrone derivatives found in a limited number of plant families. They are interesting because of their various pharmacological properties and especially because of their role as potent inhibitors of monoamine oxidase (MAO).[34] They occur as free aglycones as well as glycosides. O-Glycosides are found in two families only (Gentianaceae and Polygalaceae) while C-glycosides are widely distributed among angiosperms, ferns and fungi.[35] To obtain rapidly a precise idea of the xanthone composition of various Gentianaceae species, the combined use of LC-UV and TSP LC-MS has proved to be a method of choice.

Extracts of different *Chironia* (Gentianaceae) species from southern Africa have been screened. HPLC coupled to on-line UV diode array detection gave the traces shown for the dichloromethane and methanol extracts of the roots of *Chironia krebsii* (Fig. 8). Separation was achieved on RP-18 columns with an acetonitrile-water gradient system. Characteristic UV spectra for xanthones associated with peaks **22-39** were observed.[35] Peaks **40** and **41** corresponded to secoiridoid glycosides (similar spectra to those of **13** and **14**, Fig. 8). The identified compounds are shown[31] (Fig. 9).

As xanthones usually appear as aglycones or relatively small glycosides (mono- or diglycosides), the TSP interface was chosen for their ionization. After careful tuning of the TSP interface, the peaks recorded by UV detection (254 nm) gave a clearly discernible MS response in the total ion current trace (TIC) (Fig. 10). The TSP mass spectra of the xanthone aglycones recorded on-line after HPLC separation exhibited only $[M+H]^+$ ions as the main peak (**36** in Fig. 10). Corresponding mass spectra of the xanthone glycosides usually showed two weak ions due to $[M+H]^+$ and $[M+Na]^+$ adducts, and a main peak for the protonated aglycone moiety $[A+H]^+$ (Fig. 10). In the case of diglycosides, the successive losses of the monosaccharide units were marked by the corresponding peaks in the spectrum. Thus, for the diglycosides of *Chironia* species, a loss of 132 amu was first observed, corresponding to a pentose residue, followed by a loss of 162 amu, corresponding to a hexose moiety, leading to the aglycone ion $[A+H]^+$ (**35** in Fig. 10). The disaccharide was shown to be primeverose, consisting of a xylose moiety linked to C-6 of the glucose unit attached to the aglycone. The ion trace of *m/z* 657 allows the precise assignment of the peak to compound **35** in the extract. Similarly, the specific ion trace at *m/z* 363 corresponds to the HPLC peak of compound **36** (the ion is the main fragment of **36**) and also of compound **35**, a slower-running component of the extract. From the UV and TSP mass spectra of **35**, this compound was readily identified as the corresponding free aglycone of

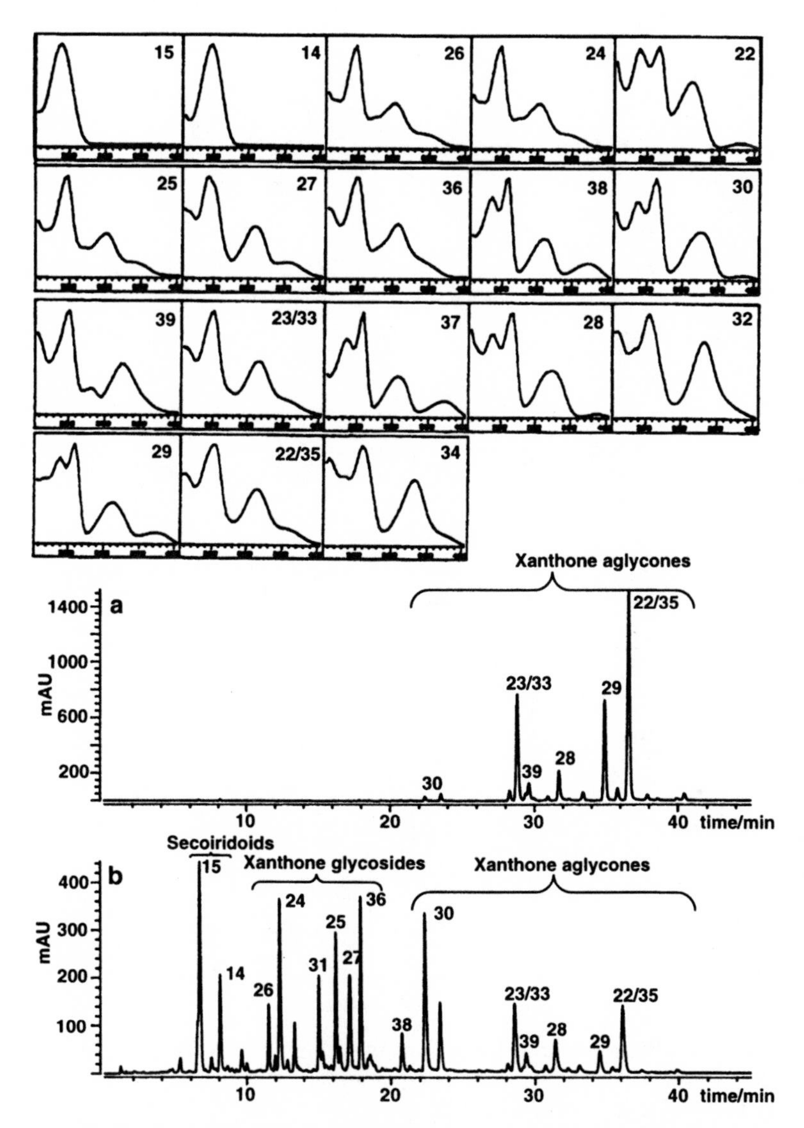

Figure 8. Comparison of the LC-UV trace of (a) the root dichloromethane and (b) the root methanolic extract of *Chironia krebsii* (Gentianaceae)[31]. The UV spectra were recorded from the chromatogram B. UV traces were recorded at 254 nm and UV spectra from 200-400 nm. For identities of peak see figure 9. Reprinted from Ref. 31 with permission.

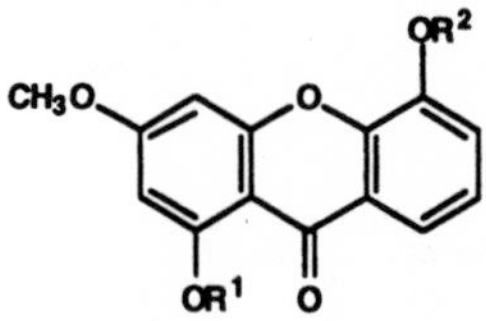

	R^1	R^2
22	H	CH_3
23	H	H
24	Glc	H
25	Prim	CH_3
26	Prim	H
27	H	Prim

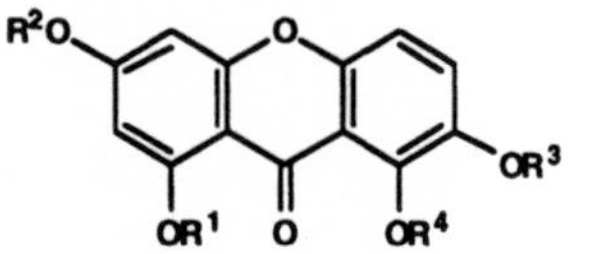

	R^1	R^2	R^3	R^4
28	H	CH_3	H	H
29	H	CH_3	CH_3	CH_3
30	H	H	H	H
31	Prim	CH_3	H	H

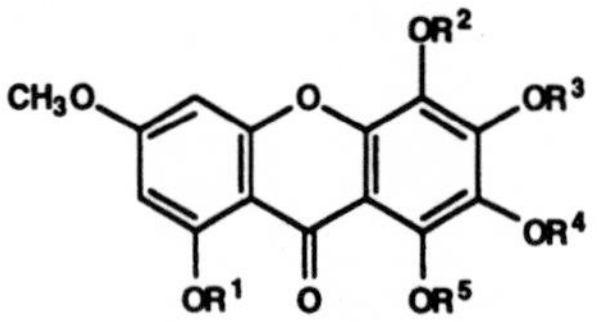

	R^1	R^2	R^3	R^4	R^5
32	H	CH_3	H	CH_3	H
33	H	CH_3	H	CH_3	CH_3
34	H	CH_3	CH_3	CH_3	H
35	H	CH_3	CH_3	CH_3	CH_3
36	Prim	CH_3	CH_3	CH_3	CH_3

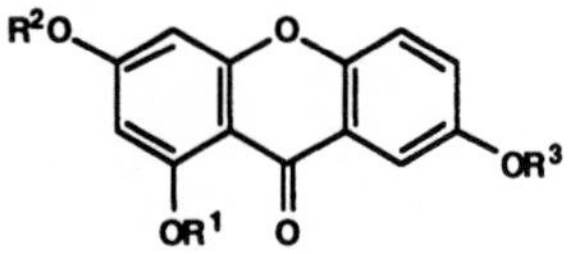

	R^1	R^2	R^3
37	H	CH_3	H
38	H	H	H

39

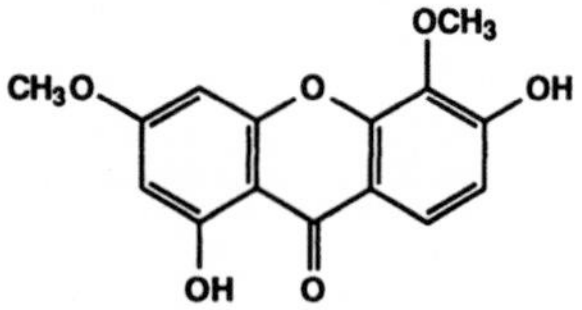

Secoiridoids:

14 Sweroside
15 Swertiamarin

Figure 9. Compounds isolated from *Chironia krebsii* (Gentianaceae).[31] Glc=glucose; Prim=primeverose [=β-D-xylopyranosyl-(1->6)-β-D-glucopyranoside.

36. Thus a simultaneous rapid identification of both aglycones and glycosides was made possible directly in the crude plant extract. In summary, the LC-UV and LC-TSP-MS results provide information on molecular mass, the number of methoxyl and hydroxyl groups, number of sugars and their sequence, and certain elementary information about the substitution patterns of the xanthones.

For more precise structural details concerning the positions of free hydroxyl groups on the xanthone nucleus, LC-UV with post-column addition of shift reagents is possible.[31,33,35] Reagents are added to the effluent leaving the HPLC column by means of a low-volume mixing tee. Diode array detection is then performed with the modified solvent. A weak base (sodium acetate) deprotonates only the more acidic phenolic groups, while a strong base (sodium methanolate,

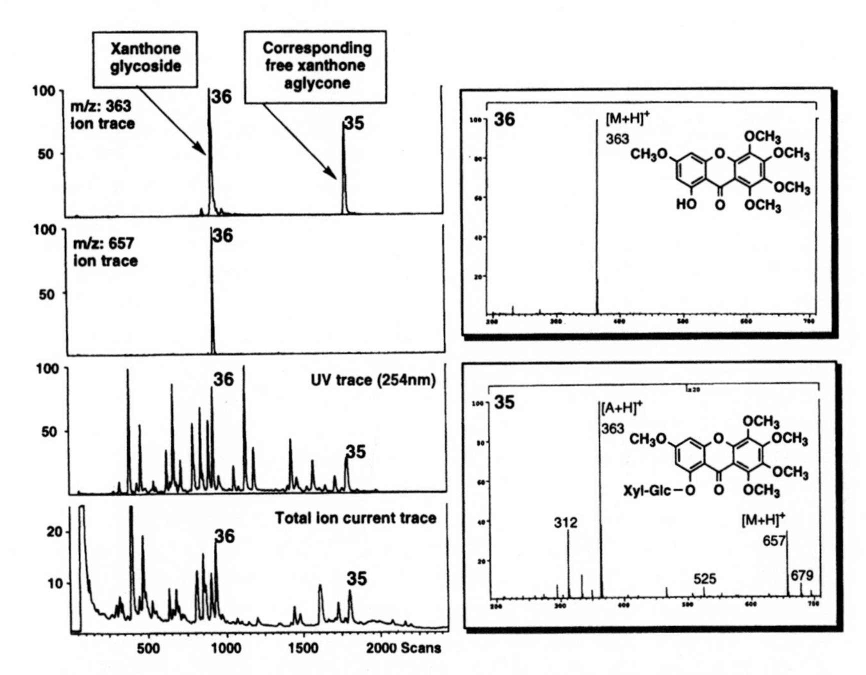

Figure 10. TSP LC-MS analysis of the root methanolic extract of *Chironia krebsii.* (Gentianaceae).[31] The ion trace m/z 657 displayed corresponds to the pseudomolecular ion $[M+H]^+$ of xanthone 36. The ion m/z 363 is the main fragment of 36 and corresponds to the pseudomolecular ion of its corresponding free aglycone 35 also present in the extract. Reprinted from Ref. 31 with permission.

potassium hydroxide) reacts with all phenolic groups. The shift reagents KOH, $AlCl_3$ and H_3BO_3 are compatible with the acidic acetonitrile-water eluent system. As disodium hydrogen phosphate is unsuitable as a weak base (due to solubility problems with the acetonitrile-water eluent), this was replaced by sodium acetate (0.5 M). However, this reagent is insufficient to deprotonate acidic phenolic groups in an aqueous solvent system and NaOH (0.01 M) was added via a second pump to obtain a pH of 8 for the solvent exiting the column. With the combination NaOAc-H_3BO_3 it is not always possible to locate *ortho*-dihydroxyl groups in the acetonitrile-water system. This problem was solved by using $AlCl_3$ in the acidified mobile phase (0.1% TFA, pH 3). Shifts were obtained similar to those with $AlCl_3$-HCl in methanol, the reagent used in the characterization of flavonoids. By comparing on-line UV spectra with the addition of $AlCl_3$ at pH 7 and pH 3, it was possible to detect labile *ortho*-dihydroxyl complexes.

The combination of LC-UV, TSP LC-MS and LC-UV with post-column derivatization is illustrated for the structure determination of three xanthone aglycones - **28, 29** and **30** in the LC-UV analysis of the methanol extract of the roots of *C. krebsii* [31] (Fig. 8). These xanthones had virtually identical UV spectra (Fig. 8), indicating that they probably have the same substitution pattern.[36] The four absorption maxima and the higher intensity of the second band were characteristic of 1,3,7,8-tetraoxygenated xanthones with a free hydroxyl group at position C-1.[36] On-line TSP mass spectra enabled determination of the molecular mass and the assignment of the number and type of substituents for the three compounds.[31] In the case of compound **28**, the $[M+H]^+$ ion (*m/z* 275) indicated a molecular weight of 274. If it is assumed, based on chemotaxonomic considerations, that simple xanthones are being dealt with, the difference between the molecular weight and the basic xanthone nucleus (*m/z* 196) gives the number and identity of substituents - in this case, 3 hydroxyl and 1 methoxyl groups. The UV spectrum of the xanthone corresponding to peak **30** (TSP MS: $[M+H]^+$ 261 -> 4xOH) was shifted on addition of NaOAc, indicating an acidic phenolic group at position C-3. A substantial shift recorded with $AlCl_3$ showed the presence of free hydroxyl groups at C-1 and C-8. Displacement of the maxima in the on-line UV spectrum, after addition of boric acid, confirmed the existence of an *ortho*-dihydroxyl group. The xanthone eluting as peak **28** (TSP MS: $[M+H]^+$ 275 -> 3xOH + 1xOMe) exhibited the same shifts as those recorded for **30**, except that the UV spectra remained unchanged after addition of NaOAc. Hence C-3 was substituted by a methoxyl group instead of a hydroxyl group, and this compound was 1,7,8-trihydroxy-3-methoxyxanthone (**28**). The existence of a chelated compound (**29**) with no free hydroxyl group in the xanthone corresponding to peak **29** (TSP MS: $[M+H]^+$ 303 -> 1xOH + 3xOMe) was indicated by a large decrease in band intensity and a small shift in the UV maxima on addition of KOH. This was confirmed by the shift measured with $AlCl_3$. The UV spectra remained unchanged with NaOAc and H_3BO_3, establishing the structure as 1-hydroxy-3,7,8-trimethoxyxanthone (**29**).

The same procedure was repeated for the other xanthones and all 18 compounds were isolated and their structures elucidated by normal spectroscopic techniques (Fig. 9). These were in good agreement with those deduced from on-line TSP LC-MS and LC-UV data. The same procedure was used to screen three other species of the genus *Chironia* and more than 40 xanthones were identified by this means.[37] This technique now is being used routinely in our laboratories to screen Gentianaceae extracts. A UV data base of all isolated xanthones also has been created, and a fully computerized UV matching of the known xanthones in crude extracts is now possible enabling early recognition.

Flavonoids

Flavonoids have long been recognized as one of the largest and most widespread classes of plant constituents, occurring in both higher and lower plants. MS data provide structural information on flavonoids and are used to determine molecular weights and to establish the distribution of substituents between the A- and B-rings. A careful study of fragmentation patterns can also be of particular value in the determination of the nature and site of attachment of the sugars in O- and C-glycosides.

For the LC-MS and LC-UV screening of extracts containing flavonoid aglycones or glycosides with a limited number of sugar units (not more than three), the same techniques as those described for xanthones have been used. The TSP LC-MS analysis of such polyphenols leads to a soft ionization of the flavonoids, providing only intense $[M+H]^+$ ions for the aglycones and weak $[M+H]^+$ ions of glycosides (mono- or disaccharide), together with intense fragment ions due to the loss of the saccharide units, leading to the aglycone moiety $[A+H]^+$ (Fig. 11). In xanthones, these spectra are often comparable to those obtained by D/CI with ammonia as reactant gas in the positive ion mode.[30]

As an example, TSP LC-MS was used successfully in combination with HPLC-UV and HPLC-UV with post-column addition of UV shift reagents[31,33] for the HPLC screening of 13 *Epilobium* species (Onagraceae) for major flavonol glycoside constituents.[38] The on-line LC-MS and LC-UV results allowed the rapid identification of only three types of flavonol aglycones in these extracts: kaempferol (MW 286); quercetin (MW 302); and myricetin (MW 312). The

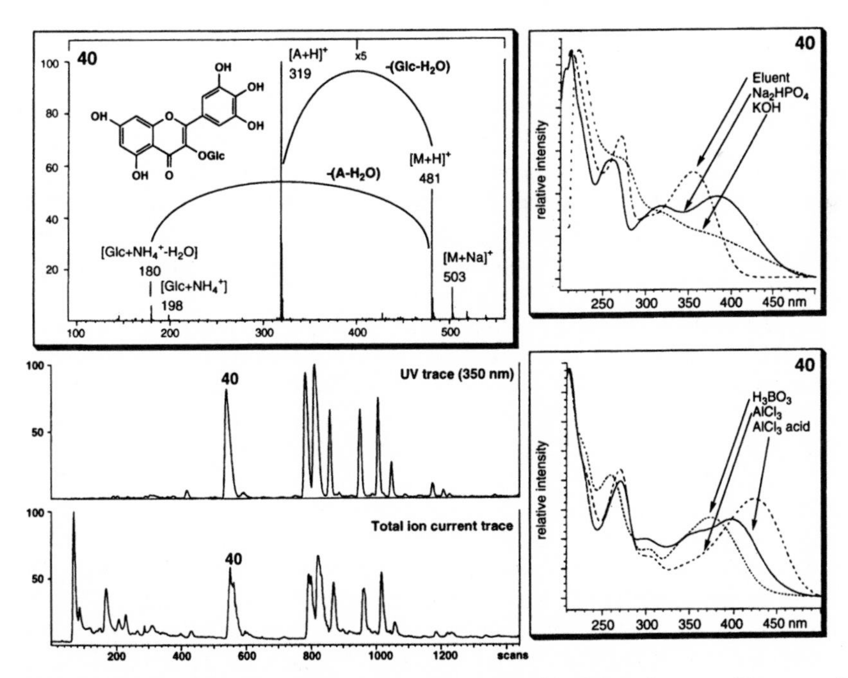

Figure 11. TSP LC-MS of the enriched methanolic extract of *Epilobium hirsutum* (Onagraceae), TSP mass spectra and on-line shifted UV spectra of isomyricitrin (40).[38]

on-line LC-UV shifted spectra showed clearly that all these metabolites were glycosylated only at position 3 (free OH on A and B ring). The TSP LC-MS spectra proved the presence of only O-monoglycosides with hexose (-162 amu), pentose (-132 amu), rhamnose (-146 amu) and glucuronic acid (-176 amu) moieties. The TSP-MS spectrum and the different UV shifted spectra obtained for isomyricitrin (**40**), together with the TIC and UV traces of the methanolic extract of Epilobium hirsutum, are presented (Fig. 11). The TSP-MS spectrum of **40** showed an intense aglycone ion $[A+H]^+$ *m/z* 319 and two distinctive pseudomolecular ions at *m/z* 481 $[M+H]^+$ and 503 $[M+Na]^+$. The loss of 162 amu between these peaks together with an ion at *m/z* 180 confirmed the presence of an hexose residue. The UV spectrum of **40** was characteristic for a flavonol. The UV shifted spectra clearly showed free hydroxyl groups at positions 5 and 7 on the A ring (shift of both $AlCl_3$ and $NaHPO_4$ spectra respectively) and three adjacent hydroxyl groups on the B ring (decomposition with KOH and shift with boric acid) (Fig. 11). This information allowed **40** to be identified as a 3-O-hexose derivative of myricetin. The nature of the hexose moiety was confirmed to be glucose, after hydrolysis of the isolated product, and **40** was confirmed to be isomyricitrin. Using the same approach, the partial on-line identification of 19 flavonols in 13 *Epilobium* extracts has been accomplished.[38]

CONCLUSION

The combination of chromatography with mass spectrometry offers the possibility of taking advantage of both chromatography as a separation method and mass spectrometry as an identification method. In the hands of a phytochemist, LC-MS can become a strategic analytical tool. Its intrinsic properties, chromatographic resolving power, high selectivity, sensitivity of detection and, capability of on-line identification allow its use for many different phytochemical applications.

As has been shown, LC-MS plays different roles in various phytochemical analyses. It can be used as a "universal" detection tool for LC, especially for compounds having weak chromophores, difficult to analyze by conventional LC-UV. Due to its ability to separate or filter ions according to their mass to charge ratio (*m/z*), MS provides high selectivities of detection. This selectivity, in addition to the resolution by chromatographic separation, permits a precise assignment or quantification of non-isomeric co-eluting compounds in extracts. LC-MS also provides important on-line MS information which is important in the early recognition of known compounds in an extract. In combination with LC-UV data, this technique avoids the time-consuming isolation of common natural products and permits the localization of interesting new ones. Hyphenated techniques can be integrated at different stages of the isolation process (extract or fraction analysis) and used to selectively follow metabolites of interest after different chromatographic steps. Under given LC conditions, the retention time, the UV and MS spectra of each peak, permit their precise assignment. This information is of great help when comparisons of LC chromatograms of different species are made for chemotaxonomic purposes.

Nevertheless, LC-MS has not become an ideal universal LC detector. Each interface presents its own restrictions of use and has a certain selectivity. The MS response is compound dependent and varies with proton affinity, polarity or melting points of the analytes as well as with the composition of the eluent and the parameters set for the interface.[39,40] In order to overcome these problems systematic studies of different sets of conditions must be undertaken for all types of constituents before their screening in crude plant extracts.

As TSP and CF-FAB are soft ionization techniques, MS results very often allow unambiguous determination of molecular weight, but fragmentation is either absent, insufficient, or insufficiently reproducible to allow definitive identification of known compounds or structure elucidation of unknown compounds. In order to compensate for this lack of on-line structural information, collision-induced dissociation (CID) can be performed on a tandem instrument in the MS-MS mode or by using another interface with electron impact (EI) capability such as particle beam (PB).

Despite these restrictions, it has been demonstrated that with the aid of TSP and CF-FAB LC-MS a rather broad range of plant metabolites can be analyzed successfully. Depending on the nature of the metabolite of interest one or the other or both techniques give molecular weight as well as some structural information. These coupled techniques allow a rapid screening and furnish useful structural information with a minute amount of material. A good estimate is obtained of the type of compounds present, and a targeted isolation is subsequently possible, saving time and costs. Although the use of LC-MS in phytochemistry currently is expensive and relatively new, we expect it to become an increasingly important tool for studying secondary metabolites in the complex mixtures encountered in most plant extracts.

REFERENCES

1. HAMBURGER, M., HOSTETTMANN, K. 1991. Bioactivity in plants: The link between phytochemistry and medicine. Phytochemistry 30: 3864-3874.
2. NIESSEN, W. M. A, VAN DER GREEF, J. 1992. Liquid chromatography-mass spectrometry. Principles and applications, Chromatogr. Sci. Ser. 58, Marcel Dekker Inc., New York. p. 479.
3. MASADA, Y. 1976. Analysis of Essential Oils by Gas Chromatography and Mass Spectrometry, John Whiley & Sons Inc., New York. p. 311.
4. KARASEK, F. W., CLEMENT, R. E. 1988. Basic Gas Chromatography-Mass Spectrometry, Elsevier, Amsterdam. p. 432.
5. DAVID, F., SANDRA, P. 1992. Capillary gas chromatography-spectroscopic techniques in natural product analysis. Phytochem. Anal. 3: 145-152.

6. SCHAUFELBERGER, D., HOSTETTMANN, K. 1987. High performance liquid chromatography analysis of secoiridoid and flavone glycosides in closely related *Gentiana* species. J. Chromatogr. 389: 450-455.
7. MARSTON, A., POTTERAT, O., HOSTETTMANN, K. 1988. Isolation of biologically active plant constituents by liquid chromatography. J. Chromatogr. 450: 3-11.
8. GARTEIZ, D. A., VESTAL, M. L. 1985. Thermospray LC/MS interface: principle and applications. LC Mag. 3: 334-346.
9. VERGEY, A. L., EDMONDS, C. G., LEWIS, I. A. S., VESTAL, M. L. 1990. Liquid Chromatography/Mass Spectrometry. Techniques and Application, Modern Anal., Chem. Ser., (D. Hercules., ed.), Plenum Press, New York. p. 306.
10. BLAKLEY, C. R., VESTAL, M. L. 1983. Thermospray interface for liquid chromatography/mass spectrometry. Anal. Chem. 55: 750-754.
11. HEEREMANS, C. E. M., VAN DER HOEVEN, R. A. M., NIESSEN, W. M. A., TJADEN, U. R., VAN DER GREEF, J. 1989. Development of optimisation strategies in thermospray liquid chromatography mass spectrometry. J. Chromatogr. 474: 149-162.
12. CAPRIOLI, R. M., TAN, F., COTRELL, J. S. 1986. Continuous-flow sample probe for fast atom bombardment mass spectrometry. Anal. Chem. 58: 949-2954.
13. CAPRIOLI, R. M. 1990. Continuous flow fast atom bombardment mass spectrometry. Anal. Chem. 62: 477A-485A.
14. COREY, E. J., KANG, M. C., DESAI, M. C., GOSH, A. K., HOUPIS, I. N. 1988. Total synthesis of (+)-ginkgolide B. J. Am. Chem. Soc. 110: 649-650.
15. BRAQUET, P., TOUQUI, L., SHEN, T. S., VARGAFTIG, B. B. 1987. Perspectives in platelet activating factor research. Pharmacol. Rev. 39: 97-210.
16. CAMPONOVO, F. F., WOLFENDER, J. L., MAILLARD, M.P., POTTERAT, O., HOSTETTMANN, K. 1995. ELSD and TSP-MS: two alternative methods for detection and quantitative LC determination of ginkgolides and bilobalide in *Ginkgo biloba* leaf extracts and phytopharmaceuticals. In press, Phytochem. Anal.
17. CHAURET, N., CARRIER, J., MANCHINI, M., NEUFELD, R., WEBER, M., ARCHAMBAULT, J. 1991. Gas chromatographic-mass spectrometric analysis of ginkgolides produced by Ginkgo biloba cell culture. J. Chromatogr., 588: 281-287.
18. FISCHER, N. H. 1991. Sesquiterpenoid Lactones. In Methods in Plant Biochemistry, vol. 7. , (B.V. Charlwood, D.V. Banthorpe, eds.), Academic Press, London. pp.187-211
19. MAILLARD, M. P., WOLFENDER, J. L., HOSTETTMANN, K. 1993. Use of liquid chromatography thermospray mass spectrometry in phytochemical analysis of crude plant extracts. J. Chromatogr. 647: 147-154.
20. BAKER, J. K., YARBER, R. H., HUFFORD, C. D., LEE, I. S., ELSOHLY, H. N., MCCHESNEY, J. D. 1988. Thermospray mass spectroscopy/high performance liquid chromatographic identification of the metabolites formed from arteether using rat liver microsome preparation. Biomed. Environ. Mass Spectrom. 18: 337-351.
21. HAMBURGER, M., WOLFENDER, J. L., HOSTETTMANN, K. 1993. Search for chlorinated lactones in the neurotoxic thistle *Centaurea solstitialis* by liquid chromatography-mass spectrometry, and model studies on their possible artifactual formation. Nat. Toxins 1: 315-327.
22. CASSADI, J. M., HOKANSON, G. C., 1978. 3 a,16a-Dihydroxytaraxene-3-acetate: a new triterpene from Centaurea solstitialis. Phytochemistry 17: 324-325.
23. FUZZATI, N., WOLFENDER, J. L., HOSTETTMANN, K., MSONTHI, J. D., MAVI, S., MOLLEYRES L. D. 1995. Isolation of antifungal valepotriates from *Valeriana capense* and search for valepotriates in other *Valeriana* species. Submitted, Planta Med.
24. WOLFENDER, J. L., MAILLARD, M., MARSTON, A., HOSTETTMANN, K. 1992. Mass spectrometry of underivatised naturally occurring glycosides. Phytochem. Anal. 3: 193-214.

25. MARSTON, A. and HOSTETTMANN, K. 1985. Plant molluscicides. Phytochemistry 24: 639-652.
26. BOREL, C., HOSTETTMANN, K. 1987. Molluscicidal saponins from *Swartzia madagascariensis* Desvaux. Hel. Chim. Acta 70: 570-576.
27. MAILLARD, M.P., HOSTETTMANN, K. 1993. Determination of saponins in crude plant extracts by liquid chromatography mass spectrometry. J. Chromatogr. 647: 137-146.
28. WAGNER, H. and MUENZIG-VASIRIAN, K. 1975. Ene Chemische Wertbestimmung der Enziandroge. Dtsch. Apoth. Ztg. 115: 1233-1239.
29. WOLFENDER, J. L., HAMBURGER, M., HOSTETTMANN, K. 1993. Search of bitter principles in Chironia species by LC-MS and isolation of a new secoiridoid diglycoside from *Chironia krebsii*. J. Nat. Prod. 56: 682-689.
30. WOLFENDER, J. L., MAILLARD, M., HOSTETTMANN, K. 1993. Liquid chromatographic-thermospray mass spectrometric analysis of crude plant extracts containing phenolic and terpene glycosides. J. Chromatogr. 647: 183-190.
31. WOLFENDER, J. L., HOSTETTMANN, K. 1993. Liquid chromatographic-UV detection and liquid chromatographic-thermospray mass spectrometric analysis of Chironia (Gentianaceae) species. J. Chromatogr. 647: 191-202
32. WEI-GUANG, M., FUZZATI, N., WOLFENDER, J. L., HOSTETTMANN, K., CHONG-REN, Y. Rhodenthoside A, a new type of acetylated secoiridoid glycoside from *Gentiana rhodentha*. Helv. Chim. Acta. 77: 1660-1671.
33. HOSTETTMANN, K., DOMON, B., SCHAUFELBERGER, D., HOSTETTMANN, M., 1984. On-line high performance liquid chromatography ultraviolet-visible spectroscopy of phenolic compounds in plant extracts using post-column derivatisation. J. Chromatogr. 283: 137-147.
34. SUZUKI O., KATSUMATA, Y., OYA, M., CHARI, V. M., KLAPFENBERGER, R, WAGNER, H., HOSTETTMANN, K. 1978. Inhibition of monoamine oxidase by isogenitsin and its 3-O-glucoside. Biochem. Pharmac. 27: 2075-2078.
35. HOSTETTMANN, K., HOSTETTMANN, M. 1989. Xanthones. In: Methods in Plant Biochemistry, Vol. 1 Plant Phenolics, (J. B. Harborne, ed.), Academic Press, London, pp. 493-508.
36. KALDAS, M. 1977. Identification des composÄs polyphÄnoliques dans *Gentianacampestris* L., *Gentiana germanica* Willd. et *Gentiana ramosa* Hegetschw. Thesis, University of Neuchëtel, Switzerland. p. 130.
37. WOLFENDER, J.L. 1993. Investigation phytochimique et analyse par chromatographie liquide-spectrometrie de masse de quatre especes du genre *Chironia* (Gentianaceae). Thesis, University of Lausanne, Switzerland. p. 342.
38. DUCREY, B., WOLFENDER, J. L., MARSTON, A., HOSTETTMANN, K. 1995. Analysis of flavonol glycosides of thirteen *Epilobium* species (Onagraceae) by LC-UV and thermospray LC-MS. Phytochemistry 38: 129-137.
39. ARPINO, P. J. 1992. Combined liquid chromatography mass spectrometry. Part III of thermospray. Mass Spectrom. Rev. 11: 3-40.
40. WOLFENDER, J. L., MAILLARD, M. P. AND HOSTETTMANN, K. (1994). Thermospray liquid chromatography-mass spectrometry in phytochemical analysis. Phytochem. Anal. 5: 153-182.

Chapter Ten

ROOT CULTURE AS A SOURCE OF SECONDARY METABOLITES OF ECONOMIC IMPORTANCE

Víctor M. Loyola-Vargas and María de Lourdes Miranda-Ham

Centro de Investigación Científica de Yucatán
Plant Biology Division
Apdo. Postal 87, 97310 Cordemex, Yuc. México

INTRODUCTION

Plants produce more than 80,000 different compounds through their secondary metabolic pathways. Some are used as pharmaceuticals, agrochemicals, dyes, flavors, pesticides, fragrances, etc., and represent multibillion dollar industries. Enormous amounts of plant material are needed for the extraction of these metabolites. Many of these compounds are obtained by direct extraction from plants that are cultivated in the field or sometimes growing in their original habitats. Several factors can alter the yield of products of economic importance. The quality of the raw material can vary widely, and some plants need to grow for several years before they are ready for harvesting. In addition, almost nothing is known about the control of pests and diseases of these plants, or of the

Phytochemistry of Medicinal Plants
Edited by John T. Arnason et al., Plenum Press, New York, 1995

postharvest procedures which are essential to preserve the active compounds until their extraction.

To overcome the inconvenience of manipulating plants or parts of them, a great effort has been directed towards obtaining *in vitro* systems to produce compounds. During the past three decades, plant cell cultures have developed into useful experimental tools to study the metabolic pathways of higher plants, in particular those leading to natural products. There are several examples of scaled-up processes based on plant cell cultures which have led to the commercial production of bioactive plant metabolites. Plant cell cultures also have permitted new insights into the physiology and biochemistry of plant secondary metabolism. For example, using plant cell cultures of *Thalictrum*, Zenk's laboratory elucidated the enzymology of berberine biosynthesis.[1] A similar approach led to the purification of several enzymes of the pathway of the terpenoid indole alkaloid biosynthesis in *Catharanthus roseus*.[2]

The above and other examples support the potential of plant cell cultures for basic and applied research. However, the potential of plant cell cultures for the production of secondary metabolites is limited still by two major factors. The first is the low productivity of the desired compounds. This problem may be partly overcome by manipulating the components of the culture medium,[3] altering the physical environment,[4] and screening for cell clones with high productivities.[5,6] The use of these strategies, alone or as a combination, has resulted in over forty examples of plant cell cultures, that are able to produce secondary metabolites at higher levels than in the whole plant. However, there are many cases in which the above strategies have failed. By and large, the problem of low productivity is still approached empirically. The second factor is the genetic instability of plant cell lines. This has been extensively documented. The instability, known as somaclonal variation, has been associated with phenomena ranging from aneuploidy and polyploidy to intra-chromosomal rearrangements,[7] and even single gene mutations. Only with a clear understanding of the underlying mechanisms that lead to the genetic instability of cultures will we be able to use somaclonal variation to our advantage.[8]

During the last decade, the need for cell organization for the biosynthesis of secondary metabolites in plant tissue cultures has been recognized to be fundamental.[9] In general, dedifferentiated plant cell cultures do not produce secondary metabolites, but will do so when they are induced to differentiate into tissues/organs. Some examples of this effect are the synthesis of tropane alkaloids in roots differentiated from callus of *Atropa belladonna*,[10] the production of morphinane alkaloids during somatic embryogenesis of *Papaver somniferum*,[11] the synthesis of indole alkaloids in root and shoot cultures of *C. roseus*,[12] the synthesis of the flavonoid naringin in shoot cultures of *Citrus paradisi*,[13] and the accumulation of cardenolides in somatic embryos of *Digitalis lanata*.[14]

The expression of secondary metabolic pathways in redifferentiated cell cultures is actually not surprising because it mimics exactly what the plant does. Furthermore, in most cases the synthesis of plant natural products is under strict temporal and spatial control. Taking this rationale as a working hypothesis, we believe that organized cultures can make a significant contribution to our understanding of secondary metabolism. Examples of this potential are already part of the plant tissue culture literature. This paper explores the use of root culture for production of phytochemicals.

ROOTS AS A SOURCE OF VALUABLE METABOLITES

The use of plant roots as a source of different compounds has long been common knowledge. Roots were used principally as medicines and food seasonings. Most of the active principles have been isolated during the past two centuries. This knowledge has provided a more rational exploitation of roots, since now they can be utilized for a reason completely different from their original use. Licorice root (*Glycyrrhiza glabra*) once used for its demulcent properties, now serves primarily as a flavoring in the food and tobacco industries.

Given the strong bias against the use of artificial chemicals to control pests, the roots of some species that contain insecticidal compounds currently are receiving more attention. Rotenoids from several tropical legumes (*Derris*, *Lonchocarpus*, *Tephrosia*, *Mundulea,* etc.), pellitorine (an heterogeneous mix of unsaturated isobutylamines) from the African Asteraceae *Anacyclus pyrethrum*, affinin from *Heliopsis longipes,* and thiophenes from *Tagetes patula* are some examples of these compounds.[15]

The commercial potential of dyes extracted from roots may increase in coming years, as some of the common synthetic dyes that represent health hazards are withdrawn from the market. The roots of madder (*Rubia tinctoria*) provided the red dye used for British military uniforms of yore. Bloodroot (*Sanguinaria canadensis*) has also been used as a dye, and more recently the alkaloid sanguinarine, derived from the root and rhizome, has found use as an antibacterial agent in toothpaste and mouthwash formulations.[16] Plant natural products also remain important in the fragrance industry. Essential oils are obtained from the roots of angelica (*Angelica archangelica*), *Saussurea lappa* (costus oil), and several species of the genus *Ferula*.[16] Many of the above mentioned compounds are costly, for they are not available in large quantities, e.g., the content of tropane alkaloids in the roots of *A. belladonna* is only 0.4 - 0.6%.[10]

The diversity of secondary metabolites found in roots shows the immense biochemical potential contained in this organ. The knowledge in this area, however, is mostly confined to accounts of empirical observations or results of chemical separations and analyses. Our understanding of the biosynthetic pathways and biological functions of root secondary metabolites is still meager. Some of the reasons for this lack of information may be: the underground location of roots; a perception of root functions as limited to a physical or mechanical role; difficulties in working in the complex chemical and biological milieu of soil; and lack of convenient *in vitro* experimental systems. As we shall discuss, however, a system applicable to the study of root secondary metabolism has been available for over half a century.

ROOT CULTURE AS A SOURCE OF CHEMICALS

In 1934 Phillip White, the pioneer of modern plant tissue culture, developed the first system that allowed indefinite proliferation of cells in isolation from the plant. It consisted of root tips which underwent elongation and branching in a medium with mineral nutrients, sugar, and vitamins. Root cultures became a standard experimental system in studies of inorganic nutrition, nitrogen metabolism, plant growth regulation, and root development.[17] The advent of rapidly growing, albeit disorganized, plant callus and suspension cultures in the late 1950's displaced the relatively slow growing root cultures as experimental systems and almost relegated them to the role of a historical curiosity.

The genetic and biochemical changes which occur when plant cells shift from being part of an organized whole to growing as a disorganized cell mass are largely unknown. However, one point is clear. Many pathways involved in secondary metabolism are developmentally regulated.[18] With a few notable exceptions, undifferentiated plant cultures rarely show the level or pattern of secondary metabolite production characteristic of their parent plants. Even in those cases where cell cultures can produce appreciable levels of compounds, their accumulation is usually incompatible with cell growth. Secondary metabolite production in these systems thus becomes a two-stage process with separate protocols for the growth and the production phases. Changes in culture performance as a result of somaclonal variation are of common occurrence in suspension cultures, and require periodic screening for the maintenance of desirable characteristics.

HAIRY ROOTS

In order to overcome the problem of slow growth of normal root cultures, a system that involves the generation of fast growing adventitious roots or hairy roots, which are the product of the infection of different tissues with *Agrobacterium rhizogenes,* has been developed. The induction of morphological changes is preceded by the stable integration of a portion of the Ri (root-inducing) plasmid into the plant genome, a characteristic it shares with the causal agent of the crown gall tumor disease, *A. tumefaciens*.[19] As far as it is known these hairy root cultures have the same metabolic features as normal root cultures, yet are as fast growing as suspension cultures.

Several protocols have been developed to carry out the transformation with different strains of *A. rhizogenes.* The infection can be done in a plant grown in a field as well as in aseptic plants grown *in vitro*, or on detached leaves, leaf discs or stem segments from greenhouse kept plants, after sterilization of the excised tissue. The scratching of the midrib of a leaf or the stem of a plantlet, with a needle of an hypodermic containing a thick bacterial suspension allows inoculation with small (about 5-10 μl) droplets of the suspension. An alternative approach for transformation is to cocultivate plant protoplasts with the bacteria (Fig. 1). Depending on the plant species, a profusion of roots may appear at the site of inoculation. In others, a tumor will form first and roots will then emerge.[20-22] In either case, hairy roots normally appear within 1 - 4 weeks (Fig. 2).

After the roots have emerged, they are separated from the infection site and placed in a semisolid medium with antibiotics (generally ampicillin or carbenicillin at a concentration of 1 mg ml^{-1}) in order to eliminate the bacteria from the transformed tissue. For most species, the establishment of axenic cultures in suspension is achieved by simply placing them in a liquid growth medium with sucrose as the energy source and without hormones. However, some species present difficulties for their establishment as independent cultures. Among the various techniques used to establish hairy root cultures, reduction of the medium strength has been very useful.[23-25] After optimization of the medium, roots can grow in culture at rates much faster than normal untransformed roots[26-32] (Fig. 3) (Table 1). This high growth rate is due mainly to the profusion of lateral roots. Both the growth rate and the extent of branching vary depending on the plant species and the culture conditions, such as ionic strength of the medium.[23-25,33]

Initial work by Flores' laboratory on *Hyoscyamus muticus* produced hairy root clones which grew significantly faster than normal root cultures or roots *in planta* while producing similar concentrations per unit dry weight of hyoscyamine. The growth parameters and alkaloid production of two selected clones have remained stable in culture for over eight years.[34,35] Similar results have been obtained in our laboratory with *D. stramonium* (6 years)[26,34,35] and *C. roseus* (4 years)[23] (Fig. 4). Apparently the fact that chromosome numbers are the same as those of the parent plants has led to this stable growth and alkaloid production.[36,37] However, some minor morphological changes have been found in one 15-month-old line of *Beta vulgaris*.[38]

While early studies on different aspects of the production of secondary metabolites by hairy root cultures were concentrated on members of the Solanaceae family, basic and applied research using this culture system now spans 26 families and 78 species of higher plants (Table 2).

FACTORS THAT AFFECT GROWTH AND SECONDARY METABOLITE YIELD

The growth of hairy root cultures is influenced by factors such as pH.[39-42] The initial lag phase of the growth cycle of cultures of *N. rustica*[39] and *D. stramonium*[41] was longer when the initial pH was raised from 5.8 to 7.0, and in the case of *N. rustica* it even affected the subsequent growth rate. In *Scopolia japonica*[40] and in *N. glauca*,[43] however, raising the pH to 7 stimulated growth.

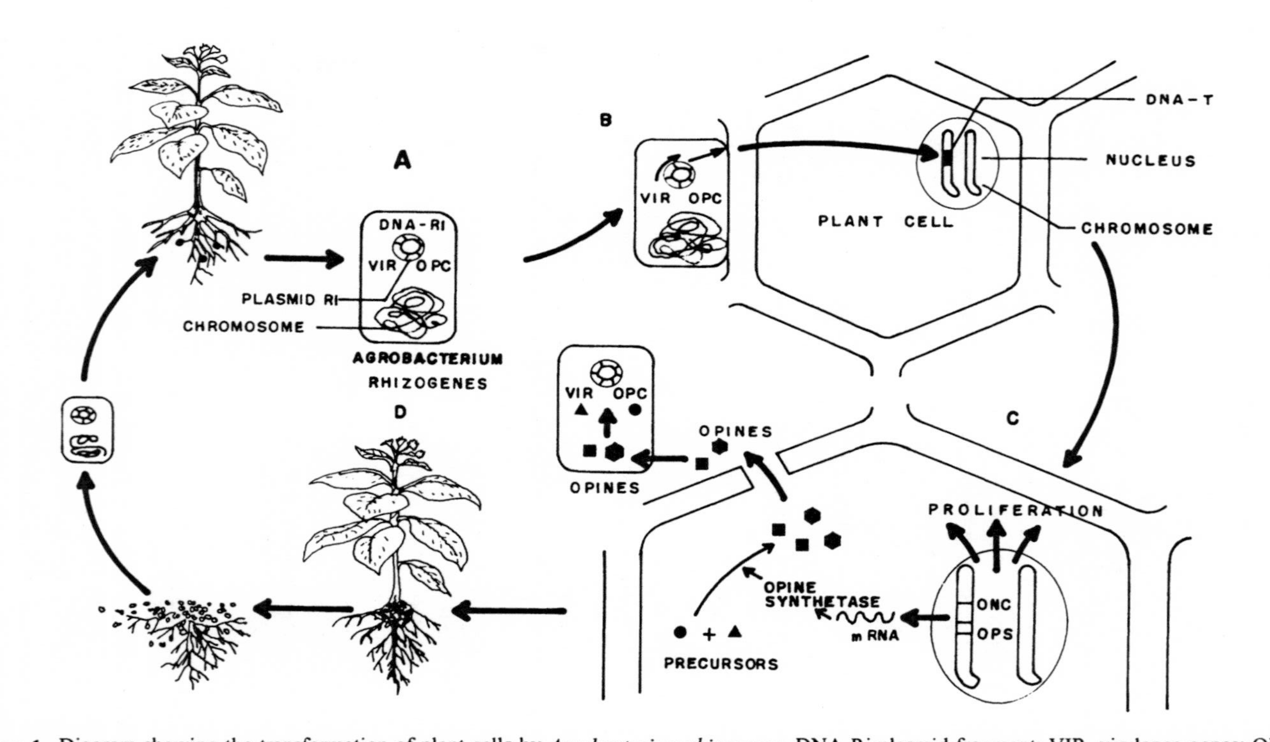

Figure 1. Diagram showing the transformation of plant cells by *Agrobacterium rhizogenes*. DNA-Ri plasmid fragment; VIR, virulence genes; OPC, genes for the catabolism of the opines; OPS, genes for the biosynthesis of the opines; ONC, oncogenes. (A) *Agrobacterium rhizogenes*; (B) infection and integration of the fragment of DNA of the bacteria to the plant genome; (C) opine synthesis; (D) formation of the hairy roots. Redrawn from Tempé and Schell.[125]

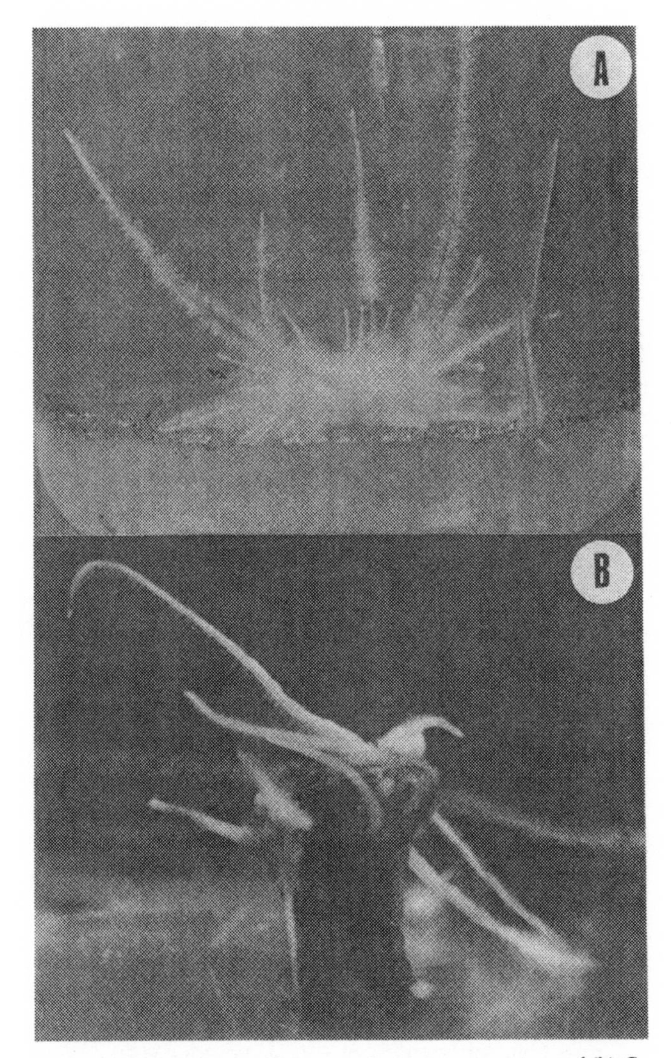

Figure 2. Photograph of the induction of hairy roots in (a) *D. stramonium* and (b) *C. roseus* hairy roots.

Hairy roots of *Tagetes patula* were grown in modified MS liquid medium at different initial pH, ranging from 4.0 to 7.0. After 12 days, the final pH of the spent medium was 4.5, irrespective of initial pH. The biomass yield was lowest when the initial pH was 4.0, but similar patterns of thiophene accumulation were observed regardless of the pH.[42] In contrast, in *C. roseus* hairy root cultures, growth was independent of the initial pH of the culture medium.[44]

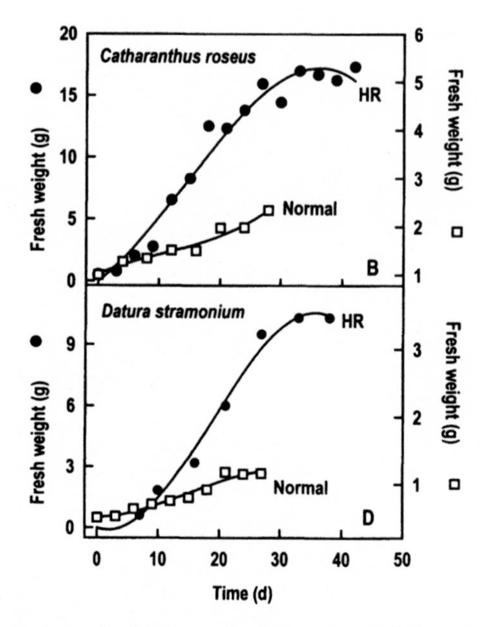

Figure 3. Time course of growth of hairy roots of *C. roseus* and *D. stramonium* compared to normal roots cultured *in vitro*.

The effects induced by different culture media,[40,43,45,46] as well as changes in the medium composition observed in cell suspension cultures,[3] have also been studied in hairy root cultures. In general, the change of one culture medium for another produces both important changes in growth patterns and the type of secondary metabolites produced by the root cultures.[43,47,48] The best results have been obtained using simple media or by reducing the ionic strength of the original medium.[23,49,50]

Various sugars and their concentrations affect growth patterns. The use of glucose, instead of sucrose, was deleterious not only for the growth of roots but also for the accumulation of hyoscyamine in *D. stramonium*.[41] In contrast, an increase in sucrose concentration (from 3 to 5%) produced an increase of 32% in hyoscyamine content, but reduced growth to 26%.[41] In another *D. stramonium* line the response to change in the sucrose concentration was even more marked, the hyoscyamine content increased eigthfold.[51] The effect of decreasing sucrose from 3 to 2% produced a tenfold decrease in scopolamine content in a hairy root line of *D. stramonium* which does not produce hyoscyamine.[52] In *C. roseus* hairy roots, an increase in sucrose concentration from 3 to 7% induced a 55% decrease in fresh weight, while the contents of ajmalicine and serpentine increased 215 and 530%, respectively.[50] Similar results have been obtained in *Rubia tinctorum*, in which 12% sucrose resulted in maximal growth and anthraquinone production.[53]

Table 1. Hairy root culture growth indices and duplication times in days(*)

Family	Genus and specie	Growth (times/days)	Reference
Apocynaceae	*Amsonia elliptica*	4	60
	Catharanthus roseus	28/17 52/40 1.6* 2.8* 3*	25 30 126 23 127
	Catharanthus trichophyllus	20/45, 25/66	128
	Rauwolfia serpentina	6.2/35	22
Araliaceae	*Panax ginseng*	3.07/21	59
Asteraceae	*Tagetes patula*	81/14	129
Campanulaceae	*Lobelia inflata*	20-60/28	48
Gentianaceae	*Swertia japonica*	150/56	29
Labiatae (Lamiaceae)	*Ajuga reptans*	230/45	101
Linaceae	*Linum flavum*	9*	21
Pedaliaceae	*Sesamum indicum*	33.3*	130
Polygonaceae	*Fagopyrum esculentum*	70/21	31
	Cinchona leddgeriana	6-8/28	131
Solanaceae	*Atropa belladonna*	60/28	32
	Datura candida X	20/28	132
	Datura stramonium	0.95* 55/28	26 41
	Duboisia leichhardtii	64/28	45
	Hyoscyamus albus	366/21	28
	Hyoscyamus niger	3.86/22	52
	Nicandra physaloides	28/26	133
Valerianaceae	*Valeriana officinalis*	112/50	24

When a two stage process was employed using sucrose in the first stage and fructose in the second one, there was an improvement in the volumetric yields of catharanthine of about twofold.[54] Changes in sucrose, phosphate, nitrate and ammonia concentrations produced contradictory results on growth and indole alkaloid production in *C. roseus* hairy root cultures.[55] However, when the content

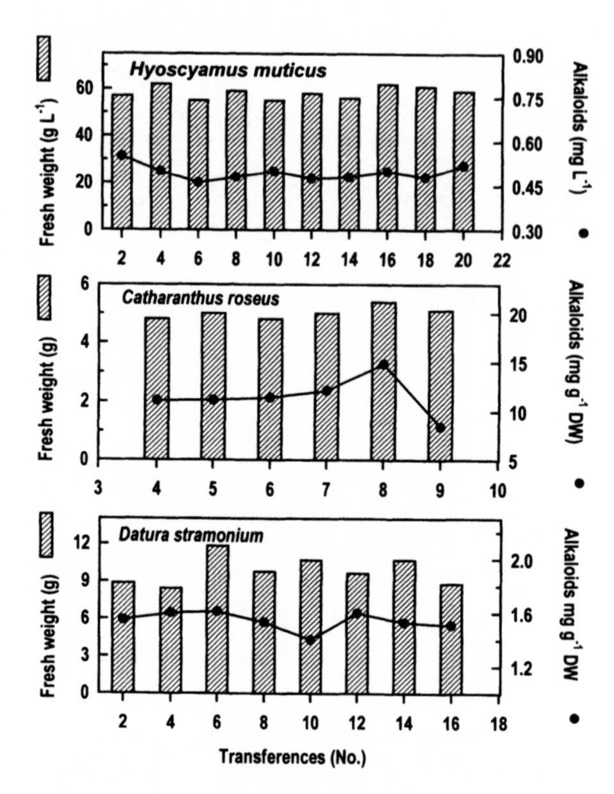

Figure 4. Stability of fresh weight (▨) and alkaloids (●) within plant tissue over time.The period of time elapsed between one transference and the next one is approximately one month. Data were redrawn from A[35] B[23] and C[26].

of individual alkaloids was analyzed, a pattern emerged. There was a preferential accumulation of ajmalicine and catharanthine when 4.5% sucrose was employed.[33]

A decrease of both phosphate and nitrate produced an increase in the synthesis of hyoscyamine in hairy roots of *D. stramonium*.[41] Similar results have been observed for the accumulation of ajmalicine and serpentine in *C. roseus* hairy roots.[50] Addition of 1% casein hydrolysate to the culture medium as a substitute for $NaNO_3$ increased threefold the content of scopolamine in *S. japonica* hairy roots.[40]

The addition of growth regulators, such as naphthaleneacetic acid (NAA), 2,4-dichlorophenoxyacetic acid (2, 4-D) and benzyladenine (BA), decreased the hyoscyamine content in *Duboisia myoporoides* hairy roots, while indoleacetic acid (IAA) and indolebutyric acid (IBA) did not affect it.[56] The exogenous

Table 2. Hairy root cultures generated for production of secondary metabolites

Family	Genus and specie	Major metabolite	Reference
Apocynaceae	*Amsonia elliptica*	Indole alkaloids	60
	Catharanthus roseus	Ajmalicine, serpentine, catharanthine, vindoline, vindolinine	23, 25, 30, 33, 44, 50, 55, 99, 126, 127, 134
	Catharanthus trichophyllus	Indole alkaloids	128, 135
	Rauwolfia serpentina	Ajmalicine, serpentine	22, 136
Araliaceae	*Panax ginseng*	Biotransformation, polyacetylenes, saponins	59, 137, 138
Asteraceae	*Acmella oppositifolia*	Polyacetylenes	81
	Artemisia absinthium	Volatile oils	102
	Ambrosia artemisiifolia	Thiarubrine A & B	120
	Ambrosia trifida	Thiophenes, thiarubrines	139
	Bidens alba	Polyacetylenes	140
	Chaenactis douglasii	Thiophenes, thiarubrines	61, 118
	Coreopsis tinctoria	Phenylpropanoids	141
	Echinacea purpurea	Alkalmides	142
	Echinops pappii	Thiophenes	143

application of gibberellic acid (GA_3), between 10^{-8} and 10^{-3} g L^{-1}, accelerated the increase in fresh weight of *Datura innoxia* hairy roots.[57] In other members of the *Solanaceae*, the highest content of tropane alkaloids was detected when hairy roots were cultured in the presence of kinetin and IAA (both at 1 mg L^{-1}) in the dark.[58] The addition of IBA (2 mg L^{-1}) to hairy roots of *Panax ginseng* produced an 84% increase in the accumulation of saponins. When kinetin was also added, this increase reached 179% (0.36 to 0.95% DW).[59] In *C. roseus* the addition of methyl jasmonate increased the content of ajmalicine and catharanthine by 57% and 108%, respectively.[33] The addition of NAA (0.5 mg L^{-1}) to hairy roots of *Amsonica elliptica* increased their growth in the light, whereas, when cultivated in the dark, there were important decreases.[60] In contrast, NAA increased growth in *Chaenactis douglassi*, but reduced the amount of thiarubrine almost to zero.[61] In *Lippia dulcis,* the use of NAA (1 mg L^{-1}) induced an increase of 100% in hernandulcin production.[62] In *Rubia tinctorum,* 5μM IAA produced the highest

Table 2. *Continued*

	Stevia rebaudiana	Steviol glucosides	144
	Tagetes erecta	Thiophenes	145
	Tagetes patula	Thiophenes	42, 63, 67-70, 84, 95, 129, 146-148
Boraginaceae	*Lithospermum erythrorhizon*	Shikonin	78, 95, 149
Campanulaceae	*Lobelia inflata*	Lobetyolinin, lobetyol, lobeline	48, 150, 151
	Lobelia sessilifolia	Lobetyolin	83
Chenopodiaceae	*Beta vulgaris*	Betalain	38, 152
Convolvulaceae	*Calystegia sepium*	Alkaloids	153
Cruciferae (Brassicaceae)	*Armoracia rusticana*	Peroxidases	79, 154
Curcurbitaceae	*Cucurbita pepo*	Peroxidases	155
	Trichosanthes kirilowii	Bioactive proteins	156
Gentianaceae	*Swertia japonica*	Phenyl glucosides, amarogentin, amaroswerin	29, 157
Geraniaceae	*Geranium thunbergii*	Tannins	49
Labiatae (Lamiaceae)	*Ajuga reptans*	Phytoecdysteroids	101
	Salvia miltiorrhiza	Diterpenes	158
Leguminosae	*Lotus corniculatus*	Tannins, phenylpropanoids	73, 93
	Trigonella foenum-graecum	Diosgenin	100

anthraquinone production and maximal growth of all conditions tested, while kinetin had no effect at all.[53] In *T. patula,* auxins could be playing a direct or indirect role in the regulation of thiophene levels in the root tips.[63]

USE OF ELICITORS IN ROOT CULTURE

Plant cells can undergo dramatic shifts in primary and secondary metabolism when exposed to plant pathogens or other stresses. This shift often involves the *de novo* induction of a metabolic pathway resulting in high concentrations of antimicrobials which are either undetectable or present at very low levels in the

Table 2. *Continued*

	Spartium junceum	Quinolizidine alkaloids	159
Linaceae	*Linum flavum*	5-mehoxypodophyl-lotoxin	21
Menispermaceae	*Stephania cepharantha*	Bisbenzylisoquinoline alkaloids	119
Papaveraceae	*Papaver somniferum*	Sanguinarine	160, 161
Pedaliaceae	*Sesamum indicum*	Anthraquinone, naphthoquinone	130
Polygonaceae	*Fagopyrum esculentum*	Flavanol	31
Rosaceae	*Sanguisorba officinalis*	Tannins	162
Rubiaceae	*Cephaelis ipecacuanha*	Emetic alkaloids	163
	Cinchona leddgeriana	Quinoline alkaloids	131
	Rubia tinctorum	Anthraquinone	53
Solanaceae	*Atropa belladonna*	Hyoscyamine, scopolamine	32, 92, 98, 113, 153, 164
	Datura candida X	Hyoscyamine, scopolamine	108, 132, 165
	Datura ferox	Hyoscyamine, scopolamine	164, 166
	Datura innoxia	Hyoscyamine, scopolamine	28, 57, 164, 166, 167
	Datura mentel	Hyoscyamine, scopolamine	164
	Datura stramonium	Hyoscyamine, scopolamine	26, 41, 51, 52, 76, 81, 95, 107, 110, 112, 113, 164, 166, 168- 170

unchallenged plant tissue. Compounds fulfilling these criteria have been termed phytoalexins. A strong correlation exists between the amount and timing of phytoalexin production and resistance to a plant pathogen.[64] The initial signaling event, known as elicitation, can be reproduced with extracts, such as autoclaved mycelial preparations derived from the cell walls of bacterial or fungal pathogens grown in axenic culture, or with hydrolytic enzymes, such as macerozyme and cellulase.[33]

Fungal elicitors have been used to study the biosynthesis of secondary metabolites or to increase the amount of these compounds in root cultures of Asteraceae,[65-70] Solanaceae,[71,72] Apocynaceae,[33] and Leguminosae.[73] Root cul-

Table 2. *Continued*

Datura wrightii	Hyoscyamine, scopolamine	166
Duboisia leichhardtii	Hyoscyamine, scopolamine	45, 96, 171-173
Duboisia myoporoides	Hyoscyamine, scopolamine	56
Hyoscyamus albus	Hyoscyamine, scopolamine, hyalbidone	28, 46, 47, 58, 116, 164, 166, 174-178
Hyoscyamus canariensis	Hyoscyamine, scopolamine	177
Hyoscyamus desertorum	Hyoscyamine, scopolamine	166
Hyoscyamus gyorffi	Hyoscyamine, scopolamine	177
Hyoscyamus muticus	Hyoscyamine, scopolamine	34, 35, 166, 167, 177
Hyoscyamus niger	Scopolamine, solavetivone	28, 34, 35, 52, 75, 109, 164, 177, 179
Hyoscyamus pusillus	Hyoscyamine, scopolamine	177
Nicandra physaloides	Hygrine	133
Nicotiana africana	Nicotine	124
Nicotiana cavicola	Nicotine	124
Nicotiana glauca	Anabasine, nicotine	43, 90
Nicotiana hesperis	Nicotine	124, 180

tures of *Bidens* elicited with mycelial extracts of *Phytophthora* spp. and *Phythium* spp. showed a dramatic increase in the concentration of a specific polyacetylene.[65,66] Treatment of *Hyoscyamus muticus* root cultures with a mycelial extract of *Rhizoctonia solani* resulted in the production and release into the surrounding medium of the sesquiterpenes lubimin and solivetivone.[72] The amount of compound produced represented nearly 0.5% of the root dry weight. The concentration of sesquiterpenes found in the elicited medium after two days matched that reported to cause a 25% inhibition of *Phytophthora infestans* germ tube growth.[74] This suggests that, at least in *H. muticus*, the induction and release of phytoalexins is physiologically significant in a host/pathogen interaction.

Table 2. *Continued*

	Nicotiana rustica	Nicotine, N'-ethyl-S-nornicotine	36, 38, 88, 94, 95, 124, 152, 181- 183
	Nicotiana tabacum	Nicotine, cadaverine, anabasine	34, 35, 91, 124, 159, 184
	Nicotiana umbratica	Nicotine	124
	Nicotiana velutina	Nicotine	124
	Scopolia carniolica	Hyoscyamine, scopolamine	164
	Scopolia japonica	Hyoscyamine, scopolamine	40
	Scopolia stramonifolia	Hyoscyamine, scopolamine	166
	Scopolia tangutica	Hyoscyamine, scopolamine	28
Umbelliferae (Apiaceae)	*Daucus carota*	Flavonoids	185
Valerianaceae	*Valeriana officinalis*	Valepotriates	24
Verbenaceae	*Lippia dulcis*	Hernandulcin	62
Zygophyllaceae	*Peganum harmala*	Serotonine	121
Ephedraceae (Gymnosperm)	*Ephedra gerardiana*	Ephedrine	20
	Ephedra minima	Ephedrine	20
	Ephedra minima hybrid	Ephedrine	20
	Ephodra saxatilis	Ephedrine	20

Another example using this strategy is that of *T. patula* which was treated with fungal, yeast or bacterial extracts. Thiophenes accumulated at a higher level (more than 200%) than in untreated control cultures.[67-70] The kinetics of thiophene formation varied in relation to the pathogen from which the elicitor was prepared, the elicitor concentration, and the duration of the exposure. A complementary approach also has been used. In *H. muticus*, the limitation of the original phosphate supply produced a fourfold increase in solivetivone. When fungal elicitors were then added, production was enhanced about 200-fold over nonelicited cultures.[75] The combination of phosphate limitation and fungal elicitation are therefore synergistic, which suggests that these factors are acting at different levels in the production of solivetivone.

Scaled-up experiments in a 10 L airlift reactor have shown that growing root mass is capable of responding to repeated elicitation, yielding consistent amounts of phytoalexins per unit of fresh weight as long as the medium is replenished.[72] The release of phytoalexins from an elicited root into the surrounding culture medium may mirror events which occur in a natural setting. This excretion of metabolites into the medium has an important practical application. Root cultures can be induced to increase this release by 50%, if compounds such as resin XAD-7 are used.[67] This fact suggests that there is a feedback inhibition process on steps of the biosynthesis of thiophenes.

In *C. roseus,* the use of an *Aspergillus* spp. homogenate increased the amount of ajmalicine and catharanthine by 66 and 19%, respectively.[33] However, the most important increases (94 and 41%) were found using 0.1% macerozyme and 1.0% cellulase.[33] These increases are preceded by an increase in tryptophan decarboxylase activity (Moreno-Valenzuela and Loyola-Vargas, unpublished).

As a result of the treatment of root cultures of *Lotus corniculatus* with glutathione, isoflavan phytoalexins accumulated in both the roots and culture medium.[73] The accumulation of these molecules was preceded by a transient increase in the activity of phenylalanine ammonia lyase. The addition of a low concentration of chitosan (0.2 -10 mg L^{-1}) enhanced the production of hernandulcin fivefold in root cultures of *Lippia dulcis*.[62] The biosynthesis of secondary metabolites in root tissue culture is induced not only by biotic stress but also by heavy metals, such as copper or cadmium. In *D. stramonium* the addition of those ions (1 mM) induced rapid accumulation of high levels of lubimin and 3-hydroxylubimin.[76] Copper has also been used to increase growth and alkaloid accumulation in *H. albus.*[46] Another abiotic elicitor, temperature, has been used to increase the alkaloid content in *D. stramonium*[51](Escalante, Maldonado-Mendoza and Loyola-Vargas, unpublished). Culturing roots at 20°C resulted in higher hyoscyamine contents than at 25 or 30°C.[51]

As a general rule, hairy root cultures do not excrete their secondary metabolites into the culture medium. However, under stress conditions, such as low pH[77] or in the presence of fungal homogenates, they may release large quantities of these compounds. The response depends on the plant species. Hairy root cultures of *Beta vulgaris* retain their betalain content inside the tissues, while *Nicotiana rustica* releases a fair amount of nicotine into the medium.[38] The excretion and even the production of certain compounds can be enhanced by the use of resins, such as XAD-2, XAD-4, XAD-7 or XAD-16, as in *Nicotiana glauca* and *Liyhospermum erythrorhizon*, where levels of production increased ten and thirteenfold, respectively.[43,78] Another strategy commonly used to induce the excretion of compounds is the use of ions, especially Ca^{2+} and Mg^{2+}, as has been demonstrated for the excretion of peroxidase from root cultures of horseradish (*Armoracia rusticana*).[79] In our laboratory we found that release of alkaloids, from root cultures of *D. stramonium* and *C. roseus,* and thiophenes, from root cultures of *T. erecta*, increased up to 75% in some cases, when the pH of the

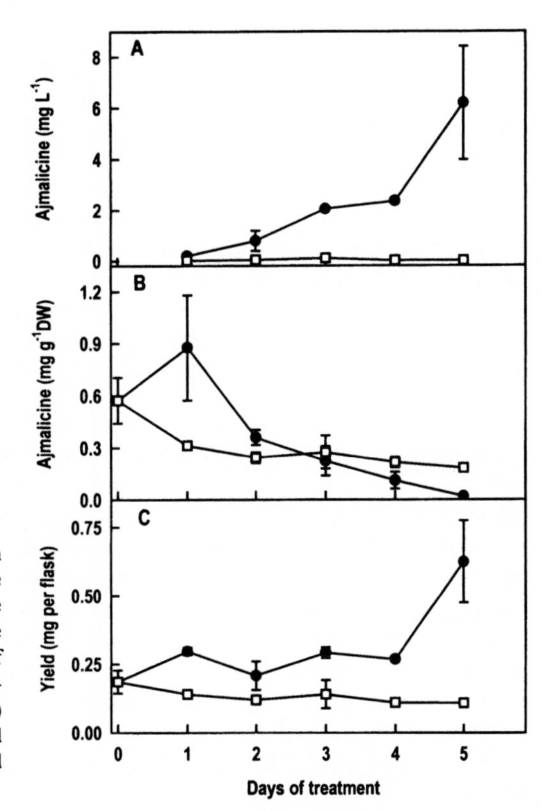

Figure 5. Effect of pH treatment on a line of hairy roots of *C. roseus*. The treatment consisted of lowering the pH of the medium to 3.5 when the roots have reached the 20th day of the culture cycle. (A) Alkaloid release into the culture medium; (B) alkaloids in tissue; (C) total alkaloid production. Control (□) and treated (●). Redrawn from Sáenz et al.[77]

culture media was reduced to 3.5[77](Fig. 5). The overall alkaloid content also was enhanced.

PHOTOSYNTHETIC ROOTS

Roots normally grow underground; therefore, root cultures are usually cultured in the dark. Some labs have found, however, that root cultures of certain species, such as *Bidens*, *Datura* and *Tagetes*, will green if grown under light. These green cultures retain the normal root anatomy. They show morphologically normal chloroplasts in the cortical cells, ribulose-1,5-biphosphate carboxylase activity, and light-dependent incorporation of CO_2, all typical of photosynthetic tissue.[65,80,81] The effect of photosynthesis on the secondary metabolism of an organ which normally lacks chloroplasts is not known.

Chloroplast-dependent reactions are known to be a vital part of certain secondary metabolic pathways,[82] and they could conceivably result in a novel pattern of compounds produced by roots. For instance, a *Datura stramonium* clone that under normal culture conditions (darkness and presence of sucrose) cannot produce scopolamine, when transferred to photoautotrophic conditions, produces scopolamine as a direct result.[81] In *Lobelia sessilifolia* only the greenish hairy roots, which produce chlorophylls a and b, showed better growth patterns and increased polyacetylene production.[83] Hairy root cultures of *Lippia dulcis* maintained in the dark do not produce detectable levels of hernandulcin or other mono- and sesquiterpenes. When placed under 16 h light day^{-1} treatment, the hairy roots acquire a light green coloration and produce several terpenes in addition to hernandulcin.[62] The effect of light also enhances the contents of alkaloids in *A. elliptica*[60] and oxygenated alkaloids (6ß-hydroxyhyoscyamine, 7ß-hydroxyhyoscyamine and scopolamine) in *Hyoscyamus albus*.[46,58] It, however, reduces drastically both growth and thiophene content in cultures of *T. patula*.[84]

MODIFICATION OF ROOT BIOSYNTHESIS

The genetic stability of hairy root clones, as opposed to cell suspensions, in which variability is common, may pose problems in attempts to modify the yield of a particular metabolite. Selection of high-yielding cells from suspension cultures, although labor intensive, has made possible dramatic increases in the production of shikonins.[85] Hairy root clones can be easily induced to form callus by hormonal treatment. Roots are readily regenerated upon withdrawal of hormones from the culture medium.[86] Suspension cultures obtained from root-derived callus show a high rate of somaclonal variation, which facilitates selection protocols.[36] Through screening or selection of suspension lines resistant to phytotoxic precursors and/or amino acid analogs, the biosynthetic potential of regenerated root tissue may be shifted to higher levels of metabolic production.[87] Any selected changes are likely to be permanent. For example, aneuploidy and associated phenotypic characteristics have been maintained stably in hairy root clones of *N. rustica* regenerated from variant cell lines.[36] This same technique has been used in transformed root cultures of *N. rustica*.[88] Among 100 regenerants, the nicotine content varied between 0.042 to 0.653 mg g^{-1}, compared to 0.160 mg g^{-1} in the original culture. Hairy roots regenerated from *H. muticus* suspension cultures resistant to p-fluorophenylalanine showed significantly higher levels of hyoscyamine than the parental hairy roots.[89] We are currently using this same approach, dedifferentiation and redifferentiation of hairy root cultures, to select 4- and 5-methyltryptophan resistant lines of *C. roseus* hairy roots.

A new approach, the introduction of "foreign" genes into hairy roots, has been employed to enhance the levels of rate-limiting intermediates. For example, the introduction of a bacterial lysine decarboxylase gene under the regulation of the 35S CaMV promoter in hairy roots of *N. glauca*[90] and *N. tabacum*[91] has resulted in a two and threefold increase of the anabasine alkaloids, respectively. This experiment also demonstrated that the corresponding bacterial enzymes can be used for bioconversion in plant tissues. Furthermore, this approach permitted cultures that were fed with lysine, even those with low lysine decarboxylase activity, to enhance drastically their levels of cadaverine and anabasine. The feeding of these amino acids to control or negative lysine decarboxylase lines showed little or no effect at all.[91] In another dramatic example, in *H. niger* the gene for hyoscyamine 6ß-hydroxylase was placed under the control of the 35S CaMV promoter and introduced to hyoscyamine-rich *A. belladonna* by a binary vector system using *A. rhizogenes*. The resulting hairy roots showed increased levels of hydroxylase activity and had a fivefold increase in hyoscyamine content compared to the wild-type hairy roots.[92]

In order to reduce the presence of undesirable compounds, another strategy has been used as in the case of tannins in *Lotus corniculatus*, where an antisense dehydroflavonol reductase (*dfr*) gene construct was made using the cDNA for *dfr* from *Antirrhinum majus*. This construct down-regulated the tannin biosynthesis in two of the recipient genotypes.[93] During stress, several metabolic pathways, in addition to the one that must be elicited, are activated simultaneously resulting in a waste of carbon skeletons and energy. Thus, the use of genetic engineering may be a suitable solution for species like *L. corniculatus*. We suggest that root clones showing higher stable production of secondary metabolites can be obtained using such an approach. This may be a major advantage over cell suspensions which show a frequent tendency to revert back from high to low producing cultures.

THE SCALING UP OF HAIRY ROOT CULTURES

Hairy root cultures present both advantages and disadvantages for scaling up. The production of nicotine from hairy roots of *N. rustica* cultivated *in vitro* in a two step system (batch/continuous flux) showed a fast growth rate, a final high root density, and the release of the majority of the nicotine produced into the culture medium.[94] Growth of hairy roots of *D. stramonium* in a column fermentor (60 cm tall), with an aspect ratio in the main compartment of about 5.7 and a total volume of 1.5 L, reached 1.12 Kg after 60 days with an initial inoculum of 1.7 g.[95] Nicotine was released into the medium. When the bioreactor was operated continuously, a fourfold stimulation in nicotine accumulation was obtained.[95] Using a 2 L airlift bioreactor and a 25 ml column packed with Amberlite XAD-2, scopolamine production was about five times higher in the column-combined reactor than in the one without it.[96] This can be explained in terms of the resin entrapping the alkaloid that is being released, and hence, possibly eliminating a feedback inhibition process.

Recently there has been a renewed interest in scaling up processes. Using *N. tabacum* as a model, several types of bioreactors have been tested: airlift, trickling film and mist. For all of them, the doubling time of the culture remained the same (approximately 2.8 days) as did the relative growth rate (24% day^{-1}).[97] Another system of bioreactors, the two-liquid phase bioreactor, also has been under study. It allows the induction of thiophene excretion (up to 70%) from *T. patula* hairy roots and shikonin from *L. erythrorhizon* (95%) into the organic phase.

Other examples of hairy root cultures that have been scaled up are: *A. belladonna,*[98] with a doubling time of 60 h in an airlift reactor; *C.roseus*[99]; *Trigonella foenum-graecum*[100] in a 9 L airlift fermentor; *Ajuga reptans,*[101] also in an airlift with a final yield of 230-fold after 45 days; and *Artemisia absinthium,*[102] in a 1 L column bioreactor for the production of volatile oils.

A prototype of a large-scale reactor has been developed to grow *H. muticus* hairy roots to high densities in a bubbling column of 9 L (Flores, personal communication). Rhodes and collaborators reported the scaling up of cultures of *D. stramonium* hairy roots to 14 L.[103] In our laboratory, a similar result has been obtained with the same species (Ayora-Talavera and Loyola-Vargas, unpublished). Recently, Furuya in the University of Kitasato (Japan) has successfully completed the scaling up of *P. ginseng* hairy roots in a culture tank of 20 tons with a yield of one ton of biomass. This would be (in terms of productivity) equivalent to 600 Ha cultivated with ginseng plants for a period of 5 to 7 years.[104]

BIOCHEMICAL STUDIES

Roots are organs where the biosynthesis of a large number of compounds takes place. These processes are difficult to study in whole plants due to the technical problems that studying such underground organs pose. Hence root cultures, normal as well as transformed, provide us with a good experimental system to begin to study different aspects of the biosynthesis of compounds. They are currently being used to elucidate enzymatic pathways, identify key intermediates, and explore the production of new root compounds.

Several enzymes have been determined and purified from root cultures. For example, hyoscyamine-6-ß-hydroxylase, a key enzyme in the biosynthesis of the tropane alkaloids, has been characterized in root cultures of *Hyoscyamus,*[105] along with other enzymes in this pathway: tropinone reductase from *D. stramonium,*[106,107] *D. candida* x *D. aurea,*[108] and *H. niger*;[109] acetyltransferase from *D. stramonium;*[110] diamine oxidase from *H. niger;*[111] putrescine N-methyltransferase from *D. candida* x *D. aurea,*[108] *D. stramonium,*[112] *A. belladonna,*[113] *N. tabacum*[114] and *H. albus.*[115-117]

Other pathways currently under study are: thiophenes in *C. douglasii,*[118] bisbenzylisoquinoline alkaloids in *Stephania cepharantha,*[119] thiarubrine A in *Ambrosia artemisiifolia,*[120] and serotonin and ß-carbolide alkaloids in *Peganum harmala.*[121] In our laboratory, efforts are directed to the study of indole alkaloid biosynthesis in hairy root cultures of *C. roseus*. We have purified tryptophan decarboxylase to homogenity,[122] and have partially purified strictosidine synthase (Muñoz, Moreno-Valenzuela and Loyola-Vargas, unpublished data), 3-hydroxymethylglutaryl CoA reductase (Galaz and Loyola-Vargas, unpublished results), and geraniol 10-hydroxylase (Canto and Loyola-Vargas, unpublished data). These data will provide us with a better understanding of the function of roots in the biosynthesis of secondary metabolites and will provide us with tools to manipulate the yield of these compounds. For the most part, however, the biosynthesis of root specific compounds remains an unexplored area.

FUTURE PROSPECTS

The recognition that the functions of plant roots involve more than water and nutrient uptake or, in some cases, the establishment of symbiotic associations, opens fascinating prospects in plant biology. Roots must also be reckoned with as the site of unique metabolic pathways and, in many cases, as major contributors to the makeup of secondary metabolites in the whole plant. The formation of biologically active metabolites by roots also has important implications for the study of plant/microbe interactions in the rhizosphere. The use of transformed root cultures presents many new opportunities for systems of commercial significance. Alternatives such as root immobilization and mist culture must be explored.

Hairy roots of tobacco have been shown to release a major portion of the total nicotine produced into the culture medium.[123,124] The ability to control this release of root metabolites into the surrounding medium has major implications for plant cell/organ culture scale-up. The use of macroreticular adsorbents[94] would further increase this release, making them suitable for industrial applications. The ability to grow root cultures from many plant species in isolation, and the ability to manipulate root metabolism will allow us to examine biosynthetic pathways specific to this organ, and eventually to isolate the relevant enzymes, leading us to the study of their corresponding genes and the control of their expression.

The use of photosynthetic root cultures must be further explored. They will surely contribute to the understanding of the biosynthetic processes that take place first in the root and are then continued in the leaves. We also will be able to address the biological significance of root chemicals, and relate the information to the whole plant. This basic information will provide the necessary background to manipulate predictably root biosynthetic potential in the whole plant as well as in scaled-up root cultures.

ACKNOWLEDGMENTS

The authors thank: Dr. Ingrid Olmsted for the revision of the English version of the manuscript; Dr. Rafael Durán for help with the classification of the plants; and Rosa M. Escobedo and Marcela Méndez for preparing the photographic material. The work of the authors has been supported by CONACYT, México, grants P2220CCOR903502, P122CCOT894672 and 0429N9108.

REFERENCES

1. ZENK, M. H. 1991. Chasing the enzymes of secondary metabolism: Plant cell cultures as a pot of gold. Phytochemistry 30: 3861-3863.
2. MEIJER, A. H., VERPOORTE, R., HOGE, J. H. C. 1993. Regulation of enzymes and genes involved in terpenoid indole alkaloid biosynthesis in *Catharanthus roseus*. J. Plant Res. 3: 145-164.
3. LOYOLA-VARGAS, V. M., MIRANDA-HAM, M. L. 1990. Aspects about the obtention of secondary metabolites from plant tissue culture. In: Production of Secondary Metabolites from Plant Tissue Cultures and its Biotechnological Perspectives.(V. M. Loyola-Vargas, ed.), CICY, Merida, Yucatan. pp. 31-79.
4. KOOIJMAN, R., DE WILDT, P., VAN DEN BRIEL, W., TAN, S., MUSGRAVE, A., VAN DEN ENDE, H. 1990. Cyclic AMP is one of the intracellular signals during the mating of *Chlamydomonas eugametos*. Planta. 181: 529-537.
5. FUJITA, Y., TAKAHASHI, S., YAMADA, Y. 1985. Selection of cell lines with high productivity of shikonin derivatives by protoplast culture of *Lithospermum erythrorhizon* cells. Agric. Biol. Chem. 49: 1755-1759.
6. SATO, F., YAMADA, Y. 1984. High berberine-producing cultures of *Coptis japonica* cells. Phytochemistry 23: 281-285.
7. BAIZA, A. M. 1990. Cytogenetics and transformed hairy root cultures. In: Production of Secondary Metabolites from Plant Tissue Cultures and its Biotechnological Perspectives.(V. M. Loyola-Vargas, ed.), CICY, Merida, Yucatan. pp. 244-261.
8. DOUGALL, D. K. 1990. Somaclonal variation as a tool for the isolation of elite cell lines to produce secondary metabolites. In: Production of Secondary Metabolites from Plant Tissue Cultures and its Biotechnological Perspectives.(V.M. Loyola-Vargas, ed.), CICY, Merida, Yucatan. pp. 122-137.
9. LINDSEY, K. , YEOMAN, M. M. 1983. The relationship between growth rate, differentiation and alkaloid accumulation in cell cultures. J. Exp. Bot. 34: 1055-1065.
10. COLLINGE, M. A., YEOMAN, M. M. 1986. The relationship between tropane alkaloid production and structural differentiation in plant cell cultures of *Atropa belladonna* and *Hyoscyamus muticus*. In: Secondary metabolism in plant cell cultures. (P. Morris, A. H. Scragg, A. Stafford, M. W. Fowler, eds.), Cambridge University Press, Cambridge. pp. 82-88.
11. GALEWSKY, S. , NESSLER, C.L. 1986. Synthesis of morphinane alkaloids during opium poppy somatic embryogenesis. Plant Sci. 45: 215-222.
12. ENDO, T., GOODBODY, A. E., MISAWA, M. 1987. Alkaloid production in root and shoot cultures of *Catharanthus roseus*. Planta Med. 53: 479-482.
13. BARTHE, G. A., JOURDAN, P. S., MCINTOSH, C. A., MANSELL, R. L. 1987. Naringin and limonin production in callus cultures and regenerated shoots from Citrus sp.. J. Plant Physiol. 127: 55-65.
14. GREIDZIAK, N., DIETTRICH, B., LUCKNER, M. 1990. Batch cultures of somatic embryos of *Digitalis lanata* in gaslift fermenters. Development and cardenolide accumulation. Planta Med. 56: 175-178.
15. JACOBSON, M. 1971. The unsaturated isobutylamines. In: Naturally Occurring Insecticides.(M. Jacobson, D. G. Crosby, eds.), Marcel Dekker, Inc. New York. pp. 137-176.
16. SIGNS, M. , FLORES, H. E. 1990. The biosynthetic potential of plant roots. BioEssays 12: 7-13.
17. BUTCHER, D. N., STREET, H. E. 1964. Excised root culture. Bot. Rev. 30: 513-586.
18. MIRANDA-HAM, M. L. 1990. Biosynthesis of secondary metabolites in plant cell and tissue cultures. In: Production of Secondary Metabolites from Plant Tissue Cultures and its

Biotechnological Perspectives.(V. M. Loyola-Vargas, ed.), CICY, Merida, Yucatan. pp. 80-90.

19. CHILTON, M. D., TEPFER, D. A., PETIT, A., DAVID, C., CASSE-DELBART, F., TEMPE, J. 1982. *Agrobacterium rhizogenes* inserts T-DNA into the genomes of the host plant root cells. Nature 295: 432-434.
20. O'DOWD, N. A. , RICHARDSON, D. H. S. 1994. Production of tumours and roots by *Ephedra* following *Agrobacterium rhizogenes* infection. Can. J. Bot. 72: 203-207.
21. OOSTDAM, A., MOL, J. N. M., VAN DER PLAS, L. H. W. 1993. Establishment of hairy root cultures of *Linum flavum* producing the lignan 5-methoxypodophyllotoxin. Plant Cell Rep. 12: 474-477.
22. BENJAMIN, B. D., ROJA, G., HEBLE, M. R. 1994. Alkaloid synthesis by root cultures of *Rauwolfia serpentina* transformed by *Agrobacterium rhizogenes*. Phytochemistry 35: 381-383.
23. CIAU-UITZ, R., MIRANDA-HAM, M. L., COELLO-COELLO, J., CHI, B., PACHECO, L. M., LOYOLA-VARGAS, V. M. 1994. Indole alkaloid production by transformed and non-transformed root cultures of *Catharanthus roseus*. In Vitro Cell. Dev. Biol. 30: 84-88.
24. GRäNICHER, F., CHRISTEN, P., KAPETANIDIS, I. 1992. High-yield production of valepotriates by hairy root cultures of *Valeriana officinalis* L. var. *sambucifolia* Mikan. Plant Cell Rep. 11: 339- 342.
25. TOIVONEN, L., BALSEVICH, J., KURZ, W. G. W. 1989. Indole alkaloid production by hairy root cultures of *Catharanthus roseus*. Plant Cell Tiss. Org. Cult. 18: 79-93.
26. MALDONADO-MENDOZA, I. E., AYORA-TALAVERA, T., LOYOLA-VARGAS, V.M . 1993. Establishment of hairy root cultures of *Datura stramonium*. Characterization and stability of tropane alkaloid production during long periods of subculturing. Plant Cell Tiss. Org. Cult. 33: 321-329.
27. MALDONADO-MENDOZA, I. E., AYORA-TALAVERA, T., LOYOLA-VARGAS, V. M. 1992. Tropane alkaloid production in *Datura stramonium* root cultures. In Vitro Cell. Dev. Biol. 28: 67-72.
28. SHIMOMURA, K., SAUERWEIN, M., ISHIMARU, K. 1991. Tropane alkaloids in the adventitious and hairy root cultures of solanaceous plants. Phytochemistry 30: 2275-2278.
29. ISHIMARU, K., SUDO, H., SATAKE, M., MATSUNAGA, Y., HASEGAWA, Y., TAKEMOTO, S., SHIMOMURA, K. 1990. Amarogentin, amaroswerin and four xanthones from hairy root cultures of *Swertia japonica*. Phytochemistry 29: 1563-1565.
30. BHADRA, R., VANI, S., SHANKS, J. V. 1993. Production of indole alkaloids by selected hairy root lines of *Catharanthus roseus*. Biotechnol. Bioeng. 41: 581-592.
31. TROTIN, F., MOUMOU, Y., VASSEUR, J. 1993. Flavanol production by *Fagopyrum esculentum* hairy and normal root cultures. Phytochemistry 32: 929-931.
32. KAMADA, H., OKAMURA, N., SATAKE, M., HARADA, H., SHIMOMURA, K. 1986. Alkaloid production by hairy root cultures in *Atropa belladonna*. Plant Cell Rep. 5: 239-242.
33. VAZQUEZ-FLOTA, F., MORENO-VALENZUELA, O., MIRANDA-HAM, M.L., COELLO-COELLO, J., LOYOLA-VARGAS, V. M. 1994. Catharanthine and ajmalicine synthesis in *Catharanthus roseus* hairy root cultures in an induction medium and during elicitation. Plant Cell Tiss. Org. Cult. (in press)
34. FLORES, H.E. , FILNER, P. 1985. Hairy roots of Solanaceae as a source of alkaloids. Plant Physiol. 77: 12s
35. FLORES, H. E., FILNER, P. 1985. Metabolic relationships of putrescine, GABA and alkaloids in cell and root cultures of Solanaceae. In: Primary and Secondary Metabolism in Plant Cell Cultures.(K. H. Neumann, W. Barz, E. Reinhard, eds.), Springer-Verlag, Heidelberg. pp. 174-186.
36. AIRD, E. L. H., HAMILL, J. D., ROBINS, R. J., RHODES, M. J. C. 1988. Chromosome stability in transformed hairy root cultures and the properties of variant lines of *Nicotiana*

rustica hairy roots. In: Manipulating Secondary Metabolism in Culture. (R.J. Robins, M.J.C. Rhodes, eds.), Cambridge University Press, Cambridge. pp. 137-144.

37. BANERJEE-CHATTOPADHYAY, S., SCHWEMMIN, A. M., SCHWEMMIN, D. J. 1985. A study of karyotypes and their alterations in cultured and *Agrobacterium* transformed roots of *Lycopersicon peruvianum* mill. Theor. Appl. Genet. 71: 258-262.
38. HAMILL, J. D., PARR, A. J., ROBINS, R. J., RHODES, M. J. C. 1986. Secondary product formation by cultures of *Beta vulgaris* and *Nicotiana rustica* transformed with *Agrobacterium rhizogenes*. Plant Cell Rep. 5: 111-114.
39. RHODES, M. J. C., ROBINS, R. J., HAMILL, J. D., PARR, A. J., WALTON, N. J. 1987. Secondary product formation using *Agrobacterium rhizogenes*-transformed "hairy-root" cultures. News Lett. 53: 2-15.
40. MANO, Y., NABESHIMA, S., MATSUI, C., OHKAWA, H. 1986. Production of tropane alkaloids by hairy root cultures of *Scopolia japonica*. Agric. Biol. Chem. 50: 2715-2722.
41. PAYNE, J., HAMILL, J. D., ROBINS, R. J., RHODES, M. J. C. 1987. Production of hyoscyamine by "hairy root" cultures of *Datura stramonium*. Planta Med. 53: 474-478.
42. MUKUNDAN, U. , HJORTSO, M. A. 1991. Growth and thiophene accumulation by hairy root cultures of *Tagetes patula* in media of varying initial pH. Plant Cell Rep. 9: 627-630.
43. GREEN, K. D., THOMAS, N. H., CALLOW, J. A. 1992. Product enhancement and recovery from transformed root cultures of *Nicotiana glauca*. Biotechnol. Bioeng. 39: 195-202.
44. HO, C. H. , SHANKS, J. V. 1992. Effects of initial medium pH on growth and metabolism of *Catharanthus roseus* hairy root cultures. A study with ^{31}P and ^{13}C NMR spectroscopy. Biotechnol. Lett. 14: 959-964.
45. MANO, Y., OHKAWA, H., YAMADA, Y. 1989. Production of tropane alkaloids by hairy root cultures of *Duboisia leichhardtii* transformed by *Agrobacterium rhizogenes*. Plant Sci. 59: 191-201.
46. CHRISTEN, P., AOKI, T., SHIMOMURA, K. 1992. Characteristics of growth and tropane alkaloid production in *Hyoscyamus albus* hairy roots transformed with *Agrobacterium rhizogenes* A4. Plant Cell Rep. 11: 597-600.
47. SAUERWEIN, M. , SHIMOMURA, K. 1991. Alkaloid production in hairy roots of *Hyoscyamus albus* transformed with *Agrobacterium rhizogenes*. Phytochemistry 30: 3277-3280.
48. YONEMITSU, H., SHIMOMURA, K., SATAKE, M., MOCHIDA, S., TANAKA, M., ENDO, T., KAJI, A. 1990. Lobeline production by hairy root culture of *Lobelia inflata* L. Plant Cell Rep. 9: 307-310.
49. ISHIMARU, K. , SHIMOMURA, K. 1991. Tannin production in hairy root culture of *Geranium thunbergii*. Phytochemistry 30: 825-828.
50. PARR, A. J., PEERLESS, A. C. J., HAMILL, J. D., WALTON, N. J., ROBINS, R. J., RHODES, M. J. C. 1988. Alkaloid production by transformed root cultures of *Catharanthus roseus*. Plant Cell Rep. 7: 309-312.
51. HILTON, M. G. , RHODES, M. J. C. 1993. Factors affecting the growth and hyoscyamine production during batch culture of transformed roots of *Datura stramonium*. Planta Med. 59: 340-344.
52. JAZIRI, M., LEGROS, M., HOMES, J., VANHAELEN, M. 1988. Tropine alkaloids production by hairy root cultures of *Datura stramonium* and *Hyoscyamus niger*. Phytochemistry 27: 419-420.
53. SATO, K., YAMAZAKI, T., OKUYAMA, E., YOSHIHIRA, K., SHIMOMURA, K. 1991. Anthraquinone production by transformed root cultures of *Rubia tinctorum*. Influence of phytohormones and sucrose concentration. Phytochemistry 30: 1507-1509.
54. BECKER, T. W., CABOCHE, M., CARRAYOL, E., HIREL, B. 1992. Nucleotide sequence of a tobacco cDNA encoding plastidic glutamine synthetase and light inducibility, organ

specificity and diurnal rhythmicity in the expression of the corresponding genes of tobacco and tomato. Plant Mol. Biol. 19: 367-379.

55. TOIVONEN, L., OJALA, M., KAUPPINEN, V. 1991. Studies on the optimization of growth and indole alkaloid production by hairy root cultures of *Catharanthus roseus*. Biotechnol. Bioeng. 37: 673-680.
56. DENO, H., YAMAGATA, H., EMOTO, T., YOSHIOKA, T., YAMADA, Y., FUJITA, Y. 1987. Scopolamine production by root cultures of *Duboisia myoporoides*. II. Establishment of a hairy root culture by infection with *Agrobacterium rhizogenes*. J. Plant Physiol. 131: 315-323.
57. OHKAWA, H., KAMADA, H., SUDO, H., HARADA, H. 1989. Effects of gibberellic acid on hairy root growth in *Datura innoxia*. J. Plant Physiol. 134: 633-636.
58. SAUERWEIN, M., WINK, M., SHIMOMURA, K. 1992. Influence of light and phytohormones on alkaloid production in transformed root cultures of *Hyoscyamus albus*. J. Plant Physiol. 140: 147-152.
59. YOSHIKAWA, T. , FURUYA, T. 1987. Saponin production by cultures of *Panax ginseng* with *Agrobacterium rhizogenes*. Plant Cell Rep. 6: 449-453.
60. SAUERWEIN, M., ISHIMARU, K., SHIMOMURA, K. 1991. Indole alkaloids in hairy roots of *Amsonia elliptica*. Phytochemistry 30: 1153-1155.
61. CONSTABEL, C. P., TOWERS, G. H. N 1988. Thiarubrine accumulation in hairy root cultures of *Chaenactis douglasii*. J. Plant Physiol. 133: 67-72.
62. SAUERWEIN, M., YAMAZAKI, T., SHIMOMURA, K. 1991. Hernandulcin in hairy root cultures of *Lippia dulcis*. Plant Cell Rep. 9: 579-581.
63. CROES, A. F., VAN DEN BERG, A. J. R., BOSVELD, M., BRETELER, H., WULLEMS, G.J. 1989. Thiophene accumulation in relation to morphology in roots of *Tagetes patula*. Effects of auxin and transformation by *Agrobacterium*. Planta 179: 43-50.
64. ERSEK, T. , KIRALY, Z. 1986. Phytoalexins: warding off compounds in plants? Physiol. Plant. 86: 343-346.
65. FLORES, H. E., PICKARD, J. J., HOY, M. W. 1988. Production of polyacetylenes and thiophenes in heterotrophic and photosynthetic root cultures of Asteraceae. In: Proceedings 1st. International Conference Naturally Occurring Acetylenes and Related Compounds.(J. Lam, H. Breteler, T. Arnason, L. Hansen, eds.), Elsevier, Amsterdam. pp. 233-254.
66. FLORES, H. E., PICKARD, J. J., SIGNS, M. 1988. Elicitation of polyacetylene production in hairy root cultures of Asteraceae. Plant Physiol. 86: 108s
67. BUITELAAR, R. M., LEENEN, E. J. T.M ., GEURTSEN, G., DE GROOT, Æ., TRAMPER, J. 1993. Effects of the addition of XAD-7 and of elicitor treatment on growth, thiophene production, and excretion by hairy roots of *Tagetes patula*. Enzyme Microb. Technol. 15: 670-676.
68. BUITELAAR, R. M., CESáRIO, M. T., TRAMPER, J. 1992. Elicitation of thiophene production by hairy roots of *Tagetes patula*. Enzyme Microb. Technol. 14: 2-7.
69. MUKUNDAN, U. , HJORTSO, M. A. 1990. Thiophene accumulation in hairy roots of *Tagetes patula* in response to fungal elicitors. Biotechnol. Lett. 12: 609-614.
70. MUKUNDAN, U. , HJORTSO, M.A. 1990. Effect of fungal elicitor on thiophene production in hairy root cultures of *Tagetes patula*. Appl. Microbiol. Biotechnol. 33: 145-147.
71. LOYOLA-VARGAS, V. M., COSGAYA, E., QUINTERO, C., AYORA, T. 1988. Modificación del perfil de poliaminas por inductores fúngicos en cultivos *in vitro* de raíces transformadas de *Datura stramonium*. Soc. Mex. Bioquím. 101
72. SIGNS, M. W. , FLORES, H. E. 1989. Elicitation of sesquiterpene phytoalexin biosynthesis in transformed root cultures of *Hyoscyamus muticus* L. Plant Physiol. 89: 135s
73. ROBBINS, M. P., HARTNOLL, J., MORRIS, P. 1991. Phenylpropanoid defence responses in transgenic *Lotus corniculatus*. 1. Glutathione elicitation of isoflavan phytoalexins in transformed root cultures. Plant Cell Rep. 10: 59-62.

74. SATO, N., YOSHIZAWA, Y., MIYAZAKI, H., MURAI, A. 1985. Antifungal activity to *Phytophthora infestans* and toxicity to tuber tissue of several popato phytoalexins. Ann. Phytopathol. Soc. (Japan). 51: 494-497.
75. DUNLOP, D. S. , CURTIS, W. R. 1991. Synergistic response of plant hairy-root cultures to phosphate limitation and fungal elicitation. Biotechnol. Prog. 7: 434-438.
76. FURZE, J. M., RHODES, M. J. C., PARR, A. J., ROBINS, R. J., WITHEHEAD, I. M., THRELFALL, D. R. 1991. Abiotic factors elicit sesquiterpenoid phytoalexin production but not alkaloid production in transformed root cultures of *Datura stramonium*. Plant Cell Rep. 10: 111-114.
77. SAENZ-CARBONELL, L., MALDONADO-MENDOZA, I. E., MORENO, V., CIAU-UITZ, R., LOPEZ-MEYER, M., OROPEZA, C., LOYOLA-VARGAS, V. M. 1993. Effect of the medium pH on the release of secondary metabolites from roots of *Datura stramonium*, *Catharanthus roseus* and *Tagetes patula* cultured *in vitro*. Appl. Biochem. Biotechnol. 38: 257-267.
78. SHIMOMURA, K., SUDO, H., SAGA, H., KAMADA, H. 1991. Shikonin production and secretion by hairy root cultures of *Lithospermum erythrorhizon*. Plant Cell Rep. 10: 282-285.
79. UOZUMI, N., KATO, Y., NAKASHIMADA, Y., KOBAYASHI, T. 1992. Excretion of peroxidase from horseradish hairy root in combination with ion supplementation. Appl. Microbiol. Biotechnol. 37: 560-565.
80. MALDONADO-MENDOZA, I. E. LOYOLA-VARGAS, V. M. 1990. Effect of photoautotrophy on tropane alkaloids contents in *Datura stramonium* hairy root cultures. Plant Physiol. 93: 21s
81. FLORES, H. E., YAO-REM, D., CUELLO, J. L., MALDONADO-MENDOZA, I. E., LOYOLA-VARGAS, V. M. 1993. Green roots: photosynthesis and photoautotrophy in an underground plant organ. Plant Physiol. 101: 363-371.
82. DE LUCA, V. , CUTLER, A.J. 1987. Subcellular localization of enzymes involved in indole alkaloid biosynthesis in *Catharanthus roseus*. Plant Physiol. 85: 1099-1102.
83. ISHIMARU, K., ARAKAWA, H., YAMANAKA, M., SHIMOMURA, K. 1994. Polyacetylenes in *Lobelia sessilifolia* hairy roots. Phytochemistry 35: 365-369.
84. MUKUNDAN, U. , HJORTSO, M.A. 1991. Effect of light on growth and thiophene accumulation in transformed roots of *Tagetes patula*. J. Plant Physiol. 138: 252-255.
85. FUJITA, Y.TABATA, M. 1987. Secondary metabolites from plant cells. Pharmaceutical applications and progress in commercial production. In: Plant Biology Vol. 3. Plant Tissue and Cell Culture. (C. E. Green, D. A. Somers, W. P. Hackett, D. D. Biesboer, eds.), Alan R. Liss, Co. New York. pp. 169-185.
86. FLORES, H. E. 1987. Use of plant cells and organ culture in the production of biological chemicals. In: Biotechnology in Agricultural Chemistry. Symposium Series 334 .(H.M. LeBaron, R.O. Mumma, R.C. Honeycutt, J.H. Duesing, J.F. Phillips, M.J. Haas, eds.), American Chemical Society, Washington, D. C. pp. 66-86.
87. RHODES, M. J. C., HAMILL, J. D., PARR, A. J., ROBINS, R. J., WALTON, N. J. 1988. Strain improvement by screening and selection techniques. In: Manipulating Secondary Metabolism in Culture.(R.J. Robins, M.JC. Rhodes, eds.), Cambridge University Press, Cambridge. pp. 83-93.
88. FURZE, J. M., HAMILL, J. D., PARR, A. J. , ROBINS, R. J., RHODES, M. J. C. 1987. Variations in morphology and nicotine alkaloid accumulation in protoplasts derived hairy root cultures of *N. rustica*. J. Plant Physiol. 131: 237-246.
89. MEDINA-BOLíVAR, F., FLORES, H. E. 1993. A novel approach for tropane alkaloid overproduction from hairy roots of *Hyoscyamus muticus*. Plant Physiol. 102: 98s
90. FECKER, L. F., HILLEBRANDT, S., RUGENHAGEN, C., HERMINGHAUS, S., LANDSMANN, J., BERLIN, J. 1992. Metabolic effects of a bacterial lysine decarboxylase gene expressed in hairy root cultures of *Nicotiana glauca*. Biotechnol. Lett. 14: 1035-1040.

91. FECKER, L. F., RüGENHAGEN, C., BERLIN, J. 1993. Increased production of cadaverine and anabasine in hairy root cultures of *Nicotiana tabacum* expressing a bacterial lysine decarboxylase gene. Plant Mol. Biol. 23: 11-21.
92. HASHIMOTO, T., YUN, D.-J., YAMADA, Y. 1993. Production of tropane alkaloids in genetically engineered root cultures. Phytochemistry 32: 713-718.
93. DOOLITTLE, R. F. 1994. Protein sequence comparisons: Searching databases and aligning sequences. Curr. Opin. Biotechnol. 5: 24-28.
94. RHODES, M. J. C., HILTON, M., PARR, A. J., HAMILL, J. D., ROBINS, R. J. 1986. Nicotine production by "hairy root" cultures of *Nicotiana rustica*: fermentation and product recovery. Biotechnol. Lett. 8: 415-420.
95. WILSON, P. D. G., HILTON, M. G., ROBINS, R. J., RHODES, M. J. C. 1989. Fermentation studies of transformed root cultures. In: Bioreactors and Biotransformations.(G. W. Moody, P. B. Baker, eds.), Elsevier Applied Sci. Pub. New York. pp. 38-51.
96. MURANAKA, T., OHKAWA, H., YAMADA, Y. 1992. Scopolamine release into media by *Duboisia leichhardtii* hairy root clones. Appl. Microbiol. Biotechnol. 37: 554-559.
97. WHITNEY, P .J. 1992. Novel bioreactors for the growth of roots transformed by *Agrobacterium rhizogenes*. Enzyme Microb. Technol. 14: 13-17.
98. SHARP, J. M. , DORAN, P. M. 1990. Characteristics of growth and tropane alkaloid synthesis in *Atropa belladonna* roots transformed by *Agrobacterium rhizogenes*. J. Biotechnol. 16: 171-186.
99. TOIVONEN, L., OJALA, M., KAUPPINEN, V. 1990. Indole alkaloid production by hairy root cultures of *Catharanthus roseus*: growth kinetics and fermentation. Biotechnol. Lett. 12: 519-524.
100. RODRIGUEZ-MENDIOLA, M. A., STAFFORD, A., CRESSWELL, R., ARIAS-CASTRO, C. 1991. Bioreactors for growth of plant roots. Enzyme Microb. Technol. 13: 697-702.
101. MATSUMOTO, T. , TANAKA, N. 1991. Production of phytoecdysteroids by hairy root cultures of *Ajuga reptans* Var Atropurpurea. Agric. Biol. Chem. 55: 1019-1025.
102. KENNEDY, A. I., DEANS, S. G., SVOBODA, K. P., GRAY, A. I., WATERMAN, P. G. 1993. Volatile oils from normal and transformed root of *Artemisia absinthium*. Phytochemistry 32: 1449-1451.
103. WILSON, P. D. G., HILTON, M. G., ROBINS, R. J., RHODES, M. J. C. 1987. Fermentation studies of transformed root cultures. In: Bioreactors and Biotransformations.(G.W. Moody, P. B. Baker, eds.), Elsevier Applied Science, London. pp. 38-51.
104. SCHEIDEGGER, A. 1990. Plant biotechnology goes commercial in Japan. Trends Biotechnol. 8: 197-198.
105. YAMADA, Y., HASHIMOTO, T. 1988. Biosynthesis of tropane alkaloids. In: Applications of Plant Cell and Tissue Culture.(G. Bock, J. Marsh, eds.), John Wiley & Sons, New York. pp. 199-212.
106. PORTSTEFFEN, A., DRAEGER, B., NAHRSTEDT, A. 1992. Two tropinone reducing enzymes from *Datura stramonium* transformed root cultures. Phytochemistry 31: 1135-1138.
107. DRAGER, B., PORTSTEFFEN, A., SCHAAL, A., MCCABE, P. H., PEERLESS, A. C. J., ROBINS, R. J. 1992. Levels of tropinone-reductase activities influence the spectrum of tropane esters found in transformed root cultures of *Datura stramonium* L. Planta 188: 581-586.
108. ROBINS, R. J., PARR, A. J., PAYNE, J., WALTON, N. J., RHODES, M. J. C. 1990. Factors regulating tropane-alkaloid production in a transformed root culture of a *Datura candida* x *D. aurea* hybrid. Planta 181: 414-422.
109. HASHIMOTO, T., NAKAJIMA, K., ONGENA, G., YAMADA, Y. 1992. Two tropinone reductases with distinct stereospecificities from cultured roots of *Hyoscyamus niger*. Plant Physiol. 100: 836- 845.

110. ROBINS, R. J., BACHMANN, P., ROBINSON, T., RHODES, M. J. C., YAMADA, Y. 1991. The formation of 3a- and 3ß-acetoxytropanes by *Datura stramonium* transformed root cultures involves two acetyl-CoA-dependent acyltransferases. FEBS Lett. 292: 293-297.
111. HASHIMOTO, T., MITANI, A., YAMADA, Y. 1990. Diamine oxidase from cultured roots of *Hyoscyamus niger*. Plant Physiol. 93: 216-221.
112. WALTON, N. J., PEERLESS, A. C. J., ROBINS, R. J., RHODES, M. J. C., BOSWELL, H. D., ROBINS, D. J. 1994. Purification and properties of putrescine *N*-methyltransferase from transformed roots of *Datura stramonium* L. Planta 193: 9-15.
113. WALTON, N. J., ROBINS, R. J., PEERLESS, A. C. J. 1990. Enzymes of N-methylputrescine biosynthesis in relation to hyoscyamine formation in transformed root cultures of *Datura stramonium* and *Atropa belladonna*. Planta 182: 136-141.
114. MCLAUCHLAN, W. R., MCKEE, R. A., EVANS, D. M. 1993. The purification and immunocharacterisation of N-methylputrescine oxidase from transformed root cultures of *Nicotiana tabacum* L. cv SC58. Planta 191: 440-445.
115. HIBI, N., FUJITA, T., HATANO, M., HASHIMOTO, T., YAMADA, Y. 1992. Putrescine *N*-methyltransferase in cultured roots of *Hyoscyamus albus*. *n*-Butylamine as a potent inhibitor of the transferase both in vitro and in vivo. Plant Physiol. 100: 826-835.
116. HASHIMOTO, T., YUKIMUNE, Y., YAMADA, Y. 1989. Putrescine and putrescine N-methyltransferase in the biosynthesis of tropane alkaloids in cultured roots of *Hyoscyamus albus*. Planta 178: 123-130.
117. HASHIMOTO, T., YUKIMUNE, Y., YAMADA, Y. 1989. Putrescine and putrescine N-methyltransferase in the biosynthesis of tropane alkaloids in cultured roots of *Hyoscyamus albus* II. Incorporation of labeled precursors. Planta 178: 131-137.
118. CONSTABEL, C. P., TOWERS, G. H. N. 1989. Incorporation of ^{35}S into dithiacyclohexadiene and thiophene polyines in hairy root cultures of *Chaenectis douglasii*. Phytochemistry 28: 93-95.
119. SUGIMOTO, Y., SUGIMURA, Y., YAMADA, Y. 1990. Biosynthesis of bisbenzylisoquinoline alkaloids in cultured roots of *Stephania cepharantha*. FEBS Lett. 273: 82-86.
120. GOMEZ-BARRIOS, M. L., PARODI, F. J., VARGAS, D., QUIJANO, L., HJORTSO, M. A., FLORES, H. E., FISCHER, N. H. 1992. Studies on the biosynthesis of thiarubrine A in hairy root cultures of *Ambrosia artemisiifolia* using ^{13}C-labelled acetates. Phytochemistry 31: 2703-2707.
121. BERLIN, J., RUGENHAGEN, C., GREIDZIAK, N., KUZOVKINA, I. N., WITTE, L., WRAY, V. 1993. Biosynthesis of serotonin and ß-carboline alkaloids in hairy root cultures of *Peganum harmala*. Phytochemistry 33: 593-597.
122. ISLAS, I., LOYOLA-VARGAS, V. M., MIRANDA-HAM, M. L. 1994. Tryptophan decarboxylase activity in transformed roots from Catharanthus roseus and its relationship to tryptamine, ajmalicine, and catharanthine accumulation during the culture cycle. In Vitro Cell. Dev. Biol. 30P: 81-83.
123. HAMILL, J. D., PARR, A. J., RHODES, M. J.C ., ROBINS, R. J., WALTON, N. J. 1987. New routes to plant secondary products. BioTechnol. 5: 800-804.
124. PARR, A. J. , HAMILL, J. D. 1987. Relationship between *Agrobacterium rhizogenes* transformed hairy roots and intact, uninfected *Nicotiana* plants. Phytochemistry 26: 3241-3245.
125. TEMPE, J. , SCHELL, J. 1987. La manipulation des plantes. La Recherche 188: 696-709.
126. VAZQUEZ-FLOTA, F., COELLO, J., LOYOLA-VARGAS, V.M. 1992. Growth kinetics and alkaloid production in hairy root cultures of *Catharanthus roseus*. Plant Physiol. 99: 49s
127. BHADRA, R., HO, C. H., SHANKS, J. V. 1991. Growth characteristics of hairy root lines of *Catharanthus roseus*. In Vitro Cell. Dev. Biol. 27: 109A-109A.

128. DAVIOUD, E., KAN, C., HAMON, J., TEMPE J., HUSSON, H. P. 1989. Production of indole alkaloids by *in vitro* root cultures from *Catharanthus trichophyllus*. Phytochemistry. 28: 2675-2680.
129. KYO, M., MIYAUCHI, Y., FUJIMOTO, T., MAYAMA, S. 1990. Production of nematocidal compounds by hairy root cultures of *Tagetes patula* L. Plant Cell Rep. 9: 393-397.
130. OGASAWARA, T., CHIBA, K., TADA, M. 1993. Production in high-yield of a naphthoquinone by a hairy root culture of *Sesamum indicum*. Phytochemistry 33: 1095-1098.
131. HAMILL, J. D., ROBINS, R. J., RHODES, M. J. C. 1989. Alkaloid production by transformed root cultures of *Cinchona ledgeriana*. Planta Med. 55: 354-357.
132. CHRISTEN, P., ROBERTS, M. F., PHILLIPSON, J. D., EVANS, W. C. 1989. High-yield production of tropane alkaloids by hairy-root cultures of a *Datura candida* hybrid. Plant Cell Rep. 8: 75-77.
133. PARR, A. J. 1992. Alternative metabolic fates of hygrine in transformed root cultures of Nicandra physaloides. Plant Cell Rep. 11: 270-273.
134. JUNG, K. H., KWAK, S. S., KIM, S. W., LEE, H., CHOI, C. Y., LIU, J. R. 1992. Improvement of catharanthine productivity in hairy root cultures of *C.roseus* by using monosaccharides as carbon source. Biotechnol. Lett. 14: 695-700.
135. DAVIOUD, E., KAN, C., QUIRION, J. C., DAS, B. C., HUSSON, H. P. 1989. Epiallo-yohimbine derivatives isolated from *in vitro* hairy-root cultures of *Catharanthus trichophyllus*. Phytochemistry 28: 1383-1387.
136. FALKENHAGEN, H., STOCKIGT, J., KUZOVKINA, I. N., ALTERMAN, I. E., KOLSHORN, H. 1993. Indole alkaloids from "hairy roots" of *Rauwolfia serpentina*. Can. J. Chem. 71: 2201-2203.
137. ASADA, Y., SAITO, H., YOSHIKAWA, T., SAKAMOTO, K., FURUYA, T. 1993. Biotransformation of 18ß-glycyrrhetinic acid by ginseng hairy root culture. Phytochemistry 34: 1049-1052.
138. HIRAKURA, K., MORITA, M., NAKAJIMA, K., IKEYA, Y., MITSUHASHI, H. 1991. Three acetylated polyacetylenes from the roots of *Panax ginseng*. Phytochemistry 30: 4053-4055.
139. LU, T., PARODI, F. J., VARGAS, D., QUIJANO, L., MERTOOETOMO, E. R., HJORTSO, M. A., FISCHER, N. H. 1993. Sesquiterpenes and thiarubrines from *Ambrosia trifida* and its transformed roots. Phytochemistry 33: 113-116.
140. NORTON, R. A. , TOWERS, G. H. N. 1986. Factors affecting synthesis of polyacetylenes in root cultures of *Bidens alba*. J. Plant Physiol. 122: 41-53.
141. HORZ, K. H. , REICHLING, J. 1993. Allylphenol biosynthesis in a transformed root culture of *Coreopsis tinctoria*: Side-chain formation. Phytochemistry 33: 349-351.
142. TRYPSTEEN, M., VANLIJSEBETTENS, M., VANSEVEREN, R., VANMONTAGU, M. 1991. *Agrobacterium-rhizogenes*-mediated transformation of *Echinacea purpurea*. Plant Cell Rep. 10: 85-89.
143. ABEGAZ, B. M. 1991. Polyacetylenic thiophenes and terpenoids from the roots of *Echinops pappii*. Phytochemistry 30: 879-881.
144. YAMAZAKI, T., FLORES, H. E., SHIMOMURA, K., YOSHIHIRA, K. 1991. Examination of steviol glucosides production by hairy root and shoot cultures of *Stevia rebaudiana*. J. Nat. Prod. 54: 986-992.
145. MUKUNDAN, U., HJORTSO, M. A. 1990. Thiophene content in normal and transformed root cultures of *Tagetes erecta*: A comparison with thiophene content in roots of intact plants. J. Exp. Bot. 41: 1497-1501.
146. UESATO, S., OGAWA, Y., INOUYA, H., SAIKI, K., ZENK, M.H. 1986. Synthesis of iridodial by cell free extracts from *Rauwolfia serpentina* cell suspension cultures. Tetrahedron Lett. 13: 2893-2896.

147. PARODI, F. J., FISCHER, N. H., FLORES, H. E. 1988. Benzofuran and bithiophenes from root cultures of *Tagetes patula*. J. Nat. Prod. 51: 594-595.
148. WESTCOTT, R. J. 1988. Thiophene production from "hairy roots" of *Tagetes*. In: Manipulating Secondary Metabolism in Culture. (R.J. Robins, M.J.C. Rhodes, eds.), Cambridge University Press, Cambridge. pp. 233-237.
149. SIM, S .J., CHANG, H. N. 1993. Increased shikonin production by hairy roots of *Lithospermum erythrorhizon* in two phase bubble column reactor. Biotechnol. Lett. 15: 145-150.
150. ISHIMARU, K., SADOSHIMA, S., NEERA, S., KOYAMA, K., TAKAHASHI, K., SHIMOMURA, K. 1992. A polyacetylene gentiobioside from hairy roots of *Lobelia inflata*. Phytochemistry 31: 1577-1579.
151. ISHIMARU, K., YONEMITSU, H., SHIMOMURA, K. 1991. Lobetyolin and lobetyol from hairy root culture of *Lobelia inflata*. Phytochemistry 30: 2255-2257.
152. BENSON, E. E., HAMILL, J. D. 1991. Cryopreservation and post freeze molecular and biosynthetic stability in transformed roots of *Beta vulgaris* and *Nicotiana rustica*. Plant Cell Tiss. Org. Cult. 24: 163-172.
153. JUNG, G. , TEPFER, D. 1987. Use of genetic transformation by the Ri T-DNA of *Agrobacterium rhizogenes* to stimulate biomass and tropane alkaloid production in *Atropa belladonna* and *Calystegia sepium* roots grown *in vitro*. Plant Sci. 50: 145-151.
154. PARKINSON, M., COTTER, T., DIX, P. J. 1990. Peroxidase production by cell suspension and hairy root cultures of horseradish (*Armoracia rusticana*). Plant Sci. 66: 271-277.
155. KATAVIC, V., JELASKA, S., BAKRAN-PETRICIOLI, T., DAVID, C. 1991. Host-tissue differences in transformation of pumpkin (*Cucurbita pepo* L.) by *Agrobacterium rhizogenes*. Plant Cell Tiss. Org. Cult. 24: 35-42.
156. SAVARY, B. J., FLORES, H. E. 1993. Characterization of bioactive proteins produced in root cultures of *Trichosanthes kirilowi* var. *Japonica*. Plant Physiol. 102: 31s
157. ISHIMARU, K., SUDO, H., SATAKE, M., SHIMOMURA, K. 1990. Phenyl glucosides from a hairy root culture of *Swertia japonica*. Phytochemistry 29: 3823-3825.
158. HU, Z. B., ALFERMANN, A. W. 1993. Diterpenoid production in hairy root cultures of *Salvia miltiorrhiza*. Phytochemistry 32: 699-703.
159. WINK, M., WITTE, L. 1987. Alkaloids in stem roots of *Nicotiana tabacum* and *Spartium junceum* transformed by *Agrobacterium rhizogenes*. Z. Naturforsch. 42c: 69-72.
160. WILLIAMS, R. D., ELLIS, B. E. 1988. Patterns of benzylisoquinoline alkaloids in *Papaver somniferum* plants and *Agrobacterium rhizogenes* transformed tissues. Phytochem. Soc. Newsletter. 28: 12
161. WILLIAMS, R. D. , ELLIS, B. E. 1993. Alkaloids from *Agrobacterium rhizogenes*-transformed *Papaver somniferum* cultures. Phytochemistry 32: 719-723.
162. ISHIMARU, K., HIROSE, M., TAKAHASHI, K., KOYAMA, K., SHIMOMURA, K. 1990. Tannin production in root culture of *Sanguisorba officinalis*. Phytochemistry 29: 3827-3830.
163. JHA, S., SAHU, N. P., SEN, J., JHA, T. B., MAHATO, S. B. 1991. Production of emetine and cephaeline from cell suspension and excised root cultures of *Cephaelis ipecacuanha*. Phytochemistry 30: 3999-4003.
164. KNOPP, E., STRAUSS, A., WEHRLI, W. 1988. Root induction on several Solanaceae species by *Agrobacterium rhizogenes* and the determination of root tropane alkaloid content. Plant Cell Rep. 7: 590-593.
165. CHRISTEN, P., ROBERTS, M. F., PHILLIPSON, J. D., EVANS, W. C. 1990. Alkaloids of hairy root cultures of a *Datura candida* hybrid. Plant Cell Rep. 9: 101-104.
166. PARR, A. J., PAYNE, J., EAGLES, J., CHAPMAN, B. T., ROBINS, R .J., RHODES, M. J. C. 1990. Variation in tropane alkaloid accumulation within the solanaceae and strategies for its exploitation. Phytochemistry 29: 2545-2550.

167. FLORES, H. E., PROTACIO, C. M., SIGNS, M. W. 1989. Primary and secondary metabolism of polyamines in plants. In: Plant Nitrogen Metabolism. Recent Advances in Phytochemistry. Vol. 23.(J.E. Poulton, J.T. Romeo, E.E. Conn, eds.), Plenum Press, New York and London. pp. 329-391.
168. ROBINS, R. J., PARR, A. J., BENT, E. G., RHODES, M. J. C. 1991. Studies on the biosynthesis of tropane alkaloids in *Datura stramonium* L. transformed root cultures. 1. The kinetics of alkaloid production and the influence of feeding intermediate metabolites. Planta 183: 185-195.
169. ROBINS, R. J., BENT, E. G., RHODES, M. J. C. 1991. Studies on the biosynthesis of tropane alkaloids by *Datura stramonium* L. transformed root cultures. 3. The relationship between morphological integrity and alkaloid biosynthesis. Planta 185: 385-390.
170. RHODES, M. J. C., ROBINS, R. J., LINDSAY, E., AIRD, H., PAYNE, J., PARR, A. J., WALTON, N. J. 1989. Regulation of secondary metabolism in transformed roots cultures. In: Primary and Secondary Metabolism of Plant Cell Cultures II. (W.G.W. Kurz, ed.). Springer-Verlag, Berlin Heidelberg. pp. 58-72.
171. KITAMURA, Y., TAURA, A., KAJIYA, Y., MIURA, H. 1992. Conversion of phenylalanine and tropic acid into tropane alkaloids by *Duboisia leichhardtii* root cultures. J. Plant Physiol. 140: 141-146.
172. KITAMURA, Y., NISHIMI, S., MIURA, H., KINOSHITA, T. 1993. Phenyllactic acid in *Duboisia leichhardtii* root cultures by feeding of phenyl[1-^{14}C]alanine. Phytochemistry 34: 425-427.
173. LEETE, E., ENDO, T., YAMADA, Y. 1990. Biosynthesis of nicotine and scopolamine in a root culture of *Duboisia leichhardtii*. Phytochemistry 29: 1847-1851.
174. DOERK-SCHMITZ, K., WITTE, L., ALFERMANN, A. W. 1994. Tropane alkaloid patterns in plants and hairy roots of *Hyoscyamus albus*. Phytochemistry 35: 107-110.
175. SAUERWEIN, M., SHIMOMURA, K., WINK, M. 1993. Incorporation of 1-^{13}C-acetate into tropane alkaloids by hairy root cultures of *Hyoscyamus albus*. Phytochemistry 32: 905-909.
176. SAUERWEIN, M., WINK, M. 1993. On the role of opines in plants transformed with *Agrobacterium rhizogenes*: Tropane alkaloid metabolism, insect-toxicity and allelopathic properties. J. Plant Physiol. 142: 446-451.
177. HASHIMOTO, T., YUKIMUNE, Y., YAMADA, Y. 1986. Tropane alkaloid production in *Hyoscyamus* root culture. J. Plant Physiol. 124: 61-75.
178. SAUERWEIN, M., ISHIMARU, K., SHIMOMURA, K. 1991. A piperidone alkaloid from *Hyoscyamus albus* roots transformed with *Agrobacterium rhizogenes*. Phytochemistry 30: 2977- 2978.
179. CORRY, J. P., REED, W. L., CURTIS, W. R. 1993. Enhanced recovery of solavetivone from *Agrobacterium* transformed root cultures of *Hyoscyamus muticus* using integrated product extraction. Biotechnol. Bioeng. 42: 503-508.
180. WALTON, N. J. , BELSHAW, N. J. 1988. The effect of cadaverine on the formation of anabasine from lysine in hairy root cultures of *Nicotiana hesperis*. Plant Cell Rep. 7: 115-118.
181. BOSWELL, H. D., WATSON, A.B ., WALTON, N. J., ROBINS, D. J. 1993. Formation of *N'*-ethyl-*S*-nornicotine by transformed root cultures of *Nicotiana rustica*. Phytochemistry 34: 153-155.
182. ROBINS, R. J., HAMILL, J. D., PARR, A. J., SMITH, K., WALTON, N. J., RHODES, M. J. C. 1987. Potential for use of nicotinic acid as a selective agent for isolation of high nicotine-producing lines of *Nicotiana rustica* hairy root cultures. Plant Cell Rep. 6: 122-126.
183. WALTON, N. J., ROBINS, R. J., RHODES, M. J. C. 1988. Perturbation of alkaloid production by cadaverine in hairy root cultures of *Nicotiana rustica*. Plant Sci. 54: 125-131.

184. WALTON, N. J. , MCLAUCHLAN, W. R. 1990. Diamine oxidation and alkaloid production in transformed root cultures of *Nicotiana tabacum*. Phytochemistry 29: 1455-1457.
185. BEL RHLID, R., CHABOT, S., PICHé, Y., CHêNEVERT, R. 1993. Isolation and identification of flavonoids from Ri T-DNA-transformed roots (*Daucus carota*) and their significance in vesicular-arbuscular mycorrhiza. Phytochemistry 33: 1369-1371.

Chapter Eleven

ANNONACEOUS ACETOGENINS

Potent Mitochondrial Inhibitors with Diverse Applications

Zhe-Ming Gu, Geng-Xian Zhao, Nicholas H. Oberlies, Lu Zeng, and Jerry L. McLaughlin

Department of Medicinal Chemistry and Pharmacognosy
School of Pharmacy and Pharmacal Sciences
Purdue University
West Lafayette, Indiana 47907, U.S.A.

Phytochemistry of Medicinal Plants
Edited by John T. Arnason et al., Plenum Press, New York, 1995

INTRODUCTION

Annonaceous acetogenins are a relatively new, but rapidly growing, class of compounds. Since the first of them, uvaricin, was reported in 1982,[1] more than 160 compounds have been published. These natural products have attracted considerable interest due to their diversity of bioactivities. Most acetogenins are potently bioactive and offer exciting potential as new antitumor, immunosuppressive, pesticidal, antiprotozoal, antifeedant, anthelmintic, and antimicrobial agents.[2-7] Chemically, the Annonaceous acetogenins seem to be derivatives of C_{32} or C_{34} fatty acids which are combined with a 2-propanol unit at C-2; the propanol is incorporated into a 2,4-disubstituted γ-lactone which can assume several different forms. Usually, one, two, or three tetrahydrofuran (THF) rings exist in the middle of the long hydrocarbon chain, which often contains a number of oxygenated moieties (hydroxyls, acetoxyls, ketones, epoxides), and/or a double bond.

All of the Annonaceous acetogenins have been reported from twenty species of six genera, *Annona, Asimina, Goniothalamus, Rollinia, Uvaria*, and *Xylopia*, existing in bark, seeds, twigs, roots, and leaves. About one third found to this date (June, 1994) were reported in the last two years. Several new skeletons, such as tri-THF rings, triepoxy rings, and adjacent bis-THF rings having only one flanking hydroxyl, or new features, like the *erythro* diol and the *cis* mono-THF ring, continue to be found in the acetogenins. The major mode of action of acetogenins, inhibition of complex I of mitochondrial oxidative phosphorylation, has been supported by additional experimental data.

All Annonaceous acetogenins have multiple stereocenters. Their waxy nature limits the direct application of X-ray crystallographic analyses to resolve their stereochemistry. In our first review,[8] we described the methodologies of Hoye and Born to solve the relative stereochemistries of THF moieties, by comparisons of the NMR data of the acetogenins with those of model compounds having known relative stereochemistries. In our second review, we encouraged other workers to resolve the absolute configurations of the stereogenic centers by the use of the advanced Mosher ester methodology.[9] Since then, the absolute configurations of many acetogenins have been defined. Two additional reviews have been published but are more limited in scope.[10,11] The time seems appropriate to summarize, once again, this class of compounds. In addition, certain advances concerning the bioactivities need to be updated and emphasized in order to encourage the development of commercial uses. In the Appendix are listed the structures of new Annonaceous acetogenins reported for 1993-1994 (June) including some not covered in our first two reviews.[8,9]

STRUCTURAL CLASSIFICATION

The Annonaceous acetogenins are conveniently classified, based on the number and arrangements of the THF rings, into mono-THF, adjacent bis-THF, non-adjacent bis-THF, and tri-THF subclasses.[8-11] The number of the THF-flanking hydroxyls and the configurations of the stereogenic centers of the THF moieties may change and, thus, may offer further subclassification. The absolute configurations of over 70 acetogenins, representative of most of the THF ring and relative stereochemistry systems, have now been defined. Acetogenins are optically pure. No pairs of mirror images are found, and we are able further to classify the acetogenins, named "type", by their THF ring systems and their stereochemistries.

Some compounds, reported by different research groups or from different plants, such as bullatalicin, cherimolin-1, and squamostatin-B, obviously represent the same structures. Usually, no direct comparisons of such compounds have been made. The comparisons of two pairs of compounds, *i.e.*, bullatacin *vs.* rolliniastatin-2 and rolliniastatin-1 *vs.* 4-hydroxy-25-deoxyneorollinicin, apparently showed different R_f values on thin layer chromatography (TLC),[12] although the spectral data suggest the same structures. Future workers are encouraged to determined absolute stereochemistries to prevent publication of such ambiguities.

EXTRACTION AND ISOLATION

Most of the pure acetogenins and the extracts containing them are easily detected in the brine shrimp lethality test (BST).[13,14] Thus, the BST is routinely used in our group and in Cavé's group for monitoring the purification. Other bioassays, *e.g.*, cytotoxicities,[15-17] do not favorably compete with the BST in speed, low cost, reproduciblity, and the quantitative utility of the results. Visualization of acetogenins on TLC plates is made possible by using phosphomolybdic acid in 95 % ethanol followed by heating, and Keddy's reagent is helpful for identifying only the α,β-unsaturated γ-lactone acetogenins.[8]

The pure Annonaceous acetogenins are quite soluble in methanol, ethanol, acetone, dichloromethane, chloroform, DMSO, and some other organic solvents, but they do not readily dissolve in cold hexane or water. Cold methanol or ethanol percolation is commonly used in the extraction of all sorts of plant parts and materials, and, sometimes, cold hexane percolation or Soxhlet extraction in hexane is applied to seeds. After extraction, partitioning of the alcoholic extract residue, between chloroform (or dichloromethane) and water, and then the further partitioning of the chloroform residue between 90 % aqueous methanol and hexane, lead to a concentration of the BST activity in the aqueous methanol residue (F005). F005 is repeatedly resolved by silica gel chromatography using open or flash columns and/or chromatotrons to give the white wax-like acetogenins; in many cases, recent workers have found that HPLC purification is necessary to separate closely related diastereomers. Some less polar epoxide acetogenins have been isolated from the hexane residue (F006).

The α,β-unsaturated γ-lactone acetogenins usually exhibit ultraviolet (UV) absorption around 210 nm and can be detected by a UV detector. In the normal phase

HPLC separation (silica gel column, 21 x 250 mm), a hexane-methanol-THF (90:9:1 to 80:8:2) solvent system is commonly used, and the detector is set at 220 - 235 nm (depending on the amount loaded). Good results have been achieved in the separation of numerous structurally similar compounds, *e.g.*, bullatalicin (C-24-*S*) and bullatanocin (C-24-*R*), differing only in the configurations of C-24, and 10-, 12-, 28-, 29-, 30-, 31-, and 32-hydroxybullatacinones, differing only in the positioning of one aliphatic hydroxyl. Good separation has also been obtained using the preparative reverse phase (RP) HPLC method, applying a C_{18} column eluted by acetonitrile:water (70:30 to 90:10). Four epimers of non-adjacent bis-THF acetogenins, bullatalicin (C-12-*R*, C-24-*S*), 12,15-*cis*-bullatalicin (C-12-*S*, C-24-*S*), bulla-tanocin (C-12-*R*, C-24-*R*), and 12,15-*cis*-bullatanocin (C-12-*S*, C-24-*R*), are separated in this way. The separation of such compounds appears to be better on RP-HPLC than on normal phase HPLC, since the C-12,15-*cis* and C-12,15-*trans* non-adjacent acetogenins did not show good separation on the normal phase HPLC in our experiments. The solvents (acetonitrile, methanol, and water) used in the RP-HPLC have UV absorptions shorter than 205 nm; therefore, another advantage of RP-HPLC is that it can be used to purify the ketolactone acetogenins; such acetogenins exhibit very weak UV absorption at short wave lengths, but they do appear as weak peaks if the UV detector is set at 205 nm.

Qualitative and quantitative analyses of acetogenins by HPLC were performed by Gromek *et al.*[18] Both normal phase and RP-HPLC methods were applied under isocratic or gradient conditions, using UV, refractometric, or evaporative light scattering detections. Their results showed that separations of certain acetogenins were better on RP-HPLC, and six acetogenins from *Annona muricata* were well-resolved. The quantitative evaluation of annonacin in the seed extract of *A. muricata* was also carried out using their HPLC methods.

STRUCTURAL ELUCIDATION STRATEGIES

Basic strategies for the structural elucidation of Annonaceous acetogenins were summarized in previous reviews.[8,9] Usually, the types of the acetogenins can be identified by the respective diagnostic ^{1}H and ^{13}C NMR spectra, and the placements of the subunits and functional groups are achieved by the electron impact mass spectrometer (EIMS) and fast atom bombardment mass spectrometer (FABMS) spectra of the acetogenins and/or their derivatives, especially the EIMS of the trimethylsilyl (TMSi) and perdeutero-TMSi derivatives.[19-22] Several special techniques are found to be useful in their structural recognition and elucidation, and these are described below.

The Terminal γ-Lactone

One of the major functional subunits in Annonaceous acetogenins is the terminal γ-lactone, and this group, including carbon atoms 1 - 4 and 35 (33) - 37

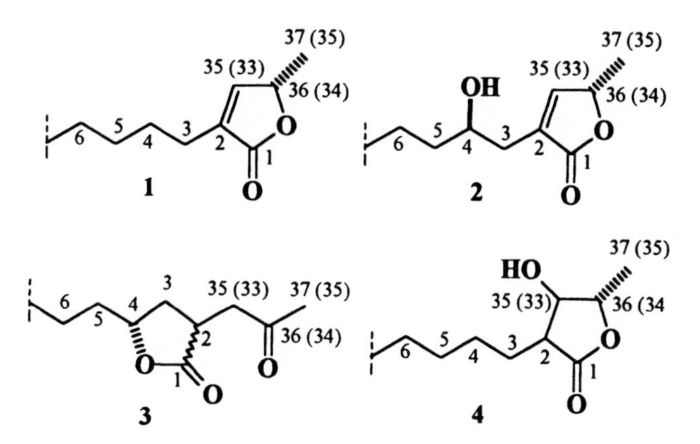

Figure 1. Annonaceous acetogenin γ-lactone subunits (1-4).

(35), assumes one of four arrangements (**1-4**, Fig. 1). The methods for identification of these different γ-lactones were summarized in previous reviews,[8,9] and, usually, they are not difficult to differentiate.

The most common functional group found in the Annonaceous acetogenins is the terminal α,β-unsaturated γ-methyl γ-lactone (**1**, Fig. 1). However, near this moiety, a C-4 hydroxyl group is quite often present (**2**, Fig. 1). Rarely, C-5 or C-6 hydroxyl-substitutions also exist instead of the C-4 hydroxyl. The placements of these hydroxyls, at C-4, 5, or 6, were originally and unambiguously established by characteristic EIMS fragment ions of the natural compounds and/or their TMSi and perdeutero-TMSi derivatives. However, today, we realize that the chemical shifts of the protons and carbons around the lactone moieties, as summarized by Gu *et al.*,[23] show characteristic values (Fig. 2) which are quite applicable in the rapid recognition of these substitutions.

Placement of Functional Groups

Much of the recent effort in the structural elucidation of the Annonaceous acetogenins has been directed to the resolution of the stereochemistries around the THF ring(s). However, additional diastereomers are created by the substitution positions of a number of additional functional groups, *e.g.*, single hydroxyls, vicinal diols, double bonds, *etc.* The positions of such groups are usually determined by the EIMS of the natural acetogenins and/or their TMSi and perdeutero-TMSi derivatives, and, in a few cases, by the FABMS of lithium-cationized acetogenins.[24] However, it is evident that some structures which were determined by MS alone are questionable.[25-27] By direct comparisons of various series of these compounds, we have found that ^{1}H and ^{13}C NMR data are also quite helpful in

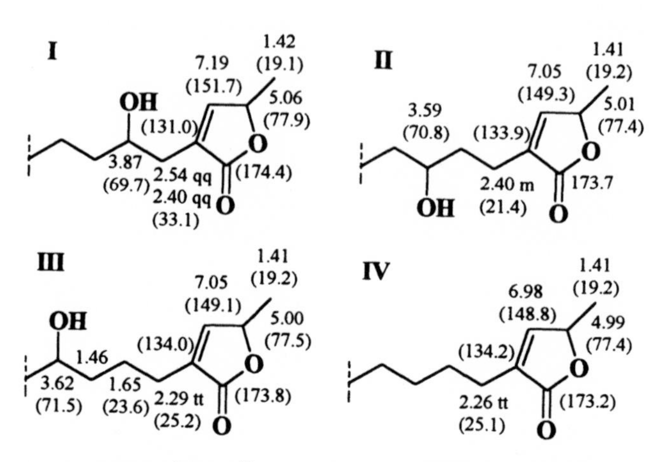

Figure 2. Comparisons of the ^{1}H and ^{13}C (in parentheses) NMR spectral data of the α,β-unsaturated γ-lactones have a 4-OH group (I), a 5-OH group (II), a 6-OH group (III), and those without any nearby hydroxyl group (IV).

placement of the functional groups. Structures elucidated by a combination of MS and NMR data, to determine the positions of the functional groups along the hydrocarbon chain, are much more trustworthy than MS data alone.

The ^{1}H NMR Spectral Patterns of Goniothalamicin, Annonacin, and Annomontacin. The structures of goniothalamicin, annonacin, and annomontacin are very similar, and all of them have an hydroxyl group at C-10. However, the proton spectra of these three compounds show patterns that are quite distinguishably different. In annomontacin, the C-10 and C-17 protons are separated by six methylenes, and the ^{1}H NMR signals of H-17 and H-22 appear as a pseudo-quartet; in annonacin, the C-10 and C-15 protons are separated by four methylenes, and the signals of H-15 and H-20 show a pseudo-quintet. In goniothalamicin, H-10 and H-13 are separated by only two methylenes, their signals are shifted further downfield, and the signals for H-13 and H-18, at *ca.* δ 3.4, appear to be two nearly separated multiplet signals. Similar ^{1}H NMR spectra were also observed in xylopianin, xylopiacin, xylopien (similar to the annomontacin pattern), (2,4-*cis* and *trans*)-isoannonacins, xylomaticin, xylomatenin (similar to the annonacin pattern), and *cis*-goniothalamicin[28] (similar to the goniothalamicin pattern).

The Chain Hydroxyl Substitution Position and the Proton Chemical Shifts of the Terminal Methyl Groups. (2,4-*cis* and *trans*)-Bullatacinones and seven pairs of hydroxylated (2,4-*cis* and *trans*)-bullatacinones having an extra hydroxyl group located at C-10, C-12, C-28, C-29, C-30, C-31, or C-32, respectively, have been isolated and characterized.[29-32] Although all of the positions of the aliphatic chain hydroxyls in these compounds were unambiguously determined by the EIMS fragmentation of TMSi derivatives, in the structural elucidation of compounds hydroxylated at C-28 to C-30, we also found that the proton signals of the terminal methyls at C-34, corresponding to different locations of the aliphatic hydroxyl group, showed sequential downfield shifts as the hydroxyl group approaches the C-34 position.[31] Although some of these differences are very small, they are quite stable and reproducible and are practically useful to the discovery and characterization of this series of compounds. The protons at C-34 of squamocin and asiminacin (both having a 28-OH), motrilin and asiminecin (both having a 29-OH), and bullanin (having a 30-OH) resonate at predictable and discernible frequencies.[31] Observations of the proton chemical shifts of the terminal methyl groups are also helpful in the determination of the absolute stereochemistries of the isolated aliphatic hydroxyl groups by the use of Mosher ester methodology (for details see the section on Absolute Stereochemistry).

Beta-Effects of the Oxygenated Carbon Centers in the ^{13}C NMR Spectra. β-Effects of the oxygenated carbon centers, usually upfield shifts of *ca.* four ppm, are universally observed in the ^{13}C NMR spectra of the Annonaceous acetogenins (Fig. 3); these are often helpful in determining the positions of such substitutions. The hydrocarbon chain methylenes normally resonate at *ca.* δ 29, but the signals

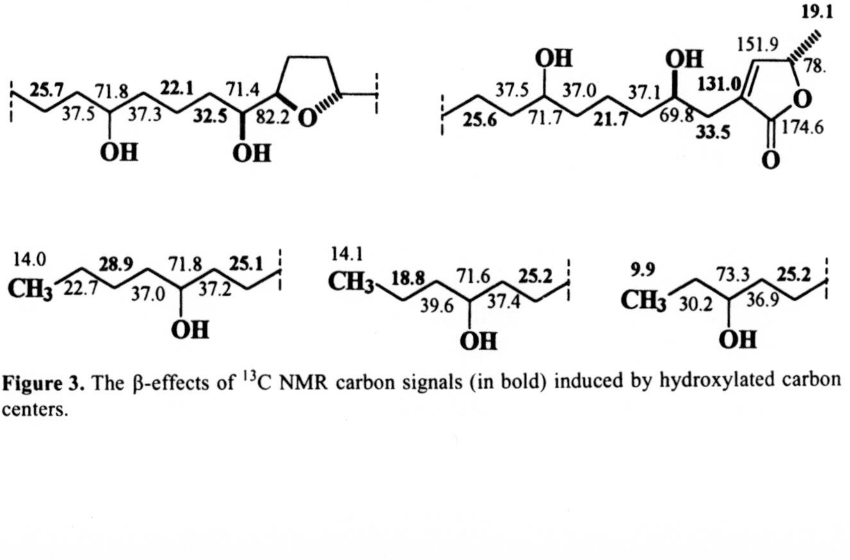

Figure 3. The β-effects of ^{13}C NMR carbon signals (in bold) induced by hydroxylated carbon centers.

of methylenes β to an hydroxylated carbon appear at *ca.* δ 25. The α-methylene carbons of an isolated aliphatic hydroxylated carbon resonate at *ca.* δ 37, while those of a THF flanking hydroxylated carbon resonate at *ca.* δ 33 (β-effected by the oxygenated C-2 or C-5 positions of the THF ring). Further changes of the chemical shifts of the hydrocarbon methylenes usually indicate an extra functional group located close by.

The hydroxyl groups at C-11 and C-13 in gonionenin form a 1,4-diol moiety, and such β-effects were observed on both methylenes between the two hydroxyl groups. The C-11, β-effected by the hydroxylated C-13, resonated at δ 33.5, not as normally at *ca.* δ 37. Thus, in the ^{13}C NMR spectra of gonionenin only two signals appeared at *ca.* δ 37 (C-5 and C-9), while in the ^{13}C NMR spectrum of annonacin there were three signals at *ca.* δ 37 (C-5, C-9, and C-11), since there is no β-effect on C-11 of annonacin. The C-12 of goniothalamicin, double-β-effected by the oxygenated C-10 and C-14 positions, shifted even further upfield *ca.* four ppm and appeared at *ca.* δ 29. Such effects have been commonly found in the acetogenins having an aliphatic hydroxyl group located only two methylenes away from a THF flanking hydroxyl group (one methylene at *ca.* δ 29 and another at *ca.* δ 33), *e.g.*, in goniothalamicin and (2,4-*cis* and *trans*)-12-hydroxybullatacinones,[33] in those having a 1,4-diol moiety between two THF rings (both methylenes at *ca.* δ 29), including all of the non-adjacent bis-THF acetogenins, and in those having a 1,2,5-triol moiety (both methylenes at *ca.* δ 29), *e.g.*, gigantriocin, gigantetrocins A and B, and muricatetrocins A and B.[34]

The double-β-effects in the ^{13}C NMR are also observed in (2,4-*cis* and *trans*)-28-hydroxybullatacinones,[32] the signals of the middle methylenes (C-26) of the 1,5-diols (24,28-OHs) appeared at δ 22.1, shifted upfield *ca.* eight ppm from the normal value, of *ca.* δ 29, of the hydrocarbon chain methylenes. A signal at *ca.* δ 22 for C-26 was also found in asiminacin, squamocin, rioclarin, panalicin, and squamostatin A; all of these possess C-24 and C-28 hydroxyl groups. Indeed, the observation of double-β-effects helped us in the placements of the fourth hydroxyl group of xylopianin, xylopiacin, and xylopien; a carbon signal, at *ca.* δ 22, indicated this hydroxyl group should be located three carbons away from C-4 which placed it at C-8. Subsequently, these placements were demonstrated to be correct by EIMS of their TMSi derivatives.[35,36]

The β-effects are also helpful to the identification of terminal hydroxylation (Fig. 3). The terminal methyl and α- and β-methylenes of the long hydrocarbon chain normally resonate at *ca.* δ 14, 22, and 32, respectively. Upfield shifts of about four ppm, aroused by the β-effects of an hydroxyl group, were observed in the carbon signals of the terminal methyls at δ 9.9 of (2,4-*cis* and *trans*)-32-hydroxybullatacinones, the α-methylenes at δ 18.8 of (2,4-*cis* and *trans*)-31-hydroxybullatacinones, and the β-methylenes at δ 28.9 of (2,4-*cis* and *trans*)-30-hydroxybullatacinones.[30,32]

The α-methylenes of a double bond normally resonate at *ca.* δ 27. A four ppm upfield shift to *ca.* δ 23 is predictable if the double bond is located two carbons away from an hydroxyl group. Thus, the carbon signal at δ 23.3 for C-20 further confirmed the location of the double bond at C-21/22 in gonionenin. In fact, a carbon signal at *ca.* δ 23 has been observed in all of the acetogenins having a double bond, and the double bonds have been found unambiguously located two carbons away from a hydroxyl group; this positioning is conveniently located, biogenetically, to form another THF ring.

Location of Double Bonds. Since the first acetogenin possessing a double bond was isolated,[37] eight such compounds, all having a C-37 skeleton, have been reported. As the ion fragments from the double bond and its adjacent carbons usually are not stable enough to give good fragment peaks, it is not sufficient to determine the position of the double bond only by MS. The placement of the double bond was first established by Fang *et al.*,[9] using 2D NMR technology with single- and double-relayed COSY spectra, and this method has been extensively demonstrated as useful in the placement of the double bond. For example, in the double-relayed COSY spectrum of gonionenin, the correlation cross peaks of C-18 to C-21 were clearly presented, and, thus, placed the double bond at C-21/22.[38]

So far, the double bond has always been found to be located two carbons away from an hydroxyl group. The unequivalence of the two pairs of protons of the methylenes α to the double bond in the ^{1}H NMR spectra of such compounds at δ 2.19 and 2.04, respectively, helps in this recognition. Further confirmation of the positions of the double bonds was performed chemically by Gu *et al.*[38] C-21/22-Double bonds in gonionenin and gigantetronenin were oxidized and then cyclized with an hydroxyl two carbons away to give new THF rings; the placements of these THF rings were then ascertained by EIMS of their TMSi derivatives. This chemical conversion unambiguously determined the positions of the double bonds in the acetogenins, and, as well, the derivatives also showed enhanced bioactivities. Fujimoto *et al.*[39] chemically cleaved the double bond of squamosten-A by oxidization and then converted the derivatives into *p*-bromophenacyl esters. A peak possessing the same retention time as that of an authentic sample of *p*-bromophenacyl undecanoate on HPLC placed the double bond at C-23. Although the result is reliable, their method requires possession of a series of the appropriate esters for HPLC identification of the ester derivatives of the acetogenins.

Relayed-Cosy NMR Technology. Single- (4 bonds) and double- (5 bonds) relayed COSY methods were first introduced into the structural elucidation of Annonaceous acetogenins by Fang *et al.*[37] In addition to the placements of the double bonds (as described above), these methods are also commonly used to place other functional groups, like hydroxyls, and they always supply quite reliable data. Usually, the τ value is a crucial factor in the relayed COSY experiments. The cross peaks between C-10 and C-13 could be clearly observed and, thus, placed the isolated aliphatic hydroxyl group in gonionenin at C-10. The placements of an hydroxyl group at C-32 in (2,4-*cis* and *trans*)-32-hydroxybullatacinones and at C-31 in (2,4-*cis* and *trans*)-31-hydroxybullatacinones were also suggested by the single- or double-relayed COSY spectra. The cross peaks between terminal methyls (H-34, at δ 0.940) and the proton on the aliphatic hydroxyl-bearing carbon (H-32, at δ 3.52) in the single-relayed COSY spectra of (2,4-*cis* and *trans*)-32-hydroxybullatacinones and the cross peaks between H-34 and H-31 (at δ 3.60 and 0.928) in the double-relayed COSY spectra of (2,4-*cis* and *trans*)-31-hydroxybullatacinones were clearly observed.[30,32]

Threo **and** ***Erythro*** **Vicinal Diols**

The formation of epoxides is one of the major steps in the proposed biogenetic pathways leading to the THF ring(s) of the acetogenins.[8] Presumably, if an epoxide opens but does not cyclize with another epoxide or an hydroxyl group and leaves two free hydroxyl groups, a 1,2-diol occurs; a *cis* epoxide (from a *cis* double bond) gives a *threo* diol and a *trans* epoxide (from a *trans* double bond) gives an *erythro* diol. Indeed, such vicinal diols exist in several of the new acetogenins, *e.g.*, gigantriocin,[40] gigantetrocins A and B,[34,40] muricatetrocins A and B,[34] *etc*. Usually, the relative stereochemistries of such vicinal diols are found to be *threo*. However, recently, *erythro* diols have been identified in (2,4-*cis* and *trans*)-bulladecinones[33] and annomuricin B.[41] The ^{1}H NMR data of an *erythro* diol (at *ca.* δ 3.60) and its acetate and acetonide derivatives are quite distinguishable from those of a *threo* diol (at *ca.* δ 3.40) and its respective derivatives (Table 1). Thus, the relative stereochemistries of such vicinal diols are easily determined by ^{1}H NMR, and future workers are encouraged to provide such determinations when they report new acetogenins that bear vicinal diols.

When Gu *et al.* tried to solve the relative configurations of 1,4- and 1,5-diols existing in certain acetogenins by converting the diols into cyclic intramolecular formaldehyde acetals (for details see sections on Syntheses),[42] they found that the distinct acetal proton signals of the *trans*-formaldehyde acetal ring (at *ca.* δ 4.96 as a singlet, corresponding to a *threo* vicinal diol) and the *cis*-formaldehyde acetal ring (at *ca.* δ 5.26 and 4.63 as two doublets, corresponding to an *erythro* vicinal diol) were also very useful in determining the relative

Table 1. ^{1}H NMR (500 MHz) signals for the protons of *threo* and *erthro* diols in Annonaceous acetogenins and their acetyl and acetonide derivatives ($CDCL_3$, δ)

	Methine Protones		Acetyl Methyls		Acetonyl Methyls	
	threo	*erythro*	*threo*	*erythro*	*threo*	*erythro*
Diols	3.45 (2H)	3.62, 3.58	-	-	-	-
Acetates	5.00 (2H)	4.97, 4.92	2.06, 2.05	2.04 (6H)	-	-
Acetonides	3.58 (2H)	4.03, 4.00	-	-	1.37 (6H)	1.43, 1.33

stereochemistry of 1,2-diols. In addition, the formaldehyde acetal derivatives are relatively stable compared to acetonides, and the acetal proton signals, located downfield from the aliphatic proton signals, can be more easily observed.

STEREOCHEMISTRY

All Annonaceous acetogenins have multiple chiral (stereogenic) centers. A number of them share the same skeleton and are differentiated from each other only by their stereochemistries around the THF subunits, *e.g.*, bullatacin (C-24 *S*) and asimicin (C-24 *R*), bullatalicin (C-24 *S*) and bullatanocin (C-24 *R*), and annonacin A (C-20 *S*) and annonacin (C-20 *R*), *etc*. Consequently, the determination of the relative and absolute stereochemistries of these stereocenters has become a major concern in the structural elucidation of new, as well as previously reported, acetogenin compounds; in addition, the stereochemistries, in many cases, influence the relative potencies and biological specificities.[8,9] Because of their waxy nature, the acetogenins and their derivatives do not readily produce crystals suitable for X-ray crystallographic analysis. Determination of relative stereochemistries around the THF ring(s) and those of the ketolactone moieties, typically by comparisons with series of synthetic model compounds of known relative stereochemistry,[43-46] and the absolute stereochemistry of the Annonaceous acetogenins defined by the application of the Mosher ester methodology[47] have been summarized.[8,9] Some newly developed methods for the determination of stereo-chemistries are discussed below.

Relative Stereochemistry of the THF Subunits

The relative stereochemistries of the THF units are now commonly determined by comparisons of the ^{1}H and ^{13}C NMR spectra of the acetogenins and their acetates with those of model compounds.[43-46] Usually, it is not difficult to determine *threo* or *erythro* relative configurations between THF ring(s) and the flanking hydroxyls by the characteristic ^{13}C NMR signals for the flanking hydroxylated carbons at δ 71.8 (71.2 to 72.5 in acetogenins) for *erythro* and at δ73.9 (73.7 to 75.5 in acetogenins) for *threo* and the corresponding ^{1}H NMR signals at *ca.* δ3.80 for *erythro* and at *ca.* δ3.40 for *threo*.[43,44] However, in some cases, it may not be easy to determine the relative configurations across the THF rings by simply analyzing the NMR data of the acetates of acetogenins. We observed that the ^{1}H NMR resonances of the THF methylenes of the natural acetogenins were characteristic in *cis* or *trans* THF rings. In Fig. 4 are illustrated the partial ^{1}H NMR spectra of gigantecin (C-18/21-*trans*) and the semi-synthetic compounds, C-18/21-*cis*-gigantecin, cyclogonionenin T (C-18/21-*trans*), and cyclogonionenin C (C-18/21-*cis*), showing that one of the methylene protons of the *cis*-THF rings resonates more upfield while the other one resonates more downfield (those of cyclogonionenin C and C-18/21-*cis*-gigantecin), by comparisons with those of

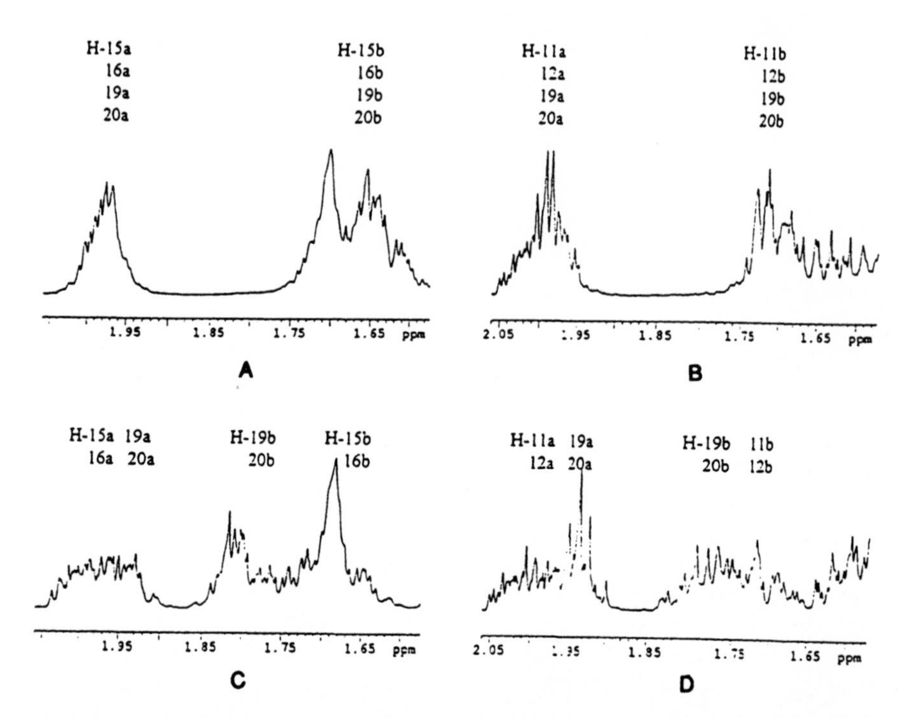

Figure 4. Partial ^{1}H NMR spectra of cyclogonionenin T (A), cyclogonionenin C (C), gigantecin (B), and C-18/21-*cis*-gigantecin (D), showing the striking differences of the proton chemical shifts of the methylenes of *cis*-THF rings from those of *trans*-THF rings.

the methylene protons of the *trans*-THF rings (those of gigantecin and cyclogonionenin T), respectively.

Recently, the Fujimoto group synthesized a series of convenient model mono-THF ring compounds with all the possible relative stereochemistries (Figs. 5 and 6) and recorded their ^{1}H and ^{13}C resonances (Tables 2, 3).[48] We compared these NMR data with those of the mono-THF and non-adjacent bis-THF acetogenins available and found they matched well. Thus, their work will be quite helpful in the determination of the relative stereochemistries of mono-THF and non-adjacent bis-THF units, or even adjacent bis-THF units, since no ^{13}C NMR data of such model compounds are available for direct comparisons.

Another useful contribution to the determination of partial relative stereochemistry has been given by Gu *et al.*; 1,2-, 1,4- and/or 1,5-diols of appropriate acetogenins have been converted into cyclic intramolecular formaldehyde acetals using chlorotrimethylsilane (Me_3SiCl) and dimethyl sulfoxide (Me_2SO).[42] The acetal moiety which is formed connects the diols but does not change the stereochemistries of their carbinol centers. Significant

Table 2. ^{1}H and ^{13}C NMR data of di-hydroxl flanking mono-THF ring model compounds[83]

A-B-C	*threo-trans-threo*		*threo-cis-threo*		*threo-trans-erythro*		*threo-cis-erythro*	
	^{1}H	^{13}C	^{1}H	^{13}C	^{1}H	^{13}C	^{1}H	^{13}C
4	-	29.4	-	29.3	-	29.4	-	29.3
5	-	25.5	-	25.6	-	25.5	-	25.7
6	-	33.4	-	34.0	-	33.2	-	34.2
7	3.41	74.0	3.42	74.3	3.40	74.3	3.44	74.2
8	3.79	82.7	3.82	82.8	3.82	83.3	3.83	82.3
9a	1.98	28.8	1.93	28.1	2.00	28.6	1.92	28.4
9b	1.66		1.74		1.64		1.76	
10a	1.98	28.8	1.93	28.1	1.91	25.2	1.96	24.1
10b	1.66		1.74		1.86		1.80	
11	3.79	82.7	3.82	82.8	3.82	82.2	3.90	82.8
12	3.41	74.0	3.42	74.3	3.88	71.6	3.83	72.1
13	-	33.4	-	34.0	-	32.5	-	33.1
14	-	25.5	-	25.6	-	25.9	-	25.9
15	-	29.4	-	29.3	-	29.3	-	29.3

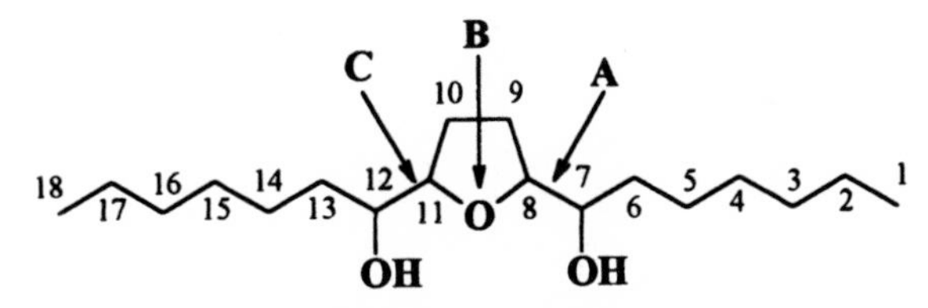

Figure 5. Di-hydroxyl flanking mono-THF rings.

differences in the ^{1}H NMR signals between the acetal protons in the *cis* (at *ca.* δ 5.26 and 4.63 as two doublets) or *trans* (at *ca.* δ 4.96 as a singlet) configurations of the cyclic formal derivatives then make it easy to assign the relative stereochemistries of the diols in the parent compounds. None of the relative configurations of the two hydroxylated carbon centers between two THF rings in the non-adjacent bis-THF rings had been defined before the application of this method. In addition, the relative stereochemical relationships with the THF subunits of some of the isolated aliphatic hydroxyl groups were also determined. Furthermore, the *erythro* configurations in bullatalicin, (2,4-*cis* and *trans*)-bullatalicinones, and squamostatin A were definitely assigned at C-23/24 and not at C-15/16 or C-19/20 after this

Table 3. ^1H and ^{13}C NMR data of single hydroxl flanking mono-THF ring model compounds[83]

A-B	*trans-threo*		*cis-threo*		*trans-erythro*		*cis-erythro*	
	^1H	^{13}C	^1H	^{13}C	^1H	^{13}C	^1H	^{13}C
4	-	29.3	-	29.3	-	29.2	-	29.2
5	-	29.7	-	29.7	-	29.7	-	29.7
6	-	26.2	-	26.2	-	26.1	-	26.2
7	1.64-1.53	35.7	-	36.1	1.57	36.1	1.58	35.8
8	3.88	79.3	3.86	79.9	3.95	80.2	3.79-3.90	79.6
9	2.03	32.4	1.96	31.4	2.03	32.3	1.95	31.4
10a	1.95	28.4	1.89	27.8	1.91-1.77	25.0	1.85	23.9
10b	1.64-1.53		1.66		1.91-1.77		1.73	
11	3.78	81.9	3.70	82.2	3.88	81.5	3.79-3.90	82.1
12	3.37	74.2	3.36	74.5	3.78	72.0	3.79-3.90	71.6
13	-	33.4	-	34.0	-	32.6	-	32.6
14	-	25.6	-	25.7	-	26.0	-	25.9
15	-	29.4	-	29.4	-	29.4	-	29.3

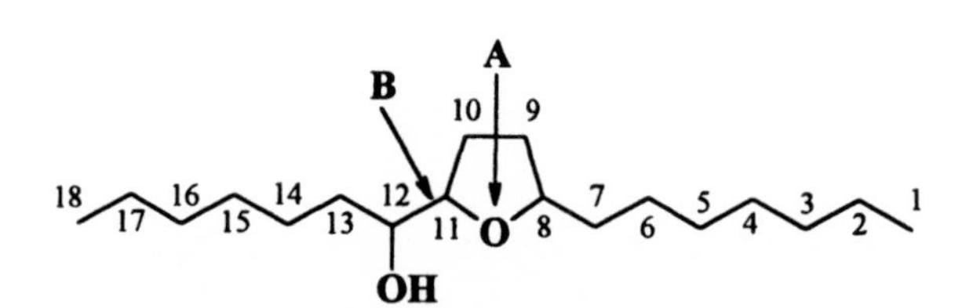

Figure 6. Single hydroxyl flanking mono-THF rings.

conversion, since, in their respective acetal derivatives, only the THF-flanking hydroxyl at C-24 is left free and the chemical shift of H-24 at δ 3.82 clearly indicates that it is *erythro*.[43,44]

Gale *et al.* synthesized a series of dimesitoate esters of α,α-dibutyl-2,5-THF-dimethanols of different relative stereochemistries and recorded their ^1H NMR data.[49] Since the relative stereochemistries can now be easily resolved by the ^1H NMR chemical shifts of the natural acetogenins, we have never tried this method. But the incorrect assignments of the relative stereochemistries of densicomacins,[50] by the authors themselves (for details see section on Structural Discussion and Revision), creates doubts about the adaptability of this method.

Absolute Stereochemistry of the THF Subunits

The absolute configurations of the acetogenins are currently being determined by the use of the advanced Mosher ester [methoxy(trifluoromethyl)phenylacetate or MTPA] methodology.[47] Details of this method were summarized previously,[9] and the method has been demonstrated to be helpful. The refined Mosher ester methodology analyzes differences between the proton chemical shifts of *S*- and *R*-MTPA esters on both sides of the chiral carbinol centers. However, this procedure cannot be directly applied to the non-adjacent THF acetogenins, such as bullatanocin, bullatalicin, and squamostatin A, because the hydroxyls between the two THF rings are only two carbons apart and the phenyl rings of the Mosher esters interfere with each other; it is also not feasible to assign accurately the complicated proton chemical shifts of the per-Mosher esters of these compounds. Similar problems are also encountered in the other subclasses with acetogenins that have hydroxyls in close proximity to each other, *e.g.*, those having a vicinal diol like gigantetrocin A, those having a 1,4-diol like goniothalamicin, and those having a non-THF-flanking hydroxyl like squamocin. Therefore, some efforts have been made to solve these problems.

As mentioned, the formaldehyde acetal derivatives of 1,2-, 1,4-, and/or 1,5-diols successfully resolve the relative stereochemistry between the diols.[42] Since this method does not affect other isolated hydroxyls, these other hydroxyls, then, are free to be converted into Mosher esters. With the relative stereochemistries around the THF ring(s) already in hand from comparisons of ^{1}H NMR spectra with those of model compounds, the absolute stereochemistries of all of the stereocenters can then be concluded by analyses of the ^{1}H NMR data of *S*- and *R*-MTPA esters of the formal derivatives.

For example, in Fig. 7 is illustrated just how the absolute configuration of gigantetrocin A was determined. Two formaldehyde acetal derivatives were obtained and separated by HPLC. One of them, the C-14/17 *cis* acetal ring derivative, revealed that the relative configuration between the C-14 and C-17 carbinol centers should be either *S*/*R* or *R*/*S*. While, the other, the C-17/18 *trans* acetal ring derivative, confirmed the *threo* relative stereochemical relationship of C-17/18, and, in addition, as it was available in more quantity than the first derivative, it was converted to the *S*- and *R*-Mosher esters. Consequently, the absolute configurations of the carbinol centers at C-4 and C-14 were determined to be *R* and *S*, respectively. Based on the already defined relative stereochemistry, the absolute configuration of gigantetrocin A was concluded to be 4*R*, 10*R*, 13*S*, 14*S*, 17*R*, 18*R* and 34*S*.[42] In this way, the absolute configurations of bullatanocin, (2,4-*cis* and *trans*)-bullatanocinones, bullatalicin, (2,4-*cis* and *trans*)-bullatalicinones, squamostatin A, squamocin, goniothalamicin, and (2,4-*cis* and *trans*)-28-hydroxybullatacinones were also determined.[42]

The absolute configurations at C-28 of squamocin, squamostatin A, and (2,4-*cis* and *trans*)-28-hydroxybullatacinones all appear to be *S*. In analyses of the ^{1}H NMR data of per-Mosher ester derivatives of these compounds, we found

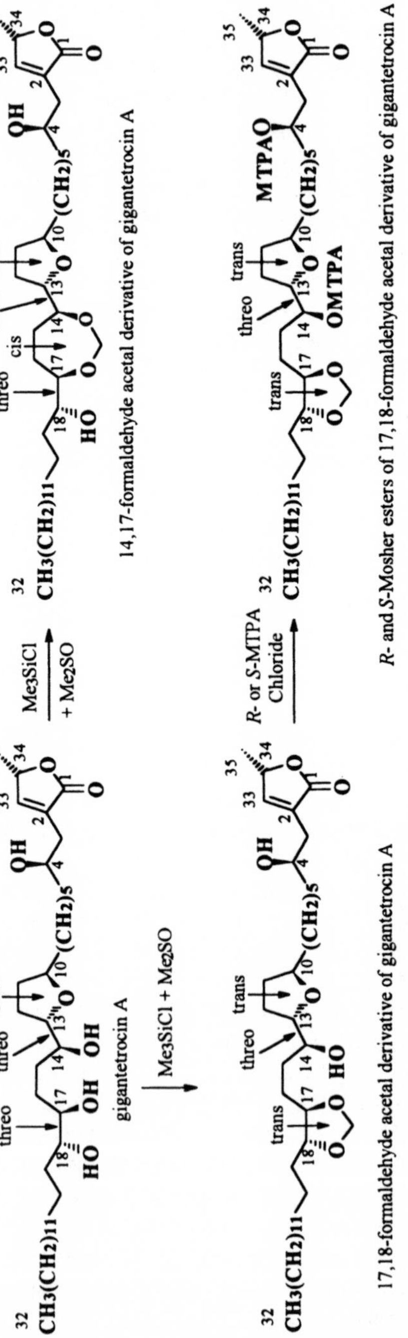

gigantetrocin A

$Me_3SiCl + Me_2SO$

17,18-formaldehyde acetal derivative of gigantetrocin A

Me_3SiCl + Me_2SO

14,17-formaldehyde acetal derivative of gigantetrocin A

R- or *S*-MTPA Chloride

R- and *S*-Mosher esters of 17,18-formaldehyde acetal derivative of gigantetrocin A

Figure 7. Gigantetrocin A and its derivatives.

that the differences [$\Delta\delta_{H\text{-}34}(\delta_S - \delta_R)$] of the chemical shifts of the terminal methyls (C-34) of the *S*- (at δ 0.860) and *R*- (at δ 0.875) Mosher esters presented negative values, which also suggested the *S* absolute configuration for C-28. This conclusion gave us the revelation that the absolute configurations of such terminal carbinol centers (C-28 through C-32) of the acetogenins can be easily defined by simply observing the chemical shift changes of the terminal methyls in the ^{1}H NMR spectra of the *S*- and *R*-Mosher ester derivatives.

Nishioka *et al.* obtained similar results.[51] From the $\Delta\delta_H(\delta_S - \delta_R)$ values of the terminal methyls of the *R*-Mosher esters of 6-undecanol (- 0.039) and 8-pentadecanol (- 0.007), they concluded that this method can be applied to a carbinol stereocenter located six, or even seven, bonds from the methyl group. They determined the absolute configurations of several acetogenins by observing the differences of the ^{1}H NMR chemical shifts of the terminal methyls (C-34 or C-32) between *S*- and *R*-Mosher ester derivatives.

Shimada *et al.* also observed that the absolute configurations of the non-adjacent bis-THF moiety could not be solved by the analyses of the $\Delta\delta_H(\delta_S - \delta_R)$ values.[52] They tried to solve this problem by comparisons of the ^{1}H NMR spectra of the *R*-Mosher esters of the non-adjacent bis-THF acetogenins with those of model mono-THF compounds. The absolute configurations of five non-adjacent acetogenins were proposed. However, the limitations of this method are obvious if an acetogenin possesses a relative stereochemistry in the non-adjacent THF moiety different from those presented in the paper, the synthesis of new model compounds becomes necessary.

Another approach to determine absolute stereochemistry was made by Rieser *et al.* through the tedious preparation of mono-Mosher esters, and the absolute configurations of four acetogenins, muricatetrocins A and B and gigantetrocins A and B, were, thus, defined.[34] After partial esterification, the products were separated by HPLC, and the positions of the Mosher esters were determined by EIMS of the TMSi derivatives. The results obtained with gigantetrocin A, as determined by this method, is exactly the same as that described above and demonstrates the accuracy of both methods.

Absolute Stereochemistry of the γ-Lactone Subunits

In the α,β-unsaturated γ-lactone moiety, the absolute configuration of C-36 (or C-34) has been found always to be *S* through several different methods.[1,39,43,47] The stereochemistry of the hydroxylation at C-4 has been determined always to be *R* by Mosher's methodology.[47] Since the ^{1}H NMR resonances of the ketolactone part of the ketolactone acetogenins agreed well with those of the ketolactone products converted from a 4-hydroxylated α,β-unsaturated γ-lactone subunit, the stereocenter at C-4 is thought to keep the *R* absolute configuration in the ketolactone acetogenins.[47] Thus, the absolute configurations of the (2,4-*cis*-) and (2,4-*trans*)-ketolactone subunits are suggested to be 2*R*,4*R* and 2*S*,4*R*, respectively.

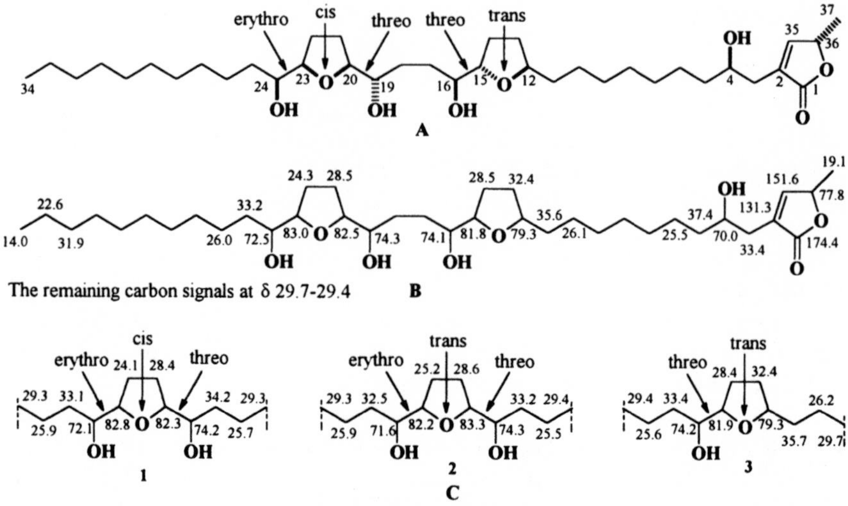

Figure 8. (A) The revised structure of sylvaticin (absolute stereochemistry is proposed based on the close similarity to bullatalicin, whose absolute configuration is known); (b) the reassignments of the ^{13}C NMR data of sylvaticin; (C) the ^{13}C NMR data of model mono-THF compounds.[83]

STRUCTURAL DISCUSSION AND REVISION

Sylvaticin and Uleicin A. Sylvaticin (**A**, Fig. 8) was the third reported member of the relatively rare subclass of non-adjacent bis-THF acetogenins.[53] The structural skeleton, established by EIMS fragmentation analyses of the TMSi derivative, has never been in doubt. Based on the ^{1}H and ^{13}C NMR data, the stereochemical relationship between the three THF flanking hydroxyl groups and the THF rings were concluded to involve two *threo* and one *erythro* arrangements; this is the same as that of bullatalicin; however, the obvious differences of both proton and carbon chemical shifts between the two compounds, due to their non-adjacent bis-THF ring moieties, demonstrated that they were different compounds.[48] In order to distinguish sylvaticin from bullatalicin, the *erythro* configuration was suggested, by the authors, to be at C-19/20. This conclusion, as well as assignment of the *trans* configuration for the THF ring at C-20 to C-23, was not correct.

As mentioned, the Fujimoto group has synthesized a series of model mono-THF ring compounds with all of the possible relative stereochemistries and has recorded their ^{1}H and ^{13}C resonances (Tables 2, 3).[48] By comparisons of the NMR data of these model compounds with those of the well-established bullatalicin and bullatanocin,[54] we have reassigned the published ^{13}C NMR chemical shifts of sylvaticin (**B**, Fig. 8). The carbon signals of C-10 to C-16 matched well with those of a model THF compound flanked by a single-hydroxyl and having a *trans-threo* configuration (**C-3**, Fig. 8) and were also very similar to those of this moiety in both bullatalicin and bullatanocin.

The other two carbinol carbon signals at δ 74.1 and 72.5 in sylvaticin, adjacent to the second THF ring with corresponding proton resonances at δ 3.48 and 3.88, respectively, indicated that one of the relative configurations of the carbon centers at C-19/20 and C-23/24 was *threo* while the other one was *erythro*. Thus, the relative stereochemistry of the second THF moiety (C-19 to C-24) was either *threo-trans-erythro* or *threo-cis-erythro*. The carbon signals for C-22 at δ 24.3, C-23 at δ 83.0, C-24 at δ 72.5, and C-25 at δ 33.2 (**B**, Fig. 8) suggested that this THF ring is more likely to be *cis* (the respective carbon signals of the model compound resonate at δ 24.1, 82.8, 72.1, and 33.1, **C-1**, Fig. 8) rather than *trans* (the respective carbon signals of the model compound resonate at δ 25.2, 82.2, 71.6, and 32.5, **C-2**, Fig. 8). The assignment of the *erythro* configuration at C-23/24 was based on the comparisons with bullatalicin, bullatacin, squamostatin A, *etc.*; in these compounds the *erythro* configuration is always located at C-23/24. This conclusion was also supported by a carbon signal at δ 33.1 (C-25). If the relative stereochemistry at C-23/24 is *threo*, a carbon signal at δ 34.2 should appear for C-26.[48] Therefore, the relative configuration of the non-adjacent bis-THF moiety of sylvaticin was concluded to be *trans-threo-threo-cis-erythro*, and the revised structure of sylvaticin, with the absolute configuration proposed similar to that of bullatalicin, is presented in **A**, Fig. 8. Sylvaticin is, thus, the first example of a non-adjacent bis-THF acetogenin which bears a *cis* C-20 to C-23 THF ring in the molecule.

Uleicin A was reported as an adjacent bis-THF acetogenin.[25] However, the published data of uleicin A, especially with a carbon signal at δ 79.3, suggest it to be a non-adjacent bis-THF acetogenin. The oxygenated carbon signals for the non-adjacent bis-THF moiety, at δ 82.9, 82.3, 81.8, 79.3, 74.2, 73.9, and 72.2, are very similar to those of sylvaticin and indicate that this moiety possesses a *trans-threo-threo-cis-erythro* configuration from C-12 to C-24. Thus, uleicin A appears to have the same structure as that of sylvaticin.

Squamostatin A and Almunequin. When Cortes *et al.* published almunequin,[55] they assigned a relative configuration of *trans-threo-threo-trans-erythro* for its non-adjacent bis-THF moiety. In order to establish this as a new compound, they also suggested the reassignment of the relative configuration of the C-19 to C-24 THF moiety of squamostatin-A to be *threo-cis-erythro*, instead of *threo-trans-erythro* as suggested by us earlier;[54] their reassignment[55] was based on only two carbon signals at δ 82.2 and 82.4 as reported for squamostatin-A by Fujimoto *et al.*[56] All of the ^{1}H and ^{13}C NMR data for the C-19 to C-26 region of squamostatin A match well with those of a *threo-trans-erythro* model bis-OH flanking THF compound (**C**-2, Fig. 8) except for the carbon signal at δ 82.2; the authors have informed us that this signal was a misprint for δ 83.2. Thus, squamostatin-A, like bullatalicin, possesses a relative configuration of *trans-threo-threo-trans-erythro* for the non-adjacent bis-THF moiety, and there are no spectral or structural differences between squamostatin A and almunequin. We have recently determined the absolute configuration of squamostatin-A, after cyclizing the C-16 and C-19 hydroxyl groups to a formaldehyde acetal, and have determined the absolute configuration of the carbinol center at C-24 to be *S* by the application of Mosher ester methodology.[42] These results also further confirmed the relative stereochemistry. Thus, almunequin is identical to squamostatin A whose absolute stereochemistry is now defined.

Yu *et al.*[57] reported a compound named squamostatin-B possessing a structure different from another squamostatin-B previously reported by Fujimoto *et al.*[48] Obviously, one of the two compounds needs to be renamed. When Yu *et al.* published their structure, they made a comparison with the planar structure of squamostatin-A, but they did not compare with the structure of squamostatin-A with its relative stereochemistry defined.[54] In fact, the published partial ^{13}C NMR data of the squamostatin-B of Yu *et al.* are similar to the refined assignment of the ^{13}C NMR of squamostatin-A, having only up to 0.1 ppm differences for respective ^{13}C NMR chemical shifts.[48] Therefore, the squamostatin-B of Yu *et al.*[57] is considered to be the same as squamostatin-A.

Giganenin. Eight mono-THF acetogenins have been reported as possessing a double bond in their structures. All of the double bonds have consistently been located between the THF ring and the terminal methyl group except in the published structure of giganenin.[37] Re-examination of the ^{1}H and ^{13}C NMR spectra of giganenin showed that, although the positional relationships of the

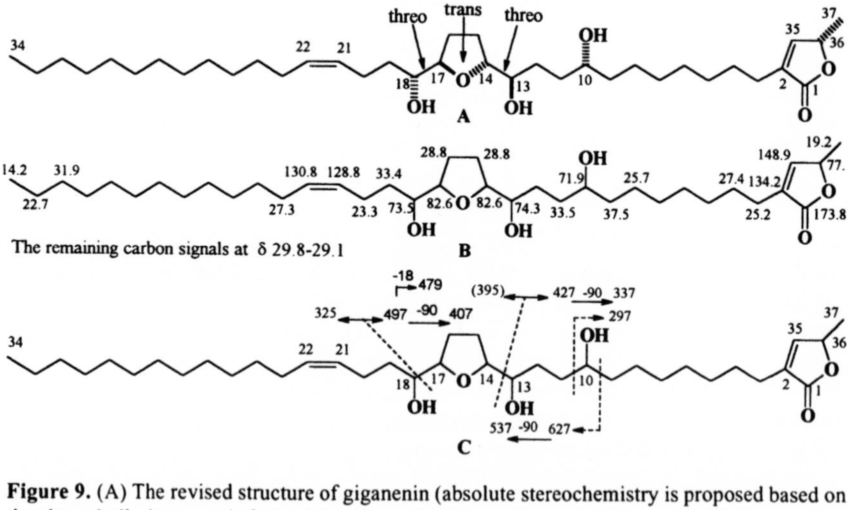

Figure 9. (A) The revised structure of giganenin (absolute stereochemistry is proposed based on the close similarity to goniothalamicin, whose absolute configuration is known); (B) the reassignments of the ^{13}C NMR data of giganenin; (C) the EIMS fragmentation of the TMSi derivative of giganenin.

mono-THF ring, the three hydroxyl groups, and the double bond were not in doubt, the placements of the isolated chain hydroxyl group and the double bond were interchangeable. The reassignments of the EIMS spectra of giganenin and its TMSi and deutero-TMSi derivatives led us to the corrected structure (**A**, Fig. 9); this structure shows a close structural relationship with gonionenin,[38] whose double bond placement was confirmed by C-18/21 cyclization, and it better integrates into the biogenetic scheme for the origin of the C-10 hydroxyl group of the acetogenins occurring in *Goniothalamus giganteus*.

Annonin XIV. In 1990, four annonins were reported by the Bayer company from the seeds of *Annona squamosa*, and all were reported as having unusual hydroxyl-substituted adjacent bis-THF rings.[37] All of these structures were wrong and have now been corrected. Three of them, annonin IV, annonin VIII, and annonin XVI, were revised and identified to be the non-adjacent bis-THF acetogenins, bullatalicin, bullatanocin, and squamostatin A, respectively;[54] however, annonin XIV was left undefined, because, at that time, the two protons at δ 3.60, with corresponding ^{13}C NMR signals at δ 74.7 and 74.6, could not be satisfactorily explained, although all of the other ^{1}H and ^{13}C NMR data indicated that it was a bullatacin type adjacent bis-THF acetogenin. An *erythro* vicinal diol, established in the bulladecinones,[33] now provides data which indicate that an *erythro* vicinal diol exists in the structure of annonin XIV (Table 1), and the other published ^{1}H and ^{13}C NMR data of annonin XIV also support this conclusion (**B**, Fig. 10). The location of the diol was suggested by the fragment ions from the published EIMS data of annonin XIV (**C**, Fig. 10). Thus, the revised structure is suggested as illustrated in **A**, Fig. 10, with the diol located at C-11/12. Recently, an *erythro*

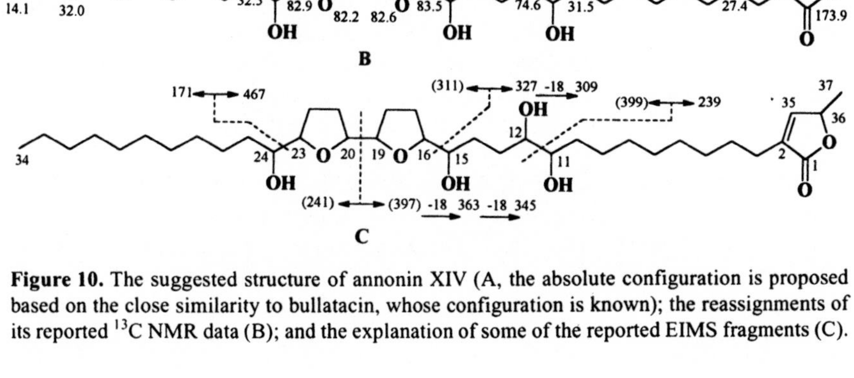

Figure 10. The suggested structure of annonin XIV (A, the absolute configuration is proposed based on the close similarity to bullatacin, whose configuration is known); the reassignments of its reported ^{13}C NMR data (B); and the explanation of some of the reported EIMS fragments (C).

vicinal diol at C-11/12 was also identified in a new mono-THF acetogenin, annomuricin B, isolated from the seeds of *Annona muricata*.[41]

Trilobacin. Trilobacin stood for a long time as the only member of the adjacent bis-THF acetogenins reported as having an *erythro* relative stereochemistry between the two THF rings,[58] and, recently, another member, trilobin, of this relatively rare type of acetogenin was isolated.[59] The relative stereochemistry of both THF rings was suggested to be *threo* when trilobacin was published.[58] Zhao *et al.* determined the absolute configuration of trilobacin and found both of the THF flanking carbinol centers to be *R*,[59] which led them to reassign the relative stereochemistries of the bis-THF rings to be one *trans* and one *cis*. In fact, the ^{1}H NMR data of the bis-THF moiety of trilobacin triacetate matches those of a model adjacent bis-THF compound with the *threo-trans-erythro-cis-threo* relative stereochemistry better than those of a model bis-THF compound with the *threo-trans-erythro-trans-threo* relative stereochemistry.[45] The presumed biogenetic pathway of the bis-THF moiety also supports this reassignment. Thus, the corrected trilobacin type of adjacent bis-THF acetogenins is proposed to have the *threo-trans-erythro-cis-threo* relative stereochemistries between C-15 and C-24.

Gigantetrocins A and B and Muricatetrocins A and B. The names of these four compounds are confused because they have been changed from gigantetrocin, gigantetrocin A, muricatetrocin, and muricatetrocin A, respectively, as described in our second review.[9] These four compounds share similar structures, possessing a THF ring with a single flanking hydroxyl and a vicinal diol. The first, gigantetrocin A, was isolated by Fang *et al.*[40] from *Goniothalamus giganteus*, and its absolute configuration was determined by Gu *et al.*[42] through formaldehyde derivatives and Mosher ester methodology. Rieser *et al.* reported, from *Annona muricata*, all of the four compounds (three were new) with absolute configurations defined as determined by the mono-Mosher ester method.[34] Unfortunately, all of the systematic chemical names which appeared in the abstract of the paper[34] were published incorrectly and have now been corrected.[60] The "m" and "n" in the same article, presenting the length of the two aliphatic chains, were also confused and should be interchanged.[60] Briefly, the gigantetrocins differ from the muricatetrocins in the position of the THF rings, at C-10 and C-12, respectively. Gigantetrocin A (17*R*, 18*R*) differs from gigantetrocin B (17*S*, 18*S*) in the stereochemistries of the C-17/18-diols, while muricatetrocin A (12*S*, 15*S*) differs from muricatetrocin B (12*R*, 15*S*) in the stereochemistries of the mono-THF rings.

Densicomacins. Two desicomacins were reported having the same skeleton as those of the gigantetrocins.[50] The reported difference between the two compounds, as defined by the authors through mesitoate derivatives, is the *threo* or *erythro* relative configurations between the mono-THF rings and their flanking hydroxyls (C-13/14). This conclusion is wrong. The ^{13}C NMR signals of the THF flanking hydroxylated carbon (C-14) in both compounds appeared at *ca.* δ 74, with corresponding ^{1}H NMR resonances at δ 3.42, and indicate that both compounds have the *threo* relative configuration.[43-45,48] The difference between the two compounds lies in the stereochemistries of the diols. Careful comparisons of the published data have led us to conclude that the densicomacins are the same as the gigantetrocins. Gigantetrocin A was published at least one year before the densicomacins were published and, thus, has precedence.[40]

Annonastatin and Epoxyrollins A and B. All Annonaceous acetogenins possess either C-37 or C-35 skeletons except these three compounds, which were reported to have either a C-38 or a C-36 skeleton; all of these proposed structures are questionable. The ^{13}C NMR chemical shifts of annonastatin,[116] given for the oxygenated THF carbons (C-16 and 19) at δ 83.2 and 81.8, do not match any of the model mono-THF compounds.[48] The molecular weight of 622, instead of $C_{38}H_{70}O_6$, can be alternatively assigned as $C_{37}H_{66}O_7$, which is the molecular formula of many of the bis-THF acetogenins. The only two carbon signals for the oxygenated THF carbons, the same as those of asimicin, can be explained by the well-known overlapping of signals created by a pseudo-symmetrical relative stereochemistry, *i.e.*, *threo-trans-threo-trans-threo*, existing in the adjacent bis-THF moiety. Thus, as suggested by Myint *et al.*,[61] the structure of annonastatin is quite likely the same as that of asimicin. Similarly, the structures of epoxyrollins A and B can be alternatively reassigned to be diepoxy compounds having a C-35 ($C_{35}H_{62}O_4$) or a C-37 ($C_{37}H_{66}O_4$) skeleton,[26] respectively.

Diepoxymontin. Another diepoxide acetogenin, diepoxymontin, appeared in the literature.[62] Although adjacent epoxides would be a new structural feature in the acetogenins, the proposed structure is doubtful; the ^{13}C NMR signals at δ 56.9 and 57.8 for the epoxymethines in diepoxymontin are similar to those (reported at δ 56.7 and 57.3) of corepoxylone,[63] a diepoxide acetogenin whose two epoxides are separated by two methylenes, and those (reported at δ 56.3 and 57.2) of epoxymurins A and B, two monoepoxide acetogenins.[64] Also, it is difficult to explain the biogenetic relationships between the proposed structure and the other acetogenins. In addition, the major EI peak at m/z 295 (100 %) of diepoxymontin can also be explained for the fragment cleaved between C-15/16 if there is an epoxide at C-15/16. Thus, we suggest that the proposed structure of diepoxymontin requires revision.

Uleicins A-E. The structures of uleicins A-E puzzled us for a long time because the published ^{1}H and ^{13}C NMR data do not match each other or agree with the skeletons proposed by the authors based on various MS spectra, particularly those of negative CI and/or MS-MS.[25] Recently, the five structures were reported as incorrect.[22] Uleicins A-D were thought to be non-adjacent acetogenins and Uleicin E was indicated to be a mixture of the ketolactone acetogenins.

Senegalene. The structure of senegalene requires revision.[65] All of its ^{1}H and ^{13}C NMR data are similar to those of gigantetronenin;[70] however, the double bond of senegalene was isolated along the hydrocarbon chain, located far away from any hydroxyl. The structure given cannot explain the unequivalence of the two pairs of methylene protons adjacent to the double bond at δ 2.19 and 2.01, respectively, in the ^{1}H NMR; these signals are exactly the same as those of gigantetronenin and all other acetogenins having a double bond, which is always located two carbons away from an hydroxyl. The position of the double bond of gigantetronenin was unambiguously determined by chemical methods (see section on Biomimic Semisythesis) and the ^{1}H NMR assignments were made by COSY and single- and double-relayed COSY spectra. Apparently, the more downfield shifts of one pair of the methylene protons adjacent to the double bond, compared to the other pair, are caused by the nearby hydroxyl. Obviously, the structure of senegalene is incorrect. The biogenetic pathway of the acetogenins also cannot explain the proposed structure. The presence of a MS fragment at m/z 281, a major fragment ion observed in the EIMS spectrum of gigantetronenin, indicates that these two compounds may be the same. The authors need to confirm the structure through epoxidation of the double bond and cyclization to form another THF ring as was done with gigantetronenin.[38]

Squamocin-I. The *erythro* relative stereochemistry in squamocin-I was assigned at C-15/16,[66] based on tiny differences in the ^{13}C NMR data compared to those of neoannonin, rather than at C-23/24 as in the bullatacin and bullatalicin types. The assignment of *erythro* at C-23/24 in bullatacin was unambiguously determined by EIMS fragmentations of the TMSi derivatives of its C-16- and C-24-mono Mosher esters.[67] As squamocin-I would represent a new type of the acetogenins, confirmation of the assigments of the relative stereochemistry seems especially necessary.

Ketolactone Acetogenins. The conversion of the 4-OH α,β-unsaturated lactone subunit to the ketolactone subunit initially revealed the structural relationship of these two kinds of acetogenins.[29] Since the first successful conversion of bullatacin to the (2,4-*cis* and *trans*)-bullatacinones, researchers have debated whether or not the ketolactone compounds are natural. The usual mixture of C-2/4 *cis* and *trans* (C-2 being either *R* or *S*) of the ketolactone subunit seems to suggest they are not. However, new ketolactone acetogenins keep being isolated and reported from extracts of Annonaceous plants. Recently, Duret *et al.*[68] experimentally demonstrated that such translactonization takes place not only in strongly basic conditions, *e.g.*, 2 % KOH in *tert*-BuOH, but also under weakly basic conditions, *e.g.*, 2 % diethylamine in CH_2Cl_2; even alkaloids existing in the same plant material may cause the translactonization. Refluxing in MeOH also leads to the translactonization, and silica gel catalyzes the translactonization. From the CH_2Cl_2 extract of plant material previously extracted by acid water to remove alkaloids, none of the ketolactone acetogenins were detected by HPLC. These results suggest that the ketolactone acetogenins are probably artifacts from the extraction and chromatographic fractionation procedures. The proposed mechanism of the translactonization is proposed.[68] Although these results may decrease the interests of future workers in these structures, their bioactivity data certainly differ from those of their parent 4-OH acetogenins and will continue to attract further biological evaluations, *e.g.*, the (2,4-*cis* and *trans*)-bullatacinones are more efficacious, but less potent than bullatacin, against L-1210 murine leukemia *in vivo*.[69]

HYPOTHESIS OF BIOSYNTHESIS

Since the first member, uvaricin, of the Annonaceous acetogenins was reported,[1] it has been, quite logically, proposed that the members of this class of natural products are derived from the polyketide pathway.[8] The THF rings are suggested to arise from diene, triene, or triene ketone groups through epoxidation and cyclization. The discovery of a double bond in certain acetogenins strongly supported this hypothesis. Thus, Fang *et al.* reasonably explained the interrelated biogenetic pathways of several acetogenins isolated from *Goniothalamus giganteus*.[70] Reports of acetogenins containing both a double bond and an epoxide and the isolation of the diepoxide acetogenins provide additional important intermediates in such a biogenetic scheme.[24,63,64,71] The determination of the absolute configurations of most types of acetogenins also helps us now to understand better the biogenetic pathways to the THF units.

In Fig. 11 are illustrated the possible biogenetic pathways to the different types of adjacent bis-THF subunits. All five types may start from similar trienes and undergo similar cyclizations. The configurations of the starting double bonds determine the final configurations between the hydroxylated carbinol centers and the THF rings, *i.e.*, *cis* double bonds become *threo* and *trans* become *erythro*, while the *cis* and *trans* types of THF rings are dependent on the epoxide direction from which the C-19/20 double bond is formed. In this way, bullatacin and all of its relative types, for example, may be derived from the same precursor (Fig. 12). The epoxidation of all three double bonds from the same face, but cyclization through three different ways (either initiating from the left side, the right side, or both sides), leads to the different types. The determination of the absolute configuration at C-20 of the bulladecinones to be *S*, as we biogenetically predicted, not *R*, as in the bullatacin and bullatalicin types (Fig. 12), gave us support for this proposal. The "right side" cyclization, as the bulladecinones probably undergo, indicates that tri-THF acetogenins should exist, if further "right side" cyclization proceeds. Indeed, just recently, the first tri-THF acetogenin, goniocin, was isolated from *Goniothalamus giganteus*.[72]

BIOMIMIC SEMISYNTHESES

The double bond in some acetogenins provides some important circumstantial evidence for the proposed biogenetic pathway to the formation of the THF ring(s). In order to demonstrate this concept further, some experiments of biomimic semisyntheses were performed. The double bonds in two mono-THF compounds, gonionenin and gigantetronenin, were oxidized to epoxides by *m*-chloroperbenzoic acid (*m*-CPBA) and then cyclized under acidic conditions, with an hydroxyl located across two methylenes away, to form another THF ring and to give pairs of adjacent bis-THF (cyclogonionenin C and cyclogonionenin T) and non-adjacent bis-THF (gigantecin and C-18/21-*cis*-gigantecin) acetogen-

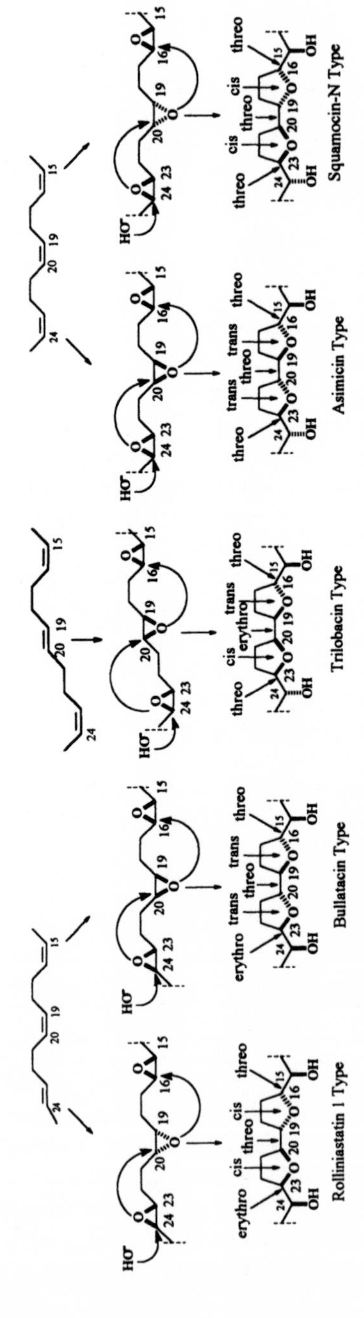

Figure 11. Hypothesis for the biogenesis of some types of the adjacent bis-THF rings of Annonaceous acetogenins.

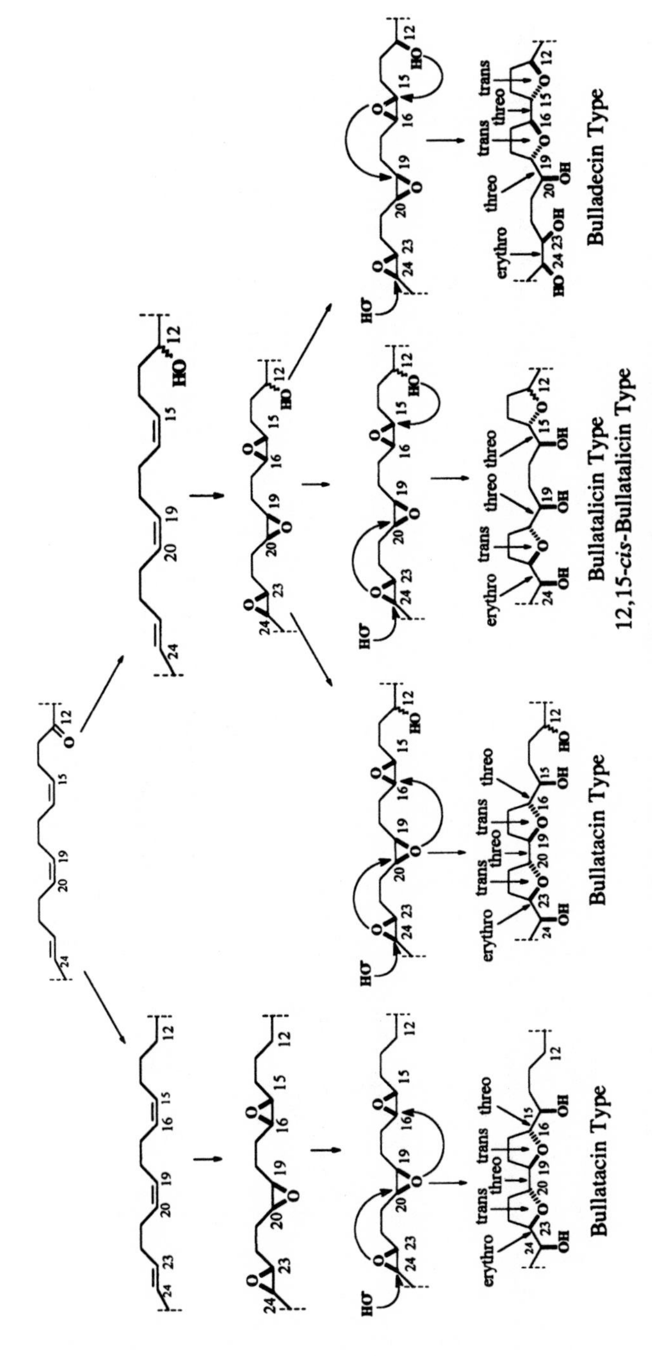

Figure 12. Hypothesis for the biogenesis of the THF-rings of bullatacin and its related types of Annonaceous acetogenins.

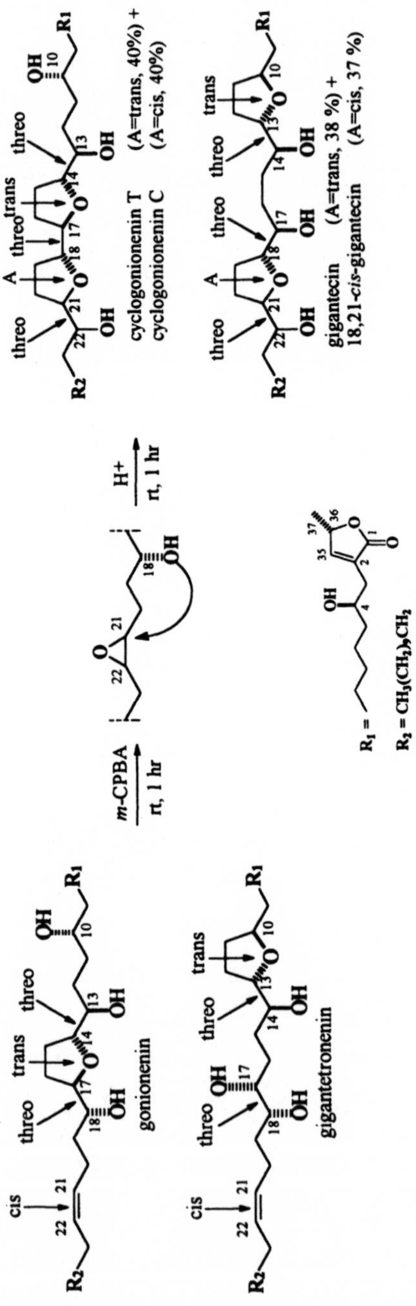

R2 cis 22 21 threo trans threo 18 17 14 13 10 OH OH OH R1
gonionenin

R2 cis 22 21 threo OH 17 threo trans 18 14 13 10 OH OH R1
gigantetronenin

$\xrightarrow[\text{rt, 1 hr}]{m\text{-CPBA}}$ 22 O 21 18 OH $\xrightarrow[\text{rt, 1 hr}]{\text{H+}}$

R2 threo A threo trans threo 22 21 18 17 14 13 10 OH OH OH R1
cyclogonionenin T (A=trans, 40%) +
cyclogonionenin C (A=cis, 40%)

R2 threo A threo threo trans 22 21 18 17 14 13 10 OH OH OH R1
gigantecin (A=trans, 38 %) +
18,21-*cis*-gigantecin (A=cis, 37 %)

R_1 = OH 4 2 35 37 36 1 O O
$R_2 = CH_3(CH_2)_9CH_2$

Figure 13. Conversion of mono-THF acetogenins to bis-THF acetogenins.

ins (Fig. 13).[38] The pairs of products, caused by the two different faces of the epoxides, demonstrated that a stereoselective oxidase probably exists naturally in the plants. This conversion of mono-THF acetogenins to bis-THF acetogenins also significantly enhanced the cytotoxicities against three human solid tumor cell lines.[38] Since gigantetronenin is a high yield acetogenin from *Goniothalamus giganteus*, this conversion provides a new, more abundant, source of gigantecin, a much more active, but rare, compound. Biomimic syntheses of mono-THF acetogenins from diepoxides was also performed by Gromek *et al*;[63] to an acetone solution of corepoxylone was added 70 % perchloric acid, and corossolone and its tetraepimer were produced. This pair of products also indicates that the initiation of cyclization is likely performed naturally by stereospecific enzymes, since the pertinent enantiomers of the THF subunits have not been found in nature. Similarly, the tri-epoxy acetogenins, tripoxyrollin and trieporeticanin, have been converted into bis-THF acetogenins.[73,74]

PARTIAL AND TOTAL SYNTHESIS

The potent and diverse bioactivities of the Annonaceous acetogenins have also recently attracted the interest of synthetic chemists. Earlier synthetic works, largely focusing on the resolution of the relative stereochemistries, have been summarized in our previous two reviews.[8,9] In the past two years, several successful total syntheses have been achieved. The determination of the absolute configurations of the stereogenic carbon centers of the acetogenins can now lead chemists to the correct natural compounds. Thus, besides the non-natural (+)-15,16,19,20,23,24-*hexepi*-uvaricin,[75] (-)-bullatacin (the enantiomer of natural bullatacin),[76] and (10ξ,15*R*,16*S*,19*S*, 20*S*,34*R*)-corossoline,[77] total syntheses of four naturally occurring acetogenins, (+)-rolliniastatin-1,[78] solamin ,[79] reticulatacin,[79] and reticulatamol,[80] have been reported. However, it seems that the multiple stereogenic carbon centers of acetogenins complicate these methods and limit the yields of the products. Interested readers are asked to refer to the specific papers to review the details of the methodologies and the yields.

BIOLOGICAL ACTIVITIES

The potent bioactivities of the Annonaceous acetogenins are certainly a major force that drives the continued and ever-expanding fractionation of plants from this family. Since the first paper, describing uvaricin as a new, *in vivo* active, antileukemic agent,[1] investigators from all over the world continue to find new and more potent derivatives of this same structural theme. Acetogenins, which have more recently been found, often possess additional THF rings, different placements of such THF rings, epoxides, double bonds, greater oxidation levels with carbonyls and hydroxyls, *etc.* Along with this great interest in new structures has come a growing desire to explain how acetogenins effect their almost catalytic bioactivities. Furthermore, researchers have been looking for ways to apply these potent bioactivities to solve modern medical as well as pesticidal problems. This section to summarize some of these recent advances.

Primary Mode of Action

One of the most important discoveries in the study of Annonaceous acetogenins has been the characterization of at least a primary mode of action. It is now well established[69,92,93] that they are inhibitors of NADH:ubiquinone oxidoreductase, which is in an essential enzyme complex (complex I) leading to oxidative phosphorylation in mitochondria. Disruption of the electron transport system (ETS) at this early stage prohibits the conversion of ADP and Pi to ATP. Hence, a type of suffocation (ATP deprivation) occurs at the cellular level.

Londershausen *et al.*[92] initially observed that the toxicities caused by Annonaceous acetogenins on insects resulted in an ever increasing lethargy, as well as a decreased mobility prior to death. They found that treated insects (larva of *Plutella xylostella*) had a substantially decreased amount of ATP (Table 4) similar to the effect of antimycin A, a known inhibitor of the ETS. Further experimentation directly measured the inhibitory effect of acetogenins on mitochondrial enzymes. In this work they found that regardless whether the mitochondrial enzymes were prepared from insects, bovine heart muscle, or *Lucilia cuprina* muscle cells, ETS was stopped (Table 5). The wide variety of organisms susceptible suggested a mode of action which inhibits an essential and common biological function.

These results were further characterized[69,92,93] by polarographically measuring oxygen consumption of mitochondrial suspensions from several different organisms. In Fig. 14A, with succinate as the substrate in rat liver mitochondria,[69]

Table 4. ATP levels in insecticide treated larvae of *Plutella xylostella*.[92]

Compound	Dose (ng)	LT_{50} (hours)	ATP [μmole (g fresh weight)$^{-1}$]
Control	--	--	1.98[a]
Parathion[c]	500	2	2.00[b]
Cyfluthrin[d]	10	2	2.25[b]
Antimycin A[e]	1×10^5	5	1.35[b]
Squamocin[f]	2×10^5	8	1.45[b]

[a]Average (n=19) with a standard deviation of +/- 0.19.

[b]Values were derived from interpolation and their standard error can be assumed to be in the range of the control standard deviation.

[c]An inhibitor of acetylcholinesterase.

[d]An effector of sodium channels.

[e]An inhibitor of the Electron Transport System.

[f]Named annonin I by the authors.

Table 5. Inhibition of electron transport of *Lucilia cuprina* muscle submitochondrial particles.[92]

Inhibitor	IC_{50} [nmole (mg protein)$^{-1}$][a]		
	NADH: ubiquinone oxidoreductase[b]	Succinate-cytochrome c reductase	alpha-Glycerophosphate-cytochrome c reductase
Squamocin[c]	4-8	>200	>200
Rotenone[d]	20	>200	>200
Antimycin A[e]	--	3-6	3-6
Cyanide[f]	>200	>200	>200

[a]Spectrophotometrical measurements of inhibitory effects on cytochrome c reductase activity.
[b]Named NADH-cytochrome c reductase by the authors.
[c]Named annonin I by the authors.
[d]An inhibitor of the Electron Transport System.
[e]An inhibitor of the Electron Transport System. A mixture of A_1 - A_4, predominantly A_1 and A_3.
[f]An uncoupler of the Electron Transport System.

bullatacin at 10 μM does not inhibit oxygen consumption; yet, in Fig. 14B, when glutamate is the substrate, bullatacin inhibits, even at 0.1 μM, and is approximately 28 times more potent than rotenone which is considered the classic complex I inhibitor. Electrons can be channeled into complex III of the ETS via either complex I with glutamate as the substrate providing NADH, or via complex II with succinate as the substrate providing $FADH_2$. The final step of electron transport is the conversion of oxygen to water (complex IV) which generates energy via a proton motive force such that ADP is converted to ATP (complex V). Since oxygen is depleted as ATP is generated, measuring the inhibition of oxygen consumption is indicative of the blockage of electron transport. Thus, it can be determined which of the two enzyme complexes (I or II) are affected by varying the substrate between glutamate and succinate. Hence, complex II is not inhibited by bullatacin as evidenced by the uninterupted consumption of oxygen when succinate serves as the substrate. However, when glutamate is the substrate, oxygen consumption is effectively blocked by bullatacin. It is, therefore, predictable that succinate could be used as an antidote for acetogenin poisoning either with respect to accidental pesticidal poisoning or as a form of rescue therapy in antitumor applications.

Recently, work has been completed which attempts to dissect exactly where the acetogenins bind within complex I. Although the data are preliminary, it has been well established that, depending on the experimental conditions, bullatacin is at least equipotent and perhaps 10 to 20 times more potent than rotenone.[94,95] Furthermore, it appears that the Annonaceous acetogenins may have a binding site which differs from that of rotenone. Friedrich *et al.*[96] as well as Esposti *et al.*[95] have found that

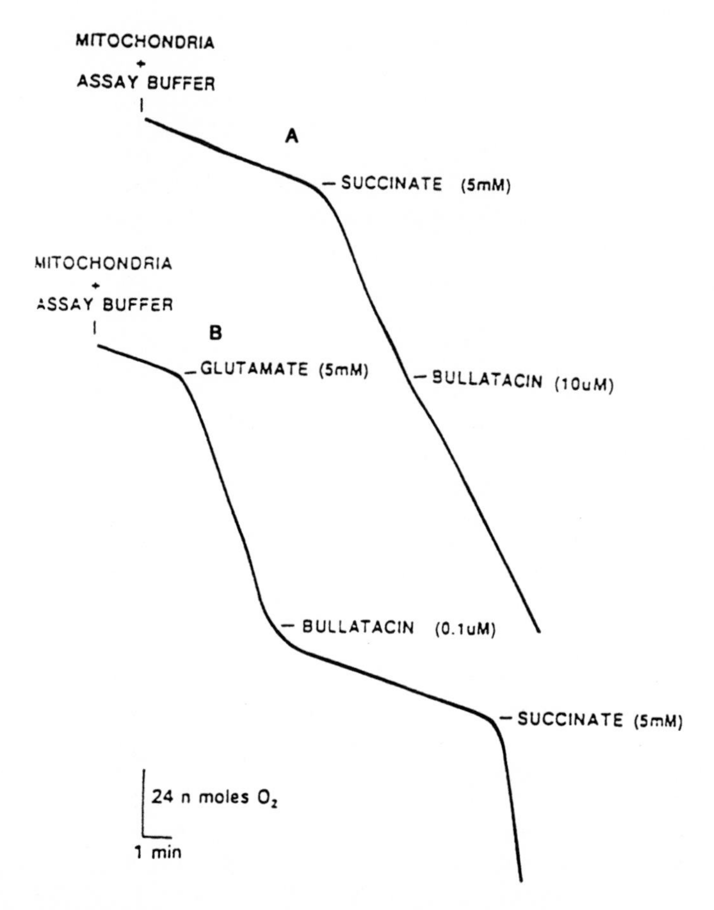

Figure 14. Effect of bullatacin on (A) succinate- and (B) glutamate dependent respiration of rat liver mitochondria.[69]

the acetogenins bind competitively with respect to the ubiquinone binding site at complex I; whereas, rotenone binds non-competitively suggesting an alternative site. More experimentation is required to characterize these separate sites of action, and, at the very least, the acetogenins are destined to become valuable as new tools for the future study of the ETS complex I. Bullatacin is the most potent of several complex I ETS inhibitors studied.[94]

Secondary Mode of Action

A new mode of action, which helps to explain the extreme potency as well as the tumor selectivity of the Annonaceous acetogenins, appears to be inhibition of NADH oxidase activity at the cell membrane. There is a fundamental difference in the NADH oxidase activity of plasma membrane vesicles derived from normal rat liver cells vs. those derived from cancerous HeLa (human cervical carcinoma) or HL-60 (human promyelocytic leukemia) cells. The former are responsive to stimuli such as hormones and growth factors; whereas, the latter two are constitutively activated. Morré *et al.*[97] have recently determined that bullatacin inhibits the NADH-oxidase activity of plasma membrane vesicles isolated from HeLa and HL-60 cells with an ED_{50} of 5 to 10 nM. More importantly, however, bullatacin does not affect the NADH oxidase activity of plasma membrane vesicles derived from normal rat liver cells. Furthermore, such activity in HL-60 cells which were resistant to adriamycin was also inhibited. This inhibition helps to explain the potency as well as the tumor selectivity of the acetogenins. It also may prove to be important as a means to combat tumors which have established multiple drug resistance. Indeed, as we predicted in our second review,[9] the dependence on ATP to drive the multiple drug resistant P-170 glycoprotein pump suggests the probable effectiveness of combination therapy, with acetogenins lowering ATP levels and thus preventing the cellular exclusion of antitumor drugs such as adriamycin.

Structure Activity Relationships

In a few recent papers, researchers have drawn conclusions about structural activity relationships (SAR) of the acetogenins which either enhance or diminish biological activities. These observations are usually presented for only a few compounds. However, Landolt *et al.*[98] have attempted to probe more thoroughly the structure activity relationships with twenty structurally diverse acetogenins. Using an oxygen electrode, they polarographically measured oxygen consumption of rat liver mitochondrial suspensions with procedures similar to those of Ahammadsahib *et al.*[69] Their results can be summarized as follows:

1. Compounds having two THF rings are more potent than those having only one, regardless if the rings are adjacent or non-adjacent;[98] however, cytotoxicity and *in vivo* antitumor data suggest that the adjacent bis-THF compounds are more potent.[69]
2. Increasing the amount of hydroxylation to three free hydroxyls increases potency, especially if an hydroxyl is present on the long aliphatic end of the compound;[98] cytotoxicity data supports this observation.[99] However, it appears as though a certain median level of polarity may be more important because the presence of four or five hydroxyls decreases biological activity, and such activity can be restored by preparing derivatives, such as formaldehyde acetals or acetonides, in which only three free hydroxyls are present.[41]

For example, the activity of bullatacin decreases in the 4-deoxy analog (desacetyluvaricin); but the activity is partially restored in the 10-hydroxy-4-deoxy analog (bullatin) and even enhanced in the 30-hydroxy-4-deoxy analog

Table 6. Mitochondrial assay IC_{50} values of the bullatacin series.[98]

Compound	Statistical IC_{50}[a] (n moles/Lt/mg of protein)
Bullatacin	71
4-Deoxybullatacin (desacetyluvaricin)	810
10-OH-4-Deoxybullatacin (bullatin)	150
30-OH-4-Deoxybullatacin (bullanin)	20
Rotenone[b]	17

[a]A standard deviation of +/- 25% should be considered when comparing this data.
[b]An inhibitor of the Electron Transport System.

(bullanin) (Table 6). This same trend holds true in the hydroxylated bullatacinone compounds[98] (Table 7) as well as, with regard to cytotoxicity, in the asimicin series.[99]

No conclusions have been drawn about how the configurations of the THF rings may affect activity.[98] Bullatacin, asimicin, and trilobacin differ only in their relative stereochemistry across the adjacent bis-THF ring system. Although asimicin has slightly greater activity, a standard deviation of +/- 25% in the mitochondrial assay makes the mitochondrial data for these compounds appear to be essentially equivalent.[98] Although this comparative examination in mitochondria[98] helped in determining some of the more essential features of the acetogenin molecules, additional SAR investigations need to be undertaken. The increased activity associated with increased oxidation is not site specific and may only be an artifact of increased water solubility or a polarity requirement. Furthermore, the mitochondrial assay is a cell free system; and, although it routinely provides results which are parallel to brine shrimp lethality as well as

Table 7. Mitochondrial assay IC_{50} values of the bullatacinone series.[98]

Compound	Statistical IC_{50}[a] (n moles/Lt/mg of protein)
Bullatacinones	49
30-OH-Bullatacinones	15
31-OH-Bullatacinones	16
Rotenone[b]	17

[a]A standard deviation of +/- 25% should be considered when comparing this data.
[b]An inhibitor of the Electron Transport System.

cell culture data, it does not take into account obstacles such as metabolism and membrane permeability. The formaldehyde acetal derivatives, acetonides, and other substitutions on the hydroxyls produce compounds which are either less active or more active than the parent compounds.[41] This, too, may be related to variations in water solubility, polarity, or membrane permeability rather than an increase or decrease in receptor binding.

In Vivo Experiments

In vivo antitumor tests of the Annonaceous acetogenins have recently given additional encouraging results. It is important to note that a great challenge has been to determine an appropriate dose since the extreme potency of some compounds may cause mortality prior to tumor efficacy. Uvaricin showed *in vivo* activity against 3PS (murine lymphocytic leukemia) (157% T/C at 1.4 mg / kg) and rollinones (147% T/C at 1.4 mg / kg) and asimicin (124% T/C at 25.0 μg / kg) were similarly active with asimicin showing fifty times the potency, but less efficacy than the other two.[8] Ahammadsahib *et al.*[69] recently reported the activity of bullatacin and (2,4-*cis* and *trans*)-bullatacinones against L1210 (murine leukemia) in normal mice, as well as in inhibiting tumors of A2780 (human ovarian carcinoma) in athymic mice (Tables 8 and 9). Bullatacin, effective at only 50.0 μg / kg, was over 300 times more potent than taxol against L1210, and bullatalicin, at 1 mg/kg, was nearly as effective as *cis*-platinum against A2780. We have

Table 8. *In vivo* activity of bullatacin and bullatacinones in conventional mice bearing i.p.-implanted L1210 murine leukemia.[69]

Compound	Dose[a] (μg/kg/day)	Survival First Trial (% T/C)	Survival Second Trial (% T/C)	5 day Weight Change (g/mouse)
Bullatacin	200.0	toxic	toxic	-2.9
(69)	100.0	106	100	-2.6
	50.0	138	138	-0.4
	25.0	125	131	+0.6
	12.5	113	138	+0.9
Bullatacinones	800.0	144	--	+0.5
	400.0	144	--	+1.1
	200.0	125	--	+1.8
	100.0	113	--	+1.6
	50.0	113	--	+2.1
	25.0	100	--	+1.7

[a] Mice were injected i.p. with L1210 cells on day zero and on days 1-9 received i.p. injections of the drug, in suitable vehicles, at the specified daily dose. % T/C refers to survival time in days of test / controls. Increases in life-span of 25% or more are considered significant.

recently patented the bullatacin compounds as new chemotherapeutic agents.[100] Previously unpublished results, obtained by the late G. Grindey at the Eli Lilly Co., showed similar effectiveness of bullatacin (67% tumor growth inhibition at 50.0 μg / kg) against X-5563 plasma cell myeloma xenografts in normal mice. Holschneider *et al.*[101] also attempted to use bullatacin in mice that were injected with MOT (murine ovarian teratocarcinoma) cells, which is a model that is believed to represent human disseminated ovarian carcinoma; they did not find an effective dose in the range of 20.0 mg / kg down to 75.0 μg / kg. However, it cannot be overemphasized how important proper dosing is in order to achieve tumor efficacy when dealing with compounds whose major mode of action may affect basic cellular functions. The lowest dose used by Holschneider *et al.*[101] is still 1.5 times greater than the effective dose reported by Ahammadsahib *et al.*,[69] and, unfortunately, the former paper fails to mention the favorable *in vivo* results, against A2780 human ovarian cancer, that had been previously reported. A favorable ratio of cytotoxic ED_{50} values vs. LD_{50} values in mice suggests that bullatacin should be less toxic *in vivo* than currently used drugs such as *cis*-platinum.[101] Presently, additional *in vivo* studies are underway which will probe the appropriate doses and suitable vehicles, and, as well, examine several other acetogenins having different structural motifs. The difficulties involved in securing the cooperation of the pharmaceutical industry or the National Cancer Institute in performing *in vivo* studies are currently the major road block in the development of these compounds as new, clinically useful, antitumor agents.

Random Results

Reports of uses of various Annonaceous plant parts in the control of lice, mosquitoes, and other insects are not uncommon in the ethnobotanical literature. Many papers have reported the pesticidal properties of the Annonaceous acetogenins,[93,94,102-104] and these uses earlier had been patented.[8,9] Potent antifungal activity against *Candida albicans* was observed with purpureacin I (0.05 μg to inhibit fungal growth on a TLC plate) but not several other acetogenins.[102] The antiparasitic activity of *Annona muricata* and *Annona cherimolia* seeds was pursued by Bories *et al.*[105] Interestingly, they found that even a crude extract could kill *Molinema dessetae*; this may be a significant result because of the need to find new filaricidal compounds. Weinberg *et al.*[106] examined the folkloric use of *Annona crassiflora* seeds as a remedy for snake venom; they found that the extract will inhibit agonist-stimulated contraction of guinea-pig ileum; the authors[106] speculate that this is due to a decrease in calcium ion permeability. Since calcium transport requires energy, the observed effect is probably caused by reduction of ATP levels. Avalos *et al.*[107] tested the allergenicity of bark extracts from *Asimina triloba*; they found only a weak irritation to guinea-pig skin; and they predicted that the amount of acetogenins used in a potential pesticidal formulation would not be enough to affect humans as an allergen.

Table 9. Antitumor efficacies of bullatacin, bullatacinones, and bullatalicin in athymic mice bearing s.c.-implanted A2780 human ovarian carcinoma xenografts.[69]

Compound	Dose[a] (μg/kg/day)	Day 10 % TGI[b]	5 day Weight Change (g/mouse)	Mortality (Dead/Total)
Bullatacin	100.0	68	-1.5	2/8
	50.0	67	-0.3	0/8
	25.0	35	+0.1	0/8
Bullatacinones	500.0	33	-0.7	1/8
	250.0	27	+0.1	0/8
	125.0	52	-0.4	0/8
Bullatalicin	2000.0	63	-0.9	2/8
	1000.0	75	-0.1	0/8
	500.0	42	-0.2	0/8
Cisplatin	5000.0	78	-1.1	0/8
Vehicle alone	--	--	-0.3	0/10

[a] Test compounds were administered i.p. once daily on days 1-9. Cisplatin (i.p.) was administered once only on day one. Bullatacin and bullatalicin were formulated in 1 % EtOH and 1% Tween 80. Bullatacinones were formulated in 1% EtOH, 1% Tween 80, and 1% DMSO. Test and positive control groups started with 8 mice per group; vehicle control group contained 10 mice. Tumor sizes were measured in two dimensions on days 3,7,10,14,17, and 21 after implantation. Tumor weights (mg) were estimated using the formula, (length x width^2)/2. % Tumor growth inhibition (TGI) was calculated as:

[b] % TGI = 100 - [(mean tumor weight of test or positive control group) / (mean tumor weight of vehicle control group)] x 100

Future Experimentation

The biological properties of the Annonaceous acetogenins continue to be an interesting pursuit. Future experiments should look for ways to characterize further their modes of action, especially in intact cellular systems. Future investigations of structure activity relationships should include the effects of stereochemistry of the THF rings on bioactivity and should lead to structures which are optimum for specific applications. In order to enter phase I human clinical trials, more *in vivo* experimentation in animal tumor models is required to determine appropriate dosing and formulation and to suggest the best candidates for antitumor development against specified tumor types. In addition, a more complete examination of the pesticidal potential of the acetogenins is necessary and should be pursued considering the present rush for more environmentally friendly formulations and the emergence of insects resistant to other pesticides. The role of ATP dependent transport in various forms of resistance (pesticidal, anticancer agents, and antibiotics) offers an exciting potential for the use of the acetogenins as adjuvants in various types of combination products. Although much has been accomplished in the biological understanding of the Annonaceous acetogenins, much more needs to be done, and the future looks bright for their development.

SUMMARY

Chemically the Annonaceous acetogenins are C32 or C34 fatty acids which are combined with a 2-propanol unit at C-2; the propanol is incorporated into a 2,4-disubstituted γ-lactone which can assume several different forms. In addition, the long hydrocarbon chain usually contains a number of oxygenated moieties (hydroxyls, acetyls, ketones, or epoxides) which originate from units of unsaturation at specific positions along the chain; some of these usually cyclize into one, two, or three tetrahydrofuran (THF) rings, creating a host of complicated stereoisomeric subclasses which are organized depending on the number and arrangement of the THF rings. To date (through June 1994), over 160 of these compounds have been isolated and identified, and details for characterization are presented herein (see Appendix) for the nearly 80 new compounds or revised structures described since the last review (through December 1992). Several chemical techniques and spectral interpretations helpful in the structural elucidations are also presented here. The Annonaceous acetogenins are extremely potent and offer exciting potential as new antitumor, pesticidal, antimalarial, anthelmintic, and antimicrobial agents. Members of all of the subclasses act biologically, at least in part, as inhibitors of ATP production through blockage of complex I in mitochondrial respiration (NADH:ubiquinone oxidoreductase) and the inhibition of NADH oxidase in the plasma membranes of cancer cells.

ACKNOWLEDGMENTS

This investigation was supported by R01 Grant No. CA 30909 from the U. S. National Cancer Institute, National Institutes of Health. N.H.O. acknowledges stipend support from the Indiana Elks Cancer Research Fund and the Purdue Research Foundation. The authors are grateful to those scientists in other laboratories who have willingly sent us their new manuscripts and, thus, have made this review as up to date as possible.

REFERENCES

1. JOLAD, S. D., HOFFMANN, J. J., SCHRAM, K. H., COLE, J. R. 1982. Uvaricin, a new antitumor agent from *Uvaria accuminata* (Annonaceae). J. Org. Chem. 47:3151-3153.
2. MOESCHLER, H. F., PFLUGER, W., WENDISH, D. 1986. Annonin insecticide. German Patent, No. DE 3438763 A1.

3. MOESCHLER, H. F., PFLUGER, W., WENDISH, D. 1987. Pure annonin and a process for the preparation. U.S. Patent, No. 4,689,232.
4. MIKOLAJCZAK, K. L., McLAUGHLIN, J. L., RUPPRECHT, J. K. 1988. Control of pests with Annonaceous acetogenins. U.S. Patent, No. 4,721,727.
5. MIKOLAJCZAK, K. L., McLAUGHLIN, J. L., RUPPRECHT, J. K. 1989. Control of pests with Annonaceous acetogenins. U.S. Patent, No. 4,855,319.
6. CAVE, A., HOCQUEMILLER, R., LAPREVOTE, O. 1989. Utilisation d'acetogenins en therapeutique en tant que substances antiparasitaires. F Patent 1048 No. 88 09 674.
7. IKEKAWA, N., FUJIMOTO, Y., IKEKAWA, T., SUMURA LIMITED CO. 1991. Novel bis-tetrahydrofuran compounds and their preparations used as anticancer agents. Japan Patent No. He 3-41076.
8. RUPPRECHT, J. K., HUI, Y.-H., McLAUGHLIN, J. L. 1990. Annonaceous acetogenins: A review. J. Nat. Prod. 53:237-278.
9. FANG, X.-P., RIESER, M. J., GU, Z.-M., ZHAO, G.-X., McLAUGHLIN, J. L. 1993. Annonaceous acetogenins: an updated review. Phytochem. Anal. 4:27-67.
10. CAVE, A., CORTES, D., FIGADERE, B., HOCQUEMILLER, R., LAPREVOTE, O., LAURENS, A., LEBOEUF, M. 1991. Recent Advances In Phytochemistry. (Downum, K. R.; Romeo, J.; Stafford, H. A., eds). Plenum Press, NewYork, 27:167-202.
11. CORTES, D., FIGADERE, R., CAVE, A. 1993. Bistetrahydrofuran acetogenins from Annonaceae. Phytochemistry 32:1467-1473.
12. ABREO, M. J., SNEDEN, A. T. 1989. 4-Hydroxy-25-desoxyneorollinicin, a new bistetrahydrofuranoid acetogenin from *Rollinia papilionella*. J. Nat. Prod. 52:822-828.
13. MEYER, B. N., FERRIGNI, N. R. PUTNAM, J. E. JACOBSON, L. B. NICHOLS, D. E. McLAUGHLIN, J. L. 1982 Planta Med. 45:31-34.
14. McLAUGHLIN, J. L. 1991. Methods in Plant Biochemistry, vol. 6. (K. Hostettmann, ed.), Academic Press, London, 1991, pp. 1-33.
15. FOGH, J., TREMPE, G. 1975. New Human Tumor Cell Lines in Human Tumor Cells, *in vitro*, J. Fogh, Ed., Plenum Press, New York, p. 115.
16. GIARD, D. J., ARONSON, S. A., TODARO, G. J., ARNSTEIN, P., KERSEY, J. H., DOSIK, H., PARKS, W. P. 1973. J. Natl. Cancer Inst. 51:1417-1423.
17. SOULE, H. D., VAZQUEZ, J., LONG, A., ALBERT, S., BRENNAN, M. 1973. J. Natl. Cancer Inst. 51:1409-1416.
18. GROMEK, D., HOCQUEMILLER, R. CAVE, A. 1994. Qualitative and quantitative evaluation of Annonaceous acetogenins by high performance liquid chromatography. Phytochem. Anal. 5:133-140.
19. LAPREVOTE, O., GIRARD, C., DAS, B. C. 1992. Formation of gas-phase lithium complexes from acetogenins and their analysis by fast atom bombardment mass spectrometry. Tetrahedron Lett. 33:5237-5240.
20. LAPREVOTE, O., GIRARD, C., DAS, B. C., LAURENS, A., CAVE, A. 1993. Desorption of lithium complexes of acetogenins by fast atom bombardment: Application to semi-quantitative analysis of crude plant extracts. Analysis 21:207-210.
21. SMITH, D. L., LIU, Y. M., WOOD, K. V. 1991. Structure elucidation of natural products by mass spectrometry. Recent Advances in Phytochemistry. (N. H. Fischer, M. B. Isman, T. H. A. Hafford, eds.) Plenum Press, New York, vol. 25, pp. 251-269.
22. LAPREVOTE, O., DAS, B. C. 1994. Structural elucidation of acetogenins from Annonaceae by fast atom bombardment mass spectrometry. Tetrahedron 50:8479-8490.
23. GU, Z.-M., FANG. X.-P., ZENG, L., WOOD, K. V., McLAUGHLIN, J. L. 1993. Bullacin: a new cytotoxic Annonaceous acetogenin from *Annona bullata*. Heterocycles 36:2221-2228.

24. LAPREVOTE, O., GIRARD, C., DAS, B. C. 1992. Location of epoxy ring in a long chain acetogenin by fast-atom bombardment and linked scan (B/E) mass spectrometry of lithium-cationized molecules. Rapid Commun. Mass Spectrom. 6:352-355.
25. LAPREVOTE, O. ROBLOT, F., HOCQUEMILLER, R., CAVE, A., CHARLES, B., TABET, J. C. 1991. Structural elucidation of five stereoisomeric acetogenins, uleicins A-E, by tandem mass spectrometry. Phytochemistry 30:2721-2727.
26. LAPREVOTE, O., ROBLOT, F., HOCQUEMILLER, R., CAVE, A. 1990. Structural elucidation of two new acetogenins, epoxyrollins A and B, by tandem mass spectrometry. Tetrahedron Lett. 31:2283-2286.
27. NONFON, M., LIEB, F., MOESCHLER, H., WENDISCH, D. 1990. Four annonins from *Annona squamosa*. Phytochemistry 29:1951-1954.
28. RIESER, M. J. 1993. Ph.D. Thesis. Annonaceous acetogenins from the seeds of *Annona muricata*. Purdue University, West Lafayette, Indiana, pp. 149-155.
29. HUI, Y. H., RUPPRECHT, J. K., ANDERSON, J. E., LIU, Y. M., SMITH, D. L., CHANG, C. J., McLAUGHLIN, J. L. 1989. Bullatacin and bullatacinone: two highly potent bioactive acetogenins from *Annona bullata*. J. Nat. Prod. 52:463-477.
30. GU, Z.-M., FANG, X.-P., MIESBAUER, L. R.; SMITH, D. L., McLAUGHLIN, J. L. 1993. 30-, 31-, and 32-Hydroxybullatacinones: bioactive terminally hydroxylated Annonaceous acetogenins from *Annona bullata*. J. Nat. Prod. 56:870-876.
31. GU, Z.-M., FANG, X.-P., HUI, Y.-H., McLAUGHLIN J. L. 1994. 10-, 12-, and 29-Hydroxybullatacinones: new cytotoxic Annonaceous acetogenins from *Annona bullata* Rich. (Annonaceae). Natural Toxins 2:49-55.
32. GU, Z. M., ZENG, L., McLAUGHLIN, J. L. 1994. Five bioactive adjacent bis-THF Annonaceous acetogenins from *Annona bullata*. Phytochemistry (accepted).
33. GU, Z.-M., FANG, X.-P., ZENG, L., KOZLOWSKI, J. F., McLAUGHLIN J. L. 1994. Novel cytotoxic Annonaceous acetogenins: (2,4-*cis* and *trans*)-bulladecinones from *Annona bullata* (Annonaceae). Bioorg. Med. Chem. Lett. 4:473-478.
34. RIESER, M. J., FANG, X. P., ANDERSON, J. E., MIESBAUER, L. R., SMITH, D. L., McLAUGHLIN, J. L. 1994. Muricatetrocins A and B and gigantetrocin B: three new cytotoxic monotetrahydrofuran-ring acetogenins from *Annona muricata*. Helv. Chim. Acta 76:2433-2444.
35. COLMAN-SAIZARBITORIA, T., ZAMBRANO, J., FERRIGNI, N. R., GU, Z. M., NG, J. H., SMITH, D. L., McLAUGHLIN, J. L. 1994. Bioactive Annonaceous acetogenins from the bark of *Xylopia aromatica*. J. Nat. Prod. 57:486-493.
36. COLMAN-SAIZARBITORIA, GU, Z. M., McLAUGHLIN, J. L. 1994. Two new bioactive monotetrahydrofuran Annonaceous acetogenins from the bark of *Xylopia aromatica*. J. Nat. Prod. 57: 1661-1669.
37. FANG, X.-P., ANDERSON, J. E., SMITH, D. L., WOOD, K. V., McLAUGHLIN, J. L. 1991. Giganenin, a highly potent monotetrahydrofuran acetogenin and 4-deoxygigantecin from *Goniothalamus giganteus*. Heterocycles 34:1075-1082.
38. GU, Z.-M., FANG, X.-P., ZENG, L., SONG, R., NG, J. H., WOOD, K. V., SMITH, D. L., cLAUGHLIN, J. L. 1994. Gonionenin: a new cytotoxic Annonaceous acetogenin from *Goniothalamus giganteus* and the conversion of mono-THF acetogenins to bis-THF acetogenins J. Org. Chem. 59:3472- 3479.
39. ARAYA, H., HARA, N., FUJIMOTO, Y., SRIVASTAVA, A., SAHAI, M. 1994. Squamostén-A, a novel mono-tetrahydrofuranic acetogenin with a double bond in the hydrocarbon chain, from *Annona squamosa* L. Chem. Pharm. Bull. 42:388-391.
40. FANG, X.-P., RUPPRECHT, J. K., ALKOFAHI, A., HUI, Y.-H., LIU, Y.-M., SMITH, D. L., WOOD, K.V., McLAUGHLIN, J. L. 1991. Gigantetrocin and gigantriocin: two novel bioactive Annonaceous acetogenins from *Goniothalamus giganteus*. Heterocycles 32:11-17.

41. WU, F. E., GU, Z. M., ZENG, L., ZHAO, G. X., ZHANG, Y., McLAUGHLIN, J. L. 1994. Two new cytotoxic monotetrahydrofuran Annonaceous acetogenins, annomuricins A and B, from the leaves of *Annona muricata*. J. Nat. Prod. (accepted).
42. GU, Z.-M., ZENG, L., FANG, X.-P., COLMAN-SAIZARBITORIA, T., HUO, M., McLAUGHLIN, J. L. 1994. Determining absolute configurations of stereocenters in Annonaceous acetogenins through formaldehyde acetal derivatives and Mosher ester methodology. J. Org. Chem. 59:5162-5172.
43. BORN, L., LIEB, F., LORENTZEN, J. P., MOESCHER, H., NONFON, M., SOLLNER, R., WENDISCH, D. 1990. The relative configuration of acetogenins isolated from *Annona squamosa*: annonin I (squamocin) and annonin VI. Planta Med. 56:312-316.
44. JOSSANG, A., DUBAELE, B., CAVE, A. 1990. Deux nouvelles acetogenins onotetrahydrofuranniques cytotoxiques: l'annomonicine et la montanacine. Tetrahedron Lett. 31:1861-1864.
45. HOYE, T. R., ZHUANG, Z. 1988. Validation of the ^{1}H NMR chemical shift method for determination of stereochemistry in the bis (tetrahydrofuranyl) moiety of uvaricin-related acetogenins from Annonaceae: Rolliniastatin 1 (and asimicin). J. Org. Chem. 53:5578-5580.
46. HOYE, T. R., HANSON, P. R. 1991. Assigning the relative stereochemistry between C(2) and C(4) of the 2-acetonyl-4-alkylbutyrolactone substructures of the appropriate Annonaceous acetogenins. J. Org. Chem. 56:5092-5095.
47. RIESER, M. J., HUI, Y.-H., RUPPRECHT, J. K., KOZLOWSKI, J. F., WOOD, K. V., McLAUGHLIN, J. L., HANSON, P. R., ZHUANG, A., HOYE, T. R. 1992. Determination of absolute configuration of stereogenic carbinol centers in Annonaceous acetogenins by ^{1}H- and ^{19}F-NMR analysis of Mosher ester derivatives. J. Am. Chem. Soc. 144:10203-10213.
48. FUJIMOTO, Y., MURASAKI, C., SHIMADA, H., NISHIOKA, S., KAKINUMA, K., SINGH, S., SINGH, M., GUPTA, Y. K., SAHAI, M. 1994. Annonaceous acetogenins from the seeds of *Annona squamosa*. Non-adjacent bis-tetrahydrofuranic acetogenins. Chem. Pharm. Bull. 42:1175-1184.
49. GALE, J. B., YU, J. G., HU, X. E., KHARE, A., HO, D. K., CASSADY, J. M. 1993. Stereochemistry of mono-tetrahydrofuranyl moiety in cytotoxic polyketides. Part B: application of proton chemical shift patterns. Tetrahedron Lett. 34:5851-5854.
50. YU, J. G., HO, D. K., Cassady, J. M., Xu, L. Z., Chang, C. J. 1992. Cytotoxic polyketides from *Annona densicoma* (Annonaceae): 10,13-*trans*-13,14-*erythro*-densicomacin, 10,13-*trans*-13,14-*threo*-densicomacin, and 8-hydroxyannonacin. J. Org. Chem. 57:6198-6202.
51. HISHIOKA, S., ARAYA, H., MURASAKI, C., SAHAI, M., FUJIMOTO, Y. 1994. Determination of absolute stereochemistry at carbinol stereocenters of tetrahydrofuranic acetogenins by advanced Mosher method. Nat. Prod. Lett. 5:117-121.
52. SHIMADA, H., NISHIOKA, S. SINGH, S., SAHAI, M., FUJIMOTO, Y. 1994. Absolute stereochemistry of non-adjacent bis-tetrahydrofuranic acetogenins. Tetrahedron Lett. 35:3961- 3964.
53. MIKOLAJCZAK, K. J., MADRIGAL, R. V., RUPPRECHT, J. K., HUI, Y. H., LIU, Y. M., SMITH, D. L., McLAUGHLIN, J. L. 1990. Sylvaticin: a new cytotoxic and insecticidal acetogenin from *Rollinia sylvatica* (Annonaceae). Experientia 46:324-327.
54. FANG, X.-P., GU, Z.-M., RIESER, M. J., HUI, Y.-H., McLAUGHLIN, J. L.1993 Structural revisions of some non-adjacent bis-tetrahydrofuran Annonaceous acetogenins. J. Nat. Prod. 56:1095-1100.
55. CORTES, D., MYINT, S. H., DUPONT, B., DAVOUST, D. 1993. Bioactive acetogenins from seeds of *Annona cherimolia*. Phytochemistry 32:1475-1482.
56. FUJIMOTO, Y., MURASAKI, C., KAKINUMA, K., EGUCHI, T., IKEKAWA, N., FURUYA, M., HIRAYAMA, K., IKEKAWA, T., SAHAI, M., GUPTA, Y. K., RAY, A. B. 1990.

Squamostatin-A: unprecedented bis-tetrahydrofuran acetogenin from *Annona squamosa*. Tetrahedron Lett. 31:535-538.

57. YU, J.-G., LUO, X.-Z., SUN, L., LIU, C.-Y., HONG, S.-L., MA, L.B. 1993. Squamostatin-B, a new polyketide from *Annona squamosa* (Annonaceae). Chin. Chem. Lett. 4:423-426.
58. ZHAO, G.-X., HUI, Y. H., RUPPRECHT, J. K., McLAUGHLIN, J. L. 1992. Additional bioactive compounds and trilobacin, a novel highly cytotoxic acetogenin, from the bark of *Asimina triloba*. J. Nat. Prod. 55:347-356.
59. ZHAO, G.-X.,GU, Z.-M., ZENG, L.,CHAO, J.-F., KOZLOWSKI, J. F., WOOD, K. V., McLAUGHLIN,J. L. 1995. The absolute configurations of trilobacin and trilobin, a novel highly potent acetogenin from the stem bark of *Asiminia triloba* (Annonaceae). Tetrahedron (submitted)
60. RIESER, M. J., FANG, X. P., ANDERSON, J. E., MIESBAUER, L. R., SMITH, D. L., McLAUGHLIN, J. L. 1994. Erratum. Helv. Chim. Acta 77:882.
61. MYINT, S. H., CORTES, D., LAURENS, A., HOCQUEMILLER, R., LEBOEUF, M., CAVE, A., COTTE, J., QUEROS, A. M. 1991. Solamin, a cytotoxic mono-tetrahydrofuranic γ-lactone acetogenin from *Annona muricata* seeds. Phytochemistry 30:3335-3338.
62. WU, Y.-C., CHANG, F.-R., CHEN, K.-S., LIANG, S.-C., LEE, M.-R. 1994. Diepoxymontin, a novel acetogenin from *Annona montana*. Heterocycles 38:1475-1478.
63. GROMEK, D., FIGADERE, B., HOCQUEMILLER, R. CAVE, A. 1993. Corepoxylone, a possible precursor of mono-tetrahydrofuran γ-lactone acetogenins: biomimetic synthesis of corossolone. Tetrahedron 49:5247-5252.
64. HISHAM, A., SREEKALA, U., PIETERS, L., BRUYNE, T. D., HEUVEL, H. V. D., CLAEYS, M. 1993. Epoxymurins A and B, two biogenetic precursors of Annonaceous acetogenins from *Annona muricata*. Tetrahedron 31:6913-6920.
65. SAHPAZ, S., LAURENS, A., HOCQUEMILLER, R., CAVE, A. 1994. Senegalene, une nouvelle acetogenine olefinique mono-tetrahydrofuranique des graines d'*Annona senegalensis*. Can. J. Chem. 72:1-4.
66. SAHAI, M., SINGH, S.,SINGH, M., GUPTA, Y. K., AKASHI, S., YUJI, R., HIRAYAMA, K., ASAKI, H., ARAYA, H., HARA, N., EGUCHI, T., KAKINUMA, K., FUJIMOTO, Y. 1994. Annonaceous acetogenins from the seeds of *Annona squamosa*. Adjacent bis-tetrahydrofuranic acetogenins. Chem. Pharm. Bull. 42:1163-1174.
67. HUI, Y.-H. 1991. Ph.D. Thesis. The search for bioactive constituents from *Annona bullata* Rich. (Annonaceae). Purdue University, West Lafayette, Indiana, pp. 136-145.
68. DURET, P., LAURENS, A., HOCQUEMILLER, R., CORTES, D., CAVE, A. 1994. Isoacetogenins, artifacts issued from translactonization from Annonaceous acetogenins Heterocycles, 39: 741-749.
69. AHAMMADSAHIB, K. I., HOLLINGWORTH, R. M., McGOVREN, J. P., HUI, Y.-H., McLAUGHLIN, J. L. 1993. Mode of action of bullatacin: A potent antitumor and pesticidal Annonaceous acetogenin. Life Sciences 53:1113-1120.
70. FANG, X.-P.ANDERSON, J. E., SMITH, D. L, McLAUGHLIN, J. L., WOOD, K. V. 1992. Gigantetronenin and gigantrionenin: novel cytotoxic acetogenins from *Goniothalamus giganteus*. J. Nat. Prod. 55:1655-1663.
71. ROBLOT, F., LAUGEL, T., LEBOEUF, M., CAVE, A., LAPREVOTE, O. 1993. Two acetogenins from *Annona muricata* seeds. Phytochemistry 34:281-285.
72. GU, Z. M., FANG, X. P., ZENG, L., McLAUGHLIN, J. L. 1994. Goniocin from *Goniothalamus giganteus*: the first tri-THF Annonaceous acetogenin. Tetrahedron Lett. 35:5367-5368.
73. TAM, V. T., HIEU, P. Q. C., CHAPPE, B., ROBLOT, F., LAPREVOTE, O., FIGADERE, B., CAVE, A. 1994. Four new acetogenins from the seeds of *Annona reticulata*. Nat. Prod. Lett. 4:255-262.

74. SAHPAZ, S., FIGADERE, B. SAEZ, J., HOCQUEMILLER, R., CAVE, A., CORTES, D. 1994. Tripoxyrollin, a new epoxy acetogenin from the seeds of *Rollinia membranacea*. Nat. Prod. Lett. 2:301-308.
75. HOYE, T. R., HANSON, P. R., KOVELESKY, A. C., OCAIN, T. D., ZHUANG, Z. P. 1991. Synthesis of (+)-(15,16,19,20,23,24)-hexepi-uvaricin: A bis (tetrahydrofuranyl) Annonaceous acetogenin analogue. J. Am. Chem. Soc. 113:9369-9371.
76. HOYE, T. R., HANSON, P. R. 1993. Synthesis of (-)-bullatacin: The enantiomer of a potent, antitumor, 4-hydroxylated, Annonaceous acetogenin. Tetrahedron Lett. 34:5043-5046.
77. YAO, Z. J., WU, Y. L. 1994. Total synthesis of (10, 15R, 16R, 19S, 20S, 34R)-corossoline. Tetrahedron Lett. 35:157-160.
78. KOERT, U. 1994. Total synthesis of (+)-rolliniastatin 1. Tetrahedron Lett. 35:2517-2520.
79. SINHA, S. C., KEINAN, E. 1993. Total synthesis of naturally occurring acetogenins: solamin and reticulatacin. J. Am. Chem. Soc. 115:4891-4892.
80. TAM, V. T., CHABOCHE, C., FIGADERE, B., CHAPPE, B., HIEU, B. C., CAVE, A. 1994. First synthesis of a new acetogenin of Annonaceae, reticulatamol: activated tin hydride with enhanced reducing ability. Tetrahedron Lett. 35:883-886.
81. HOYE, T. R., SUHADOLNIK, J. C. 1986. Stereocontrolled synthesis of 2,5-linked bistetrahydrofurans via the triepoxide cascade reaction. Tetrahedron 42:2855-2862.
82. BERTRAND, P., GESSON, J.-P. 1992. Approach to the synthesis of Annonaceous acetogenins from D-glucose. Tetrahedron Lett. 33:5177-5180.
83. HARMANGE, J. C., FIGADERE, B., CAVE, A. 1992. Stereocontrolled synthesis of 2,5-linked monotetrahydrofuran units of acetogenins. Tetrahedron Lett. 33:5749-5752.
84. CORTES, D., MYINT, S. H., HARMANGE, J. C., SAHPAZ, S., FIGADERE, B. 1992. Catalytic hydrogenation of Annonaceous acetogenins. Tetrahedron Lett. 33:5225-5226.
85. FIGADERE, B., HARMANGE, J. C., HAI, L. X., CAVE, A. 1992. Synthesis of 2,33-dihydro-4-oxo-murisolin: conjugate addition of primary alkyl iodides to α,β-unsaturated ketones. Tetrahedron Lett. 33:5189-5192.
86. JONATHAN, B. G., YU, J. G., HU, X. E., KHARE, A., HO, D. K., CASSADY, J. M. 1993. Stereochemistry of mono-tetrahydrofuranyl moiety in cytotoxic polyketides. Part A: synthesis of model compounds. Tetrahedron Lett. 34:5847-5850.
87. HOPPE, R., FLASCHE, M., SCHARF, H. D. 1994. An approach towards 2,5-disubstituted tetrahydrofurans of Annonaceous acetogenins. Tetrahedron Lett. 35:2873-2876.
88. HARMANGE, J. C.,FIGADERE, B., HOCQUEMILLER, R. 1991. Enantiospecific preparation of the lactone fragment of murisolin. Tetrahedron: Asymmetry 2:347-350.
89. BERTRAND. P., GESSON, J.-P. 1992. Stereoselective synthesis of the trans 2,4-disubstituted butyrolactone moiety of Annonaceous acetogenins. Synlett. 889-890.
90. SCHOLZ, G., TOCHTERMANN, W. 1991. Optisch aktive γ-lactone aus cyclooctin und furan-synthese von (-)-muricatacin. Tetrahedron Lett. 32:5535-5538.
91. FIGADERE, B., CHABOCHE, C., PEYRAT, J.-F., CAVE, A. 1993. Stereocontrolled synthesis of key intermediates in the total synthesis of acetogenins of Annonaceae. Tetrahedron Lett. 34:8093-8096.
92. LONDERSHAUSEN, M., LEICHT, W., LIEB, F.,MOESCHLER, H., WEISS, H. 1991. Molecular mode of action of annonins. Pesticide Sci. 33:427-438.
93. LEWIS, M A., ARNASON, J. T., PHILOGENE, B. J. R., RUPPRECHT, J. K., McLAUGHLIN, J. L. 1993. Inhibition of respiration at site I by asimicin, an insecticidal acetogenin of the pawpaw, *Asimina triloba* (Annonaceae). Pesticide Biochem. Physiol. 45:15-23.
94. HOLLINGWORTH, R. M., AHAMMADSAHIB, K. I., GADELHAK, G., McLAUGHLIN, J. L. 1994. New inhibitors of complex I of the mitochondrial electron transport chain with activity as pesticides. Biochem. Soc. Trans. 22:230-233.

95. ESPOSITI, M. D., GHELLI, A., BATTA, M., CORTES, D., ESTORNELL. E. 1994. Natural substances (acetogenins) from the family Annonaceae are powerful inhibitors of mitochondrial NADH dehydrogenase (complex I). Biochem. J. 301:161-167.
96. FRIEDRICH, T., OHNISHI, T., FORCHE, E., KUNZE, B., JANSEN, R., TROWITZSCH, W., HOFLE, G., REICHENBACH, H., WEISS, H. 1994. Two binding sites for naturally occurring inhibitors in mitochondrial and bacterial NADH: ubiquinone oxidoreductase (complex I). Biochem. Soc. Trans. 22:226-230.
97. MORRE, J. D., DECABO, R., FARLEY, C., OBERLIES, N. H., McLAUGHLIN, J. L. 1995. Mode of action of bullatacin, a potent antitumor acetogenin: inhibition of NADH oxidase activity of HELA and HL-60, but not liver, plasma membranes. Life Sciences 56:343-348.
98. LANDOLT, J. L., AHAMMADSAHIB, K. I., HOLLINGWORTH, R. M., BARR, R., CRANE, F. L., BUERCK, N. L., McCABE, G. P., McLAUGHLIN, J. L. 1995 Chemico-Biol. Interact. (in press).
99. ZHAO, G.-X., MIESBAUER, L. R., SMITH, D. L., McLAUGHLIN, J. L. 1994. Asimin, asiminacin and asiminecin: novel highly cytotoxic asimicin isomers from *Asimina triloba*. J. Med. Chem. 37:1971-1976.
100. McLAUGHLIN, J. L., HUI, Y. H. 1993. Chemotherapeutically active acetogenins. U.S. Patent, No. 5,229,419.
101. HOLSCHNEIDER, C. H., JOHNSON, M. T., KNOX, R. M., REZAI, A., RYAN, W. J., MONTZ, F. J. 1994. Bullatacin-in vivo and in vitro experience in an ovarian cancer model. Cancer Chemother. Pharmacol. 34:166-170.
102. CEPLEANU, F., OHTANI, K., HAMBURGER, M., GUPTA, M. P., SOLIS, P., HOSTETTMANN, K. 1993. Novel acetogenins from the leaves of *Annona purpurea*. Helv. Chim. Acta 76:1379-1388.
103. OHSAWA, D., ATSUZAWA, S., MITSUI, T., YAMAMOTO, I. 1991 Isolation and insecticidal activity of three acetogenins from seeds of pond apple, *Annona glabra* L. J. Pesticide Sci. 16:93-96.
104. SAHPAZ, S., BORIES, C., LOISEAU, P., CORTES, D., HOCQUEMILLER, R., LAURENS, A., CAVE, A. 1994. Cytotoxic and antiparasitic activity from A*nnona senegalensis* seeds. Planta Med. 60:538-540.
105. BORIS, C., LOISEAU, P., CORTES, D., MYINT, S. H., HOCQUEMILLER, R., GAYRAL, P., CAVE, A., LAURENS, A. 1991. Antiparasitic activity of *Annona muricata* and *Annona cherimolia* seeds. Planta Med. 57:434-436.
106. WEINBERG, M. L. D., PIRES, V., WEINBERG, J., OLIVEIRA, A., B. D. 1993. Inhibition of drug-induced contractions of guinea-pig ileum by *Annona crassiflora* seed extract. J. Pharm. Pharmacol. 45:70-72.
107. AVALOS, J., RUPPRECHT, J. K., McLAUGHLIN, J. L., RODRIGUEZ, E. 1993. Guinea pig maximization test of the bark extract from pawpaw, *Asimina triloba* (Annonaceae). Contact Dermatitis 29:33-35.
108. ALKOFAHI, A.; RUPPRECHT, J. K.; SMITH, D. L.; CHANG, C.-J.; McLAUGHLIN, J. L. 1988. Goniothalamicin and annonacin: bioactive acetogenins from *Goniothalamus giganteus* (Annonaceae). Experientia 44:83-85.
109. RIESER, M. J., FANG, X. P., RUPPRECHT, J. K., HUI, Y. H., SMITH, D. L., McLAUGHLIN, J. L. Bioactive single-ring acetogenins from seed extracts of *Annona muricata*. Planta Med. 59:91-92.
110. CHANG, F. R., WU, Y. C., DUH, C. Y., WANG, S. K. 1993. Studies on the acetogenins of Formosan Annonaceous plants, II. Cytotoxic acetogenins from *Annona reticulata*. J. Nat. Prod. 56:1688-1694.
111. MYINT, S. H., LAURENS, A., HOCQUEMILLER, R., CAVE, A. 1990. Murisolin: a new cytotoxic mono-tetrahydrofuran-γ-lactone from *Annona muricata*. Heterocycles 31:861-867.

112. CORTES, D., MYINT, S. H., LAURENS, A., HOCQUEMILLER, R. LEBOEUF, M., CAVE A. 1991. Corossolone et corossoline, deux nouvelles γ-lactones mono-tetrahydrofuraniques cytotoxiques. Can. J. Chem. 69:8-11.

113. WU, Y. C., CHANG, F. R., DUH, C. Y., WANG, S. K. 1992. Annoreticuin and isoannoreticuin: two new cytotoxic acetogenins from *Annona reticulata*. Heterocycles 34:667-674.

114. LI, X. H., HUI, Y. H., RUPPRECHT, J. K., LIU, Y. M., WOOD, K. V., SMITH, D. L., CHANG, C. J., McLAUGHLIN, J. L. 1990. Bullatacin, bullatacinone, and squamone, a new bioactive acetogenin, from the bark of *Annona squamosa*. J. Nat. Prod. 53:81-86.

115. JOSSANG, A., DUBAELE, B., CAVE, A., BARTOLI, M. H., BERIEL, H. 1991 Annomontacin: une nouvelle acetogenine γ-lactone-monotetrahydrofurannique cytotoxique de l'*Annona montana*. J. Nat. Prod. 54:967-971.

116. LIEB, F., NONFON, M., WACHENDORFF-NEUMANN, U., WENDISCH, D. 1990 Annonacin and annonastatin from *Annona squamosa*. Planta Med. 56:317-319.

117. McCLOUD, T. G., SMITH, D. L., CHANG, C. J., CASSADY, J. M. 1987. Annonacin, a novel, biologically active polyketide from *Annona densicoma*. Experientia 43:947-949.

118. ZHAO, G. X., RIESER, M. J., HUI, Y. H., MIESBAUER, L. R., SMITH, D. L., McLAUGHLIN, J. L. 1993. Biologically active acetogenins from stem bark of *Asimina triloba*. Phytochemistry 33:1065-1073.

119. XU, L. Z., CHANG, C. J., YU, J. G., CASSADY, J. M. 1989. Chemistry and selective cytotoxicity of annonacin-10-one, isoannonacin, and isoannonacin-10-one. Novel polyketides from *Annona densicoma* (Annonaceae). J. Org. Chem. 54:5418-5421.

120. HISHAM, A., PIETERS, L. A. C., CLAEYS, M., ESMANS, E., DOMMISSE, R., VLIETINCK, A. J. 1990. Uvariamicin-I, II and III: three novel acetogenins from *Uvaria narum*. Tetrahedron Lett. 31:4649-4652.

121. HUI, Y. H., WOOD, K. V., McLAUGHLIN, J. L. 1992. Bullatencin, 4-deoxyasimicin, and the uvariamicins: additional bioactive Annonaceous acetogenins from *Annona bullata* Rich. (Annonaceae). Natural Toxins 1:4-14.

122. CORTES, D., MYINT, S. H., LEBOEUF, M., CAVE A. 1991. A new type of cytotoxic acetogenins: the tetrahydrofuranic β-hydroxy methyl γ-lactones. Tetrahedron Lett. 32:6133-6134.

123. SAAD, J. M., HUI, Y. H., RUPPRECHT, J. K., ANDERSON, J. E., KOZLOWSKI, J. F., ZHAO, G. X., WOOD, K. V., McLAUGHLIN, J. L. 1991. Reticulatacin: a new bioactive acetogenin from *Annona reticulata* (Annonaceae). Tetrahedron 47:2751-2756.

124. RATNAYAKE, S., GU, Z. M., MIESBAUER, L. R. SMITH, D. L., WOOD, K. V., EVERT, D. R., McLAUGHLIN, J. L. 1994. Parvifloracin and parviflorin: cytotoxic bistetrahydrofuran acetogenins with 35 carbons from *Asimina parviflora* (Annonaceae). Can. J. Chem. 72:287-293.

125. DURET, P., GROMEK, D., HOCQUEMILLER, R., CAVE., A. 1995. Isolation and structure of three new bis-tetrahydrofuranic acetogenins from the roots of *Annona cherimolia*. J. Nat. Prod. 58: (in press).

126. HISHAM, A., PIETERS, L. A. C., CLAEYS, M., ESMANS, E., DOMMISSE, R., VLIETINCK, A. J. 1991. Acetogenins from root bark of *Uvaria narum*. Phytochemistry 30:2373-2377.

127. RIOS, J. L., CORTES, D., VALERDE, S. 1989. Acetogenins, aporphinoids, and azaanthraquinone from *Annona cherimolia* seeds. Planta Med. 55:321-323.

128. RUPPRECHT, J. K., CHANG, C. J., CASSADY, J. M., McLAUGHLIN, J. L. 1986. Asimicin, a new cytotoxic and pesticidal acetogenin from the pawpaw, *Asimina triloba* (Annonaceae). Heterocycles 24:1197-1201.

129. KAWAZU, K., ALCATARA, J. P., KOBAYASHI, A. 1989. Isolation and structure of Neoannonin, a novel insecticidal compound from the seeds of *Annona squamosa*. Agric. Biol. Chem. 53:2719-2722.

130. CORTES, D., MYINT, S. H., HOCQUEMILLER, R. 1991. Molvizarin and motrilin: two novel cytotoxic bis-tetrahydrofuranic γ-lactone acetogenins from *Annona cherimolia*. Tetrahedron 47:8195-8202.

131. GU, Z.-M., FANG, X.-P., RIESER, M. J., HUI, Y.-H., MIESBAUER, L. R., SMITH, D. L., WOOD, K. V., McLAUGHLIN, J. L. 1993 New cytotoxic Annonaceous acetogenins: bullatanocin and *cis*- and *trans*-bullatanocinone, from *Annona bullata* (Annonaceae). Tetrahedron 49:747-754.

132. JOLAD, S. D., HOFFMANN, J. J., COLE, J. R. 1985. Desacetyluvaricin from *Uvaria accuminata*, configuration of uvaricin at C-36. J. Nat. Prod. 48:644-645.

133. ETSE J. T., WATERMAN, P. G. 1986. Chemistry in the Annonaceae, XXII. 14-hydroxy-25-desoxyrollinicin from the stem bark of *Annona reticulata*. J. Nat. Prod. 49:684-686.

134. PETTIT, G. R., RIESEN, R., LEET, J. E., POLONSKY, J., SMITH, C. R., SCHMIDT, J. M., DUFRESNE, C., SCHAUFELBERGER, D., MORETTI, C. 1989. Isolation and structure of rolliniastatin 2: a new cell growth inhibitory acetogenin from *Rollinia mucosa*. Heterocycles 28:213-217.

135. ZHAO, G.-X., NG, J. H., KOZLOWZKI, J. F., SMITH, D. L, McLAUGHLIN, J. L. 1994. Bullatin and bullanin: two novel, highly cytotoxic acetogenins from *Asimina triloba*. Heterocycles 38:1897-1908.

136. FUJIMOTO, Y., EGUCHI, T., KAKINUMA, K., IKEKAWA, N., SAHAI, M., GUPTA, Y. K. 1988. Squamocin, a new cytotoxic bis-tetrahydrofuran containing acetogenin from *Annona squamosa*. Chem. Pharm. Bull. 36:4802-4806.

137. SAEZ, J., SAHPAZ, S., VILLAESCUSA, L., HOCQUEMILLER, R., CAVE, A., CORTES, D. 1993. Rioclarine et membranacine, deux nouvelles acetogenines bis-tetrahydrofuraniques des graines de *Rollinia membranacea*. J. Nat. Prod. 56:351-356.

138. DABRAH, T. T., SNEDEN A. 1984. Rollinicin and isorollinicin, cytotoxic acetogenins from *Rollinia papilionella*. Phytochemistry 23:2013-2016.

139. HISHAM, A., PIETERS, L. A. C., CLAEYS, M., ESMANS, E., DOMMISSE, R., VLIETINCK, A. J. 1991. Squamocin-28-one and panalicin, two acetogenins from *Uvaria narum*. Phytochemistry 30:545-548.

140. PETTIT, G. R., CRAGG, G. M., POLONSKY, J., HERALD, D. L., GOSWAMI, A., SMITH, C. R., MORETTI, C., SCHMIDT, J. M., WEISLEDER, D. 1987. Isolation and structure of rolliniastatin 1 from the South American tree *Rollinia mucosa*. Can. J. Chem. 65:1433-1435.

141. ABREO, M. J., SNEDEN, A. T. 1989. Rollinone, a revision and extension of structure. J. Nat. Prod. 53:983-935.

142. DABRAH, T. T., SNEDEN A. 1984. Rollinone, a new cytotoxic acetogenin from *Rollinia papilionella*. J. Nat. Prod. 47:652-657.

143. ALKOFAHI, A., RUPPRECHT, J. K., LIU, Y.-M., CHANG, C.-J., SMITH, D. L., McLAUGHLIN, J. L. 1990. Gigantecin: a novel antimitotic and cytotoxic acetogenin, with nonadjacent tetrahydrofuran rings, from *Goniothalamus giganteus* (Annonaceae). Experientia 46:539-541.

144. HUI, Y. H., RUPPRECHT, J. K., LIU, Y. M., ANDERSON, J. E., SMITH, D. L., CHANG, C. J., McLAUGHLIN, J. L. 1989. Bullatalicin, a novel bioactive acetogenin from *Annona bullata* (Annonaceae). Tetrahedron 45:6941-6948.

145. HUI, Y. H., RUPPRECHT, J. K., ANDERSON, J. E., WOOD, K. V., McLAUGHLIN, J. L. 1991. Bullatalicinone, a new potent bioactive acetogenin, and squamocin from *Annona bullata* (Annonaceae). Phytotherapy Res. 5:124-129.

146. CORTES, D., RIOS, J. L., VILLAR, A., VALVERDE, S. 1984. Cherimoline et dihydrocherimoline: deux nouvelles γ-lactones bis-tetrahydrofuranniques possedant une activite antimicrobienne. Tetrahedron Lett. 25:3199-3202.
147. GU, Z. M., ZENG, L., McLAUGHLIN, J. L. 1995. Isolation and structural elucidation of bioactive C- 12,15-*cis*-non-adjacent bis-THF Annonaceous acetogenins. Heterocycles (accepted).
148. FANG, X.-P., RONG, S., GU, Z.-M., RIESER, M. J., MIESBAUER, L. R., SMITH, D. L., McLAUGHLIN, J. L. 1993. A new type of cytotoxic Annonaceous acetogenin: giganin from *Goniothalamus giganteus*. Bioorg. Med. Chem. Lett. 3:1153-1156.

APPENDIX: New Structures of Annonaceous Acetogenins (1993–1994)

1. Xylopianin $C_{35}H_{64}O_7$, MW 596, (35)

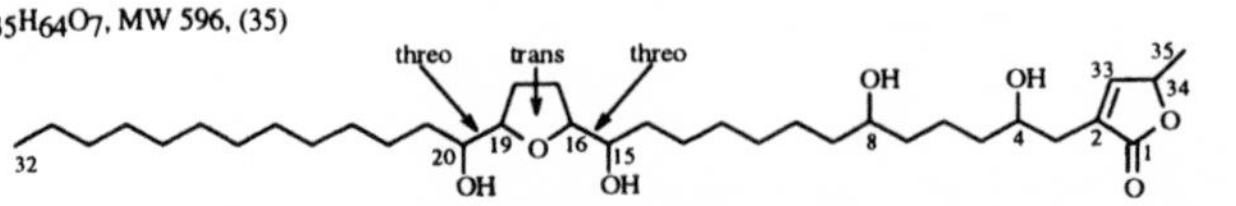

No.	1	2	4	8	15	16	19	20	32	33	34	35
$^1H(\delta)$			3.86m	3.60m	3.41m	3.80dt	3.80dt	3.41m	0.88t	7.15q	5.06qd	1.44d
$^{13}C(\delta)$	174.6	131.0	69.7	71.7	74.1	82.7	82.6	74.0	14.2	151.9	78.0	19.1

MP: 78-79 °C, white waxy solid; $[\alpha]_D$ +23.3° (*c* 0.008, MeOH); **UV** (λ_{max}, MeOH, nm): 217; **IR** (υ_{max}, film, cm^{-1}): 3426, 2920, 2837, 1737, 1455, 1314, 1073, 667; **MS**: FABMS (glycerol, *m/z*) 597, 579, 561, 543, 525; **NMR**: ^{1}H-NMR (500 MHz, $CDCl_3$), ^{13}C-NMR (125 MHz, $CDCl_3$); **Derivatives**: tetra-TMSi (EIMS); **Biological activities**: BST LC_{50}=0.33 μg/mL, A-549 ED_{50}=4.26 x 10^{-2} μg/mL, MCF-7 ED_{50} =18.47 μg/mL, HT-29 ED_{50}=6.63 μg/mL; **Source**: *Xylopia aromatica*, bark.

2. Annoreticuin $C_{35}H_{64}O_7$, MW 596, (113)

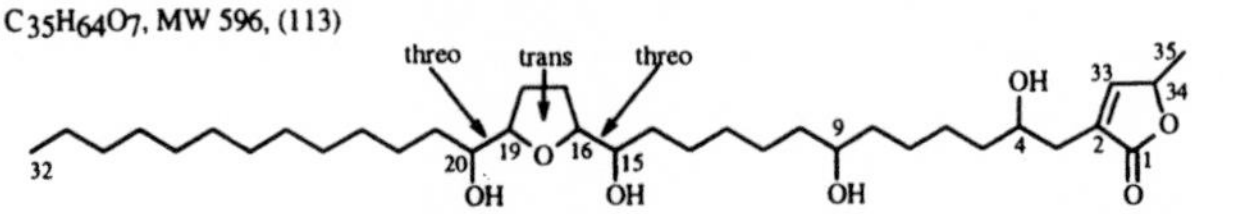

No.	1	2	4	9	15	16	19	20	32	33	34	35
$^1H(\delta)$			3.82m	3.60m	3.41m	3.79m	3.79m	3.41m	0.88t	7.20d	5.07qd	1.44d
$^{13}C(\delta)$	175.4	131.4	70.08	72.01	74.29	83.4	83.55	74.29	14.27	152.6	79.89	19.23

White amorphous powder; $[\alpha]_D$ +10.5° (*c* 0.02, $CHCl_3$); **UV** (λ_{max}, MeOH, nm): 210 (ε=4010); **IR** (υ_{max}, film, cm^{-1}): 3450, 1745; **MS**: CIMS (isobutane, *m/z*) 597, EIMS (*m/z*) 327, 227, 197, 169, 141, 123, 97, FABMS (*m/z*) 597, 579, 561, 543, 525; **NMR**: ^{1}H-NMR (200 MHz, $CDCl_3$), ^{13}C-NMR (50 MHz, $CDCl_3$); **Derivatives**: acetate (^{1}H-NMR), tetra-TMSi (EIMS); **Biological activities**: P388 ED_{50}=3.60 μg/mL, A-549 ED_{50}=0.34 μg/mL, KB ED_{50}=3.37 μg/mL, HT-29 ED_{50}=2.28 μg/mL; **Source**: *Annona reticulata*, leaves.

3. Isoannoreticuin $C_{35}H_{64}O_7$, MW 596, (113)

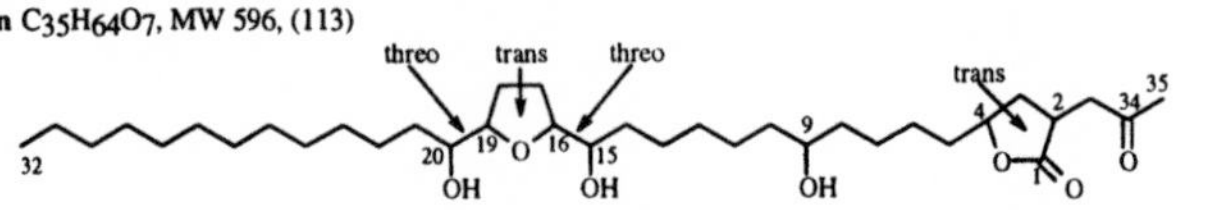

No.	1	2	4	9	15	16	19	20	32	34	35
$^1H(\delta)$		3.03ddd	4.53m	3.58m	3.41m	3.79m	3.79m	3.41m	0.88t		2.20s
$^{13}C(\delta)$	180.37	34.76	79.53	72.27	74.69	83.39	83.44	74.73	14.24	207.36	29.61

White amorphous powder; $[\alpha]_D$ +9.7° (*c* 0.33, $CHCl_3$); **UV** (λ_{max}, MeOH, nm): 190; **IR** (υ_{max}, film, cm^{-1}): 1760, 1700; **MS**: CIMS (isobutane, *m/z*) 597, EIMS (*m/z*) 397, 379, 361, 327, 309, 297, 227, 199, 197, 169, 141; **NMR**: ^{1}H-NMR (200 MHz, $CDCl_3$), ^{13}C-NMR (50 MHz, $CDCl_3$); **Derivatives**: tetra-TMSi (EIMS); **Biological activities**: P-388 ED_{50}=1.02 μg/mL, A-549 ED_{50}=0.41 μg/mL, KB ED_{50}=6.69 μg/mL, HT-29 ED_{50}=3.06 μg/mL; **Source**: *Annona reticulata*, leaves.

4. Annoreticuin-9-one $C_{35}H_{62}O_7$, MW 594, (110)

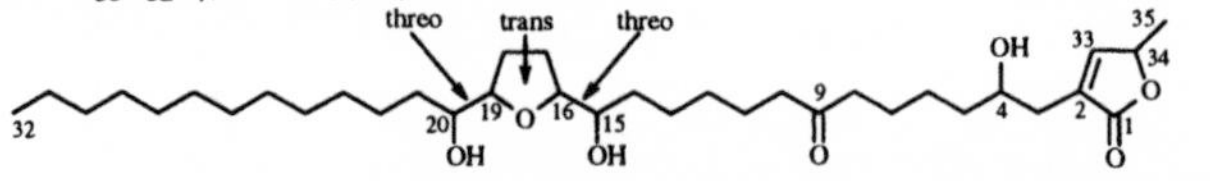

No.	1	2	4	9	15	16	19	20	32	33	34	35
$^1H(\delta)$			3.88m		3.40m	3.79m	3.79m	3.40m	0.88t	7.21d	5.08qd	1.44d
$^{13}C(\delta)$	176.31	132.22	70.19	213.46	74.86	83.59	74.71	83.49	14.34	153.45	78.82	19.34

White amorphous powder; $[\alpha]_D$ +11.7° (*c* 0.02, $CHCl_3$); **UV** (λ_{max}, MeOH, nm): 215 (ε=3400); **IR** (υ_{max}, film, cm^{-1}): 3420, 2915, 2850, 1750, 1700, 1470, 1320, 1080, 1030, 960; **MS**: EIMS (*m/z*) 385, 377, 359, 325, 307, 289, 269, 225, 207, 197, 179, 169, 141, 123, 111, 97; **NMR**: ^{1}H-NMR (200 MHz, $CDCl_3$), ^{13}C-NMR (50 MHz, $CDCl_3$); **Derivatives**: triacetate (IR, CIMS, EIMS, ^{1}H-NMR), tri-TMSi (EIMS, FABMS); **Biological activities**: P388 ED_{50}=2 x 10^{-1} μg/mL, A-549 ED_{50}=10^{-2} μg/mL, KB ED_{50}=4.66 μg/mL, HT-29 ED_{50}=1.32 μg/mL; **Source**: *Annona reticulata*, leaves.

5. *cis*-Isoannonacin $C_{35}H_{64}O_7$, MW 596, (109, 118)

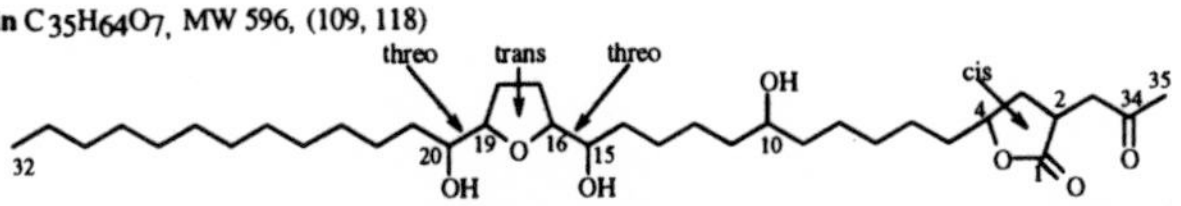

No.	1	2	4	10	15	16	19	20	32	34	35
$^1H(\delta)$		3.08 m	4.37 dddd	3.57 m	3.38 ddd	3.78 ddd	2.78 ddd	3.38 ddd	0.86 t		2.17 s
$^{13}C(\delta)$	178.2	43.42	78.86	71.84	74.04	82.63	82.57	73.98	14.17	205.5	22.73

Mp: 95-96 °C; $[\alpha]_D$ +20° (0.2 mg/ml, $CHCl_3$); **UV** (λ_{max}, MeOH, nm): 205; **IR** (υ_{max}, film, cm^{-1}): 3414, 2916, 2845, 1752, 1717, 1454, 1391, 1371, 1160, 1062, 864, 744; **MS**: EIMS (*m/z*) 596, 578, 560, 542, 397, 379, 361, 327, 309, 291, 241, 141; FAB HR-MS (glycerol); **NMR**: ^1H-NMR (500 MHz, $CDCl_3$), ^{13}C-NMR (125.75 MHz, $CDCl_3$); **Derivatives**: triacetate (IR, CIMS, ^1H-NMR), triTMSi (EIMS); **Biological activities**: BST LC_{50}=0.33 μg/mL, A-549 ED_{50}=4.42 x 10^{-5} μg/mL, MCF-7 ED_{50} < 10^{-3} μg/mL, HT-29 ED_{50}=1.70 x 10^{-1} μg/mL; **Source**: *Asimina triloba*, stem bark; *Annona muricata*, seeds.

6. 8-Hydroxyannonacin $C_{35}H_{64}O_8$, MW 612, (55)

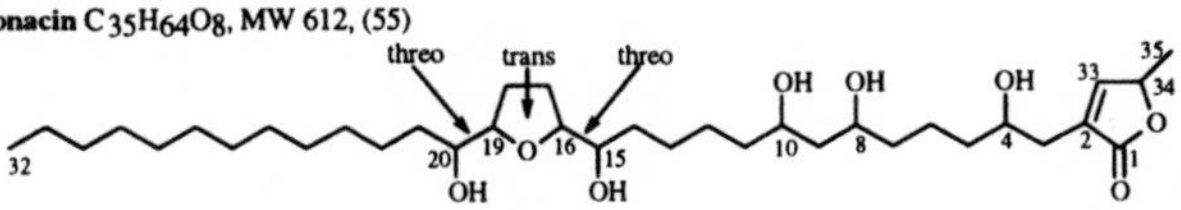

No.	4	8	10	15	16	19	20	32	33	34	35
$^1H(\delta)$	3.84 m	3.84 m	3.61 m	3.38 m	3.77 dd	3.77 dd	3.38 m	0.88 t	7.18 d	5.05 qdd	1.44 d

Waxy solid; $[\alpha]_D$ +6.1° (*c* 0.12, MeOH); **UV** (λ_{max}, MeOH, nm): 209.5 (log ε=4.22); **IR** (υ_{max}, film, cm^{-1}): 3421, 1465, 1404, 1216, 1120, 1080; **MS**: HR-FABMS, EIMS (*m/z*) 413, 395, 377, 359, 343, 325, 307, 289, 369, 257, 239, 213, 199, 195, 141; **NMR**: ^1H-NMR (500 MHz, $CDCl_3$); **Derivatives**: pentaacetate (EIMS, ^1H-NMR), penta-TMSi (EIMS); **Biological activities**: P388 ED_{50}=2 x 10^{-1} μg/mL, A-549 ED_{50}=5 x 10^{-2} μg/mL, HT-29 ED_{50}=3 x 10^{-1} μg/mL; **Source**: *Annona densicoma*, stem bark.

7. Giganenin $C_{37}H_{60}O_6$, MW 606, (37)

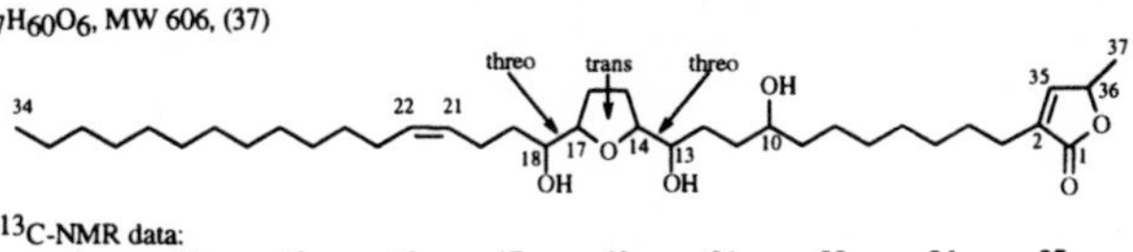

Reassignments of ^{13}C-NMR data:

No.	1	2	10	13	14	17	18	21	22	34	35	36	37
$^{13}C(\delta)$	173.8	134.2	71.9	74.3	82.6	82.6	73.5	128.8	130.8	14.2	148.9	77.4	19.2

Other data were reported in the previous review (9)

8. Gonionenin $C_{37}H_{66}O_7$, MW 622, (38)

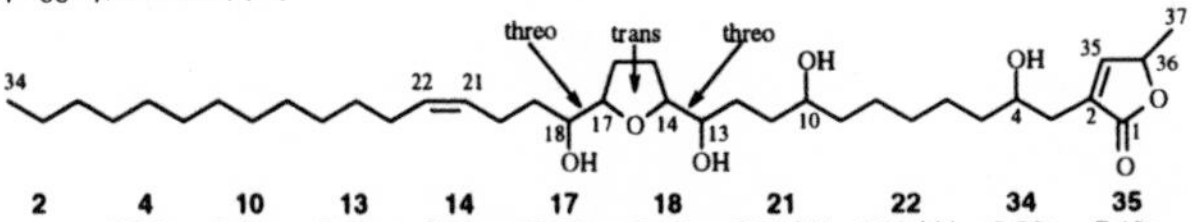

No.	1	2	4	10	13	14	17	18	21	22	34	35	36	37
$^1H(\delta)$			3.84m	3.63m	3.45m	3.82m	3.82m	3.43m	5.36ddd	5.39ddd	0.88t	7.19q	5.06qq	1.44d
$^{13}C(\delta)$	174.6	131.0	69.8	71.5	74.3	82.6	82.7	73.6	128.8	130.7	14.2	151.8	78.0	19.1

Mp: 87-88 °C; $[\alpha]_D$ +19.5° (*c* 0.22, MeOH); **UV** (λ_{max}, MeOH, nm): 209 (logε=4.05); **IR** (υ_{max}, film, cm^{-1}): 3454, 2910, 2852, 2361, 1732, 1468, 1317, 1108; **MS**: FAB HR-MS (glycerol); **NMR**: ^1H-NMR (500 MHz, $CDCl_3$), ^{13}C-NMR (125 MHz, $CDCl_3$); **Derivatives**: tetraacetate, 21/22-Epoxide (FABMS, ^1H-NMR), cyclogonionenin T (UV, IR, HR-FABMS, ^1H-, ^{13}C-NMR), cyclogonionenin C (UV, IR, HR-FABMS, ^1H-, ^{13}C-NMR), tetra-TMSi (EIMS); **Biological activities**: BST LC_{50}=21.7 μg/mL, A-549 ED_{50}=1.34 x 10^{-3} μg/mL, MCF-7 ED_{50}=4.54 x 10^{-3} μg/mL, HT-29 ED_{50}=1.12 x 10^{-4} μg/mL; **Source**: *Goniothalamus giganteus*, bark.

9. Xylopiacin $C_{37}H_{69}O_7$, MW 624, (35)

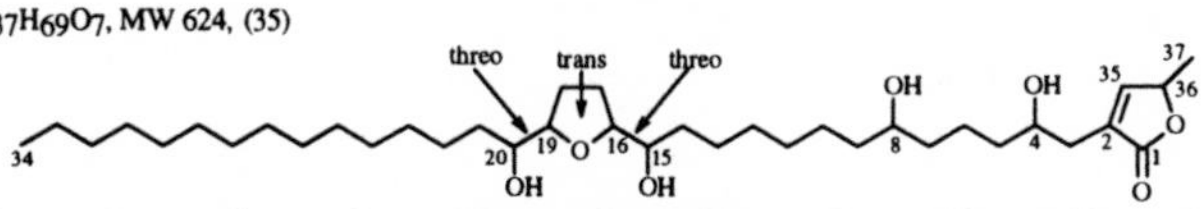

No.	1	2	4	8	15	16	19	20	34	35	36	37
$^1H(\delta)$			3.86m	3.60m	3.41m	3.80dt	3.80dt	3.41m	0.88t	7.19q	5.07qd	1.44d
$^{13}C(\delta)$	174.6	131.0	69.8	71.7	74.1	82.7	82.6	74.0	14.2	151.9	78.0	19.2

MP: 90-91 °C, white wax; $[\alpha]_D$ +24° (*c* 0.006, MeOH); **UV** (λ_{max}, MeOH, nm): 220; **IR** (υ_{max}, film, cm^{-1}): 3426, 2920, 2847, 1737, 1314, 1071; **MS**: FABMS (glycerol, *m/z*) 625, 607, 589, 571, 553; **NMR**: ^1H-NMR (500 MHz, $CDCl_3$), ^{13}C-NMR (125 MHz, $CDCl_3$); **Derivatives**: tetra-TMSi (EIMS); **Biological activities**: BST LC_{50}=0.49 μg/mL, A-549 ED_{50}=1.36 x 10^{-6} μg/mL, MCF-7 ED_{50}=1.39 x 10^{-7} μg/mL, HT-29 ED_{50}=1.11 x 10^{-1} μg/mL; **Source**: *Xylopia aromatica*, bark.

10. Xylopien $C_{37}H_{66}O_7$, MW 622, (36)

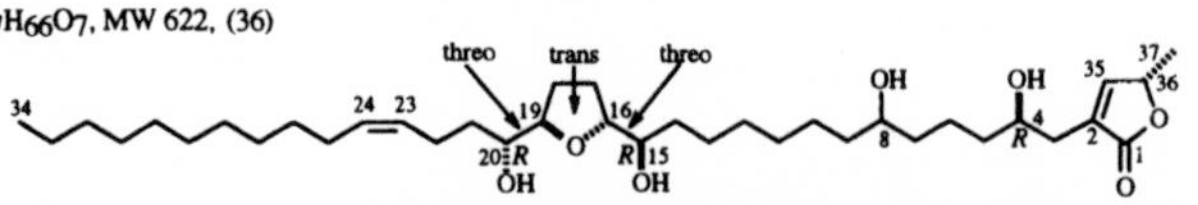

No.	1	2	4	8	15	16	19	20	23	24	34	35	36	37
$^1H(\delta)$			3.86m	3.62m	3.42m	3.80dt	3.80dt	3.42m	5.35ddd	5.40ddd	0.88t	7.19q	5.06qq	1.44d
$^{13}C(\delta)$	174.7	131.2	69.8	71.7	74.0	82.6	82.6	73.5	129.0	130.9	14.1	151.9	78.0	19.1

Mp: 48-49 °C, white wax; $[\alpha]_D$ +15° (*c* 0.001, MeOH); **UV** (λ_{max}, MeOH, nm): 225 (log ε=3.16); **IR** (υ_{max}, film, cm^{-1}): 3431, 2924, 2854, 1747, 1652, 1456, 1318, 1076, 668; **MS**: EIMS (*m/z*) 395, 379, 361, 343, 327, 309, 295, 225, 213, 195, 141; **NMR**: ^{1}H-NMR (500 MHz, $CDCl_3$), ^{13}C-NMR (125.75 MHz, $CDCl_3$); **Derivatives**: tetraacetate, tetra-TMSi (EIMS); **Biological activities**: BST LC_{50}=0.871 μg/mL, A-549 ED_{50}=3.64 x 10^{-3} μg/mL, MCF-7 ED_{50}=7.11 x 10^{-3} μg/mL, HT-29 ED_{50}=7.26 x 10^{-3} μg/mL; **Source**: *Goniothalamus giganteus*, bark.

11. Xylomaticin $C_{37}H_{69}O_7$, MW 624, (35)

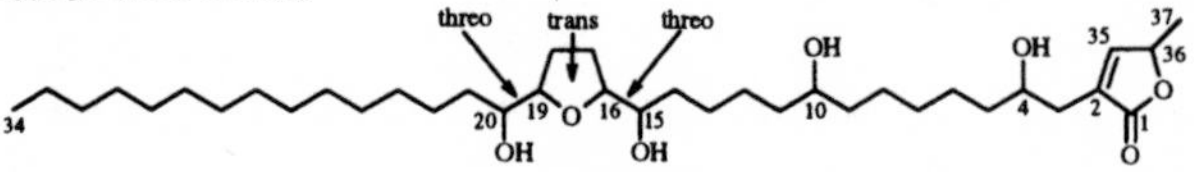

No.	1	2	4	10	15	16	19	20	34	35	36	37
$^1H(\delta)$			3.85m	3.60m	3.41m	3.80dt	3.80dt	3.41m	0.88t	7.19q	5.06qd	1.44d
$^{13}C(\delta)$	174.6	131.1	69.9	71.8	74.1	82.6	82.6	74.0	14.2	151.8	78.0	19.2

MP: 67-68 °C, white wax; $[\alpha]_D$ +5.3° (*c* 0.006, MeOH); **UV** (λ_{max}, MeOH, nm): 219; **IR** (υ_{max}, film, cm^{-1}): 3408, 2934, 2841, 1744, 1460, 1315, 1067, 845; **MS**: FABMS (glycerol, *m/z*) 625, 607, 589, 571, 553; **NMR**: ^{1}H-NMR (500 MHz, $CDCl_3$), ^{13}C-NMR (125 MHz, $CDCl_3$); **Derivatives**: tetra-TMSi (EIMS); **Biological activities**: BST LC_{50}=0.11 μg/mL, A-549 ED_{50}=1.54 x 10^{-4} μg/mL, MCF-7 ED_{50}=1.47 x 10^{-6} μg/mL, HT-29 ED_{50}=3.04 x 10^{-1} μg/mL; **Source**: *Xylopia aromatica*, bark.

12. Xylomatenin $C_{37}H_{66}O_7$, MW 622, (36)

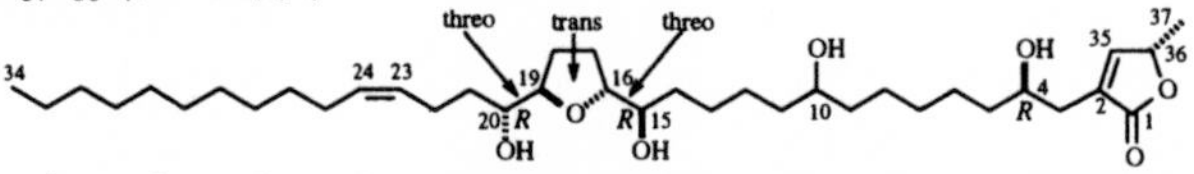

No.	1	2	4	10	15	16	19	20	23	24	34	35	36	37
$^1H(\delta)$			3.85m	3.60m	3.42m	3.80dt	3.80dt	3.42m	5.35ddd	5.40ddd	0.88t	7.19q	5.06qq	1.44d
$^{13}C(\delta)$	174.6	131.2	69.9	71.8	74.0	82.6	82.6	73.5	128.9	130.8	14.1	151.9	77.9	19.1

Mp: 52-53 °C, white wax; $[\alpha]_D$ +19° (*c* 0.001, MeOH); **UV** (λ_{max}, MeOH, nm): 222 (log ε=2.886); **IR** (υ_{max}, film, cm^{-1}): 3422, 2923, 2855, 1734, 1650, 1457, 1318, 1079, 668; **MS**: EIMS (*m/z*) 397, 379, 361, 343, 327, 309, 295, 225, 241, 141; **NMR**: ^{1}H-NMR (500 MHz, $CDCl_3$), ^{13}C-NMR (125 MHz, $CDCl_3$); **Derivatives**: tetraacetate, tetra-TMSi (EIMS); **Biological activities**: BST LC_{50}=0.77 μg/mL, A-549 ED_{50}<10^{-3} μg/mL, MCF-7 ED_{50}=6.31 x 10^{-1} μg/mL, HT-29 ED_{50}<10^{-3} μg/mL; **Source**: *Goniothalamus giganteus*, bark.

13. Squamosten-A $C_{37}H_{66}O_7$, MW 622, (39)

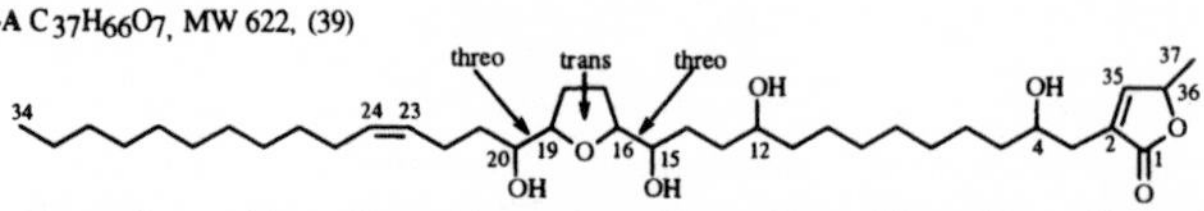

No.	1	2	4	12	15	16	19	20	23	24	34	35	36	37
$^1H(\delta)$			3.83m	3.63m	3.44m	3.83m	3.83m	3.44m	5.36m	5.39m	0.88t	7.18q	5.05qq	1.39d
$^{13}C(\delta)$	174.58	131.24	69.99	71.70	74.37	82.63	83.65	73.50	128.92	130.87	14.10	151.78	77.96	19.12

Mp: 64-67 °C; $[\alpha]_D$ +9° (*c* 0.1, MeOH); **UV** (λ_{max}, MeOH, nm): 210 (ε=7000); **IR** (υ_{max}, $CHCl_3$, cm^{-1}): 3665, 3575, 3450, 1745; **MS**: EIMS (*m/z*) 397, 379, 361, 343, 327, 325, 309, 291, 273, 269, 251, 141; FAB HR-MS; **NMR**: ^{1}H-NMR ($CDCl_3$), ^{13}C-NMR ($CDCl_3$); **Derivatives**: tetra-(*R*)-MTPA ester; **Source**: *Annona squamosa*, seeds.

14. Annomuricin A $C_{35}H_{64}O_8$, MW 612, (41)

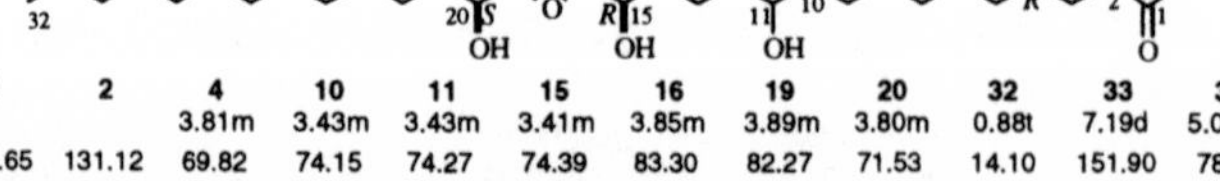

No.	1	2	4	10	11	15	16	19	20	32	33	34	35
$^1H(\delta)$			3.81m	3.43m	3.43m	3.41m	3.85m	3.89m	3.80m	0.88t	7.19d	5.05dq	1.42d
$^{13}C(\delta)$	174.65	131.12	69.82	74.15	74.27	74.39	83.30	82.27	71.53	14.10	151.90	78.01	19.08

White powder; $[\alpha]_D$ -6.4° (*c* 0.025); **UV** (λ_{max}, CH_2Cl_2, nm): 232 (ε=3100); **IR** (υ_{max}, film, cm^{-1}): 3431, 2919, 2849, 1734, 1699, 1073; **MS**: CIMS (*m/z*) 613, 595, 577, 559, 541, 395, 353, 325, 271, 269, 253, 241, 223, 205, 199, EIMS (*m/z*) 341, 325, 271, 269, 241, 223, 205, 199, 141; **NMR**: ^{1}H-NMR (500 MHz, $CDCl_3$), ^{13}C-NMR (125 MHz, $CDCl_3$); **Derivatives**: pentaacetate (^{1}H-NMR), penta-TMSi (EIMS), per-Mosher esters (^{1}H-NMR), acetonide (^{1}H-NMR); **Biological activities**: BST LC_{50}=0.625 μg/mL, A-549 ED_{50} =3.30 x 10^{-1} μg/mL, MCF-7 ED_{50}>1.0 μg/mL, HT-29 ED_{50}>1.0 μg/mL; **Source**: *Annona muricata*, leaves.

15. Annomuricin B $C_{35}H_{64}O_8$, MW 612, (41)

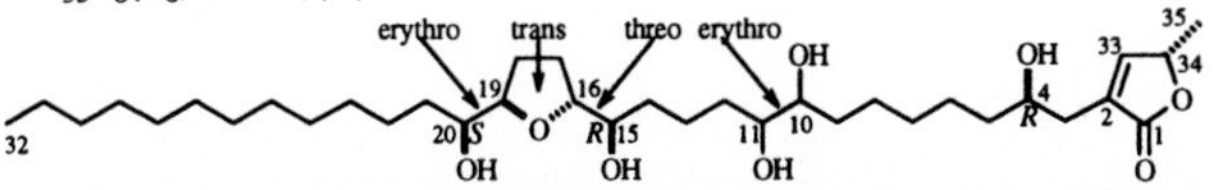

No.	1	2	4	10	11	15	16	19	20	32	33	34	35
$^1H(\delta)$			3.81m	3.61m	3.61m	3.41m	3.85m	3.89m	3.80m	0.88t	7.19d	5.05dq	1.42d
$^{13}C(\delta)$	174.65	131.12	69.82	74.23	74.35	74.53	83.30	82.27	71.53	14.10	151.90	78.01	19.08

White powder; $[\alpha]_D$ -11.7° (c 0.064); **UV** (λ_{max}, CH_2Cl_2, nm): 232 (ε=3000); **IR** (υ_{max}, film, cm^{-1}): 3430, 2921, 2851, 1733, 1684, 1075; **MS**: CIMS (m/z) 613, 595, 577, 559, 541, EIMS (m/z) 271, 269, 253, 241, 223, 199, 141; **NMR**: ^1H-NMR (500 MHz, $CDCl_3$), ^{13}C-NMR (125 MHz, $CDCl_3$); **Derivatives**: pentaacetate (^1H-NMR), penta-TMSi (EIMS), per-Mosher esters (^1H-NMR), acetonide (^1H-NMR); **Biological activities**: BST LC_{50}=0.69 μg/mL, A-549 ED_{50}=1.59 x 10^{-1} μg/mL, MCF-7 ED_{50}>1.0 μg/mL, HT-29 ED_{50}=0.435 μg/mL; **Source**: *Annona muricata*, leaves.

16. Gigantetrocin B $C_{35}H_{64}O_7$, MW 596, (50, 34, 60)

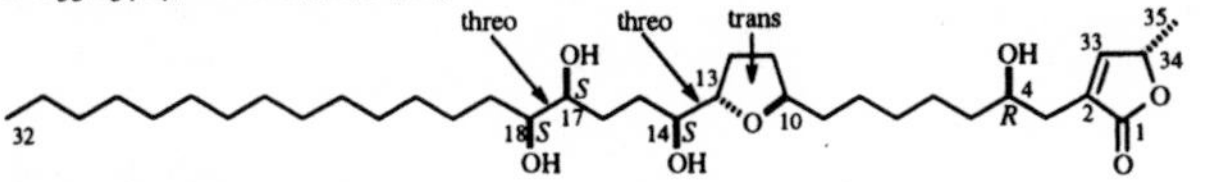

Other data were reported in the previous review (9).

17. Muricatetrocin B $C_{35}H_{64}O_7$, MW 596, (34, 41, 60)

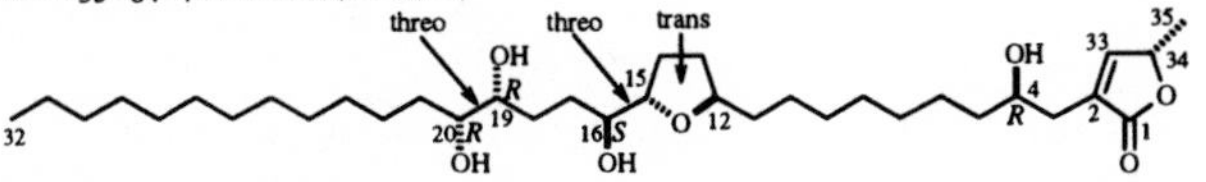

Other data were reported in the previous review (9).

18. Senegalene $C_{37}H_{66}O_7$, MW 622, (65)

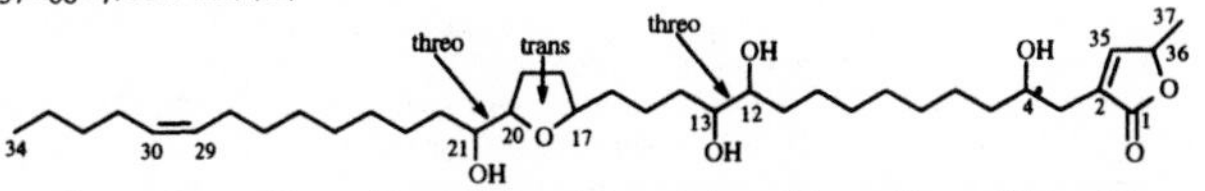

No.	1	2	4	12	13	17	20	21	29	30	34	35	36	37
$^1H(\delta)$			3.81m	3.41m	3.41m	3.90m	3.81m	3.41m	5.35m	5.35m	0.86t	7.18d	5.04dq	1.42d
$^{13}C(\delta)$	174.6	130.8	69.48	74.06	74.20	79.20	81.70	73.54	130.4	128.9	13.92	151.9	77.88	18.84

Amorphous powder; $[\alpha]_D$ +16° (c 0.21, $CHCl_3$); **IR** (υ_{max}, film, cm^{-1}): 3350, 2840, 1750, 1450, 1310, 1050, 748; **MS**: CIMS (NH_3, m/z) 640, 623, 605, 587, 569, 551; **NMR**: ^1H-NMR (200 MHz, $CDCl_3$), ^{13}C-NMR (50 MHz, $CDCl_3$); **Derivatives**: tetraacetate (FABMS, ^1H-NMR), 2,35,29,30-tetrahydro-senegalene (CIMS), 2,35,29,30-tetrahydro-tetraacetyl-senegalene (CIMS), 12,13-acetonide (CIMS), 29,30-epoxy-tetraacetyl-senegalene (CIMS); **Biological activities**: KB ED_{50}=1.5 x 10^{-4} μg/mL, VERO ED_{50}=2 x 10^{-2} μg/mL; **Source:** *Annona senegalensis*, seeds.

19. Muricatetrocin A $C_{35}H_{64}O_7$, MW 596, (34, 41, 60)

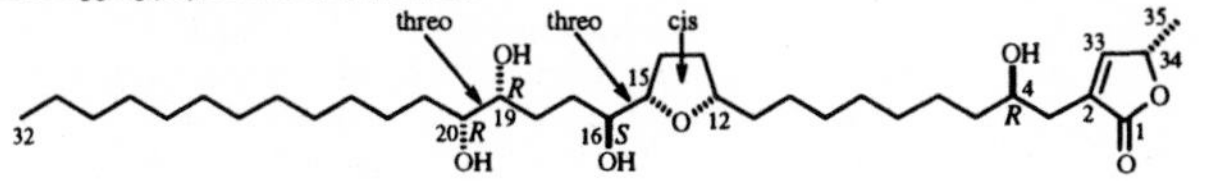

Other data were reported in the previous review (9).

20. Squamocin-K $C_{35}H_{62}O_6$, MW 578, (66)

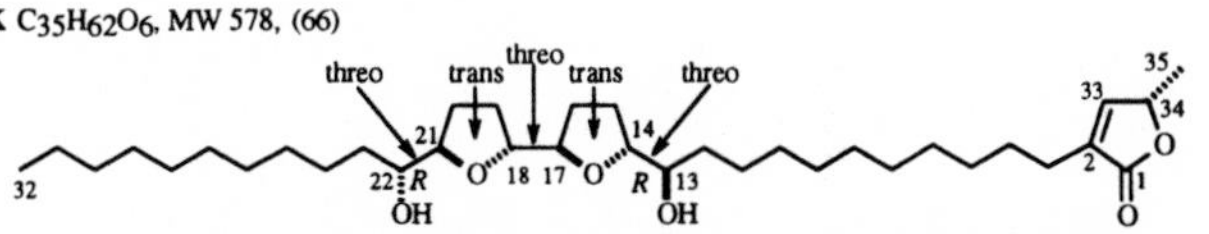

No.	1	2	13	14	21	17	18	22	32	33	34	35
$^1H(\delta)$			3.39m	3.81-3.87q		3.87-3.93 m		3.39m	0.88t	6.98d	4.99q	1.40d
$^{13}C(\delta)$	173.8	134.3	74.0	83.1	83.1	81.7	81.7	74.0	14.0	148.8	77.3	19.2

White wax, $[\alpha]_D$ +20.5° (c 0.53, MeOH); **IR** (υ_{max}, cm^{-1}): 3560, 3450, 1750; **NMR**: ^1H-NMR (500 MHz, $CDCl_3$), ^{13}C-NMR (125 MHz, $CDCl_3$); **Derivatives**: diacetate (^1H-NMR), di-(R)-MTPA ester; **Source**: *Annona squamosa*, seeds.

21. Parviflorin $C_{35}H_{62}O_7$, MW 594, (23, 66, 124)

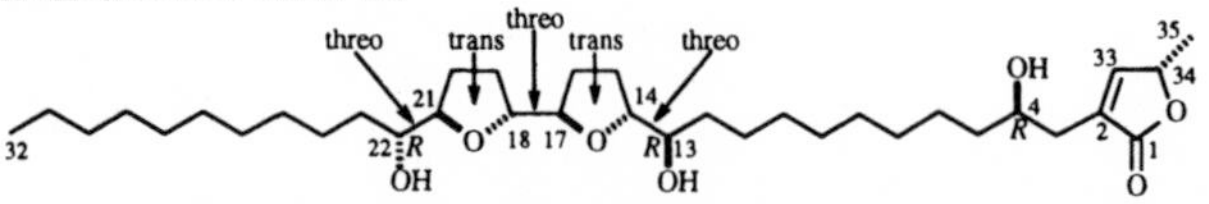

No.	1	2	4	13	14	17	18	21	22	32	33	34	35
$^1H(\delta)$			3.87m	3.39m	3.84m	3.84m	3.84m	3.84m	3.39m	0.87t	7.18q	5.06qq	1.43d
$^{13}C(\delta)$	174.52	131.13	69.99	74.08	83.15	81.79	81.77	83.12	74.05	14.17	151.72	77.97	19.17

Whitish wax; $[\alpha]_D$ +18.33° (*c* 0.06, EtOH); **UV** (λ_{max}, EtOH, nm): 223 (ε=8405); **IR** (υ_{max}, film, cm^{-1}): 3452, 1753, 1648; **MS**: CIMS (isobutane, *m/z*) 595, 577, 559, 541, 283, EIMS (*m/z*) 483, 423, 353, 311, 283, 265, 241, 171, 141, 111; **NMR**: ^{1}H-NMR (500 MHz, $CDCl_3$), ^{13}C-NMR (125 MHz, $CDCl_3$); **Derivatives**: triacetate (EIMS, CIMS, ^{1}H-NMR), tri-TMSi (EIMS), and tri-d_9-TMSi (EIMS), Mosher esters; **Biological activities**: BST LC_{50}=8.8 x 10^{-2} μg/mL, A-549 ED_{50}<10^{-12} μg/mL, MCF-7 ED_{50}=1.72 μg/mL, HT-29 ED_{50}=5.49 x 10^{-1} μg/mL; **Source**: *Asimina parviflora*, twigs; *Annona bullata*, bark; *Annona squamosa*, seeds.

Isomolvizarin-2s $C_{35}H_{62}O_7$, MW 594, (125)
(reported as a *cis* and *trans* mixture and mixed with isomolvizarin-1s)

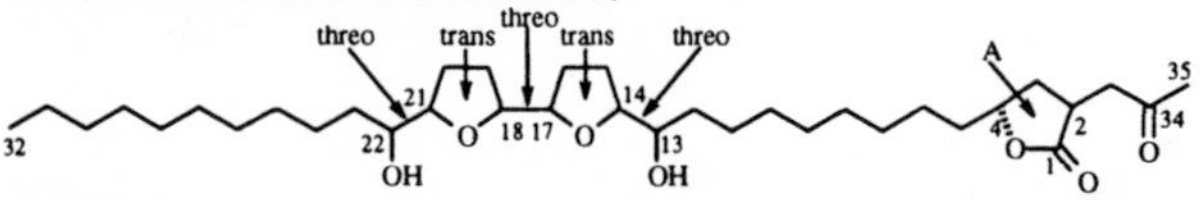

22. (2,4-*cis*)-Isomolvizarin-2 (A=*cis*)

No.	1	2	4	13	14	17	18	21	22	32	34	35
$^1H(\delta)$		3.00m	4.38m	3.38m	3.85m	3.85m	3.85m	3.85m	3.38m	0.86t		2.20s
$^{13}C(\delta)$	178.81	34.43	79.36	74.07	83.21	81.73	81.73	83.21	74.07	14.06	205.50	29.92

23. (2,4-*trans*)-Isomolvizarin-2 (A=*trans*)

No.	1	2	4	13	14	17	18	21	22	32	34	35
$^1H(\delta)$		3.00m	4.54m	3.38m	3.85m	3.85m	3.85m	3.85m	3.38m	0.86t		2.20s
$^{13}C(\delta)$	178.81	36.69	78.90	74.07	83.21	81.73	81.73	83.21	74.07	14.06	205.50	29.92

Colorless wax; $[\alpha]_D$ +33° (*c* 0.19, MeOH); **UV** (λ_{max}, MeOH, nm): 204.0 (ε=3674); **IR** (υ_{max}, film, cm^{-1}): 3432, 2929, 2856, 1771, 1721, 1570, 1467, 1408, 1372, 1310, 1165, 1063, 722; **MS**: CIMS (isobutane, *m/z*) 595, 557, 559, 481, 463, 311, 281, 241, EIMS (*m/z*) 576, 558, 551, 423, 405, 387, 353, 335, 317, 293, 283, 265, 241, 141, Li-FAB-MS (*m*-NBA, *m/z*) 601, 557, 487, 493, 441, 429, 404, 244; **NMR**: ^{1}H-NMR (200 MHz, $CDCl_3$), ^{13}C-NMR(50 MHz, $CDCl_3$); **Biological activities**: KB ED_{50}=10^{-4} μg/mL, HeLa ED_{50}=10^{-4}, VERO ED_{50}=10^{-3} μg/mL; **Source**: *Annona cherimolia*, roots.

24. Bullacin $C_{35}H_{62}O_7$, MW 594, (23)

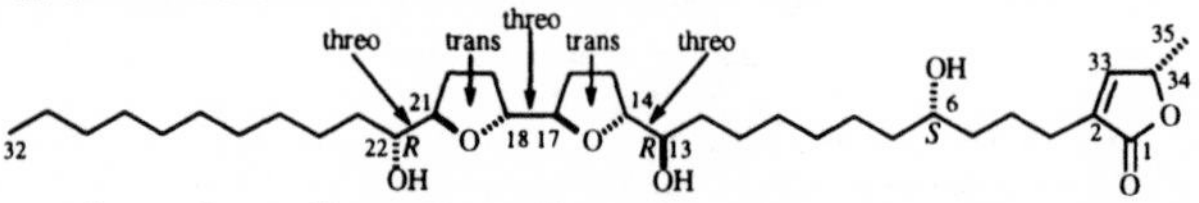

No.	1	2	6	13	14	17	18	21	22	32	33	34	35
$^1H(\delta)$			3.62m	3.38m	3.87m	3.87m	3.84m	3.83m	3.38m	0.88t	7.05q	5.00qq	1.40d
$^{13}C(\delta)$	173.8	134.0	71.5	74.0	83.1	81.8	81.8	83.2	74.1	14.2	149.1	77.5	19.2

Whitish wax; $[\alpha]_D$ +15.6° (*c* 0.3, $CHCl_3$); **UV** (λ_{max}, MeOH, nm): 213.3 (log ε=3.75); **IR** (υ_{max}, film, cm^{-1}): 1755; **MS**: CIMS (isobutane, *m/z*) 577, 559, 541, EIMS (*m/z*) 353, 311, 283, 241, 169; **NMR**: ^{1}H-NMR (500 MHz, $CDCl_3$), ^{13}C-NMR (125 MHz, $CDCl_3$); **Derivatives**: triacetate (^{1}H-NMR), tri-TMSi (EIMS), and tri-d_9-TMSi (EIMS), Mosher esters; **Biological activities**: BST LC_{50}=7.0 x 10^{-2} μg/mL, A-549 ED_{50}=1.79 x 10^{-5} μg/mL, MCF-7 ED_{50}=1.0 x 10^{-5} μg/mL, HT-29 ED_{50}=5.23 x 10^{-3} μg/mL; **Source**: *Annona bullata*, bark.

25. Asimin $C_{37}H_{66}O_7$, MW 622, (99)

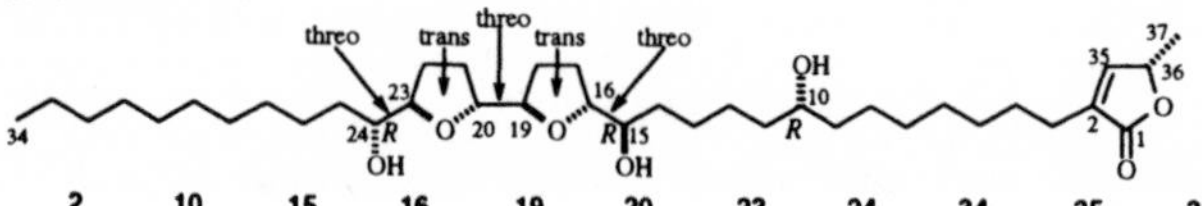

No.	1	2	10	15	16	19	20	23	24	34	35	36	37
$^1H(\delta)$			3.59m	3.40m	3.85m	3.85m	3.85m	3.85m	3.40m	0.878t	6.99q	5.00qq	1.41d
$^{13}C(\delta)$	173.80	134.20	71.84	74.08	83.17	81.81	81.76	83.05	73.97	14.18	148.82	77.41	19.27

Colorless wax; $[\alpha]_D$ +26° (1 mg/ml, $CHCl_3$); **UV** (λ_{max}, MeOH, nm): 215; **IR** (υ_{max}, film, cm^{-1}): 3438, 2925, 2855, 1752, 1457, 1318, 1200, 1069, 954, 870; **MS**: CIMS (isobutane, *m/z*) 623, 605, 587, 569, 225, 311, 293, 275, 241, 171, FAB HR-MS (glycerol); **NMR**: ^{1}H-NMR (500 MHz, $CDCl_3$), ^{13}C-NMR (125 MHz, $CDCl_3$); **Derivatives**: triacetate (IR, CIMS, ^{1}H-NMR), tri-TMSi (EIMS), and tri-d_9-TMSi (EIMS); **Biological activities**: BST LC_{50}=4.6 x 10^{-3} μg/mL, A-549 ED_{50}=7.99 x 10^{-9} μg/mL, MCF-7 ED_{50}=9.57 x 10^{-9} μg/mL, HT-29 ED_{50}<10^{-12} μg/mL; **Source**: *Asimina triloba*, stem bark.

26. Squamocin F $C_{37}H_{66}O_7$, MW 622 (66)

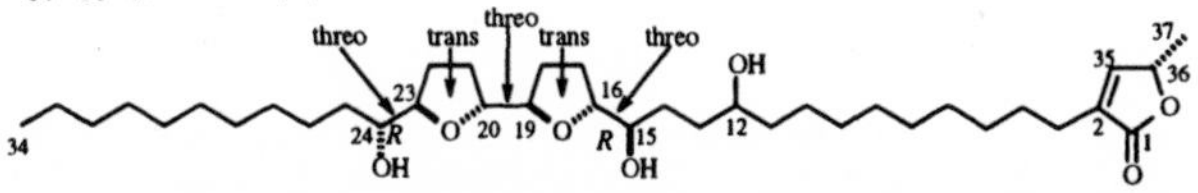

No.	1	2	12	15	16	19	20	23	24	34	35	36	37
$^1H(\delta)$			3.60m	3.44m	3.80-3.94 m		3.80-3.94 m		3.40m	0.88t	6.99s	4.99q	1.41d
$^{13}C(\delta)$	173.9	134.3	71.7	74.2	83.2	81.8	81.7	83.2	74.0	14.2	149.0	77.5	19.3

White wax, $[\alpha]_D$ +21.0° (*c* 0.58, MeOH); **IR** (υ_{max}, cm^{-1}): 3560, 3450, 1750; **MS**: EIMS (*m/z*) 445, 433, 415, 397, 363, 345, 327, 293, 275, 253, FAB HR-MS; **NMR**: ^{1}H-NMR (500 MHz, $CDCl_3$), ^{13}C-NMR (125 MHz, $CDCl_3$); **Derivatives**: triacetate (^{1}H-NMR), tri-(*R*)-MTPA ester; **Source**: *Annona squamosa*, seeds.

27. Asiminacin $C_{37}H_{66}O_7$, MW 622, (99)

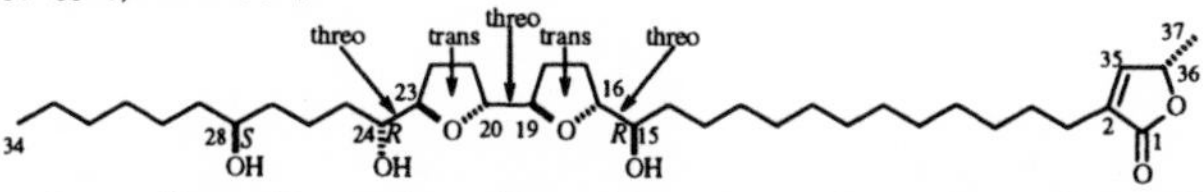

No.	1	2	15	16	19	20	23	24	28	34	35	36	37
$^1H(\delta)$			3.40m	3.86m	3.86m	3.86m	3.86m	3.40m	3.60m	0.882t	6.99q	5.00qq	1.41d
$^{13}C(\delta)$	173.83	134.25	73.89	83.05	81.78	81.85	83.19	74.07	71.73	14.15	148.78	77.40	19.27

Colorless wax; $[\alpha]_D$ +21.1° (3.8 mg/ml, $CHCl_3$); **UV** (λ_{max}, MeOH, nm): 215; **IR** (υ_{max}, film, cm^{-1}): 3418, 2925, 2856, 1753, 1457, 1319, 1200, 1069, 953, 872; **MS**: CIMS (isobutane, *m/z*) 623, 605, 587, 569, 295, 365, 347, 309, 257, 239, 115; FAB HR-MS (glycerol); **NMR**: ^{1}H-NMR (500 MHz, $CDCl_3$), ^{13}C-NMR (125 MHz, $CDCl_3$); **Derivatives**: triacetate (IR, CIMS, ^{1}H-NMR), tri-TMSi (EIMS), and tri-d_9-TMSi (EIMS); **Biological activities**: BST LC_{50}=5.7 x 10^{-3} µg/mL, A-549 ED_{50}=3.58 x 10^{-9} µg/mL, MCF-7 ED_{50}<10^{-12} µg/mL, HT-29 ED_{50}<10^{-12} µg/mL; **Source**: *Asimina triloba*, stem bark; *Annona squamosa*, seeds.

28. Squamocin-D $C_{37}H_{66}O_7$, MW 622, (51, 66)

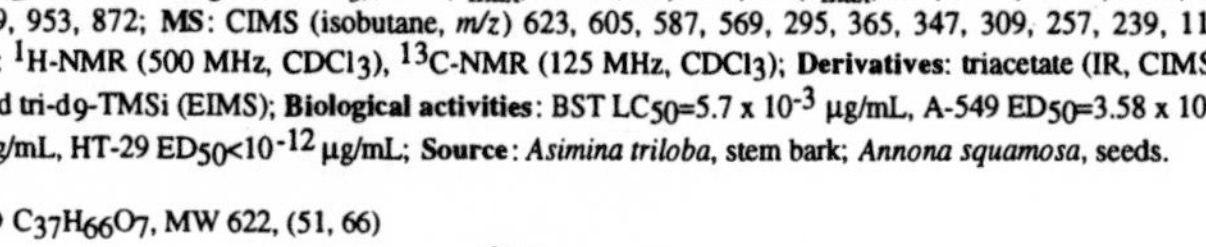

No.	1	2	15	16	19	20	23	24	28	34	35	36	37
$^1H(\delta)$			3.40m	3.81-3.93 m		3.81-3.93 m		3.40m	3.60m	0.88t	6.99s	4.99q	1.41d
$^{13}C(\delta)$	173.9	134.3	74.1	83.1	81.8	81.8	83.2	73.9	71.7	14.1	148.8	77.4	19.2

Colorless oil, $[\alpha]_D$ +30.1° (*c* 0.58, MeOH); **IR** (υ_{max}, cm^{-1}): 3560, 3450, 1745; **MS**: EIMS (*m/z*) 519, 501, 483, 465, 435, 417, 399, 365, 347, 329, 295; FAB HR-MS; **NMR**: ^{1}H-NMR (500 MHz, $CDCl_3$), ^{13}C-NMR (125 MHz, $CDCl_3$); **Derivatives**: triacetate (^{1}H-NMR), tri-(*R*)-MTPA ester; **Source**: *Annona squamosa*, seeds.

29. Asiminecin $C_{37}H_{66}O_7$, MW, 622, (99)

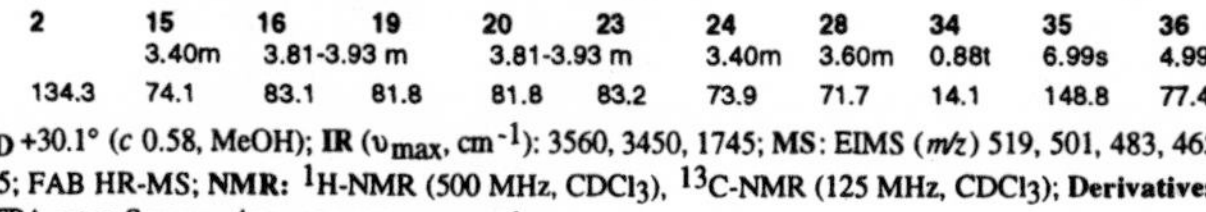

No.	1	2	15	16	19	20	23	24	29	34	35	36	37
$^1H(\delta)$			3.40m	3.86m	3.86m	3.86m	3.86m	3.40m	3.60m	0.891t	6.99q	5.00qq	1.41d
$^{13}C(\delta)$	173.83	134.25	73.87	83.05	81.78	81.82	83.17	74.07	71.81	14.11	148.79	77.41	19.26

Colorless wax; $[\alpha]_D$ +22° (1 mg/ml, $CHCl_3$); **UV** (λ_{max}, MeOH, nm): 215; **IR** (υ_{max}, film, cm^{-1}): 3419, 2925, 2856, 1751, 1456, 1318, 1199, 1069, 953, 872; **MS**: CIMS (isobutane, *m/z*) 623, 605, 587, 569, 551, 435, 417, 365, 347, 311, 295, 241, 239; FAB HR-MS (glycerol); **NMR**: ^{1}H-NMR (500 MHz, $CDCl_3$), ^{13}C-NMR (125 MHz, $CDCl_3$); **Derivatives**: triacetate (IR, CIMS, ^{1}H-NMR), tri-TMSi (EIMS), and tri-d_9-TMSi (EIMS); **Biological activities**: BST LC_{50}=4.9 x 10^{-3} µg/mL, A-549 ED_{50}=3.29 x 10^{-7} µg/mL, MCF-7 ED_{50}=2.74 x 10^{-9} µg/mL, HT-29 ED_{50}<10^{-12} µg/mL; **Source**: *Asimina triloba*, stem bark.

Isomolvizarin-1s (reported as a *cis* and *trans* mixture), $C_{35}H_{62}O_7$, MW 594, (125)

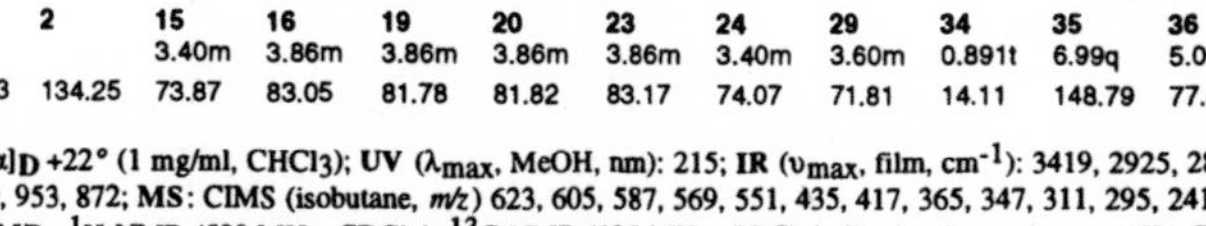

30. (2,4-*cis*)-Isomolvizarin-1 (A=*cis*)

No.	1	2	4	13	14	17	18	21	22	32	34	35
$^1H(\delta)$		3.00m	4.38m	3.38m	3.85m	3.85m	3.85m	3.85m	3.85m	0.86t		2.20s
$^{13}C(\delta)$	178.81	34.43	79.36	74.07	83.13	82.48	82.80	82.22	71.27	14.06	205.50	29.92

31. (2,4-*trans*)-Isomolvizarin-1 (A=*trans*)

No.	1	2	4	13	14	17	18	21	22	32	34	35
$^1H(\delta)$		3.00m	4.54m	3.38m	3.85m	3.85m	3.85m	3.85m	3.85m	0.86t		2.20s
$^{13}C(\delta)$	178.81	36.69	78.90	74.07	83.13	82.48	82.80	82.22	71.27	14.06	205.50	29.92

White wax; $[\alpha]_D$ +24° (*c* 0.32MeOH); **UV** (λ_{max}, MeOH, nm): 206.1 (ε=3674); **IR** (υ_{max}, film, cm^{-1}): 3434, 2922, 2851, 1767, 1715, 1465, 1360, 1313, 1167, 1120, 1075, 1011, 960, 722; **MS**: CIMS (isobutane, *m/z*) 595, 557, 559, 481, 463, 311, 281, 241, EIMS (*m/z*) 576, 558, 551, 423, 405, 387, 353, 335, 317, 293, 283, 265, 241, 141, Li-FAB-MS (*m*-NBA, *m/z*) 601, 557, 487, 493, 441, 429, 404, 244; **NMR:** ^{1}H-NMR (200 MHz, $CDCl_3$), ^{13}C-NMR (50 MHz, $CDCl_3$); **Biological activities**: KB $ED_{50}=10^{-4}$ μg/mL, HeLa $ED_{50}=10^{-4}$, VERO $ED_{50}=10^{-3}$ μg/mL; **Source**: *Annona cherimolia*, roots.

32. Squamocin-B $C_{35}H_{62}O_7$, MW 594, (51, 66)

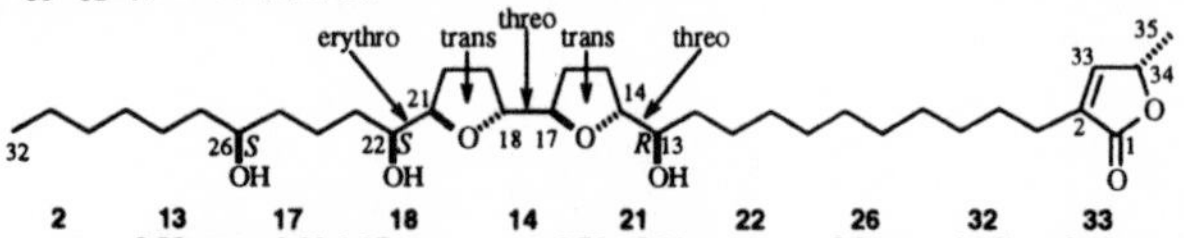

No.	1	2	13	17	18	14	21	22	26	32	33	34	35
$^1H(\delta)$			3.38m	3.89-3.97 m		3.78 - 3.89 m			3.60m	0.88t	6.99 s	4.99 q	1.41d
$^{13}C(\delta)$	173.9	134.3	74.1	82.2	82.5	83.3	82.8	71.4	71.7	14.0	148.8	77.4	19.2

White wax, $[\alpha]_D$ +27.6° (*c* 0.2, MeOH); **IR** (υ_{max}, cm^{-1}): 3660, 3575, 3450, 1745; **MS**: EIMS (*m/z*) 491, 473, 445, 427, 407, 389, 371, 353, 337, 319, 301, 267, 249, FAB HR-MS; **NMR**: ^{1}H-NMR (500 MHz, $CDCl_3$), ^{13}C-NMR (125 MHz, $CDCl_3$); **Derivatives**: tri-(*R*)-MTPA ester; **Source**: *Annona squamosa*, seeds.

33. Bullatin $C_{37}H_{66}O_7$, MW 622, (135)

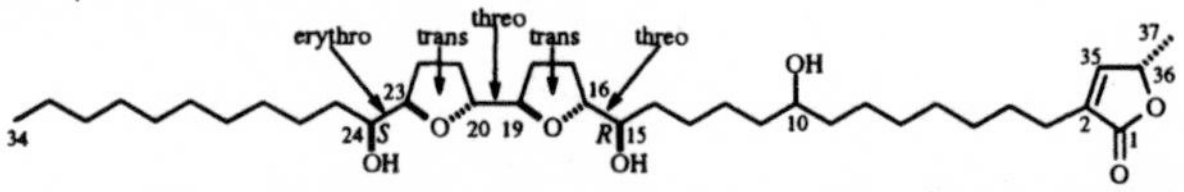

No.	1	2	10	15	16	19	20	23	24	34	35	36	37
$^1H(\delta)$			3.59m	3.41m	3.87m	3.87m	3.87m	3.93m	3.87m	0.878t	6.99q	5.00qq	1.41d
$^{13}C(\delta)$	173.83	134.21	71.85	73.99	83.14	82.33	82.52	82.81	71.31	14.18	148.85	77.43	19.27

Colorless wax; $[\alpha]_D$ +7.5° (0.4 mg/ml, EtOH); **UV** (λ_{max}, MeOH, nm): 213 (log ε= 2.34); **IR** (υ_{max}, film, cm^{-1}): 3442, 2971, 2849, 1748, 1458, 1186, 1051; **MS**: EIMS (*m/z*) 409, 391, 373, 363, 345, 327, 311, 293, 275, 225, 171, FAB HR-MS (glycerol); **NMR**: ^{1}H-NMR (500 MHz, $CDCl_3$), ^{13}C-NMR (125 MHz, $CDCl_3$); **Derivatives**: triacetate (IR, CIMS, ^{1}H-NMR), tri-TMSi (EIMS), and tri-d_9-TMSi (EIMS); **Biological activities**: BST $LC_{50}=4.0 \times 10^{-3}$ μg/mL, A-549 $ED_{50}=9.39 \times 10^{-6}$ μg/mL, MCF-7 $ED_{50}=8.33 \times 10^{-6}$ μg/mL, HT-29 $ED_{50}=3.78 \times 10^{-6}$ μg/mL; **Source**: *Asimina triloba*, stem bark.

34. Bullanin $C_{37}H_{66}O_7$, MW 622, (135)

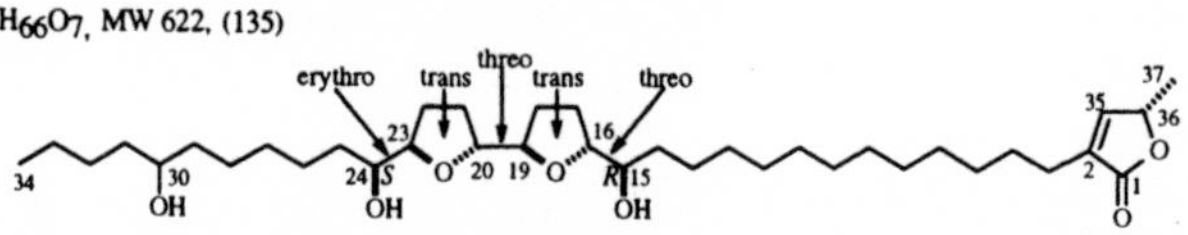

No.	1	2	15	16	19	20	23	24	30	34	35	36	37
$^1H(\delta)$			3.40m	3.85m	3.85m	3.85m	3.93m	3.85m	3.58m	0.907t	6.99q	5.00qq	1.41d
$^{13}C(\delta)$	173.83	134.23	74.11	83.25	82.26	82.52	82.75	71.28	71.87	14.14	148.79	77.41	19.26

Colorless wax; $[\alpha]_D$ +28° (0.5 mg/ml, EtOH); **UV** (λ_{max}, MeOH, nm): 220 (log ε=2.42); **IR** (υ_{max}, film, cm^{-1}): 3442, 2927, 2852, 1747, 1459, 1320, 1193, 1070; **MS**: EIMS (*m/z*) 565, 511, 417, 295, 291, 277, 239, 221, 187, 169, 151, FAB HR-MS (glycerol); **NMR**: ^{1}H-NMR (500 MHz, $CDCl_3$), ^{13}C-NMR (125 MHz, $CDCl_3$); **Derivatives**: triacetate (IR, CIMS, ^{1}H-NMR), tri-TMSi (EIMS), and tri-d_9-TMSi (EIMS); **Biological activities**: BST $LC_{50}=6.0 \times 10^{-3}$ μg/mL, A-549 $ED_{50}=3.11 \times 10^{-14}$ μg/mL, MCF-7 $ED_{50}=3.22 \times 10^{-14}$ μg/mL, HT-29 $ED_{50}=4.77 \times 10^{-12}$ μg/mL; **Source**: *Asimina triloba*, stem bark.

35. Purpureacin-2 $C_{37}H_{66}O_8$, MW 638, (102)

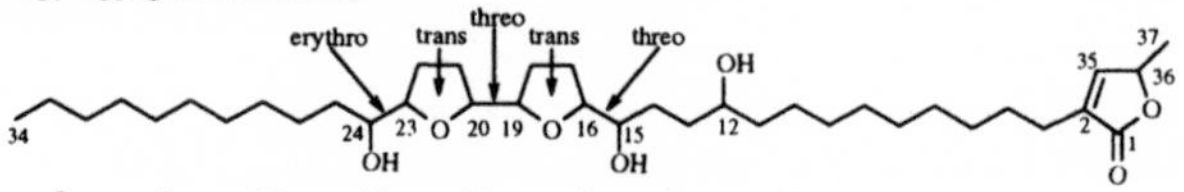

No.	1	2	4	12	15	16	19	20	23	24	34	35	36	37
$^1H(\delta)$			3.78m	3.53m	3.39m	3.82m	3.8-3.9 m		3.8-3.9 m		0.80t	7.12q	5.00qq	1.37d
$^{13}C(\delta)$	174.6	131.2	69.9	71.7	74.2	83.2	82.5	82.8	82.3	71.3	14.1	151.8	78.0	19.1

White amorphous power; $[\alpha]_D$ +6.5° (*c* 0.17, MeOH); **UV** (λ_{max}, MeOH, nm): 207 (logε=3.90); **MS**: CIMS (NH_3, *m/z*) 656 (MNH_4^+), 639, 621, 603, 585, 567, 491, 363, 311, 293, 269, 251, 241, 223, 111, EIMS (*m/z*) 467, 449, 431, 413, 379, 363, 361, 343, 341, 323, 311, 309, 293, 291, 269, 251, 241, 223, 171, 141, 111; **NMR**: ^{1}H-NMR (200 MHz, $CDCl_3$), ^{13}C-NMR (50 MHz, $CDCl_3$); **Derivatives**: tri-TMSi (EIMS); **Biological activities**: BST LC_{50}=0.38 μg/mL, Larvicidal activity LC_{100}/24h=1.0 μg/mL, antifungal activity: 1.0 μg, antibacterial activity: not active; **Source**: *Annona purpurea*, leaves.

36. Annonin XIV $C_{37}H_{66}O_8$, MW 638, (27)

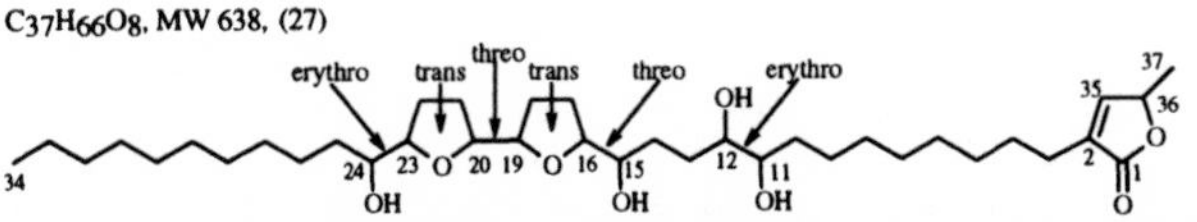

No.	1	2	11	12	15	16	19	20	23	24	34	35	36	37
$^1H(\delta)$			3.60	3.60	3.40	3.75-4.00		3.75-4.00		3.75-4.00	0.85	7.00	5.00	1.41
$^{13}C(\delta)$	173.9	134.4	74.7	74.6	74.2	83.5	82.6	82.2	82.9	71.6	14.1	148.9	77.4	19.2

Amorphous wax, $[\alpha]_D$ +15.7° (*c* 0.30, CH_2Cl_2); **MS**: EIMS (*m/z*) 467, 363, 345, 327, 309, 239, 171; **NMR**: ^{1}H-NMR ($CDCl_3$), ^{13}C-NMR ($CDCl_3$); **Derivatives**: tetraacetate (^{1}H-NMR); **Biological activities**: weak activity against *Caenorhabditis elegans* and *Plutella xylostella*; **Source**: *Annona squamosa*, seeds.

10-Hydroxybullatacinones (reported as a *cis* and *trans* mixture), $C_{37}H_{66}O_8$, MW 638, (31, 32)

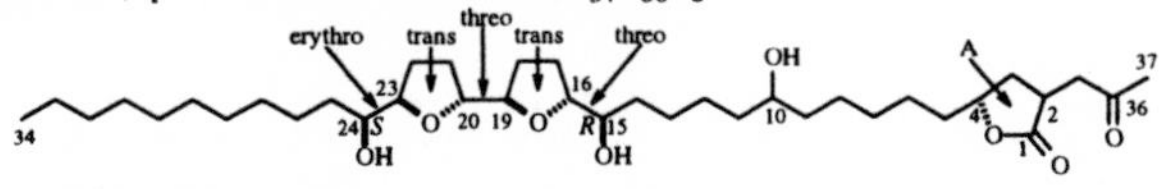

37. (2,4-*cis*)-10-Hydroxybullatacinone (A=*cis*)

No.	1	2	4	10	15	16	19	20	23	24	34	36	37
$^1H(\delta)$		3.02m	4.39m	3.59m	3.42m	3.86m	3.86m	3.93m	3.93m	3.86m	0.878t		2.20s
$^{13}C(\delta)$	178.8	44.3	78.0	71.7	73.9	83.1	82.8	82.5	82.3	71.3	14.17	205.4	24.9

38. (2,4-*trans*)-10-Hydroxybullatacinone (A=*trans*)

No.	1	2	4	10	15	16	19	20	23	24	34	36	37
$^1H(\delta)$		3.03m	4.54m	3.59m	3.42m	3.86m	3.86m	3.93m	3.93m	3.86m	0.878t		2.20s
$^{13}C(\delta)$	178.2	43.9	79.4	71.7	73.9	83.1	82.8	82.5	82.3	71.3	14.17	205.5	24.9

White powder; **IR** (υ_{max}, film, cm^{-1}): 3426, 2925, 2846, 1764, 1716, 1066; **MS**: FAB HRMS (glycerol); **NMR**: ^{1}H-NMR (500 MHz, $CDCl_3$), ^{13}C-NMR (125 MHz, $CDCl_3$), ^{1}H-^{1}H COSY, single relayed COSY; **Derivatives**: triacetate (^{1}H-NMR), tri-TMSi (EIMS); **Biological activities**: BST LC_{50}=7.0 x 10^{-2} µg/mL, A-549 ED_{50}=8.52 x 10^{-3} µg/mL, MCF-7 ED_{50}=1.86 x 10^{-4} µg/mL, HT-29 ED_{50}=6.93 x 10^{-2} µg/mL; **Source**: *Annona bullata*, bark.

12-Hydroxybullatacinones (reported as a *cis* and *trans* mixture), $C_{37}H_{66}O_8$, MW 638, (31, 32)

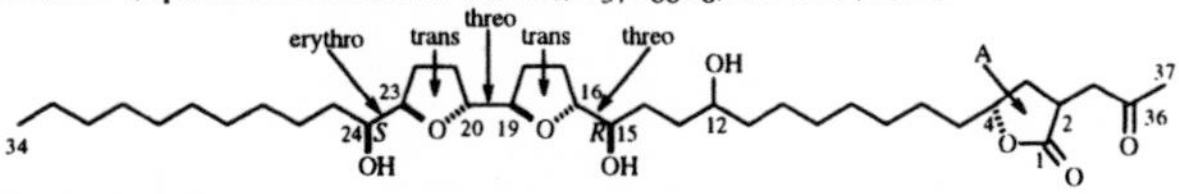

39. (2,4-*cis*)-12-Hydroxybullatacinone

No.	1	2	4	12	15	16	19	20	23	24	34	36	37
$^1H(\delta)$		3.02m	4.39m	3.62m	3.45m	3.87m	3.87m	3.93m	3.93m	3.87m	0.878t		2.20s
$^{13}C(\delta)$	178.8	44.3	78.9	71.7	74.3	83.0	82.8	82.5	82.3	71.3	14.6	205.4	24.9

40. (2,4-*trans*)-12-Hydroxybullatacinone

No.	1	2	4	12	15	16	19	20	23	24	34	36	37
$^1H(\delta)$		3.03m	4.54m	3.62m	3.45m	3.87m	3.87m	3.93m	3.93m	3.87m	0.878t		2.20s
$^{13}C(\delta)$	178.2	43.9	79.4	71.7	74.4	83.0	82.8	82.5	82.3	71.3	14.6	205.5	24.9

White powder; **IR** (υ_{max}, film, cm^{-1}): 3426, 2925, 2359, 1763, 1715, 1062; **MS**: FAB HRMS (glycerol); **NMR**: ^{1}H-NMR (500 MHz, $CDCl_3$), ^{13}C-NMR (125 MHz, $CDCl_3$), ^{1}H-^{1}H COSY, single relayed COSY; **Derivatives**: triacetate (^{1}H-NMR), tri-TMSi (EIMS); **Biological activities**: BST LC_{50}=4.43 x 10^{-2} µg/mL, A-549 ED_{50}=1.30 x 10^{-2} µg/mL, MCF-7 ED_{50}=2.15 x 10^{-6} µg/mL, HT-29 ED_{50}=1.51 µg/mL; **Source**: *Annona bullata*, bark.

28-Hydroxybullatacinones (reported as a *cis* and *trans* mixture), $C_{37}H_{66}O_8$, MW 638, (32)

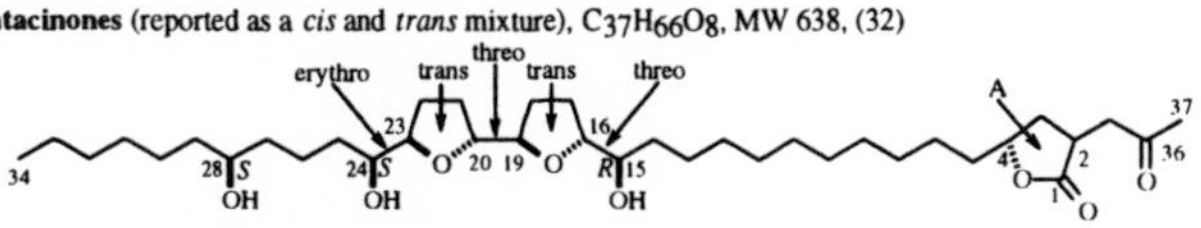

41. (2,4-*cis*)-28-Hydroxybullatacinone (A=*cis*)

No.	1	2	4	15	16	19	20	23	24	28	34	36	37
$^1H(\delta)$		3.02m	4.39m	3.42m	3.86m	3.86m	3.93m	3.93m	3.86m	3.60m	0.878t		2.20s
$^{13}C(\delta)$	178.8	44.3	78.0	73.9	83.1	82.8	82.5	82.3	71.3	71.4	14.17	205.4	24.9

42. (2,4-*trans*)-28-Hydroxybullatacinone (A=*trans*)

No.	1	2	4	15	16	19	20	23	24	28	34	36	37
^1H (δ)		3.03m	4.54m	3.42m	3.86m	3.86m	3.93m	3.93m	3.86m	3.60m	0.878t		2.20s
^{13}C (δ)	178.2	43.9	79.4	73.9	83.1	82.8	82.5	82.3	71.3	71.4	14.17	205.5	24.9

White powder; **IR** (υ_{max}, film, cm^{-1}): 3426, 2925, 2846, 1764, 1716, 1066; **MS**: FAB HRMS (glycerol); **NMR**: ^1H-NMR (500 MHz, $CDCl_3$), ^{13}C-NMR (125 MHz, $CDCl_3$); **Derivatives**: triacetate (^1H-NMR), tri-TMSi (EIMS); **Biological activities**: BST LC_{50}=7.0 x 10^{-2} μg/mL, A-549 ED_{50}=8.52 x 10^{-3} μg/mL, MCF-7 ED_{50}=1.86 x 10^{-4} μg/mL, HT-29 ED_{50}=6.93 x 10^{-2} μg/mL; **Source**: *Annona bullata*, bark.

29-Hydroxybullatacinones (reported as a *cis* and *trans* mixture), $C_{37}H_{66}O_8$, MW 638, (31, 32)

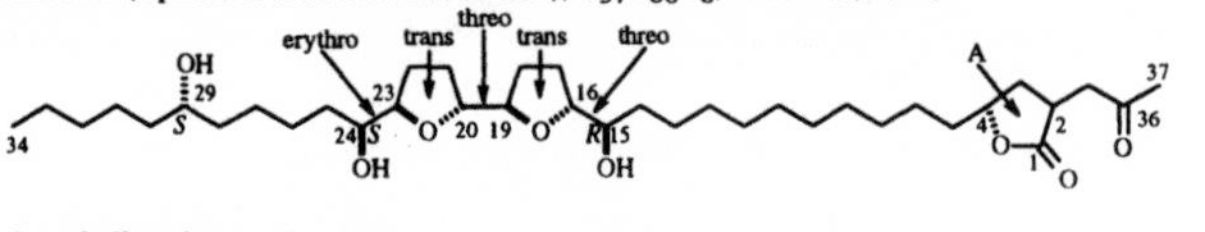

43. (2,4-*cis*)-29-Hydroxybullatacinone (A= *cis*)

No.	1	2	4	15	16	19	20	23	24	29	34	36	37
^1H (δ)		3.02m	4.39m	3.40m	3.87m	3.87m	3.93m	3.93m	3.87m	3.59m	0.892t		2.20s
^{13}C (δ)	178.8	44.3	78.0	74.1	83.2	82.8	82.5	82.3	71.3	71.9	14.12	205.4	24.9

44. (2,4-*trans*)-29-Hydroxybullatacinone (A=*trans*)

No.	1	2	4	15	16	19	20	23	24	29	34	36	37
^1H (δ)		3.03m	4.54m	3.40m	3.87m	3.87m	3.93m	3.93m	3.87m	3.59m	0.892t		2.20s
^{13}C (δ)	178.2	43.9	79.4	74.1	83.2	82.8	82.5	82.3	71.3	71.9	14.12	205.5	24.9

White powder; **IR** (υ_{max}, film, cm^{-1}): 3426, 2919, 2847, 1766, 1716, 1063; **MS**: FAB HRMS (glycerol); **NMR**: ^1H-NMR (500 MHz, $CDCl_3$), ^{13}C-NMR (125 MHz, $CDCl_3$) ^1H-^1H COSY; **Derivatives**: triacetates (^1H-NMR), tri-TMSi (EIMS); **Biological activities**: BST LC_{50}=7.53 x 10^{-2} μg/mL, A-549 ED_{50}=2.40 x 10^{-3} μg/mL, MCF-7 ED_{50}=1.35 x 10^{-4} μg/mL, HT-29 ED_{50}=1.57 x 10^{-1} μg/mL; **Source**: *Annona bullata*, bark.

45. Squamocin-I $C_{35}H_{63}O_6$, MW 578, (66)

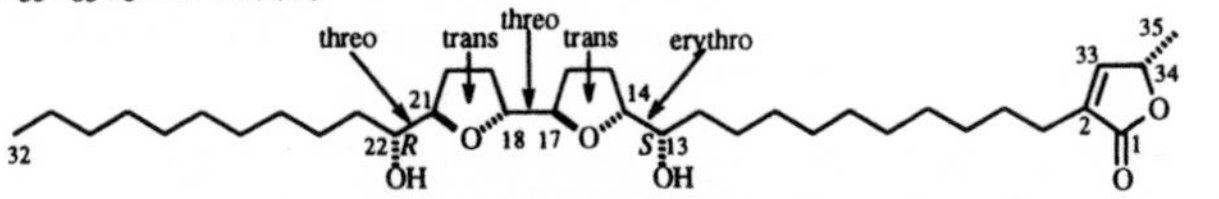

No.	1	2	13	14	21	17	18	22	32	33	34	35
^1H(δ)				3.80 - 3.90 m		3.90-3.97m		3.40m	0.88t	6.99brs	5.00q	1.41d
^{13}C(δ)	173.8	134.3	71.3	82.8	83.3	82.5	82.2	74.1	14.0	151.7	77.3	19.2

Whitish needles, **MP**: 68.5-71 °C; $[\alpha]_D$ +22.2° (*c* 0.05, EtOH); **UV** (λ_{max}, nm): 209 (log ε=3.7); **IR** (υ_{max}, film, cm^{-1}): 3560, 3450, 1750, 1722, 1645; **MS**: EIMS (*m/z*) 419, 407, 389, 371, 353, 337, 319, 311, 301, 293, 267; **NMR**: ^1H-NMR (500 MHz, $CDCl_3$), ^{13}C-NMR (125 MHz, $CDCl_3$); **Derivatives**: diacetate (^1H-NMR), Mosher esters; **Source**: *Annona squamosa*, seeds.

46. Trilobacin $C_{37}H_{66}O_7$, MW 622, (58, 59)

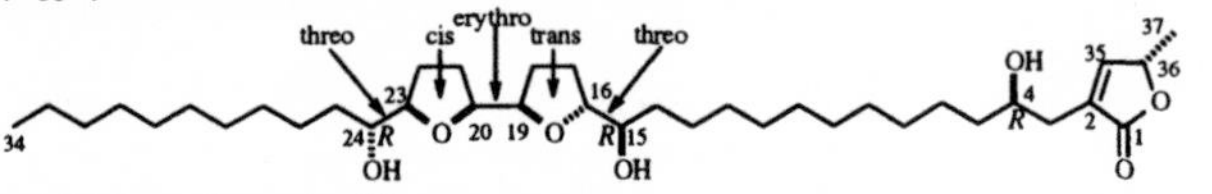

Other data were reported in the previous review (9).

47. Trilobin $C_{37}H_{66}O_7$, MW 622, (59)

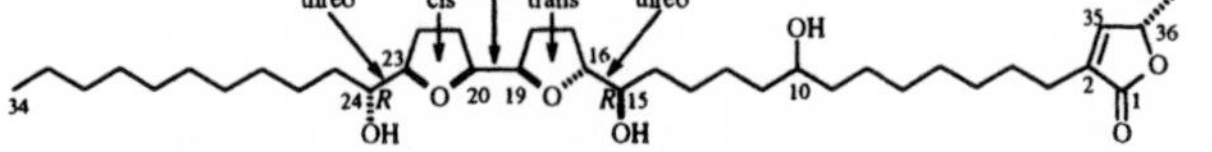

No.	1	2	10	15	16	19	20	23	24	34	35	36	37
^1H (δ)			3.59m	3.37m	3.83m	3.97q	4.05q	3.83m	3.40	0.878t	6.99q	5.00qq	1.41d
^{13}C (δ)	173.8	131.3	71.9	74.6	83.2	81.6	81.0	82.6	73.8	14.1	148.8	77.4	19.2

Colorless wax; **MS**: FABMS (glycerol, *m/z*) 623, 605, 569, 551, EIMS (*m/z*) 345, 311, 293, 275, 241, 225, 195; **NMR**: ^1H-NMR (500 MHz, $CDCl_3$), ^{13}C-NMR (125 MHz, $CDCl_3$) ^1H-^1H COSY, HMBC, HMQC; **Derivatives**: triacetate (^1H-NMR), tri-TMSi (EIMS), Mosher esters; **Biological activities**: BST LC_{50}=9.7 x 10^{-3} μg/mL, A-549 ED_{50}=5.71 x 10^{-12} μg/mL, MCF-7 ED_{50}=2.95 x 10^{-12}, HT-29 ED_{50}=1.68 x 10^{-1} μg/mL; **Source**: *Asimina triloba*, stem bark.

48. Squamocin-N $C_{37}H_{66}O_6$, MW 606, (66)

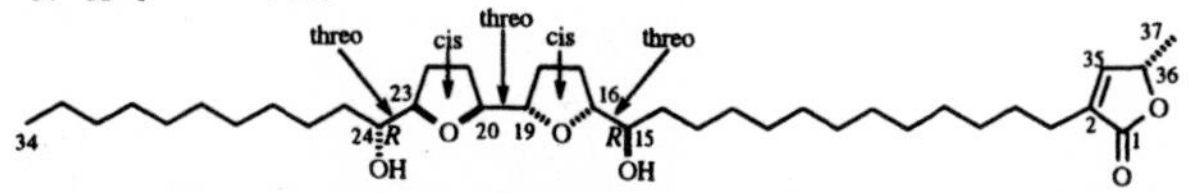

No.	1	2	15	24	16	23	19	20	34	35	36	37
^1H (δ)			3.40qui		3.81-3.86m		3.88-3.93m		0.88t	6.98d	4.99q	1.41d
^{13}C (δ)	173.9	134.4	74.0	74.0	82.2	82.8	81.0	81.0	14.1	148.8	77.4	19.2

Colorless oil; $[\alpha]_D$ +40.6° (*c* 0.43, MeOH); **NMR**: ^{1}H-NMR (500 MHz, $CDCl_3$), ^{13}C-NMR (125 MHz, $CDCl_3$); **Derivatives**: diacetate (^{1}H-NMR), Mosher esters (^{1}H-NMR); **Source**: *Annona squamosa*, seeds.

49. Rollinone $C_{37}H_{66}O_7$, 622, (142)

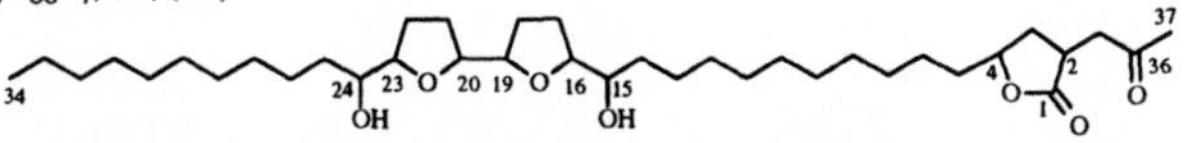

No.	1	2	4	15	16	19	20	23	24	34	36	37
^1H (δ)		3.05m	4.40m	3.42m	3.80-3.95 m			3.80 - 3.95 m		0.88t		2.18d
^{13}C (δ)	177.0s	36.8d	78.6d	74.1q	83.4d	80.8d	81.2d	83.8d	72.3d	14.4q	204.2s	23.2q

Fine colorless crystals, **MP**: 54-55 °C; $[\alpha]_D$ +25.0° (*c* 0.1371, $CHCl_3$); **UV** (λ_{max}, CH_2Cl_2, nm): 321, 280 (ε=15199); **IR** (υ_{max}, film, cm^{-1}): 3486, 2930, 1780, 1726, 1466, 1410, 1356, 1050, 715; **MS**: CIMS (*m/z*) 623, 622, 605, 587, 569, 451, 381, 363, 341, 312, 311, 293, 283, 265, 253, 241, 223, 171, 141, 113, 99, 85, 71; **NMR**: ^{1}H-NMR (360 MHz, $CDCl_3$), ^{13}C-NMR (22 MHz, C_6D_6); **Derivatives**: diacetate (IR, CIMS, ^{1}H-NMR); **Biological activities**: P388 $ED_{50}<10^{-5}$ µg/mL (*in vitro*), T/C=147% at 1.4 mg/kg (*in vivo*); **Source**: *Annona reticulata*, leaves.

Bulladecinones (reported as a *cis* and *trans* mixture) $C_{37}H_{66}O_8$, MW 638, (33, 32)

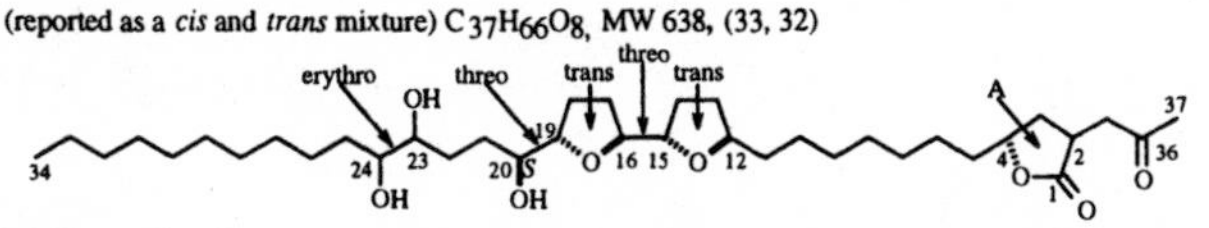

50. (2,4-*cis*)-Bulladecinone (A=*cis*)

No.	1	2	4	12	15	16	19	20	23	24	34	36	37
^1H(δ)		3.02m	4.39dddd	3.93m	3.87m	3.87m	3.87m	3.47m	3.58m	3.62m	0.88t		2.20s
^{13}C(δ)	178.2	43.8	79.4	79.9	81.3	82.2	82.7	74.1	74.4	74.7	14.2	205.53	25.3

51. (2,4-trans)-Bulladecinone (A=*trans*)

No.	1	2	4	12	15	16	19	20	23	24	34	36	37
^1H(δ)		3.03m	4.55dddd	3.93m	3.87m	3.87m	3.87m	3.47m	3.58m	3.62m	0.88t		2.20s
^{13}C(δ)	178.8	44.3	78.9	79.9	81.3	82.2	82.7	74.1	74.4	74.7	14.2	205.46	25.3

MS: EIMS (*m/z*) 467, 449, 431, 379, 329, 311, 309, 293, 259, 241, 171; FAB HRMS (glycerol); **NMR**: ^{1}H-NMR (500 MHz, $CDCl_3$), ^{13}C-NMR (125 MHz, $CDCl_3$), COSY, NOESY; **Derivatives**: triacetate (^{1}H-NMR), acetonides (COSY, NOESY), triTMSi (EIMS), and tri-d_9-TMSi (EIMS); **Biological activities**: BST LC_{50}=1.37 x 10^{-1} µg/mL, A-549 ED_{50}=3.37 x 10^{-5} µg/mL, MCF-7 ED_{50}=1.07 x 10^{-3} µg/mL, HT-29 ED_{50}=2.29 x 10^{-1} µg/mL; **Source**: *Annona bullata*, bark.

52. Parviflorацин $C_{35}H_{62}O_8$, MW 610, (134)

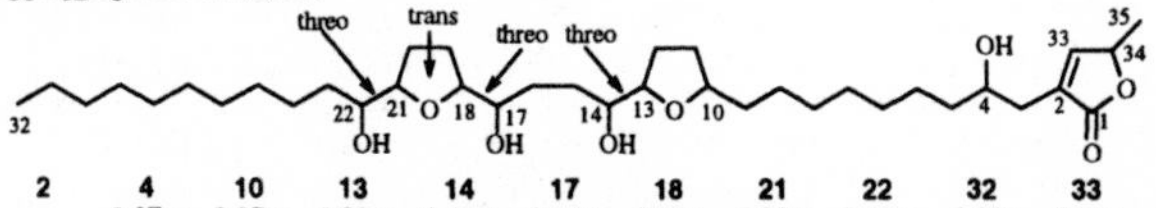

No.	1	2	4	10	13	14	17	18	21	22	32	33	34	35
^1H(δ)			3.87m	3.87m	3.80m	3.41m	3.41m	3.80m	3.80m	3.41m	0.87t	7.19q	5.06qq	1.43d
^{13}C(δ)	174.6	131.1	69.87	79.24	81.96	74.07	74.25	82.66	82.68	74.37	14.07	151.9	77.97	19.06

Whitish wax; $[\alpha]_D$ +18.75° (*c* 0.08, EtOH); **UV** (λ_{max}, EtOH, nm): 221 (ε=8967); **IR** (υ_{max}, film, cm^{-1}): 3466, 1745, 1722, 1645; **MS**: CIMS (isobutane, *m/z*) 611, 593, 575, 557, 539, 351, 333, 281, EIMS (*m/z*) 499, 439, 421, 369, 351, 333, 329, 311, 293, 281, 263, 241, 171, 141, 123, 111; **NMR**: ^{1}H-NMR (500 MHz, $CDCl_3$), ^{13}C-NMR (125 MHz, $CDCl_3$); **Derivatives**: tetraacetate (CIMS, EIMS, ^{1}H-NMR), tetra-TMSi (EIMS), tetra-d_9-TMSi (EIMS); **Biological activities**: BST LC_{50}=2.01 x 10^{-2} µg/mL, A-549 ED_{50}=2.83 x 10^{-11} µg/mL, MCF-7 $ED_{50}<10^{-12}$ µg/mL, HT-29 ED_{50}=2.15 µg/mL; **Source**: *Asimina parviflora*, twigs.

53. Squamostatin-E $C_{37}H_{66}O_7$, MW 622, (52, 36)

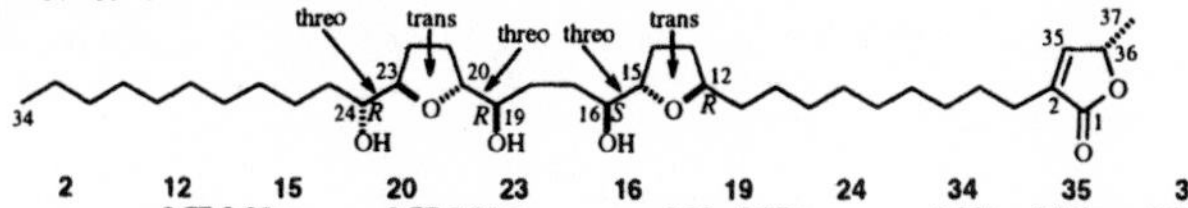

No.	1	2	12	15	20	23	16	19	24	34	35	36	37
^1H(δ)			3.77-3.92m		3.77-3.92m			3.38 - 3.57m		0.88t	6.98br s	4.98qq	1.40d
^{13}C(δ)	173.9	134.3	79.3	82.0	82.7	82.7	74.4	74.2	74.1	14.1	148.8	77.4	19.2

White crystals; **MP**: 105-106 °C; $[\alpha]_D$ +14.7° (*c* 0.51, MeOH); **IR** (υ_{max}, film, cm^{-1}): 3560, 3450, 1750; **NMR**: ^{1}H-NMR (500 MHz, $CDCl_3$), ^{13}C-NMR (125 MHz, $CDCl_3$); **Derivatives**: Mosher esters (^{1}H-NMR); **Source**: *Annona squamosa*, seeds.

54. Cherimolin-2 $C_{37}H_{66}O_8$, MW 638, (55)

No.	1	2	4	12	15	16	19	20	23	24	34	35	36	37
^{1}H(δ)			3.78	3.82	3.74	3.38	3.38	3.74	3.74	3.38	0.82	7.16	5.00	1.39
^{13}C(δ)	174.5	130.9	69.52	79.14	81.84	74.21	74.10	82.50	82.58	73.93	13.91	151.7	77.79	18.87

$[\alpha]_D$ +6°(*c* 0.31, MeOH); **IR** (υ_{max}, film, cm^{-1}): 3540, 2940, 2870, 1750; **MS**: CIMS (CH_4, *m/z*) 639, 621, 603, 585, 567; **NMR**: ^{1}H-NMR (200 MHz, $CDCl_3$), ^{13}C-NMR (50 MHz, $CDCl_3$), ^{1}H-^{1}H COSY, ^{1}H-^{13}C COSY, COSY-45, HOHAHA; **Biological activities**: antiparasitic against infective larvae of *Molinema dessetae* LD_{50}=0.04-0.25 mg/L (D7), KB ED_{50}=10^{-3}-10^{-5} µg/mL, VERO ED_{50} =10^{-2}-10^{-3} µg/mL; **Source**: *Annona cherimolia*, seeds.

55. Squamostatin-C $C_{37}H_{66}O_8$, MW 638, (52, 36)

No.	1	2	4	12	15	20	23	16	19	24	34	35	36	37
^{1}H(δ)			3.77-3.90m			3.77-3.90m			3.41m		0.88t	7.19s	5.06q	1.43d
^{13}C(δ)	173.9	134.3	70.0	79.3	82.0	82.7	82.7	74.4	74.3	74.0	14.1	151.7	77.9	19.1

White crystals; **MP**: 95-97 °C; $[\alpha]_D$ +12.0° (*c* 0.20, MeOH); **IR** (υ_{max}, film, cm^{-1}): 3685, 3585, 3540, 1755; **NMR**: ^{1}H-NMR (500 MHz, $CDCl_3$), ^{13}C-NMR (125 MHz, $CDCl_3$); **Derivatives**: Mosher esters (^{1}H-NMR); **Source**: *Annona squamosa*, seeds.

56. Squamostatin-D $C_{37}H_{66}O_7$, MW 622, (52, 36)

No.	1	2	4	12	15	20	23	16	19	24	34	35	36	37
^{1}H(δ)			3.77-3.90 m			3.77 - 3.90 m		3.41m	3.41m	3.41m	0.88t	7.19s	5.06q	1.43d
^{13}C(δ)	173.9	134.3	70.0	79.3	82.0	82.7	82.7	74.4	74.3	74.0	14.1	151.7	77.9	19.1

White crystals; **MP**: 112-113.5 °C; $[\alpha]_D$ +7.9° (*c* 0.51, MeOH); **IR** (υ_{max}, film, cm^{-1}): 3560, 3450, 1750; **MS**: EIMS (*m/z*) 622, 433, 415, 397, 381, 363, 345, 327, 293, 275; **NMR**: ^{1}H-NMR (500 MHz, $CDCl_3$), ^{13}C-NMR (125 MHz, $CDCl_3$); **Derivatives**: Mosher esters (^{1}H-NMR); **Source**: *Annona squamosa*, seeds.

57. Cherimolin-1 $C_{37}H_{66}O_8$, MW 638, (146, 102, 55)

No.	1	2	4	12	15	16	19	20	23	24	34	35	36	37
^{1}H(δ)			3.80	3.80	3.80	3.40	3.40	3.80	3.80	3.80	0.86	7.16	5.02	1.40
^{13}C(δ)	174.5	131.0	69.78	79.25	81.93	74.45	74.33	83.26	82.19	71.46	14.01	151.8	77.88	18.99

$[\alpha]_D$ +64° (*c* 0.30, MeOH); **IR** (υ_{max}, film, cm^{-1}): 3590, 2930, 2856, 1750, 1460; **MS**: CIMS (CH_4, *m/z*) 639, 621, 603, 585, 567, CIMS (NH_3, *m/z*) 656, 639; **NMR**: ^{1}H-NMR (200 MHz, $CDCl_3$), ^{13}C-NMR (50 MHz, $CDCl_3$), ^{1}H-^{1}H COSY, ^{1}H-^{13}C COSY, COSY-45, HOHAHA; **Derivatives**: tetraacetate (IR, CIMS, ^{1}H-NMR), 2,35-dihydro (CIMS, ^{1}H-NMR); **Biological activities**: antiparasitic against infective larvae of *Molinema dessetae* LD_{50}=0.04-0.25 mg/L (D7), KB ED_{50}=10^{-3}-10^{-5} µg/mL, VERO ED_{50}=10^{-2}-10^{-3} µg/mL; **Source**: *Annona cherimolia*, seeds, *Annona purpurea*.

58. Squamostatin-B $C_{37}H_{67}O_8$, MW 638, (52, 36)

No.	1	2	4	12	15	20	23	24	16	19	34	35	36	37
^{1}H(δ)			3.76-3.91m		3.76-3.91m		3.76-3.91m		3.41m	3.41m	0.88t	7.19s	5.06q	1.43d
^{13}C(δ)	174.6	131.2	70.0	79.3	82.0	83.3	82.2	71.6	74.4	74.3	14.1	151.8	78.0	19.1

White crystals; **MP**: 98-101 °C; $[\alpha]_D$ +10.5° (*c* 0.10, MeOH); **IR** (υ_{max}, film, cm^{-1}): 3590, 3450, 1745; **MS**: EIMS (*m/z*) 638, 449, 431, 413, 395, 379, 361, 343, 309, 291, 273; **NMR**: ^{1}H-NMR (500 MHz, $CDCl_3$), ^{13}C-NMR (125 MHz, $CDCl_3$); **Derivatives**: Mosher esters (^{1}H-NMR); **Source**: *Annona squamosa*, seeds.

59. Almunequin $C_{37}H_{66}O_8$, MW 638, (137, 55, 35)

erythro trans threo threo 35 37 36 34 28 24 23 O 20 19 16 15 O 12 2 O 1 OH OH OH OH O

No.	1	2	12	15	16	19	20	23	24	28	34	35	36	37
$^1H(\delta)$			3.83	3.77	3.38	3.38	3.80	3.80	3.80	3.56	0.89	6.95	4.96	1.37
$^{13}C(\delta)$	174.0	134.1	79.17	81.87	74.31	74.29	83.22	82.07	71.38	71.55	14.93	148.8	77.26	19.04

$[\alpha]_D$ +25° (*c* 0.38, MeOH); **IR** (υ_{max}, film, cm^{-1}): 3580, 2940, 2870, 1755, 1460, 1370, 1325, 1060, 1030, 960; **MS**: CIMS (NH_3, *m/z*) 639, 621, 603, 585, 567, FABMS (*m/z*) 661, 639; **NMR**: ^1H-NMR (400 MHz, $CDCl_3$), ^{13}C-NMR (100 MHz, $CDCl_3$), ^1H-^1H COSY, ^1H-^{13}C COSY, COSY-45, HOHAHA; **Derivatives**: 2,35-dihydro (CIMS, ^1H-NMR, ^{13}C-NMR); **Biological activities**: antiparasitic against infective larvae of *Molinema dessetae* LD_{50}=0.04-0.25 mg/L (D7), KB $ED_{50}=10^{-3}$-10^{-5} μg/mL, VERO $ED_{50}=10^{-2}$-10^{-3} μg/mL; **Source**: *Annona cherimolia*, seeds, *Annona reticulata*, roots.

60. C-12,15-*cis*-Bullatanocin $C_{37}H_{66}O_8$, MW 638, (147)

threo trans threo threo cis OH 35 37 36 34 24 R 23 O 20 R 19 16 S 15 O S 12 4 R 2 O 1 OH OH OH O

No.	1	2	4	12	15	16	19	20	23	24	34	35	36	37
$^1H(\delta)$			3.85m	3.85m	3.71m	3.39m	3.44m	3.81m	3.81m	3.39m	0.88t	7.19q	5.06qq	1.43d
$^{13}C(\delta)$	174.6	131.2	69.9	79.9	82.3	74.8	74.3	82.7	82.7	74.0	14.1	151.8	78.0	19.1

C-12,15-*cis*-Bullatanocinones (reported as a *cis* and *trans* mixture) $C_{37}H_{66}O_8$, MW 638, (147)

threo trans threo threo cis A 37 36 34 24 R 23 O 20 R 19 16 S 15 O S 12 4 O 2 1 O OH OH OH O

61. (2,4-*cis*)-C-12,15-*cis*-Bullatanocinone (A=*cis*)

No.	12	15	16	19	20	23	24	34
$^1H(\delta)$	3.85m	3.85m	3.39m	3.44m	3.81m	3.81m	3.38m	0.88t
$^{13}C(\delta)$	79.8	82.3	74.7	74.3	82.7	82.7	74.0	1.43

62. (2,4-*trans*)-C-12,15-*cis*-Bullatanocinone (A=*trans*)

No.	12	15	16	19	20	23	24	34
$^1H(\delta)$	3.85m	3.85m	3.39m	3.44m	3.81m	3.81m	3.38m	0.88t
$^{13}C(\delta)$	79.8	82.3	74.7	74.3	82.7	82.7	74.0	1.43

63. C-12,15-*cis*-Bullatalicin $C_{37}H_{66}O_8$, MW 638, (147)

erythro trans threo threo cis OH 35 37 36 34 24 S 23 O 20 R 19 16 S 15 O S 12 4 R 2 O 1 OH OH OH O

No.	1	2	4	12	15	16	19	20	23	24	34	35	36	37
$^1H(\delta)$			3.84m	3.84m	3.71m	3.38m	3.43m	3.86m	3.85m	3.88m	0.88t	7.19q	5.06q	1.43d
$^{13}C(\delta)$	174.6	131.2	69.9	79.9	82.3	74.8	74.5	82.2	83.3	71.5	14.1	151.8	78.0	19.1

C-12,15-*cis*-Bullatalicinones (reported as a *cis* and *trans* mixture) $C_{37}H_{66}O_8$, MW 638, (147)

threo trans threo threo cis A 37 36 34 24 R 23 O 20 R 19 16 S 15 O S 12 4 O 2 1 O OH OH OH O

64. (2,4-*cis*)-C-12,15-*cis*-Bullatalicinone (A=*cis*)

No.	12	15	16	19	20	23	24	34
$^1H(\delta)$	3.84m	3.71m	3.38m	3.43m	3.86m	3.85m	3.88m	0.88t
$^{13}C(\delta)$	79.8	82.2	74.8	74.5	82.2	83.3	71.5	14.1

65. (2,4-*trans*)-C-12,15-*cis*-Bullatalicinone (A=*trans*)

No.	12	15	16	19	20	23	24	34
$^1H(\delta)$	3.84m	3.71m	3.38m	3.43m	3.86m	3.85m	3.88m	0.88t
$^{13}C(\delta)$	79.8	82.2	74.8	74.5	82.2	83.3	71.5	14.1

66. C-12,15-*cis*-Squamostatin-A $C_{37}H_{66}O_8$, MW 638, (147)

No.	1	2	12	15	16	19	20	23	24	28	34	35	36	37
$^1H(\delta)$			3.84m	3.71m	3.38m	3.43m	3.86m	3.85m	3.88m	3.60m	0.89t	6.98q	4.99q	1.41d
$^{13}C(\delta)$	173.9	134.3	80.0	82.3	74.9	74.6	82.2	83.4	71.6	71.8	14.1	148.9	77.4	19.2

67. Sylvaticin $C_{37}H_{66}O_8$, MW 638, (25, 53, 102)

Reassignments of ^{13}C-NMR data:

No.	1	2	4	12	15	16	19	20	23	24	34	35	36	37
$^{13}C(\delta)$	174.4	131.3	70.0	79.3	81.8	74.1	74.3	82.5	83.0	72.5	14.0	151.6	77.8	19.1

Other data were reported in the previous review (8).

68. Goniocin $C_{37}H_{64}O_7$, MW 620, (72)

No.	1	2	4	10	13	14	17	18	21	22	34	35	36	37
$^1H(\delta)$			3.85m	3.93m	3.92-3.83m		3.92-3.83m		3.80m	3.37m	0.879t	7.20q	5.06qq	1.35d
$^{13}C(\delta)$	174.3	131.1	69.8	79.6	82.3-81.1		82.3-81.1		83.0	74.2	14.1	151.9	78.0	19.1

NMR: ^{1}H-NMR (500 MHz, $CDCl_3$), ^{13}C-NMR (125 MHz, $CDCl_3$), ^1H-^1H COSY; **Derivatives**: bis-TMSi derivative (EIMS), Mosher esters; **Biological activities**: BST LC_{50}=57 μg/mL, A-549 ED_{50}=9.42 x 10^{-1} μg/mL, MCF-7 ED_{50}=4.85, HT-29 ED_{50}=1.61 x 10^{-2} μg/mL; **Source**: *Goniothalamus giganteus*, bark.

69. Reticulatamol $C_{35}H_{66}O_3$, MW 534, (80)

No.	1	2	15	32	33	34	35
$^1H(\delta)$			3.58 t	0.88 t	6.98 d	5.00 qd	1.42 d
$^{13}C(\delta)$	173.9	134.4	72.2	14.2	148.9	77.5	19.3

MP: 92-94 °C (heptane); $[\alpha]_D$ +2° (*c* 1, $CHCl_3$); **UV** (λ_{max}, EtOH, nm): 206 (log ε=1.011); **IR**(υ_{max}, cm^{-1}): 3350, 1750; **MS**: CIMS (isobutane, *m/z*) 535, 517, 489, EIMS (*m/z*) 534, 516, 309, 295, 281, 266, FAB-Li MS (*m/z*) 545; **NMR**: ^{1}H-NMR (200 MHz, $CDCl_3$), ^{13}C-NMR (50 MHz, $CDCl_3$); **Source**: *Annona reticulata*, seeds.

70. Diepoxymontin $C_{35}H_{62}O_4$, MW 546, (62)

No.	1	2	11	12	13	14	32	33	34	35
$^1H(\delta)$			2.95 m	2.95 m	2.95 m	2.95 m	0.88 t	6.98 td	4.99 qtd	1.39 d
$^{13}C(\delta)$	174.3	134.8	56.9	57.8	57.8	56.9	14.5	149.3	77.8	19.4

White amorphous powder; **IR** (υ_{max}, film, cm^{-1}): 1745; **MS**: FABMS (*m/z*) 547, 546, EIMS, linked scan MS/MS (*m/z*) 295, 277, 267, 251, 233, 223; **NMR**: ^{1}H-NMR (200 MHz, $CDCl_3$), ^{13}C-NMR (50 MHz, $CDCl_3$), ^1H-^1H COSY, ^1H-^{13}C HETCOR; **Source**: *Annona reticulata*, fruits.

71. Epomuricenin-B $C_{35}H_{62}O_3$, MW 530, (71)

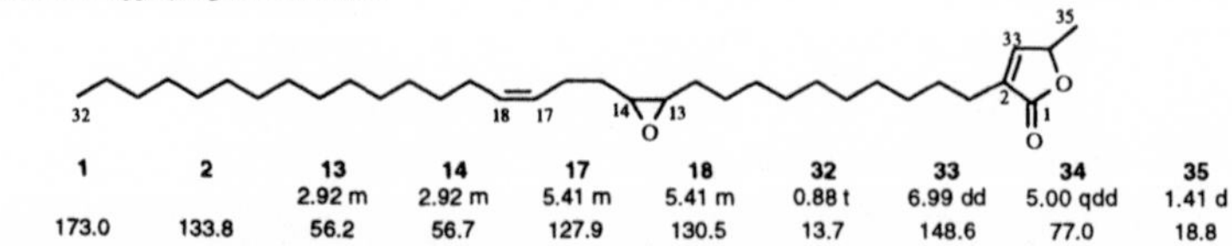

No.	1	2	13	14	17	18	32	33	34	35
$^1H(\delta)$			2.92 m	2.92 m	5.41 m	5.41 m	0.88 t	6.99 dd	5.00 qdd	1.41 d
$^{13}C(\delta)$	173.0	133.8	56.2	56.7	127.9	130.5	13.7	148.6	77.0	18.8

White wax; $[\alpha]_D$ +18°(*c* 0.22, MeOH); **UV** (λ_{max}, EtOH, nm): 210 (log ε=3.80); **IR** (υ_{max}, film, cm^{-1}): 2900, 2840, 1760, 1650, 1460, 1310, 1190, 1110, 1065, 1020, 710; **MS**: EIMS (*m/z*) 530, 512, 321, 308, 307, 295, 293, 280, 279, 267, 251, 237, 223, 209, 195, 181, 167, 153, 139, 125, 112, 111, 110; **NMR**: ^{1}H-NMR (200 MHz, $CDCl_3$), ^{13}C-NMR (50 MHz, $CDCl_3$), ^1H-^1H COSY, ^1H-^{13}C COLOC; **Derivatives**: diepomuricanin (HREIMS, linked scan *B/E* FABMS (m-NBA+LiCl), ^{1}H-, ^{13}C-NMR); **Biological activities**: KB ED_{50}=2 μg/mL, VERO ED_{50}>10 μg/mL; **Source**: *Annona muricata*, seeds.

72. Epomuricenin-A $C_{35}H_{62}O_3$, MW 530, (64, 71)

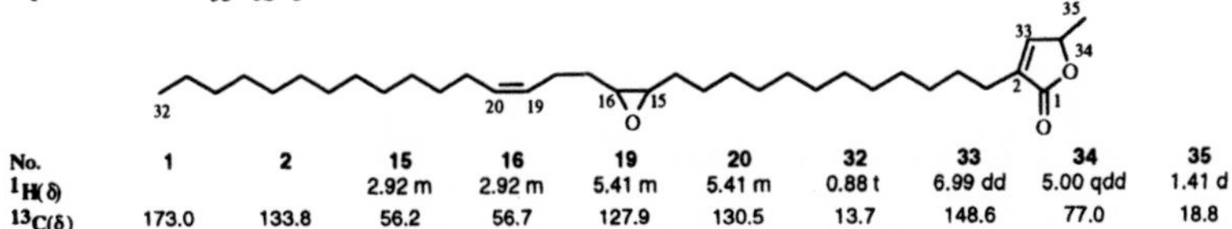

No.	1	2	15	16	19	20	32	33	34	35
$^1H(\delta)$			2.92 m	2.92 m	5.41 m	5.41 m	0.88 t	6.99 dd	5.00 qdd	1.41 d
$^{13}C(\delta)$	173.0	133.8	56.2	56.7	127.9	130.5	13.7	148.6	77.0	18.8

White wax; $[\alpha]_D$ +18°(*c* 0.22, MeOH); **UV** (λ_{max}, EtOH, nm): 210 (log ε=3.80); **IR** (υ_{max}, film, cm^{-1}): 2900, 2840, 1760, 1650, 1460, 1310, 1190, 1110, 1065, 1020, 710; **MS**: EIMS (*m/z*) 530, 512, 321, 308, 307, 295, 293, 280, 279, 267, 251, 237, 223, 209, 195, 181, 167, 153, 139, 125, 112, 111, 110; **NMR**: ^{1}H-NMR (200 MHz, $CDCl_3$), ^{13}C-NMR (50 MHz, $CDCl_3$), ^{1}H-^{1}H COSY, ^{1}H-^{13}C COLOC; **Derivatives**: diepomuricanin (HREIMS, Linked scan *B/E* FABMS (m-NBA+LiCl), ^{1}H-, ^{13}C-NMR); **Biological activities**: KB ED_{50} = 2 µg/mL, VERO ED_{50}>10 µg/mL; **Source**: *Annona muricata*, seeds, bark.

73. Epoxymurin-B $C_{35}H_{62}O_3$, MW 530, (64)

No.	1	2	15	16	19	20	32	33	34	35
$^1H(\delta)$			5.40 m	5.40 m	2.92 m	2.92 m	0.88 t	6.98 d	4.98 dq	1.40 d
$^{13}C(\delta)$	173.85	134.1	130.87	128.18	57.23	56.70	14.01	148.74	77.32	19.10

Wax; **MS**: EIMS (*m/z*) 530, FABMS (*m/z*) 531, 537, product ion spectrum (M+Li) 537, 519, 517, 509, 507, 493, 491, 179, 165, 451, 437, 425, 423, 411, 409, 397, 395, 383, 381, 369, 353, 341, 299, 283, 273, 257, 243, 229, 215, 201, 187; **NMR**: ^{1}H-NMR (400 MHz, $CDCl_3$), ^{13}C-NMR (100 MHz, $CDCl_3$), ^{13}C-NMR-APT, ^{1}H-^{1}H COSY; **Source**: *Annona muricata*, seeds.

74. Corepoxylone $C_{35}H_{60}O_5$ MW 560 (63)

No.	2	10	15	16	19	20	32	33	34	35
$^1H(\delta)$			2.96m	2.96m	2.96m	2.96m	0.89t	6.99d	4.98dq	1.40d
$^{13}C(\delta)$	134.24	210.89	57.33	56.70	56.40	57.01	14.09	148.90	77.38	19.20

Amorphous powder; $[\alpha]_D$ +36.8°(*c* 0.08, $CHCl_3$); **UV** (λ_{max}, EtOH, nm): 228 (log ε=3.09); **IR** (υ_{max}, cm^{-1}): 1750, 1705, 1465, 1270, 1080, 750; **MS**: EIMS (*m/z*) 560, 238, 223, 220, 195, L-SIMS (NBA+LiCl, *m/z*) 567, 537, 482, 455, 413, 397, 385, 327, 315, 287, 257; **NMR**: ^{1}H-NMR (200 MHz, $CDCl_3$), ^{13}C-NMR (50 MHz, $CDCl_3$), ^{1}H-^{1}H COSY, COSY-45, COSY-RELAY, ^{1}H-^{13}C COSY; **Derivatives**: conversion to corossolone (CIMS, EIMS, ^{1}H-, ^{13}C-NMR); **Biological activities**: KB ED_{50}=1.6 x 10^{-3} µg/mL, VERO ED_{50}=2.5 µg/mL; **Source**: *Annona muricata*, seeds.

75. Dieporeticanin-1 (isolated as a mixture with dieporeticanin-2), $C_{37}H_{66}O_4$, MW 574, (73)

No.	1	2	17	18	21	22	34	35	36	37
$^1H(\delta)$			2.94 m	2.94 m	2.94 m	2.94 m	0.87 t	6.96 d	4.98 dq	1.41 d
$^{13}C(\delta)$	173.80	134.36	56.71	57.31	57.31	56.71	14.06	148.77	77.31	19.19

76. Dieporeticanin-2 (isolated as a mixture with dieporeticanin-1), $C_{37}H_{66}O_4$, MW 574, (73)

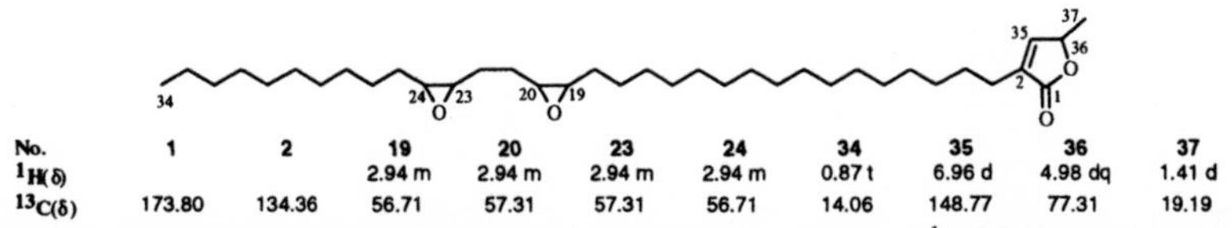

No.	1	2	19	20	23	24	34	35	36	37
$^1H(\delta)$			2.94 m	2.94 m	2.94 m	2.94 m	0.87 t	6.96 d	4.98 dq	1.41 d
$^{13}C(\delta)$	173.80	134.36	56.71	57.31	57.31	56.71	14.06	148.77	77.31	19.19

Amorphous whitish solid; $[\alpha]_D$ +12° (*c* 1, $CHCl_3$); **UV** (λ_{max}, nm): 207; **IR** (υ_{max}, film, cm^{-1}): 2930, 2850, 1750; **MS**: CIMS (isobutane, *m/z*) 575, L-SIMS (*m/z*) 581, EIMS (*m/z*) 574, 556, 538, 463, 351, 323; **NMR**: ^{1}H-NMR (200 MHz, $CDCl_3$), ^{13}C-NMR (50 MHz, $CDCl_3$); **Source**: *Annona muricata*, seeds.

77. Dieporeticenin $C_{37}H_{64}O_4$, MW, 572, (73)

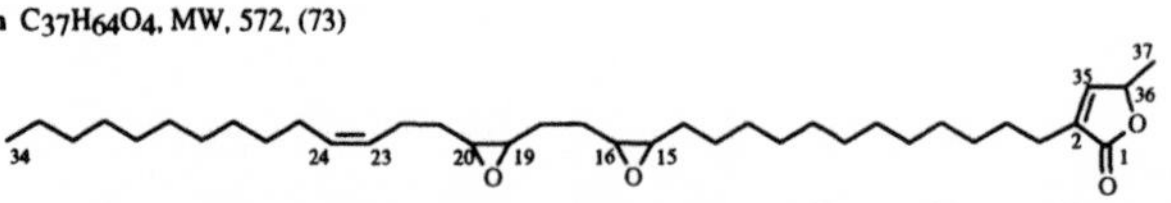

No.	1	2	15	16	19	20	23	24	34	35	36	37
$^1H(\delta)$			2.95 m	2.95 m	2.95 m	2.95 m	5.40 m	5.40 m	0.89 t	6.96 d	4.98 dq	1.41 d
$^{13}C(\delta)$	173.77	134.35	56.70	56.78	56.87	56.29	128.13	131.13	14.07	148.77	77.33	19.20

White wax, $[\alpha]_D$ +11° (*c* 1, $CHCl_3$); **UV** (λ_{max}, nm): 208; **IR** (υ_{max}, cm^{-1}): 2928, 2856, 1750; **MS**: CIMS (isobutane, *m/z*) 573, L-SIMS (*m/z*) 579, CID (*m/z*) 535, 301/313, 371/383, 271, EIMS (*m/z*) 572, 554, 530, 295; **NMR**: ^1H-NMR (200 MHz, $CDCl_3$), ^{13}C-NMR (50 MHz, $CDCl_3$); **Source**: *Annona muricata*, seeds.

78. Tripoxyrollin $C_{37}H_{64}O_5$, MW 588, (74)

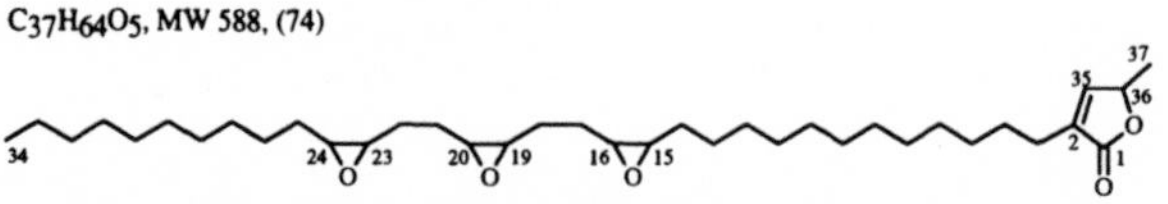

No	1	2	15	16	19	20	23	24	34	35	36	37
$^1H(\delta)$			2.95 m	2.95 m	2.95 m	2.95 m	2.95 m	2.95 m	0.85 t	6.96 d	4.95 dq	1.40 d
$^{13}C(\delta)$	173.77	134.35	57.26	56.83	56.61	56.31	57.26	56.61	14.04	148.77	77.31	19.15

Amorphous white solid, $[\alpha]_D$ +10.3° (*c* 0.68, $CHCl_3$); **UV** (λ_{max}, MeOH, nm): 205 (log ε = 4.05); **IR** (υ_{max}, film, cm^{-1}): 2900, 2840, 1740, 1460, 1315, 1200, 1070, 930, 745; **MS**: FABMS (NAB+Li, *m/z*) 595, 577, 567, 551, 453, 441, 383, 371, 329, 317, 313, 301, 259; **NMR**: ^1H-NMR (200 MHz, $CDCl_3$), ^{13}C-NMR (50 MHz, $CDCl_3$); **Source**: *Rollinia membranacea*, seeds.

79. trieporeticanin $C_{37}H_{64}O_5$, MW 588, (73)

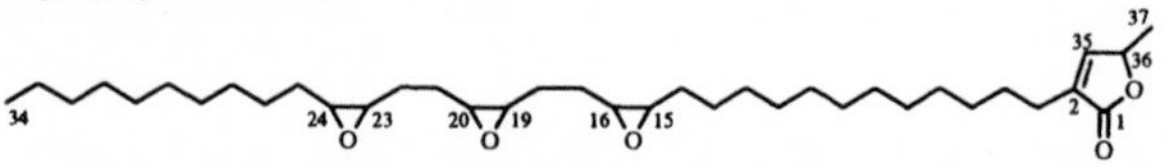

No.	1	2	15	16	19	20	23	24	34	35	36	37
$^1H(\delta)$			2.96 m	2.96 m	2.96 m	2.96 m	2.96 m	2.96 m	0.88 t	6.98 d	4.98 dq	1.41 d
$^{13}C(\delta)$	174.01	134.43	56.50	56.70	56.80	57.01	57.43	57.43	14.04	148.94	77.31	14.19

Amorphous white solid, $[\alpha]_D$ +13.0° (*c* 1, $CHCl_3$); **UV** (λ_{max}, MeOH, nm): 208; **MS**: CIMS (isobutane, *m/z*) 589, L-SIMS (*m/z*) 595, CID (*m/z*) 551, 301/313, 371/383, 441/453, 329, EIMS (*m/z*) 588; **NMR**: ^1H-NMR (200 MHz, $CDCl_3$), ^{13}C-NMR (50 MHz, $CDCl_3$); **Source**: *Annona reticulata*, seeds.

80. Purpureacin-1 $C_{37}H_{66}O_8$, MW 638, (102)

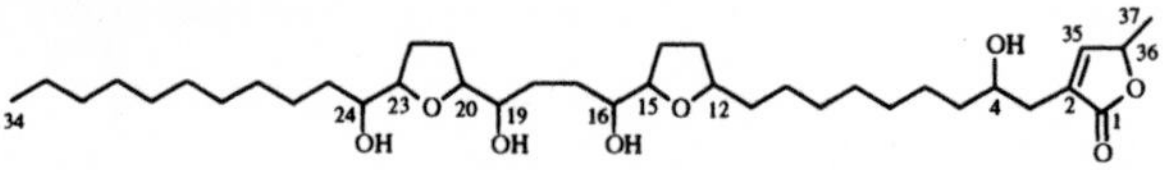

No.	1	2	4	12	15	16	19	20	23	24	34	35	36	37
$^1H(\delta)$			3.80m	3.55m	3.85m	3.42m	3.8-3.9m	3.80m	3.8-3.9 m		0.81t	7.12dd	4.99dq	1.37d
$^{13}C(\delta)$	174.6	131.2	69.9	79.9	82.7	74.7	74.4	82.2	81.3	74.1	14.1	151.8	78.0	19.1

White amorphous power; $[\alpha]_D$ -3.3°(*c* 0.12, MeOH); **UV** (λ_{max}, nm): 208 (log ε=3.93); **MS**: EIMS (*m/z*) 509, 491, 479, 461, 443, 467, 449, 431, 413, 397, 381, 379, 363, 361, 339, 329, 321, 311, 309, 293, 291, 271, 253, 241, 223, 171, 141, 111; **NMR**: ^1H-NMR (200 MHz, $CDCl_3$), ^{13}C-NMR (50 MHz, $CDCl_3$); **Derivatives**: triTMSi (EIMS); **Biological activities**: BST LC_{50}=0.53 μg/mL, Larvicidal activity $LC_{100}/24h$=2.0 μg/mL, antifungal activity: 0.05 μg, antibacterial activity: 20 μg; **Source**: *Annona purpurea*, leaves.

Bioassay abbreviations:

BST: Brine shrimp lethality, *Artemia salina*.
A-549: Human lung carcinoma.
MCF-7: Human breast carcinoma.
HT-29: Human colon adenocarcinima.
9KB: Human naopharyngeal carcinoma.
P388: Methylcholanthrene-induced murine lymphocytic leukemia.
HeLa: Human carcinoma of the cervix cells.
VERO: Monkey epitheloid renal cell.
Larvicidal activity: Activity against larvae of *Aedes aegypti*.
Antifungal activity: Activity against *Candida albicans*: minimal amount inhibiting fungal growth on TLC plate.
Antibacterial activity: Activity against *Bacillus subtilis*: minimal amount inhibiting bacterial growth on TLC plate.

Chapter Twelve

NEO-CLERODANE DITERPENOIDS FROM AMERICAN SALVIA SPECIES

Lydia Rodríguez-Hahn, Baldomero Esquivel, and
Jorge Cárdenas

Instituto de Química de la Universidad Nacional Autónoma de
México
Cd. Universitaria, Circuito Exterior
Coyoacán 04510 México D.F.

INTRODUCTION

Interest in finding biodegradable products possessing a defined biological activity, such as bactericidal, fungicidal, antitumoral, antiviral or pesticidal, has been a driving force in the study of the secondary metabolites of plants, fungi, marine organisms, and others. Phytochemical studies often make use of accumulated ethnobotanical knowledge, particularly in regard to uses of the plants in traditional medicine.[1] Knowledge of the secondary metabolites responsible for biological activity and of chemotaxonomic relationships within specific groups being studied is helpful in finding new sources of potentially economically important products or compounds which can be transformed into active principles.[2] The search for plants which contain secondary metabolites with pesticidal properties is important as they provide biodegradable and renewable resources which are potentially able to replace the more harmful synthetic ones currently being employed.

The Labiatae is a widespread and diversified family of plants with *circa* 4000 species distributed throughout the world.[3] Labiatae have been used as medicinal plants since ancient times. For example, *Salvia miltiorrhiza* (Bunge)

Phytochemistry of Medicinal Plants
Edited by John T. Arnason et al., Plenum Press, New York, 1995

(Danshen), from which a large number of abietane diperpenoids have been isolated,[4] was included in the Chinese Pharmacopoeia of traditional medicine.

The phytochemical study of plants of the genera *Ajuga, Teucrium, Scutellaria, Leonurus, Stachys* and *Salvia* of the Labiatae, has led to the isolation of a large number of neo-clerodane diterpenoids.[5] Some of these products have shown antifeedant activity against economically important insects. For example, ajugarin I (**1**), a neo-clerodane diterpenoid isolated from *Ajuga remota*, a plant collected in Nairobi (Kenya), has been thoroughly studied by Kubo et all.[6] Ajugarin I was found to be antifeedant against *Spodoptera exempta* (African army worm), *S. littoralis* (Egyptian cotton leaf worm) and *Heliothis armigera*.[7] Ajugarin IV (**2**), isolated from the same source, was inactive against *S. exempta* but was insecticidal against *Bombyx mori* (silk worm).[8] Several neo-clerodane diterpenoids isolated from *Teucrium* species also have been found to possess antifeedant activity against *S. littoralis*.[7] Jodrellin B (**3**), isolated from *Scutellaria woronowii*, is considered the most potent clerodane antifeedant yet found against *S. littoralis*.[9]

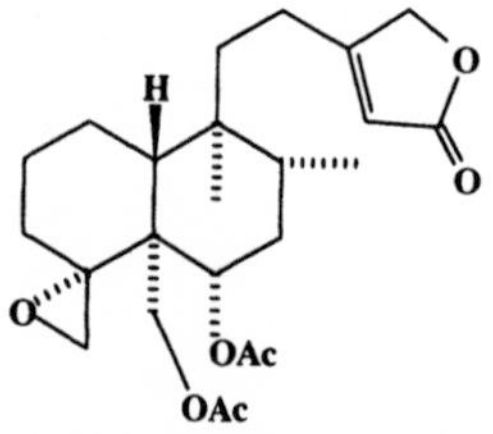

Ajugarin I **1** Ajugarin IV **2** Jodrellin B **3**

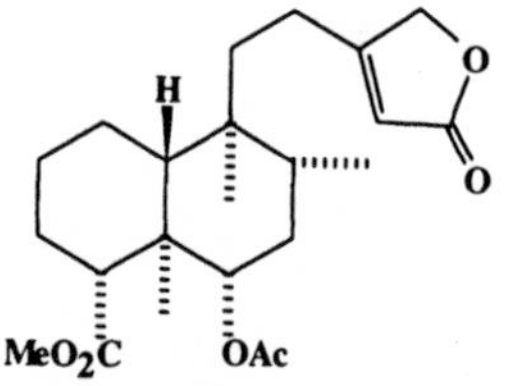

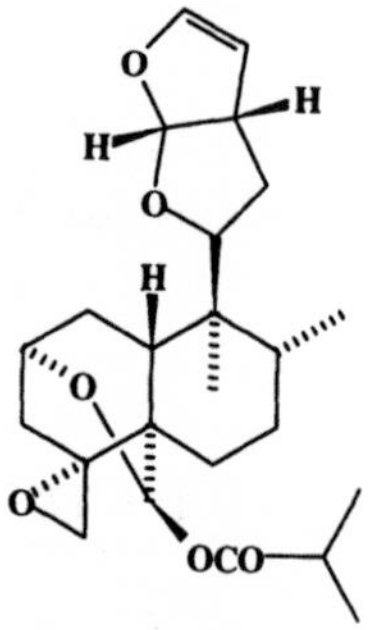

NEO-CLERODANE DITERPENOIDS IN AMERICAN SALVIA SPECIES

The genus *Salvia* L., is one of the largest genera of the Lamiaceae with 900 species widespread throughout the world, mainly in tropical and subtropical regions.[10] Bentham divided this genus into four subgenera: *Salvia*, *Sclarea*, *Leonia* and *Calosphace*.[11] Phytochemical studies of plants of this genus have established an interesting relationship among species in regard to the diterpenoid content and the geographical distribution and the botanical classification.[12] The European and Asiatic *Salvia* species (*Salvia* and *Sclarea* subgenera) contain diterpenoids with abietane and labdane skeletons.[12] *Salvia* spp. found in North

America have been included in subgenus *Leonia*, from which abietane and clerodane diterpenoids have been isolated.[12] *Salvia*, subgenus *Calosphace* is the largest subgenus with over 500 species distributed in Mexico, Central and South America.[13] Epling thoroughly studied this subgenus and organised it into 105 Sections.[14] In Mexico it is represented by over 300 species with a high degree of endemism.[13] Field observations revealed that most of them have developed an efficient defence system against insect attack.

The phytochemical studies of the American *Salvia* spp. included in the subgenus *Calosphace*, have shown a close relationship between the diterpenoid content of species and their taxonomic sections. The species classified in section *Erythrostachys* contain abietane diterpenoids commonly found in European and Asiatic *Salvia* spp.[15] Of species included in section *Tomentellae*, diterpenoids with abietane and icetexane skeletons have been isolated, some of them with a high degree of unsaturation.[16] Most of the diterpenoids isolated from *Salvia* spp. included in sections *Scorodonia*, *Fulgentes*, *Angulatae* and others are neo-clerodane diterpenoids or have a skeleton which can be biogenetically related to a clerodanic precursor.[17,18]

Some years ago we studied the diterpenoid content of a population of *S. melissodora*, Lag. *(Salvia*, sect. *Scorodonia*) collected north east of Mexico City. From this plant we isolated melisodoric acid (**4**) whose structure was established by chemical degradation and spectroscopic means.[19] The relative stereochemistry was not ascertained. Melisodoric acid exhibited antifeedant activity against *Euxoa* and *Heliothis* larvae. (Taboada, personal communication). In order to find an adequate source of this diterpenoid, we studied several populations of *S. melissodora*. From a population collected in the state of Mexico we isolated two neo-clerodane diterpenoids (**5,6**) whose structures were established by spectroscopic means.[20] The dilactone (**5**) had been previously found in *Baccharis trimera* (Asteraceae), and its structure deduced from spectroscopic data and X-ray diffraction analysis.[21] The neo-clerodane (**6**) had been isolated from *Portulaca* C.V. Jewel.[22] It can be considered a biogenetic precursor of the α - and β - substituted terminal butenolides found in many of the neo-clerodane diterpenoids isolated from American *Salvia* spp.

A population of *S. melissodora* collected in the state of Hidalgo (Mexico) yielded the dilactone (**7**) in which the terminal butenolide is a -substituted,[20] and brevifloralactone (**8**), a neo-clerodane diterpenoid found also in *S. breviflora* (*Salvia*, sect. *Scorodonia*).[23] The *S. melissodora* population collected in San Luis Potosi (Mexico) was very rich in neo-clerodane diterpenoids. The structurally related products (**7,9-16**) were isolated, and their structures established by spectroscopic means and chemical correlations.[24] All of them contained an a-substituted terminal butenolide. Manganese dioxide treatment of **16** produced the dilactone (**9**). The hydroxymethylenes bound to C4 and C5 could be the precursors of the 18,19-olide group found in the clerodane diterpenoids (**7,9-15**).

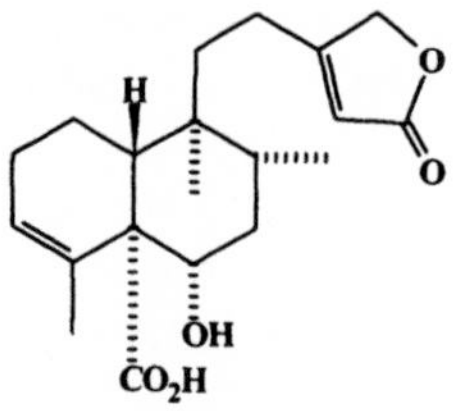

Melisodoric acid

4

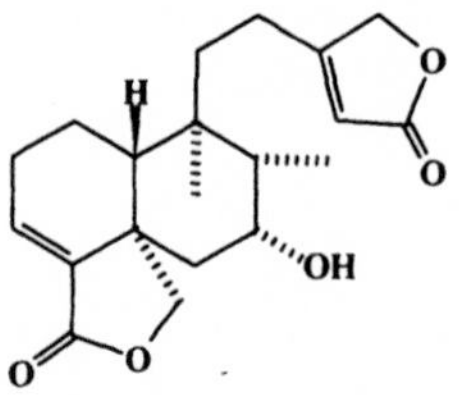

5

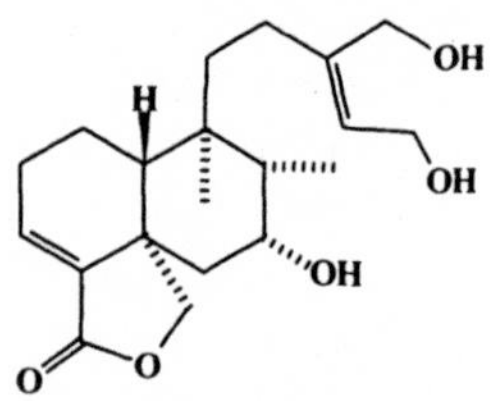

Portulide C

6

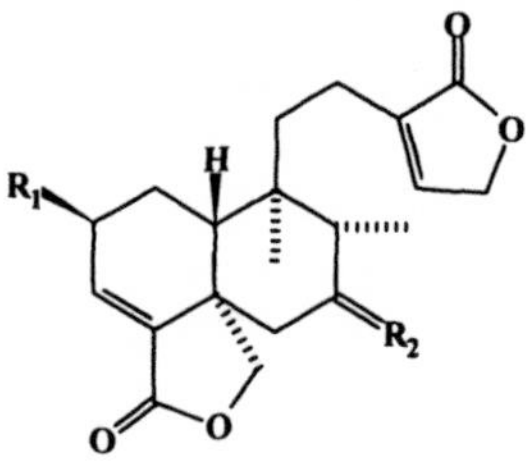

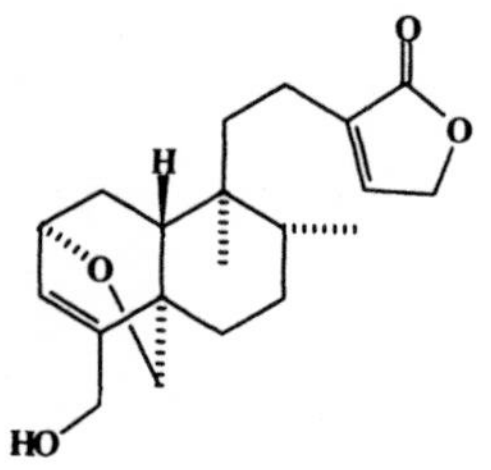

Breviflorolactone

8

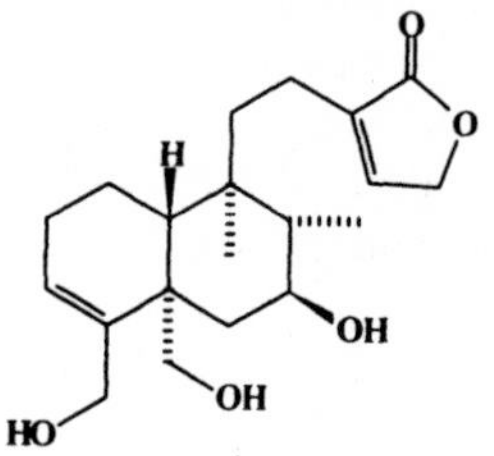

16

	R_1	R_2
7	H	α OH, H
9	H	β OH, H
10	H	α OAc, H
11	OH	α OH, H
12	OH	α OAc, H
13	OH	H_2
14	OH	O
15	H	O

Salvia keerlii also belongs to sect. *Scorodonia*. A phytochemical study of a population collected in Oaxaca (Mexico) led to the isolation of kerlin (**17**) and kerlinolide (**18**).[25] Their structures were deduced by spectroscopic means and an X-ray diffraction analysis of kerlin which established a 12*R* configuration. The relative stereochemistry shown for kerlinolide (**18**) was deduced on biogenetic grounds.[25] From a population of *S. keerlii* collected in Queretaro (Mexico), kerlinic acid (**19**) was obtained.[26] Its structure was established by spectroscopic analysis and transformation into the 3,4-epoxy-12β-butenolide derivative[26] (**20**), identical to the epoxidation product obtained from melisodoric acid (**4**). The 3,4-epoxy-6-acetyl-melisodoric methyl ester (**20**) had a moderate antifeedant activity against larvae of *Spodoptera littoralis*.[27] The different populations of *S. melissodora* and *S. keerlii* studied did not contain melisodoric acid. Apparently there is chemotypic variation in diterpenoid content found among the species studied.

Salvia semiatrata has been included in sect. *Atratae*.[14] A population of this species collected in Oaxaca, contained the dilactone **5** and the neo-clerodane diterpenoid semiatrin (**21**) whose structure was deduced on spectroscopic grounds.[28] The X-ray diffraction analysis established the relative stereochemistry shown with C-12*S*.[29] Semiatrin showed a strong antifeedant activity against *Spodoptera littoralis* larvae.[27]

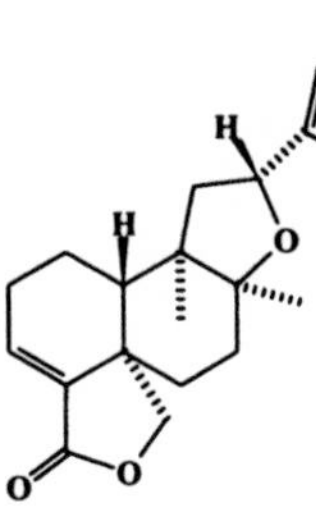

Kerlin **17**

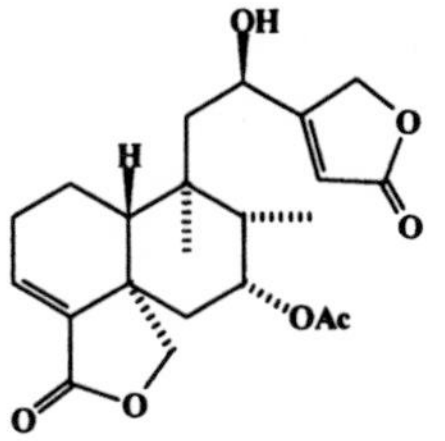

Kerlinolide **18**

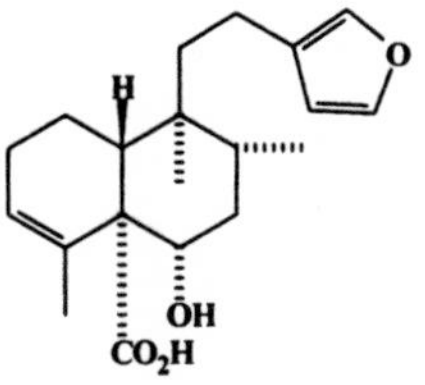

Kerlinic acid **19**

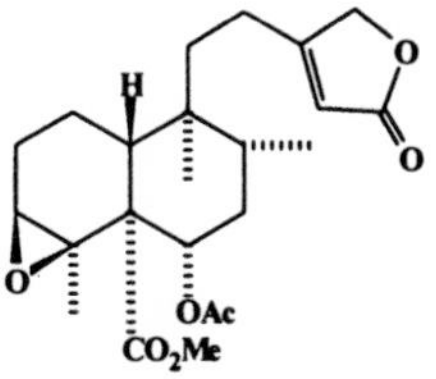

20

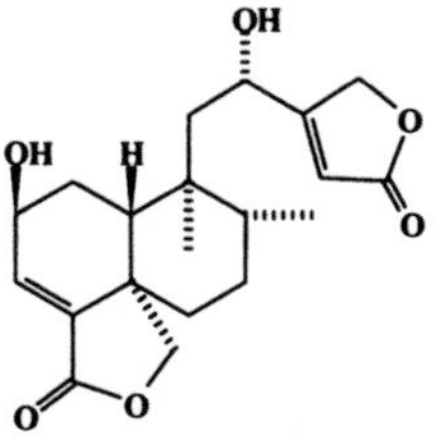

Semiatrin **21**

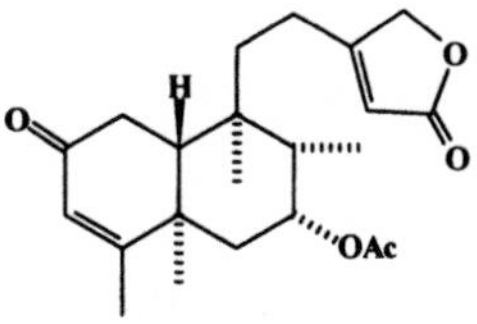

Lasianthin **22**

S. lasiantha (*Salvia*, sect *Mitratae*) contained the neo-clerodane diterpenoid lasianthin (**22**), whose structure was established by chemical transformations, spectroscopic data and X-ray diffraction analysis.[30] It is interesting to note that the 18 and 19 methyl groups are not oxidized in this diterpenoid, a feature not very common in the neo-clerodane diterpenoids isolated from American *Salvia* species.[5]

Salvia divinorum has a long tradition of use by the Mazatec Indians of Oaxaca (Mexico) in their divination rites. It belongs to the section *Dusenostachys*. From this species the neo-clerodane diterpenoids salvinorin A (**23**) and salvinorin B (**24**) were isolated, and their structures established by spectroscopic means and X-ray diffraction analysis of salvinorin A.[31] The absolute stereochemistry was unambiguously determined by exciton chirality circular dichroism of their 1α,2α-diol dibenzoate derivative (**25**).[32] The hallucinogenic properties of the plant have been attributed to the diterpenoids found in it.[33]

A phytochemical study of *S. madrensis* (*Salvia*, sect *Dusenostachys*) led to the isolation of the neo-clerodane diterpenoids salvimadrensin (**26**) and salvimadrensinone (**27**), whose structures were established by spectral analysis and chemical transformations.[34] Both diterpenoids contain a seven membered 17,19-olide function, which produces a distortion of the B ring of the *trans*-decalin to a boat or twisted chair conformation. The presence of this functionality is unusual in neo-clerodane diterpenoids.[5] Salvimadrensin (**26**) showed antifeedant activity against larvae of *S. littoralis,* [27] which could be related to the presence of the 17,19-olide group. The conformational constraint produced by the 2,19-ethereal linkage present in jodrellins A and B (**3**) has been related to the potent antifeedant activity of these clerodane diterpenoids.[9]

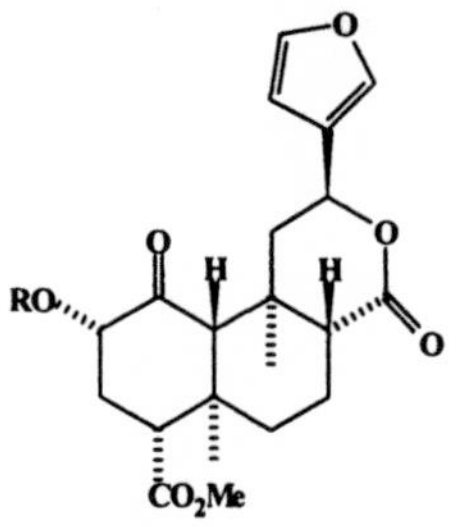

23 R=H Salvinorin A
24 R=Ac Salvinorin B

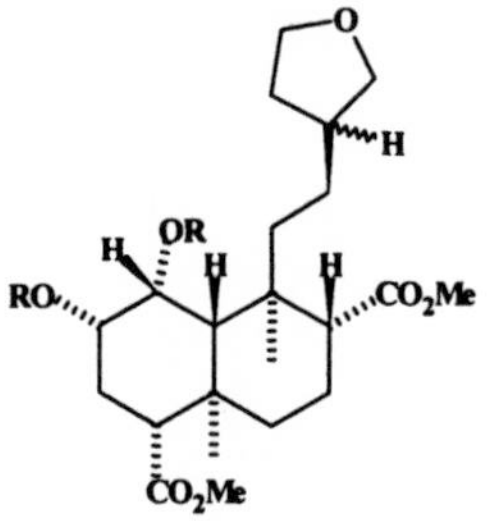

25 R=COΦ

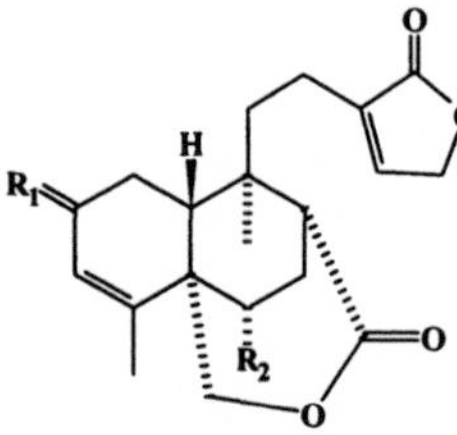

26 $R_1=H_2$; $R_2=OH$ Salvimadrensin
27 $R_1=O$; $R_2=H$ Salvimadrensinone

Only a few Salvia species from Central and South America (*Salvia*, subgenus *Calosphace*) have been analysed. *S. splendens* is endemic to Brazil, and classified in section *Secundae.*[14] A phytochemical study of this plant led to the isolation of two neo-clerodane diterpenoids, salviarin (**28**) and splendidin (**29**) whose structures also were established on spectral bases which included X-ray diffraction analysis of salviarin.[35,36]

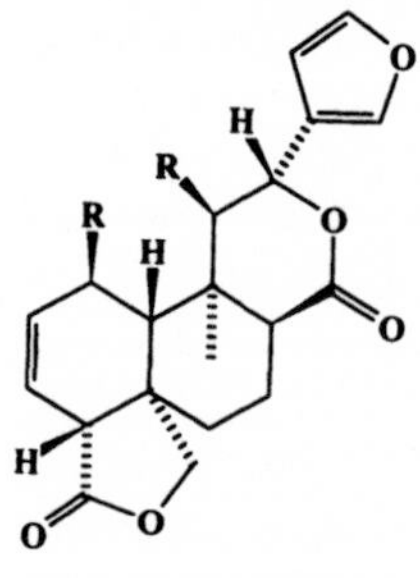

28 R=H Salviarin
29 R=OAc Splendidin

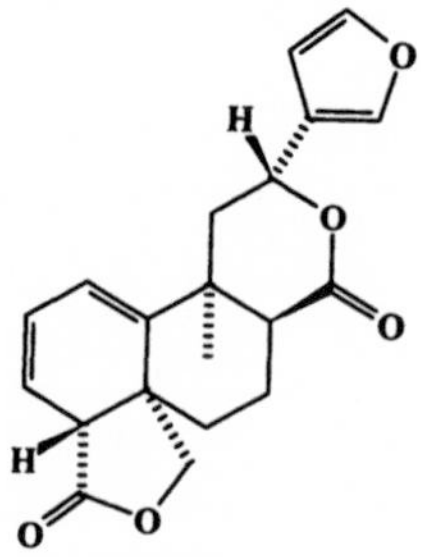

30

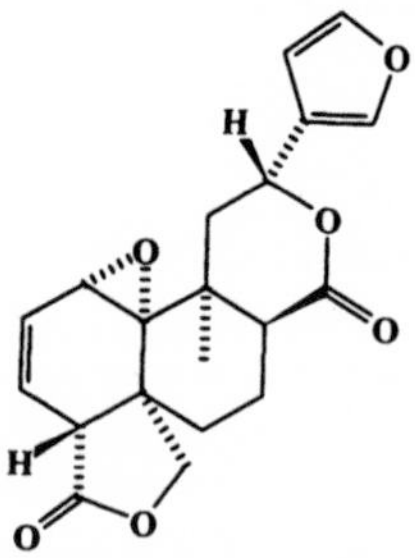

31

The phytochemical study of Salvia species classified in section *Fulgentes* deserves special mention from the ethnobotanical point of view. The name "mirto" has been assigned to several plants used in Mexican traditional medicine,[37] among them the mixture of *S. fulgens* and *S. microphylla*. The different *Salvia* species included in sect. *Fulgentes*, are widely distributed in Mexico. From a population of *S. lineata* (sect. *Fulgentes*) collected in Tehuacan (Puebla, Mexico) two neo-clerodane diterpenes were isolated, and their structures (**30, 31**) were deduced by spectral analysis and chemical transformations.[38] The structure of the diterpenoid **30** was identical to the product obtained by dehydration of the bitter principle **32** isolated from *S. rubescens* (sect. *Rubescentes*) endemic to Venezuela.[39] The relative stereochemistry shown was established through correlation with the diene **30** and comparison with similar structures. The neo-clerodane diterpenoids **30-32** are structurally related to salviarin (**28**). All of them contain a 17,12-olide function and a terminal β-substituted furan ring. A population of *S. lineata* collected in Oaxaca did not contain the 1α,10α-epoxy-salviarin (**31**), instead the clerodanic diene linearolactone (**33**) was found, and its structure was deduced by spectral analysis.[38] X-ray diffraction established a *cis*-A/B decalin stereochemistry.[40] The diene **30** showed a strong antifeedant activity against larvae of *S. littoralis*, while linearolactone (**33**) was not active.[27]

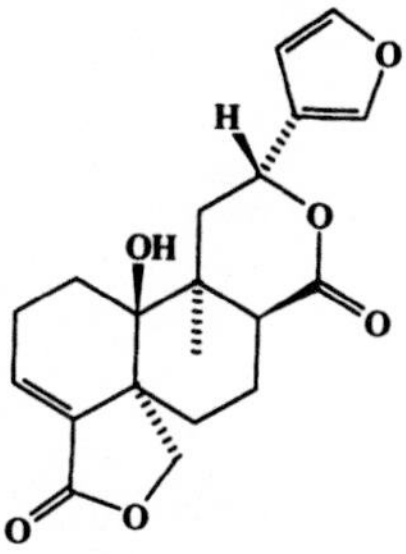

32

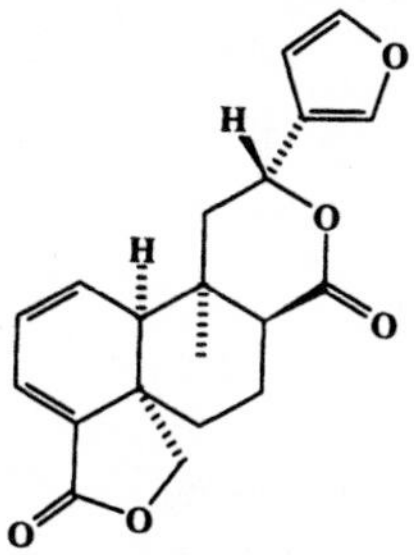

Linearolactone
33

A phytochemical study of a population of *S. fulgens* (*Salvia*, sect. *Fulgentes*) led to the isolation of sandaracopimaric acid (**34**).[42] This is the first report of the presence of a pimarane diterpenoid in a *Salvia* species. The main product isolated from *S. fulgens* was salvigenolide (**35**), a diterpenoid whose structure was deduced by spectral means and chemical transformation reactions.[41] The X-ray diffraction analysis of **35** confirmed the structure and relative stereochemistry proposed.[41] Salvigenolide (**35**) showed a rearranged neo-clerodane skeleton with a six-seven A/B ring system which was named salvigenane. This rearrangement could be produced by the loss of a C11 suitable leaving group and migration of the C8-C9 bond of a clerodanic precursor, as shown (Fig. 1). Treatment of salvigenolide (**35**) with metachloroperbenzoic acid produced the epoxy derivative **36**. Saponification of **35** led to the relactonization product **37**.[41]

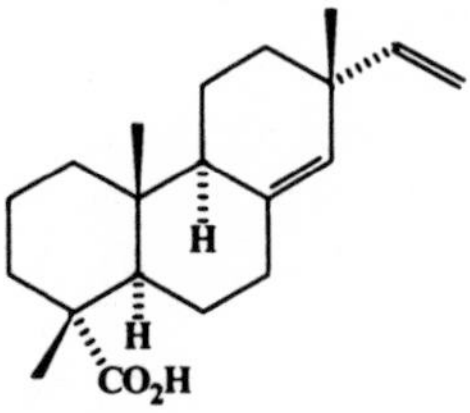

Sandaracopimaric acid
34

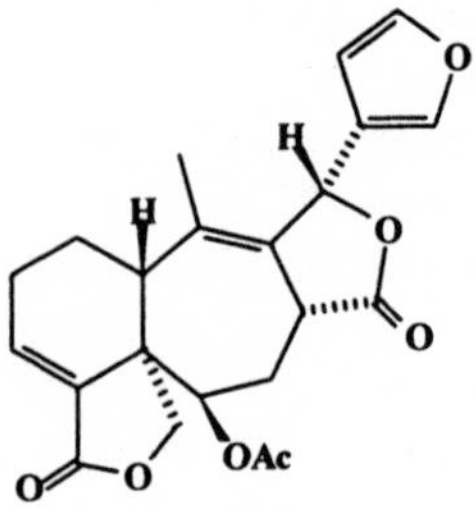

Salvigenolide
35

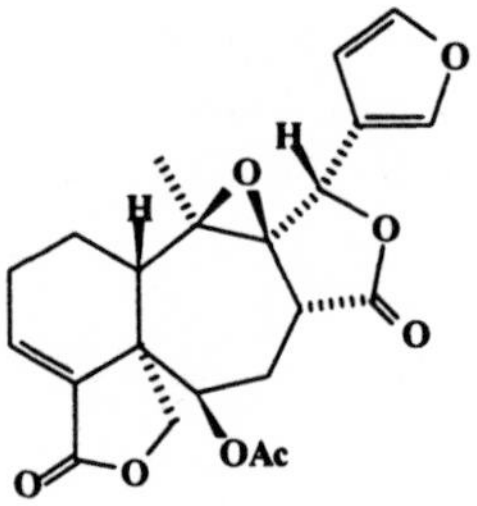

36

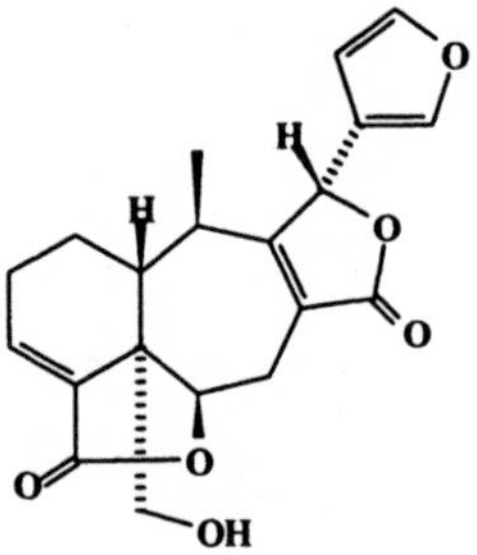

37

From different populations of *S. microphylla* the neo-clerodane dilactone **5**, together with several pimarane diterpenoids, were isolated.[42,43] The structures of the pimarane derivatives **38-40** were deduced from spectral evidence and comparison with known products. Lithium aluminum hydride reduction of the methyl ester of 7α-hydroxy-sandaracopimaric acid (**38**), yielded the diol **41**, which has been

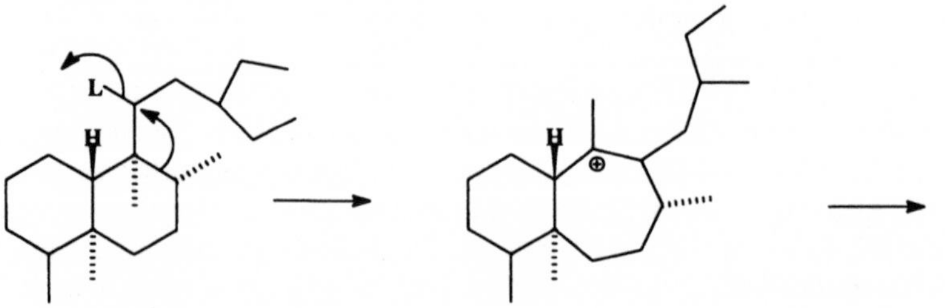

Figure 1. Biogenetic hypothesis of formation of salvigenane skeleton.

considered the active principle of *Tetradenia riparia* (Lamiaceae), a plant used in Rwanda for antimicrobial and antispasmodic activities.[44] The pimarane diterpenoids found in *S. fulgens* and *S. microphylla* could be the secondary metabolites responsible for the therapeutic properties attributed to this "mirto".

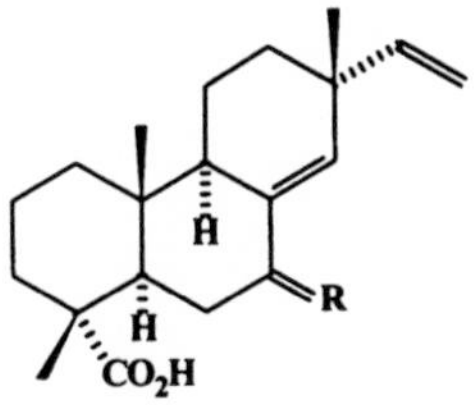

38 R = α OH, H
39 R = O

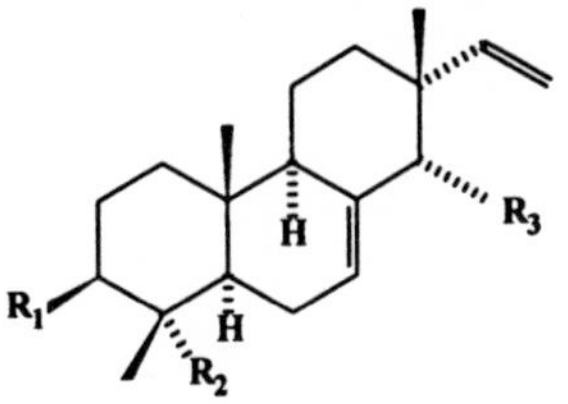

40 R_1 = H, R_2 = CH_2OH, R_3 = OH
42 R_1 = R_3 = H, R_2 = CO_2H
43 R_1 = H, R_3 = OH, R_2 = CO_2H
44 R_1 = H, R_3 = OH, R_2 = CH_2OH
45 R_1 = OH, R_2 = CO_2H, R_3 = H

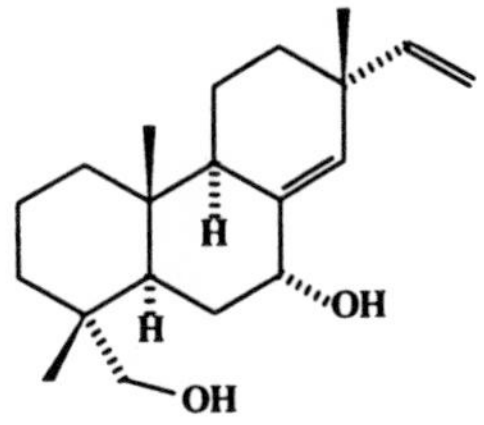

41

Pimarane diterpenoids **42-45** also were isolated from *S. greggii*, a *Salvia* species botanically related to *S. microphylla*, but included in sect. *Flocculosae*.[14] The neo-clerodane diterpenoid, 10β-hydroxy-8-epi-salviarin (**46**), was also isolated from this species,[45] and its structure established from spectral evidence and comparison with the product obtained by sodium borohydride treatment of salviacoccin (**47**), a product isolated from *S. coccinea* (sect. *Subrotundae*). The structure of salviacoccin was deduced by spectral analysis and chemical transformations.[46] Treatment with thionyl chloride and pyridine, gave the dehydroderivative (**48**). Salviacoccin also has been found in *S. plebeia*, a species included in subgenus *Leonia*.[47] The neo-clerodane diterpenoid, 2β,3β-epoxy derivative of salviacoccin (**49**), also was isolated from this species, and its structure and sterochemistry proven by treatment of salviacoccin with m-chloroperbenzoic acid.

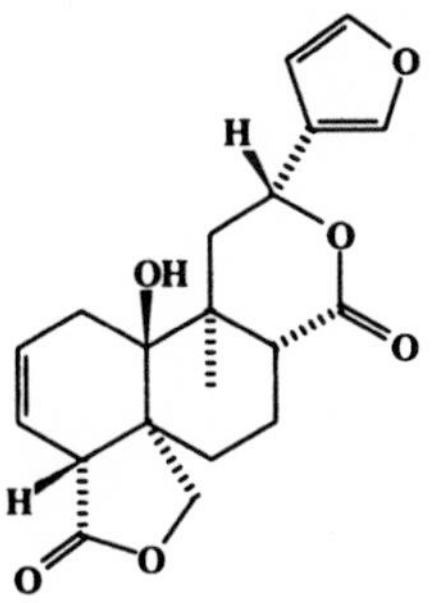

46

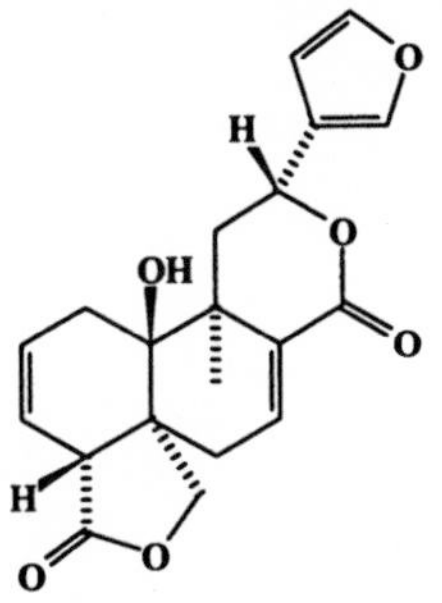

Salviacoccin
47

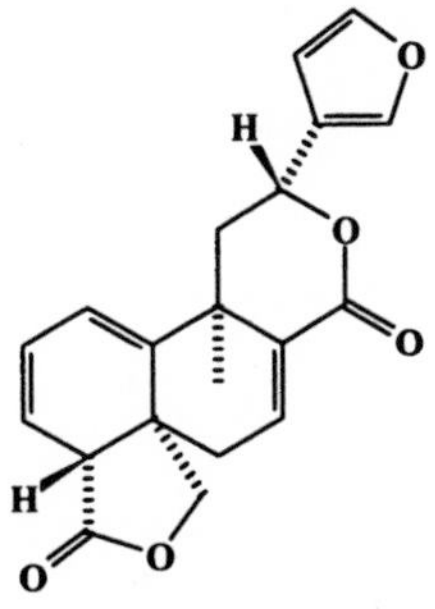

48

S. gesneraeflora is a *Salvia* species found in central Mexico, included in sect. *Nobiles* which is botanically related to sect. *Fulgentes*.[14] From this species salviarin (**28**) and the neo-clerodane diterpenoid **50** were isolated.[48] The *cis*-A/B ring fusion and the relative stereochemistry shown in **50** were deduced by comparison of the ^{1}H NMR data published with those of the spectrum of linearolactone (**33**).

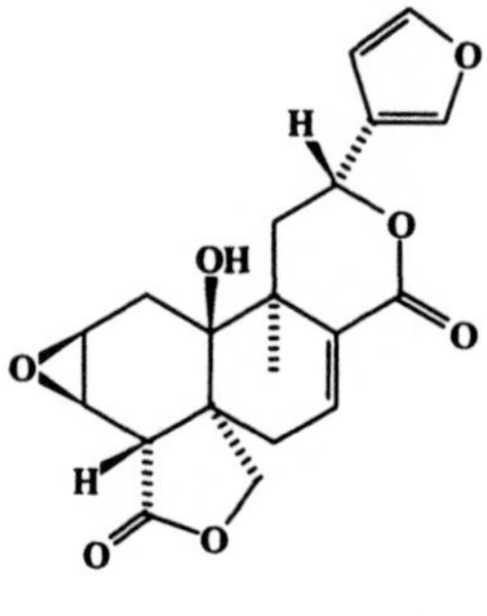

49 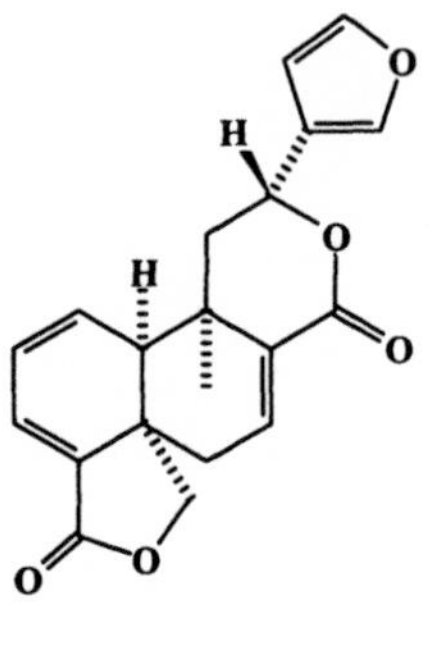 **50**

Salvia puberula, Fern. recently has been classified in sect. *Holwaya* (Ramamoorthy),[49] a section botanically related to sect. *Fulgentes*. This botanical relationship was a great help in the elucidation of the structures of the two diterpenoids isolated from this species.[50] The benzonorcaradiene structure of salvipuberulin (**51**) was deduced by spectroscopic analysis. The ^{1}H NMR spectrum indicated the presence of a disubstituted cyclopropane ring and a pentasubstituted aromatic ring. The ^{13}C NMR was consistent with the presence of the pentasubstituted aromatic ring in the molecule. The benzonorcaradiene diterpenoid (**51**) was slowly rearranged into the benzocycloheptatriene diterpenoid isosalvipuberulin (**52**) by boiling a methylene chloride solution. This transformation can be envisaged as a disrotatory electrocyclic reaction, followed by two suprafacial 1,5-sigmatropic hydrogen shifts, as shown (Fig. 2). Isosalvipuberulin (**52**) also was isolated from the acetone extract of the plant, and its structure deduced by spectroscopic analysis and chemical reactions. Catalytic reduction of

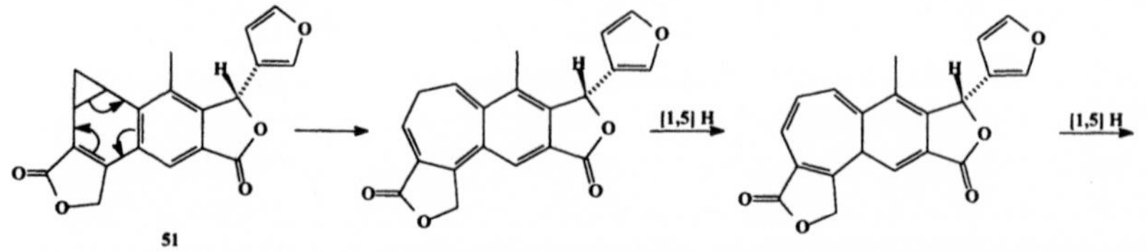

Figure 2. Proposed mechanism of rearrangement of salvipuberulin to isosalvipuberulin.

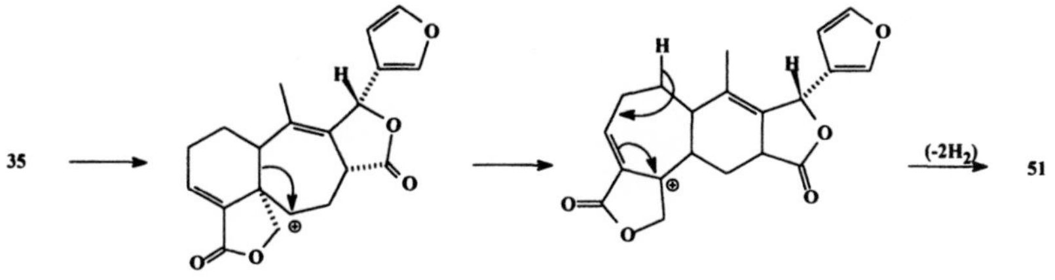

Figure 3. Biogenetic hypothesis of formation of salvipuberulin.

52 gave the 1,2-dihydroderivative which was treated with lithium aluminum hydride, followed by acetylation of the tetrol to yield the tetra-acetylated derivative **53**. The unusual structure proposed for isosalvipuberulin was confirmed by X-ray diffraction analysis.[50] The 12 *R* chirality shown in **52** is a common feature found in most of the neo-clerodane diterpenoids isolated from *Salvia* species.[5] The botanical relationship mentioned earlier suggested that salvigenolide (**35**) could be a possible biogenetic precursor of salvipuberulin (**51**) and isosalvipuberulin (**52**) as outlined (Fig.3). The skeleton of **52** has been named isosalvipuberulan. Salvipuberulin and isosalvipuberulin are the first examples of benzonorcaradiene and benzocycloheptatriene diterpenoids of clerodanic origin.

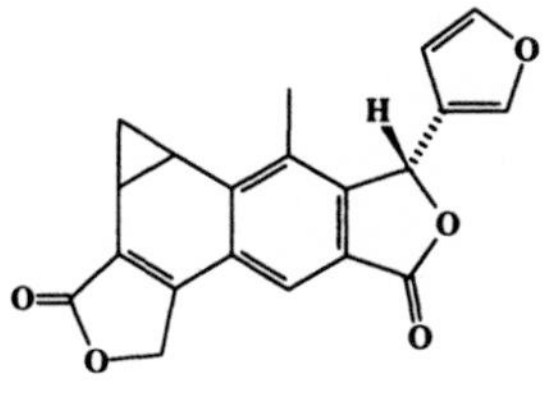

Salvipuberulin
51

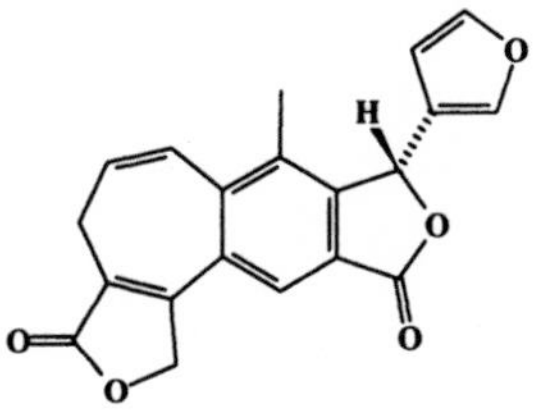

Isosalvipuberulin
52

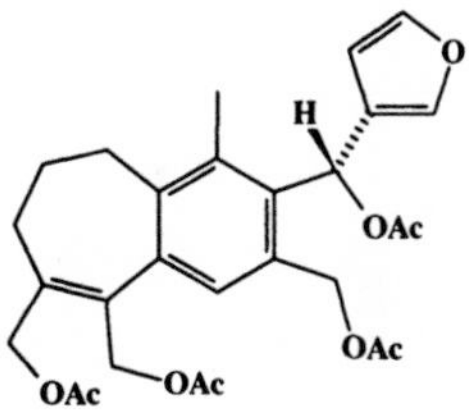

53

Isosalvipuberulin (**52**) was also found in *S. tiliaefolia*, a *Salvia* species included in section *Angulatae*, subsection *Tiliaefoliae*, botanically unrelated to sect. *Fulgentes* or *Holwaya*.[14] The tetralin diterpenoid tilifodiolide (**54**) also was isolated from this species. Its structure was established by chemical and spectroscopic means which included HETCOR and INEPT resonance experiments.[51] The ^{1}H NMR spectrum suggested the presence of a tetrasubstituted aromatic ring in **54** with two ortho hydrogens. The ^{13}C NMR spectrum confirmed this system in the molecule. The presence of a tetraline skeleton in tilifodiolide was verified by an ozonolysis reaction, followed by oxidative treatment of the ozonide. The diacid obtained was converted to the dimethyl ester **55** with ethereal diazomethane. The unusual structure proposed for tilifodiolide was confirmed by X-ray diffraction analysis.[51] This diterpenoid crystallizes with two independent molecules per

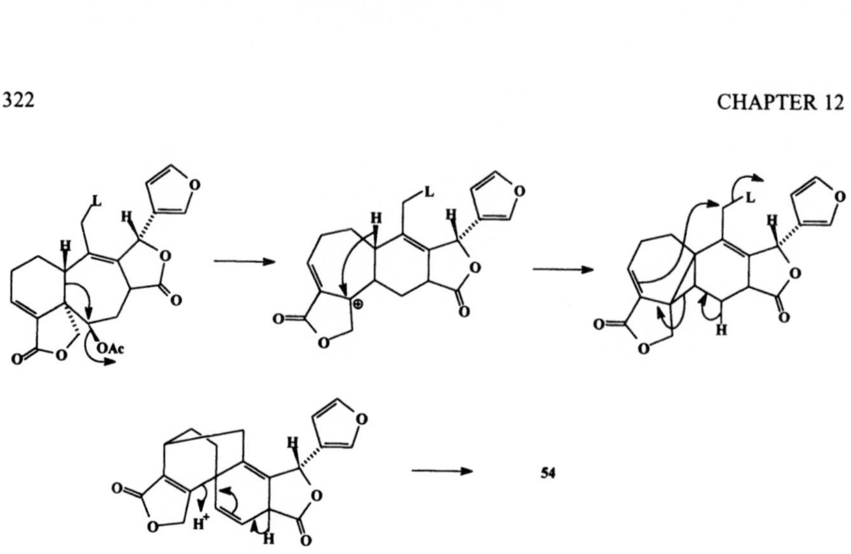

Figure 4. Biogenetic hypothesis of formation of tilifodiolide.

asymmetric unit which differ in the orientation of the α,β-unsaturated γ-lactone bound to C-3 with respect to the hydrobenzofuranone moiety. The presence of isosalvipuberulin and tilifodiolide in *S. tiliaefolia*, points to the possibility of a common biogenetic precursor, such as salvigenolide (**35**), for both diterpenoids. The presence of a good leaving group at C-20 could explain the formation of tilifodiolide as proposed (Fig. 4). From a population of *S. tiliaefolia* collected in Cuernavaca (Morelos, Mexico) the neo-clerodane diterpenoid salvifolin (**56**) was isolated together with isosalvipuberulin and tilifodiolide. The structure was deduced from spectral evidence and comparison with the spectral data of related diterpenoids.[51] Tilifodiolide (**54**) showed an interesting antifeedant activity against *S. littoralis* and *S. exiqua* larvae.[27]

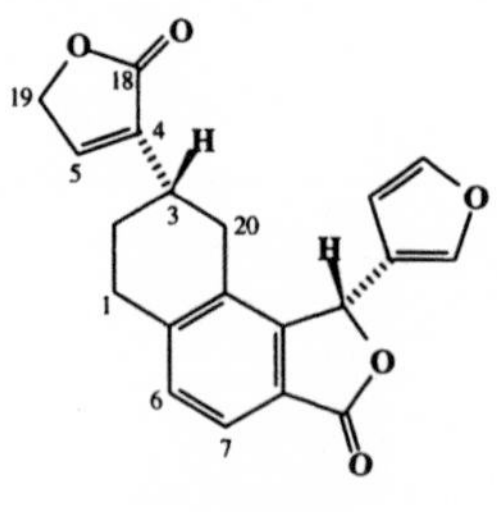

Tilifodiolide
54

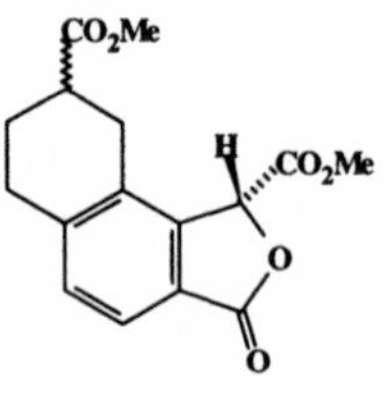

55

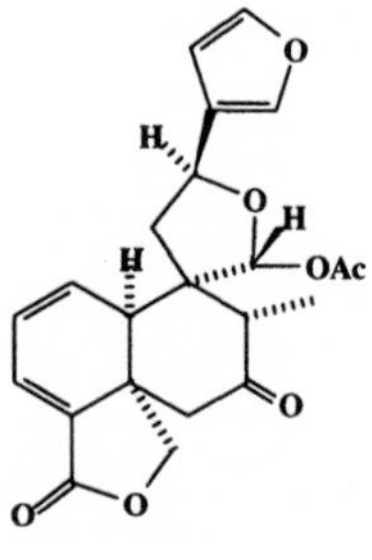

Salvifolin
56

S. rhyacophyla has been included in sect. *Angulatae*, subsect. *Tiliaefoliae*.[14] A study of the diterpenoid content of this species led to the isolation of several neo-clerodane diterpenoids.[52] Salviarin (**28**) (R=H) and the salviarin derivatives: 6β-hydroxy- (**57**), 6β-acetoxy- (**58**), 10β-acetoxy-(**59**) and 6β-hy-

Figure 5. Hypothesis of biogenesis of rhyacophane skeleton.

droxy-7,8-dehydrobacchatricuneatin (**60**), were isolated from this plant and their structures established by spectral means. Dehydrokerlin (**61**) was also isolated, its structure closely related to that of kerlin (**17**) from which it differs in the presence of a terminal β-substituted furan ring in place of the β-butenolide found in kerlin. The unusual structure of rhyacophiline (**62**), also found in *S. rhyacophyla*, was deduced from spectral data and confirmed by X-ray diffraction analysis.[52] The ^{1}H and ^{13}C nmr spectra established the presence of a trisubstituted aromatic ring with the 18,19-olide function attached to it. Both resonance spectra showed that rhyacophiline contained two secondary methyl groups assigned to carbon atoms 6 and 17. They also indicated the existence of a ketalic function which was attributed to C20. Rhyacophiline (**62**) is a 5,6-seco clerodane diterpenoid with an aromatic A ring, a new skeleton named rhyacophane, which could be biogenetically derived from a 7-keto-clerodanic precursor with a conjugated diene in the A ring and an A/B-*cis*-decaline ring fusion, as shown (Fig. 5).

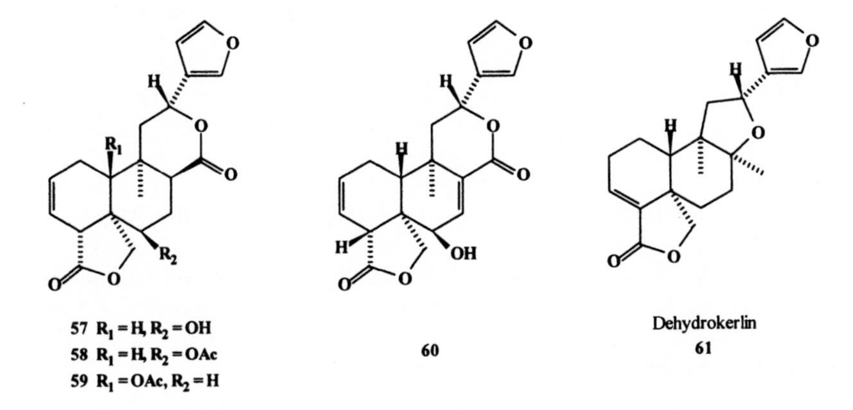

S. languidula also belongs to sect. *Angulatae* but has been included in subsect. *Glumacea*.[14] The diterpenoids isolated from this species, present two different carbon skeletons which could be derived from a common clerodanic precursor. The structure of languiduline (**63**) was deduced from spectral data and X-ray diffraction analysis.[53] Its ^{1}H and ^{13}C NMR spectra indicated the presence of a disubstituted furan ring conjugated to the C12 ketonic group. Both spectra also indicated that C16 was bound to C1 in languiduline. This new skeleton was named languidulane. The salvilanguidulines A-D (**64-67**) were also isolated from *S. languidula*, and their structures were established on spectral grounds and X-ray diffraction analysis of **64**.[54] All of them had an unusual common skeleton named salvilanguidulane. The ^{1}H and ^{13}C NMR spectra of salvilanguidulines A-D indicated the presence of an epoxy-spirolactone function in place of the disubstituted furan ring found in **63**, and a cyclohexanone group. The languidulane diterpenoid **63** and the salvilanguidulane derivatives **64-67** could be biogenetically derived from a neo-clerodane precursor in which the A ring of the decalin contains a conjugated diene group. The oxidation of the β-substituted furan ring to the epoxy-lactone could be produced prior to the nucleophilic attack of C13 on C1. The possible biogenetic relationships discussed are shown (Fig. 6).

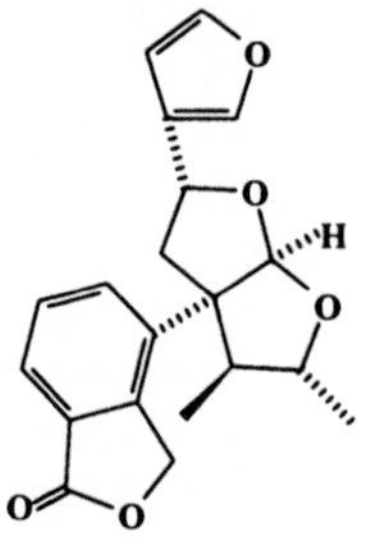

Rhyacophiline
62

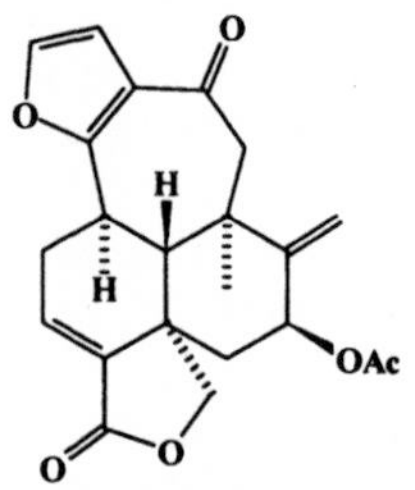

Languiduline
63

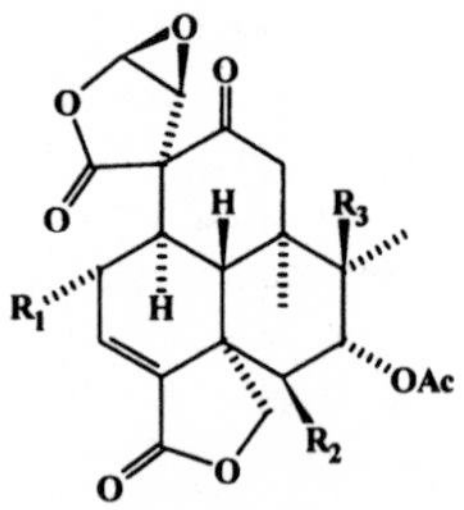

Salvilanguidulines
A: **64** $R_1=R_2=R_3=H$
B: **65** $R_2=R_3=H$, $R_1=OH$
C: **66** $R_1=R_3=H$, $R_2=OH$
D: **67** $R_1=R_2=H$, $R_3=OH$

A diterpenoid with a languidulane skeleton, salvisousolide (**68**), has been isolated from *S. urolepis* (*Salvia*, sect *Angulatae*, subsect. *Tiliaefoliae*)[55] and *S. sousae* (sect. *Polystachyae*)[56]. *S. urolepis* also contained the neo-clerodane diterpenoid **69** and the unresolved mixture of hydroxy-lactones **70** and **71**. Diterpenoid **69** could be considered a possible clerodanic precursor of the languidulane diterpenoid **68**, according to the biogenetic hypothesis proposed. The neo-clerodane derivative (**72**) was also isolated from *S. sousae*.[56] Its structure was deduced by spectral analysis and comparison with similar structures which led us to propose a *cis*-A/B-decalin fusion. The assignment of the oxirane group at 6,7 was established by double resonance experiments.[56]

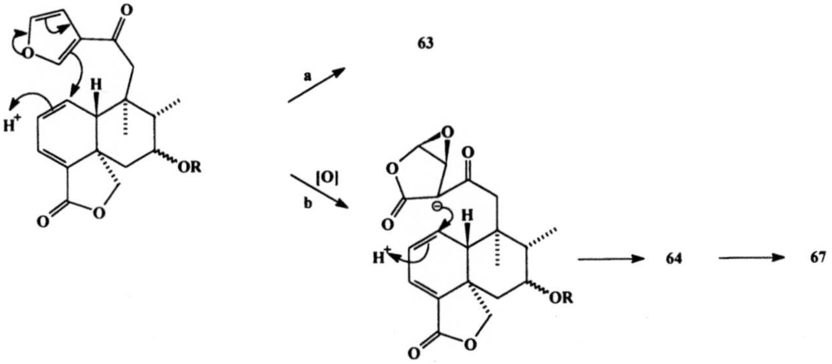

63

a

[O]

b

64

67

Figure 6. Possible biogenesis of languidulane and salvilanguidulane skeletons.

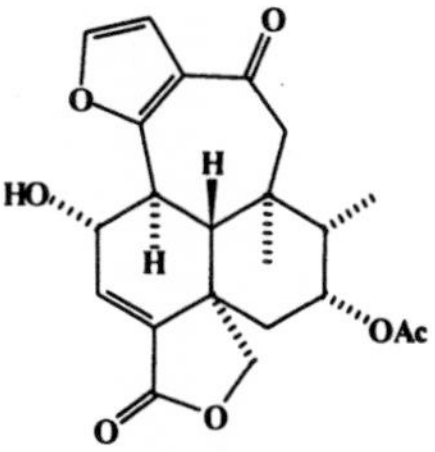

Salvisousolide
68

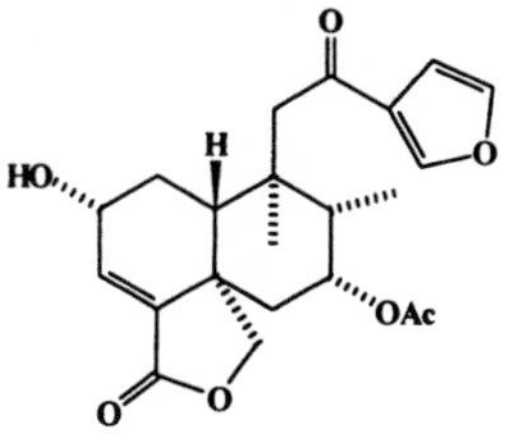

69

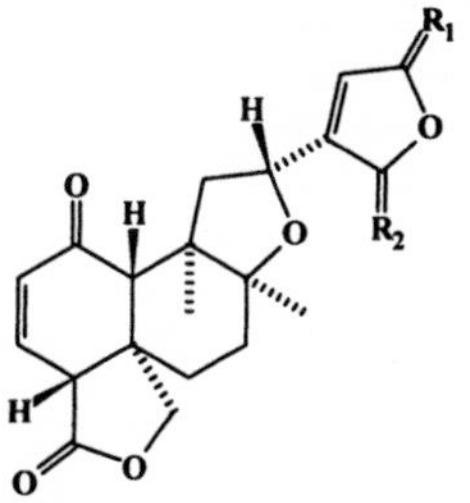

70 $R_1 = H, OH$; $R_2 = O$
71 $R_1 = O$; $R_2 = H, OH$

S. reptans (*Salvia*, sect. *Farinaceae*) contained two diterpenoids (**73**,**74**), whose structures were deduced by spectral means.[57] The X-ray diffraction analysis of both products established the relative stereochemistry as shown.[57,58] Salvireptanolide (**73**) had a rhyacophane skeleton with a phtalide A ring as in rhyacophiline (**62**). It differed in the degree of oxidation of C12 and C20. The second diterpenoid (**74**) had a neo-clerodane skeleton structurally related to linearolactone in which the A/B ring fusion of the decalin has been established as *cis*. It is interesting to draw attention to the *trans*-fused 18,19-olide group which, as shown by X-ray analysis, produces a distortion of the A ring to a half-chair conformation.[57] This constitutes an unusual feature of the saturated 18,19-olide function found in neo-clerodane diterpenoids isolated from *Salvia* spp.. It is usually found to be *cis*-fused as in salviarin (**28**) and related products.[5]

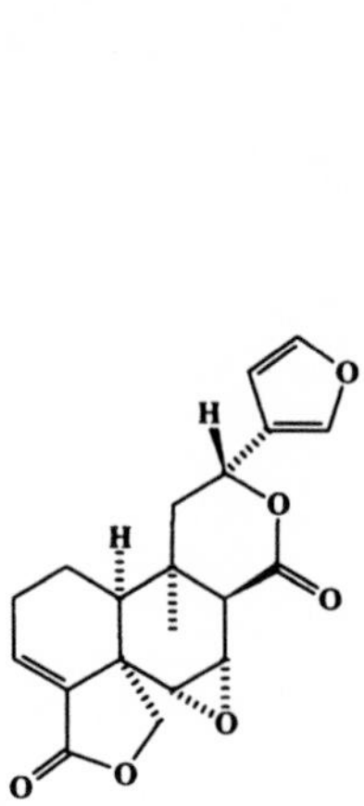

Salvireptanolide

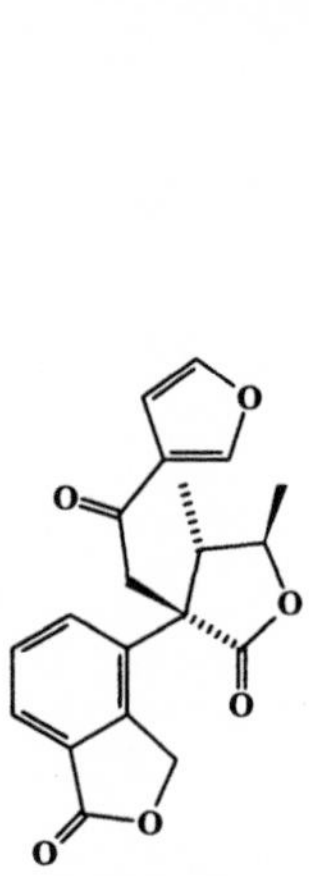

72 73

S. farinacea also has been included in sect. *Farinacea*.[14] Two structurally related neo-clerodane diterpenoids, salvifarin (**76**) and salvifaricin (**75**), were isolated from this species.[59] Their structures were established by spectroscopic studies, chemical correlation and X-ray diffraction analysis.[60] Both diterpenoids had a *cis*-A/B-decalin ring fusion and a ketalic function which involved carbon atoms 20,12 and 7, a functionality unusual in neo-clerodane diterpenoids isolated from American *Salvia* spp.. Treatment of salvifarin (**76**) with sodium iodide-p-toluensulfonic acid in acetonitrile solution, produced salvifaricin (**75**).[61]

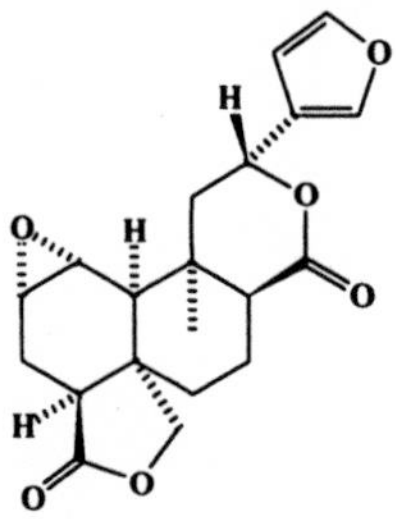

Salvifaricin Salvifarin

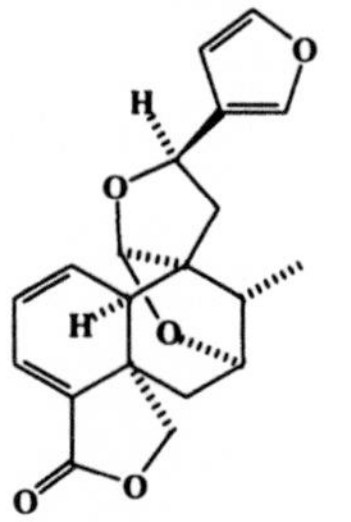

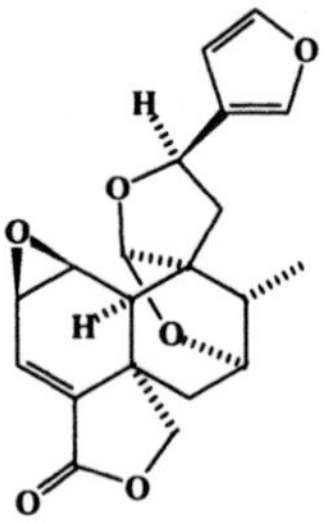

74 75 76

A rhyacophane diterpenoid, salviandulin C (**77**) was isolated from *S. lavanduloides* (*Salvia*, sect *Lavanduloideae*),[14] from which the 9,10-seco-clerodane diterpenoids salviandulins A (**78**) and B (**79**) were also obtained.[62,63] The structures of these diterpenoids were established by spectral means, chemical correlation and X-ray diffraction analysis of salviandulin B (**79**). The acetyl derivative of **79** was proved to be epimeric at C-7 to salviandulin A. Jones oxidation of salviandulin B and desacetyl salviandulin A, yielded the 7-keto derivative (**80**).[63] A possible biogenetic precursor of these diterpenoids could be the 7,8-epoxy derivative of the 10β-hydroxy neo-clerodane (**47**), as indicated (Fig. 7).

S. cardiophylla is a Paraguayan *Salvia* species included in sect. *Rudes*.[14] From the aerial parts of this species, the 5,10-seco-clerodane diterpenoid cardiophyllidin (**81**) was isolated[64] together with the abietane diterpenoid 2α-hydroxysugiol.[65] The presence of abietane and clerodane type diterpenoids in the aerial parts of American *Salvia* species is noteworthy. The structure of cardiophyllidin was established by spectroscopic means and X-ray diffraction analysis.[64] The *R* configuration ascribed to C-9 is unusual as most neo-clerodane diterpenoids isolated from American *Salvia* species have a 9-*S* absolute configuration.[5]

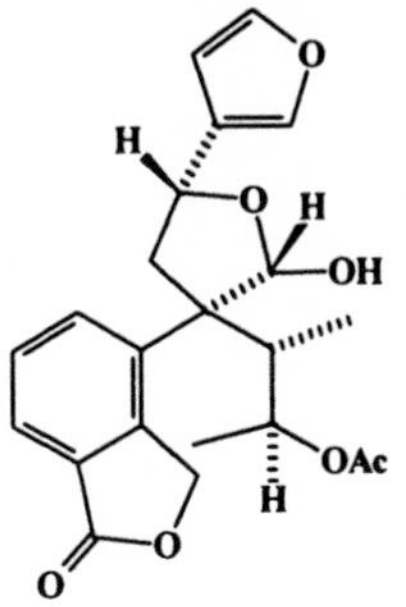

Salviandulin C
77

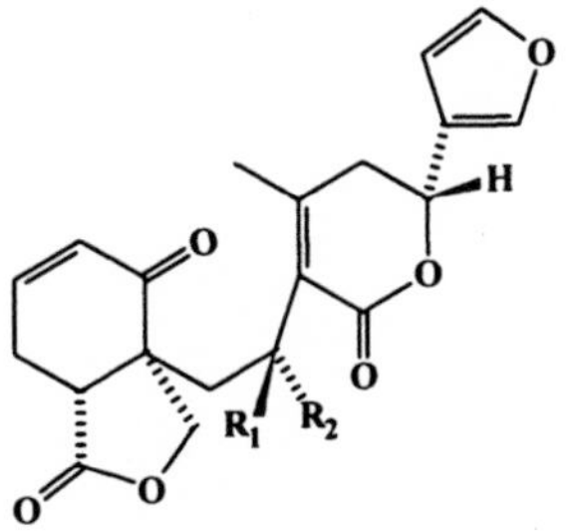

78 $R_1 = H; R_2 = OAc$ Salvaindulin A
79 $R_1 = OH; R_2 = H$ Salvaindulin B
80 $R_1, R_2 = O$

5,10-Seco-clerodane diterpenoids have been found in two Mexican *Salvia* spp.. *S. thymoides* (sect. *Flocculosae*) contained the product **82**, while *S. purpurea* (sect. *Purpureae*) yielded the terminal butenolide isomer **83**. Both 5,10-seco-clerodane diterpenoids were first found in *Pulicaria angustifolia* (Compositae).[68]

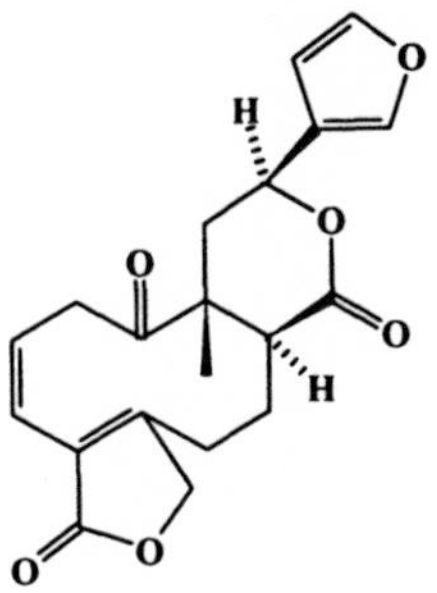

Cardiophyllidin
81

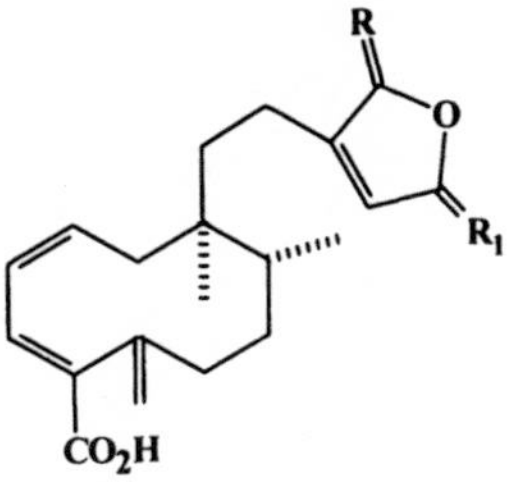

82 $R = H_2; R_1 = O$
83 $R = O; R_1 = H_2$

CONCLUSIONS

The systematic chemotaxonomic study of the Mexican *Salvia* spp. has been a useful guide for the study of *Salvia* species producing neo-clerodane

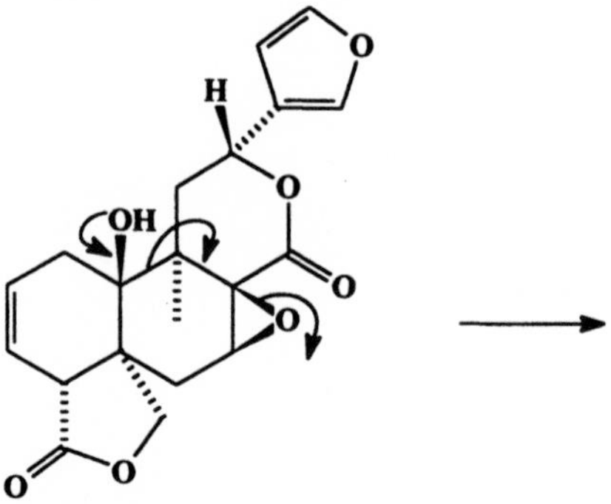

Figure 7. Possible biogenesis of 9,10-seco clerodane diterpenoids.

diterpenoids or secondary metabolites with rearranged clerodane skeletons. The presence of these compounds may explain the observed resistance to insect predation of some *Salvia*.

Some of these compounds, isolated from the American *Salvia* spp., have shown interesting antifeedant activity against lepidopteran species.[27] For example, 2β-hydroxy-7α-acetoxy-16,15-olide (**12**), semiatrin (2β,12α-dihydroxy-15,16-olide) (**21**), and 2β-acetoxy-7α-hydroxy-16,15-olide (**11**,R'=OAc) showed significant antifeedant activity against *Spodoptera littoralis*, while other dihydroxy derivatives examined were inactive.[27]

The study of the antifeedant activity of the diterpenoids structurally related to salviarin (**28**) revealed that 1(10)-dehydrosalviarin (**30**) was very active while linearolactone-1,2-3,4 diene (**33**) was completely inactive against *S. littoralis*. These data thus indicate that there is no clear relationship between the structure of the neo-clerodane diterpenoid and the antifeedant activity analyzed.

The rearranged neo-clerodane tilifodiolide (**54**) was very active against *S. littoralis* and *S. exiqua*. This diterpenoid was isolated from Salvia tiliaefolia as part of a program of the IOCD (International Organization for Chemistry in Development) to study a series of plants with insecticidal properties selected by Lagunes et al..[69]

ACKNOWLEDGMENTS

We wish to express our deep gratitude to Dr. T.P. Ramamoorthy, without whose botanical guidance, the systematic chemotaxonomic study of the Mexican *Salvia* spp. could not have been possible, and Dr. Monique Simmons for the study of the antifeedant activities. We also would like to thank all the young collaborators whose names appear in the references. The financial help of CONACyT

(Projects PCCBNA-021142 and PCECCNA-050325) and DGAPA-UNAM (Project PAPIID IN 205192) are gratefully acknowledged.

REFERENCES

1. BAI, D. 1993. Traditional Chinese medicines and new drug development. Pure and Appl. Chem. 65: 1103-1112.
2. DENIS, D.N., GREENE, A.E., GUENARD, D., GUERITTE-VOEGELEIN, F., MANGATAL, L., POTIER, P. 1988. A highly efficient, practical approach to natural taxol. J. Am. Chem. Soc. 110: 5917-5919.
3. HEDGE, I.C. 1992. A global survey of the biogeography of the Labiatae. In: Advances in Labiatae Science, (R.M. Harley, T. Reynolds, eds.), Royal Botanic Gardens Kew, pp. 7-17.
4. CHANG, H.M., CHENG, K.P., CHOANG, T.F., CHOW, H.F., CHUI, K.Y., HON, P.M., TAN, F.W.L., YANG, Y., ZHONG, Z.P., LEE, C.M., SHAM, H.L., CHAN, C.F., CUI, Y.X., WONG, N.C. 1990. Structure elucidation and total synthesis of new tanshinones isolated from *Salvia miltiorrhiza* Bunge (Danshen). J. Org. Chemistry 55: 3537-3543.
5. RODRIGUEZ-HAHN, L., ESQUIVEL, B., CARDENAS, J. 1994. Clerodane diterpenes in Labiatae. In: Progress in the Chemistry of Organic Natural Products (W. Herz, G.W. Kirby, R.E. Moore, eds.) Springer Verlag, New York, pp. 107-196.
6. KUBO , I., LEE, Y.W., BATOGH-NAIR, V., NAKANISHI, K., CHAPYA, A. 1976. Structure of ajugarins. Chem. Commun. 949-950.
7. SIMMONDS, M.S.J., BLANEY, W.M, LEY, S.V., SAVONA, G., BRUNO, M., RODRIGUEZ, B. 1989. The antifeedant activity of clerodane diterpenoids from *Teucrium*. Phytochemistry 28: 1069-1071.
8. KUBO, I., KLOCKE, J.A., MINRA , L., FUKUYAMA, Y. 1982. Structure of ajugarin IV. Chem. Commun. 618-619 .
9. ANDERSON, J.C. BLANEY, W.M., COLE, M.D., FELLOWS, L.E., LEY, S.V., SHEPPARD, R.N., SIMMONDS, M.S.J. 1989. The structure of two new clerodane diterpenoids, potent insect antifeedants from *Scutellaria woronowii* (Juz); Jodrellin A and B. Tetrahedron Letters. 30: 4737-4740.
10. STANDLEY, P., WILLIAMS, L. 1973. Flora of Guatemala. Fieldiana Bot. 24(9): 237-317.
11. BENTHAM, G. 1876. Labiatae. In: Genera Plantarum 2, (G. Bentham, J.D. Hooker, eds.),. Reeve and Co. London, pp. 1166-1223..
12. RODRIGUEZ-HAHN, L., ESQUIVEL, B., CARDENAS, J., RAMAMOORTHY, T.P. 1992. The distribution of diterpenoids in *Salvia*. In: Advances in Labiatae Science, (R.M. Harley, T. Reynolds, eds.), Royal Botanic Gardens, Kew, pp. .335-347
13. RAMAMOORTHY, T.P., LORENCE, D.H. 1987. Species vicariance in the Mexican flora and description of a new species of *Salvia* (Lamiaceae). Bull. Mus. Hist. Nat., B, Adamsonia 9(2): 167-175.
14. EPLING, C. 1939. A revision of *Salvia* subgenus Calosphace. Repert. Spec. Nov. Regni. Veg. Beih. 110: 1-383.
15. RAMAMOORTHY, T.P., ESQUIVEL, B., SANCHEZ, A.A., RODRIGUEZ-HAHN, L. 1988. Phytogeographical significance of the occurrence of abietane-type diterpenoids in *Salvia* sect. Erythrostachys (Lamiaceae). Taxon 37: 908-912.
16. RODRIGUEZ-HAHN, L., ESQUIVEL, B., SANCHEZ, A.A., SANCHEZ, C., CARDENAS, J., RAMAMOORTHY, T.P. 1989. Diterpenos abietánicos de Salvias mexicanas. Rev. Latinoamer. Quím. 20: 105-110.
17. RODRIGUEZ-HAHN, L., ESQUIVEL, B., SANCHEZ, A.A., SANCHEZ, C., CARDENAS, J., RAMAMOORTHY, T.P. 1987. Nuevos diterpenos de Salvias mexicanas. Rev. Latinoamer. Quím. 18: 104-109.

18. RODRIGUEZ-HAHN, L., ESQUIVEL, B., CARDENAS, J., 1992. New diterpenoid skeletons of clerodanic origin from Mexican *Salvia* species. Trends in Organic Chemistry. 3: 99-111
19. RODRIGUEZ-HAHN, L., MARTINEZ C.G., ROMO, J. 1973. Estructura del ácido melisodórico, un diterpeno aislado de *Salvia melissodora* Lag. Rev. Latinoamer. Quím. 4: 93-100.
20. ESQUIVEL, B., VALLEJO, A., GAVIÑO, R., CARDENAS, J., SANCHEZ, A.A., RAMAMOORTHY, T.P., RODRIGUEZ-HAHN, L. 1988. Clerodane diterpenoids from *Salvia melissodora*. Phytochemistry. 27: 2903-2905.
21. HERZ, W., PILOTTI, A.M., SODERHOLM, A.C., SHUHAMA, I.K., VICHNEWSKI, W. 1977. New ent-clerodane-type diterpenoids from *Baccharis trimera*. J. Org. Chem. 42: 3913-3917.
22. OHSAKI, A., OHNO, M., SHIBATA, K., TOKOROGAMA, T., KUBOTA, T. 1986. Clerodane diterpenoids from *Portulaca* cv Jewel. Phytochemistry 25: 2414-2416.
23. CUEVAS, G., COLLERA, O., GARCIA, F., CARDENAS, J., MALDONADO, E., ORTEGA, A. 1987. Diterpenes from *Salvia breviflora*. Phytochemistry 26: 2019-2021.
24. ESQUIVEL, B., HERNANDEZ, L.M., CARDENAS, J., RAMAMOORTHY, T.P., RODRIGUEZ-HAHN, L.1989. Further ent-clerodane diterpenoids from *Salvia melissodora*. Phytochemistry 28: 561-566.
25. ESQUIVEL, B., MENDEZ, A., ORTEGA, A., SORIANO-GARCIA, M.,TOSCANO, A., RODRIGUEZ-HAHN, L. 1985. Neo-clerodane-type diterpenoids from *Salvia keerlii*. Phytochemistry 24: 1769-1772.
26. RODRIGUEZ-HAHN, L., GARCIA, A., ESQUIVEL, B., CARDENAS, J., 1987. Structure of kerlinic acid from *Salvia keerlii*. Chemical correlation with melisodoric acid. Can. J. Chem. 65: 2687-2690.
27. ESQUIVEL, B., RODRIGUEZ-HAHN, L., SIMMONDS, M.S.J., BLANEY, W.M.1994. Antifeedant activity of some neo-clerodane diterpenoids from Mexican *Salvia spp*. Poster 577 at eight IUPAC International Congress of Pesticide Chemistry. Washington D.C. USA.
28. ESQUIVEL, B., HERNANDEZ, M., RAMAMOORTHY, T.P. CARDENAS, J., RODRIGUEZ-HAHN, L. 1986. Semiatrin, a neo-clerodane diterpenoid from *Salvia semiatratha*. Phytochemistry 25: 1484-1486.
29. SORIANO-GARCIA, M., TOSCANO, A., ESQUIVEL, B., HERNANDEZ, M., RODRIGUEZ-HAHN, L. 1987. Structure and stereochemistry of 2-hydroxy-12(S)-hydroxyneoclerodane- 3,13(14)-diene-15,16:19,20-diolide (semiatrin), a diterpene. Acta Cryst. C43: 272-274.
30. SANCHEZ, A.A., ESQUIVEL, B., PERA, A., CARDENAS, J., SORIANO-GARCIA, M., TOSCANO, A., RODRIGUEZ-HAHN, L. 1987. Lasianthin, a neo-clerodane diterpenoid from *Salvia lasiantha*. Phytochemistry 26: 479-482.
31. ORTEGA, A., BLOUNT, J.F., MANCHAND, P.S. 1982. Salvinorin, a new trans-neo-clerodane diterpene from *Salvia divinorum* (Labiatae). J. Chem. Soc. Perkin. Trans. I 2505-2508.
32.KOREEDA, M., BROWN, L., VALDES III, L.J. 1990. The absolute stereochemistry of salvinorins. Chemistry Letters 2015-2018.
33. VALDES III, L.J., BUTLER, W.M., HATFIELD, G.M. PAUL, A.G., KOREEDA, M. 1984. Divinorin A, a psychotropic terpenoid, and divinorin B from the hallucinogenic mexican mint *Salvia divinorum*. J. Org. Chem. 49: 4716-4720.
34. RODRIGUEZ-HAHN, L., ALVARADO, G., CARDENAS, J., ESQUIVEL, B., GAVIÑO, R. 1994. Neo-clerodane diterpenoids from *Salvia madrensis*. Phytochemistry 35: 447-450.
35. SAVONA, G., PATERNOSTRO, M.P., PIOZZI, F., HANSON, J.R., HITCHCOCK, P.B., THOMAS, S.A. 1978. Salviarin, a new diterpenoid from *Salvia splendens*. J. Chem. Soc. Perkin Trans. I. 643-646.

36. SAVONA, G., PATERNOSTRO, M.P., PIOZZI, F., HANSON, J.R. 1979. Splendidin, a new trans-clerodane from *Salvia splendens*. J. Chem. Soc. Perkin Trans. I. 533-534.
37. MARTINEZ, M. 1992. Las plantas medicinales de México.6th. Edition. Ediciones Botas México. pp. 22-657.
38. ESQUIVEL, B., CARDENAS, J., RAMAMOORTHY, T.P., RODRIGUEZ-HAHN, L. 1986. Clerodane diterpenoids of *Salvia lineata*. Phytochemistry 25: 2381-2384.
39. BRIESKORN, C.H., STEHLE, T. 1973. Labiaten-bitterstoffe:eine neue verbindung des clerodantyps. Chem. Ber. 106: 922-928.
40. SORIANO-GARCIA, M., ESQUIVEL, B., TOSCANO, R.A., RODRIGUEZ-HAHN, L. 1987. Structure and stereochemistry of (8S,12R)-cis-clerodane-1,3,13(16),14-tetraene-15,16-epoxy-12(17);18,19-diolide (Linearolactone), a diterpene. Acta Cryst. C43: 1565-1567.
41. ESQUIVEL, B., CARDENAS, J., TOSCANO, A., SORIANO-GARCIA, M., RODRIGUEZ-HAHN, L. 1985. Structure of salvigenolide, a novel diterpenoid with a rearranged neo-clerodane skeleton from *Salvia fulgens*. Tetrahedron 41: 3213-3217.
42. ESQUIVEL, B., CARDENAS, J., RODRIGUEZ-HAHN, L., RAMAMOORTHY, T.P. 1987. The diterpenoid constituents of *Salvia fulgens* and *Salvia microphylla*. Journal of Natural Products 50: 738-740.
43. ESQUIVEL, B., MARTINEZ, N.S., CARDENAS, J., RAMAMOORTHY, T.P., RODRIGUEZ-HAHN, L. 1989. The pimarane-type diterpenoids of *Salvia microphylla* var. *neurepia*. Planta Medica 55: 62-63.
44. DE KIMPE, N., SCHAMP, N., PUYVELDE, L.V., DUBE, S., CHAGNON-DUBE, M., BORREMANS, F., ANTEUNIS, M.J.D., DECLERCQ, J.P., GERMAIN, G., VAN MERSSCHE, M. 1982. Isolation and structural identification of 8(14),15-sandaracopimaradiene 7a,18-diol from *Iboza riparia*. J. Org. Chem. 47: 3628-3630.
45. BRUNO, M., SAVONA, G., FERNANDEZ-GADEA, F., RODRIGUEZ, B., 1986. Diterpenoids from *Salvia greggii*. Phytochemistry 25: 475-477.
46. SAVONA, G., BRUNO, M., PATERNOSTRO, M., MARCO, J.L., RODRIGUEZ, B., 1982. Salviacoccin, a neo-clerodane diterpenoid from *Salvia coccinea*. Phytochemistry 21: 2563-2566.
47. GARCIA-ALVAREZ, M.C., HASAN, M., MICHAVILA, A., FERNANDEZ-GADEA, F., RODRIGUEZ, B. 1986. Epoxysalviacoccin, a neo-clerodane diterpenoid from *Salvia plebeia*. Phytochemistry 25: 272-274.
48. JIMENEZ, M., MORENO, E.D., DIAZ.E. 1979. Diterpenos de la *Salvia gensneraefolia* I. Estructuras de las gensnerofolinas A y B. Rev. Latinoamer. Quím. 10: 166-171.
49. RAMAMOORTHY, T.P. 1984. Typifications in *Salvia* (Lamiaceae). Taxon 33: 322-324.
50. RODRIGUEZ-HAHN, L., ESQUIVEL, B., SANCHEZ, A.A., CARDENAS, J., TOVAR, O.G., SORIANO-GARCIA, M., TOSCANO, A. 1988. Puberulin and isopuberulin, benzonorcaradiene and benzocycloheptatriene diterpenoids of clerodanic origin from *Salvia puberula*. J. Org. Chem. 53: 3933-3936.
51. RODRIGUEZ-HAHN, L., O'REILLY, R., ESQUIVEL, B., MALDONADO, E., ORTEGA, A., CARDENAS,J., TOSCANO, R.A. 1990. Tilifodiolide, tetraline-type diterpenoid of clerodanic origin from *Salvia tiliaefolia*. J. Org. Chem. 55: 3522-3525.
52. FERNANDEZ, M.C., ESQUIVEL, B., CARDENAS, J., SANCHEZ, A.A., TOSCANO, R.A., RODRIGUEZ-HAHN, L. 1991. Clerodane and aromatic seco-cleroane diterpenoids from *Salvia rhyacophila*. Tetrahedron 47: 7199-7208.
53. CARDENAS, J., ESQUIVEL, B., TOSCANO, R.A., RODRIGUEZ-HAHN, L. 1988. Languiduline, a diterpenoid with an unusual structure from *Salvia languidula*. Heterocycles 27: 1809-1812.

54. CARDENAS, J., PAVON, T., ESQUIVEL, B., TOSCANO, A., RODRIGUEZ-HAHN, L. 1992. Salvilanguidulines, four new diterpenoids isolated from *Salvia languidula* with an unusual epoxy spiro g-lactone. Tetrahedron Letters 33: 581-584.
55. SANCHEZ, A.A., ESQUIVEL, B., RAMAMOORTHY, T.P., RODRIGUEZ-HAHN, L. Clerodane diterpenoids from *Salvia urolepis*. Phytochemistry in press.
56. ESQUIVEL, B., OCHOA, J., CARDENAS, J., RAMAMOORTHY, T.P., RODRIGUEZ-HAHN, L. 1988. Clerodane-type diterpenoids from *Salvia sousae*. Phytochemistry 27: 483-486.
57. ESQUIVEL, B., ESQUIVEL, O., CARDENAS, J., SANCHEZ, A.A., RAMAMOORTHY, T.P., TOSCANO, R.A., RODRIGUEZ-HAHN, L. 1991. Clerodane and seco-clerodane diterpenoids from *Salvia reptans*. Phytochemistry 30: 2335-2338.
58. TOSCANO, R.A., SANCHEZ, A.A., ESQUIVEL, B., ESQUIVEL, O., RODRIGUEZ-HAHN, L. 1994. Salvireptanolide. Acta Cryst. in press.
59. SAVONA, G., RAFFA, D., BRUNO, M., RODRIGUEZ, B., 1983. Salvifarin and salvifaricin, neo-clerodane diterpenoids from *Salvia farinacea*. Phytochemistry 22: 784-786.
60. EGUREN, L., FAYOS, J., PERALES, A., SAVONA, G., RODRIGUEZ, B. 1984. Salvifarin, X-ray structure determination of a cis neo-clerodane diterpenoid from *Salvia farinacea*. Phytochemistry 23: 466-467.
61. RODRIGUEZ, B., PASCUAL, C., SAVONA, G. 1984. The correct structure of salvifaricin, a cis -neo-clerodane diterpenoid from *Salvia farinacea*. Phytochemistry 23: 1193-1194.
62. MALDONADO, E., CARDENAS, J., SALAZAR, B., TOSCANO, R.A., ORTEGA, A., JANKOWSKI, C.K., AUMELAS, A., VAN CALSTEREN, M.R. 1992. Salvianduline C, a 5,6-secoclerodane diterpenoid from *Salvia lavanduloides*. Phytochemistry 31: 217-220.
63. ORTEGA, A., CARDENAS, J., TOSCANO, A., MALDONADO, E., AUMELAS, A., VAN CALSTEREN, M.R., JANKOWSKI, C. 1991. Salviandulines A and B. Two secoclerodane diterpenoids from *Salvia lavanduloides*. Phytochemistry 30: 3357-3360.
64. GONZALEZ, A.G., HERRERA, J.R., LUIS, J.G., RAVELO, A.G., RODRIGUEZ, M.L., FERRO, E. 1988. Cardiophyllidin, a seco-ent-neoclerodane diterpenoid from *Salvia cardiophylla*. Tetrahedron Letter 29: 363-366.
65. GONZALEZ, A.G., HERRERA, J.R., LUIS, J.G., RAVELO, A.G., FERRO, E.A. 1988. Terpenes and flavones of *Salvia cardiophylla*. Phytochemistry 27: 1540-1541.
66. FLORES, E.A. 1986. Estudio fitoquímico de *Salvia thymoides* Benth. B.Sc. Thesis. Faculty of Chemistry of the National Autonomous University of Mexico. p. 66.
67. HERNANDEZ, S. 1986. Estudio fitoquímico de *Salvia purpurea* Cav.(dos poblaciones). B.Sc. Thesis. Faculty of Chemistry of the National Autonomous University of Mexico.
68. SINGH, P., SHARMA, M.C., JOSHI, K. C., BOHLMANN, F. 1985. Diterpenes derived from clerodanes from *Pulicaria angustifolia*. Phytochemistry 24: 190-192.
69. LAGUNES, T.A., ARENAS, C., RODRIGUEZ, C. 1985. Extractos acuosos y polvos vegetales con propiedades insecticidas. Consejo Nacional de Ciencia y Tecnología. (ed.), México. p. 203.

Chapter Thirteen

SESQUITERPENE LACTONES REVISITED

Recent Developments in the Assessment of Biological Activities and Structure Relationships

Robin J. Marles,[1] Liliana Pazos-Sanou,[2] Cesar M. Compadre,[3] John M. Pezzuto,[4] Elżbieta Błoszyk,[5] and J. Thor Arnason[6]

[1] Botany Department, Brandon University, Brandon, MB, Canada R7A 6A9

[2] Laboratory of Biological Assays, School of Medicine University of Costa Rica, San José, Costa Rica

[3] Department of Biopharmaceutical Sciences, University of Arkansas for Medical Sciences Little Rock, Arkansas 72205-7122

[4] Program for Collaborative Research in the Pharmaceutical Sciences, Department of Medicinal Chemistry and Pharmacognosy, College of Pharmacy, University of Illinois at Chicago, Chicago, Illinois 60612

[5] Department of Medicinal Plants, University of Medical Sciences Mazowiecka 33, 60-623 Poznań, Poland

[6] Ottawa-Carleton Institute of Biology, University of Ottawa Ottawa, Ontario, Canada K1N 6N5

INTRODUCTION

Sesquiterpene lactones are plant constituents often possessing a bitter taste which are common in most tribes of the Asteraceae (Compositae) and also are found in at least 14 other angiosperm families and to a limited extend in gymnosperms, liverworts and fungi.[1] There are more than 4,000 known structures in this class of plant secondary metabolites, and the bioactivity of some of these compounds has been reviewed previously, e.g. by Picman,[1] Ivie and Witzel,[2] and Rodriguez, Towers, and Mitchell.[3] Planta Medica has announced the planned publication of an updated review by Rodriguez. These reviews have reported a wide variety of biological effects, some of which are listed in Table 1. Sesquiterpene lactone mechanisms of action also have been investigated extensively, and some of their reported molecular targets include acid phosphatase, aryl sulfatase, cathepsins, cyclooxygenase, DNA, DNA polymerase, glycogen synthase, 5-lipoxygenase, phosphofructokinase, phospholipase A_2, reduced glutathione, and thymidylate synthetase.[1-3]

We became interested in this class of compounds in part because of their identification as the primary active principles of the medicinal plant "feverfew" (*Tanacetum parthenium* (L.) Schultz-Bip., Asteraceae), the leaves of which have been clinically proven to reduce the incidence and severity of migraine headaches.[4,5] As part of an effort to set quality control guidelines for the registration by Health and Welfare Canada of tablets containing powdered feverfew leaf as a nonprescription migraine prophylactic drug, we developed a practical bioassay using bovine platelets to measure the ability of crude feverfew extracts or purified constituents to inhibit the release of serotonin (5-hydroxytryptamine, 5-HT), which is believed to play roles both as a vasoactive agent and neurotransmitter in the etiology of migraine.[6] We showed that while bioactivity was closely correlated (r = 0.95) to the content of one germacranolide sesquiterpene lactone, parthenolide (**1** in Fig. 1), other sesquiterpene lactones present in feverfew contributed to 5-HT release inhibition, and hence standardization should include both biological and phytochemical assays. We also showed that the content of parthenolide and bioactivity varied enormously in feverfew plants grown under identical conditions from seeds of different geographical origins. This suggested the existence of several chemical races of *Tanacetum parthenium*, further supported by the finding that plants devoid of parthenolide often had elevated levels of the biogenically-related eudesmanolides, reynosin (**19**) and santamarin (**23**).

Further molecular-based bioassays and structure-activity relationship analyses were undertaken to learn more about the mechanism of sesquiterpene lactone 5-HT release inhibition and to explore the possibility that more active and more selective (i.e. less cytotoxic) derivatives might be found. We hoped to develop predictive models of the structural features necessary for selectivity. We report here our progress along these lines of investigation.

Table 1. A selection of sesquiterpene lactone bioactivities reported in Picman's review[1]

Allelopathic	Antihyperlipidemic	Genotoxic
Allergenic	Antiinflammatory	Hypoglycemic
Analgesic	Antiprotozoan	Hypotensive
Anthelmintic	Antitumor	Insecticidal
Anticholinergic	Antiulcerogenic	Molluscicidal
Antiarthritic	Cardiotoxic	Mutagenic
Antiasthmatic	Cholinergic	Piscicidal
Antibacterial	CNS stimulant	Schistosomicidal
Antifeedant (insect, mammal)	Cytotoxic	Sedative
Antifungal	Diuretic	Toxic acutely to mammals

MATERIALS AND METHODS

Sesquiterpene Lactones

Parthenolide was obtained by extraction and isolation from an authenticated commercial bulk sample of *T. parthenium* leaf powder, using the methods of Bohlmann and Zdero.[7] It was identified spectroscopically and by comparison with an authentic sample obtained from the collection of the Chemistry Department, Louisiana State University. The exocyclic methylene of parthenolide was selectively reduced by H_2 and a Pd catalyst to afford 11β,13-dihydroparthenolide (**2**). The 1,10-dihydro derivative of parthenolide (**4**) was prepared by first protecting the exocyclic methylene through preparation of the dimethylamine adduct, then catalytic reduction of the 1,10-double bond followed by oxidative regeneration of the exocyclic methylene using MeI and Ag_2O. Other naturally-occurring and semisynthetic sesquiterpene lactones were donated by a number of colleagues who are named in the Acknowledgements.

HPLC Purity Check

Purity of the compounds was verified by hplc prior to bioassay. Ethanolic solutions were analyzed by direct injections (20 μl) onto a reversed-phase hplc system consisting of a Brownlee Spheri-10 RP-18 column (250 mm x 4.6 mm x 10 μm), an SP8700 pump (Spectra Physics, Mississauga, Ont.), a Valco CW6 injector (Valco Instruments Co. Inc., Houston, TX), and a Pharmacia-LKB 214O RSD diode-array detector (Pharmacia-LKB, Baie d'Urfe, Que.), set for detection at 210 nm. Elution was isocratic with a mobile phase of $CH_3CN{:}H_2O$ (45:55) at a flow rate 2.0 ml/min (900 psi). Quantitative results were obtained using parthenolide (t_R = 5.9 min) as an external standard in CH_3CN solutions ranging in concentration from 50 to 2000 μg/ml ($r^2 > 0.999$).[6] Quantitative data evaluation was conducted on an IBM-AT computer using Pharmacia-LKB Wavescan-EG and

Nelson Analytical 2600 Chromatography Software. All chemicals were analytical grade and all solvents were chromatography grade (BDH).

Bioassays

Serotonin release inhibition was determined as described previously.[6] Briefly, venous blood was collected from cattle during commercial slaughtering and treated with trisodium citrate anticoagulant and [^{14}C]-5-HT. The radio-labelled 5-HT is rapidly taken up into the platelets by high affinity active transport. Platelet-rich plasma (PRP) was prepared by centrifugation and diluted to a standard concentration of platelets with autologous platelet-poor plasma. Aliquots of PRP were preincubated (37°C, 5% CO_2) with a compound dissolved in phosphate-buffered saline, positive control (100 μM parthenolide), or negative control (phosphate-buffered saline), and then adenosine diphosphate was added to stimulate platelet aggregation and degranulation (5-HT release). The reaction was stopped after 6 min with ice-cold acetylsalicylic acid, the mixture centrifuged, and an aliquot of the supernatant subjected to scintillation counting. The IC_{50} of [^{14}C]-5-HT release and its 95% confidence interval were calculated by probit analysis from the percentage of release in the negative control versus log concentration. Each experiment included three treatment replicates, each experiment was repeated twice, and the values were averaged.

Cytotoxicity of a selection of sesquiterpene lactones against the KB-3 cell line was assessed by methods previously reported.[8] Parthenolide was assayed for mutagenic and coincidental antibacterial activity in a forward mutation assay utilizing *Salmonella typhimurium* Strain TM677.[9] It was tested for inhibition of cyclooxygenase from sheep seminal vesicles according to the methods of Kulmacz and Lands.[10] Parthenolide also was tested for inhibitory activity against protein kinase C type I from bovine brain,[11-13] and against membrane-associated protein kinase C prepared from S49T-lymphoma cells, as measured by the incorporation of ^{32}P into a protein kinase C-selective peptide substrate, according to the methods of Chakravarthy et al.[14]

Structure-Activity Relationship Analysis

To facilitate obtaining a correct starting conformation for structural studies on the germacranolides, x-ray crystallographic coordinates[15] for the parent compound, parthenolide (**1**), were used to generate its structure. As a measure of hydrophobicity, the logarithms of the octanol/water partition coefficients (CLOGP), and as a measure of steric bulk, the molar refractivities (CMR), were calculated using the CLOGP3 and CMR modules of the program "MedChem" version 3.54, Daylight Chemical Information Systems, Inc., on a VAX-8530 computer. Comparative molecular field analysis and molecular surface mapping of hydrophobic potential were performed with the CoMFA and MolCad modules of the program Sybyl version 6.0a on a Silicon Graphics Indigo II computer.

RESULTS AND DISCUSSION

Serotonin Release Inhibition Structure-Activity Relationships

A total of 54 sesquiterpene lactones representing all the major skeletal classes were assayed for their ability to inhibit the release of serotonin (5-HT) from bovine platelets. The names, skeletal types, hydrophobicity (CLOGP) and steric bulk (CMR) are listed in Table 2. The micromolar concentration necessary for a 50% inhibition of 5-HT release (IC_{50}) for each compound and its 95% confidence limits are provided with the chemical structure of the compound in Figures 1 to 4, arranged by structural type to facilitate structure-activity comparisons.

We previously[6] reported a comparison between the 5-HT release inhibition activity of parthenolide (**1**, IC_{50} = 3.03 μ*M*) versus two of the most widely used migraine prophylactic drugs, verapamil hydrochloride (IC_{50} = 577.5 μ*M*) and propranolol hydrochloride (IC_{50} = >939.8 μ*M*). While these drugs operate by

Table 2. Physicochemical properties of the sesquiterpene lactones studied (structures are given in Figs. 1 - 4)

No.	Class	Compound name	CLOGP	CMR
1	Germacranolide: Germacrolide	Parthenolide	2.544	6.698
2	Germacranolide: Germacrolide	Parthenolide, 11β,13-dihydro-	2.561	6.648
3	Germacranolide: Germacrolide	Parthenolide, 1,10-epoxy-11β,13-dihydro-	0.390	6.566
4	Germacranolide: Germacrolide	Parthenolide, 1,10-dihydro-	3.088	6.724
5	Germacranolide: Germacrolide	Stizolicin	-0.516	9.563
6	Germacranolide: Germacrolide	Costunolide, 1,10-epoxy-	2.140	6.698
7	Germacranolide: Germacrolide	Ursiniolide A	2.691	10.380
8	Germacranolide: Germacrolide	Ursiniolide B	1.916	11.674
9	Germacranolide: Germacrolide	Salonitenolide	0.236	7.086
10	Germacranolide: Germacrolide	Cnicin	-0.104	9.798
11	Germacranolide: Germacrolide	Alatolide	0.282	9.095
12	Germacranolide: Germacrolide	Glaucolide A	-0.363	11.087
13	Germacranolide: Leucantholide	Cinerenin	-0.323	8.018
14	Germacranolide: Leucantholide	Cinerenin acetate	0.583	8.982
15	Germacranolide: Melampolide	Schkuhriolide	0.497	7.045
16	Germacranolide: Melampolide	Melampodin A	0.286	11.001
17	Germacranolide: Melampolide	Enhydrin	1.233	11.027
18	Germacranolide: Melampolide	Tatridin B	0.136	7.086
19	Eudesmanolide	Reynosin	1.963	6.803
20	Eudesmanolide	Reynosin, 11β,13-dihydro-	1.887	6.753
21	Eudesmanolide	Reynosin-8β-O-epoxyangelate	1.525	9.287
22	Eudesmanolide	Reynosin-8β-O-2,3-dihydroxy-2-methylbutyrate	-0.117	9.617
23	Eudesmanolide	Santamarin	1.963	6.803
24	Eudesmanolide	Santamarin, 11β,13-dihydro-	1.887	6.753

(continued)

Table 2. *Continued*

No.	Class	Compound name	CLOGP	CMR
25	Eudesmanolide	Santamarin-8β-O-epoxyangelate	1.525	9.287
26	Eudesmanolide	Santamarin-8β-O-(2-hydroxyethyl)acrylate	0.671	9.515
27	Eudesmanolide	Santamarin, 3,4-*cis*-α-epoxy-8β-epoxyangeloyloxy-	-0.644	9.205
28	Eudesmanolide	Arbusculin B, 1β-hydroxy-8β-epoxyangeloyloxy-	1.525	9.287
29	Eudesmanolide	Santonin, α-	1.611	6.762
30	Eudesmanolide (non-lactone)	Vachanic acid	2.974	7.089
31	Eudesmanolide (non-lactone)	Vachanic acid methyl ester	3.560	7.553
32	Eudesmanolide (linear)	Alantolactone, iso-	4.050	6.650
33	Eudesmanolide (linear)	Asperilin	1.963	6.803
34	Eudesmanolide (linear)	Pulchellin C	0.729	6.956
35	Eudesmanolide (linear)	Telekin	1.963	6.803
36	Guaianolide	Grossheimin	-0.586	6.839
37	Guaianolide	Grandolide, 3-oxo-	-0.662	6.788
38	Guaianolide	Lippidiol, 8-epi-iso-	-0.200	6.906
39	Guaianolide	Repin, 15-deoxy-	0.462	8.798
40	Guaianolide	Repin	-0.872	8.799
41	Guaianolide	Centaurepensin	-0.381	10.136
42	Guaianolide	Artecanin	-1.579	6.697
43	Guaianolide	Xerantholide	1.901	6.762
44	Pseudoguaianolide: Helenanolide	Helenalin	0.536	6.915
45	Pseudoguaianolide: Helenanolide	Linifolin A	1.408	7.878
46	Pseudoguaianolide: Helenanolide	Tenulin	-0.031	7.221
47	Pseudoguaianolide: Helenanolide	Aromaticin, 6α-hydroxy-2,3-dihydro-	0.772	6.864
48	Pseudoguaianolide: Helenanolide	Geigerinin	1.273	6.982
49	Pseudoguaianolide: Ambrosanolide	Burrodin	0.772	6.864
50	Pseudoguaianolide: Ambrosanolide	Inuchinenolide C	0.145	9.062
51	Pseudoguaianolide: Ambrosanolide	Parthenin	0.892	6.915
52	Pseudoguaianolide: Ambrosanolide	Coronopilin	1.188	6.864
53	Pseudoguaianolide: Ambrosanolide	Confertiflorin	1.280	7.828
54	Pseudoguaianolide: Secoambrosanolide	Psilostachyin A	1.665	7.101

different mechanisms, i.e. verapamil is a calcium channel blocker and propranolol is a β-adrenergic blocker, they also have been reported to inhibit platelet aggregation and 5-HT release.[16] As can be seen from the IC_{50} values given in Figures 1 to 4, many of the sesquiterpene lactones studied here are 100 - 200 times more active in this bioassay than verapamil hydrochloride. The good range of values seen for our series of sesquiterpene lactones facilitates the determination of structure-activity relationships.

For our series of compounds, the germacranolides were in general the most active, although good levels of activity (same order of magnitude as parthenolide **1** = 3.03 μ*M*) were found for some eudesmanolides (**22, 25**), guaianolides (**39 - 41**), and pseudoguaianolides (**44, 53**). The lack of an α,β-unsaturated moiety in

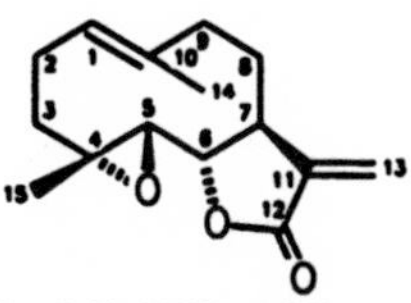

1 = 3.03 (1.80 - 4.27) 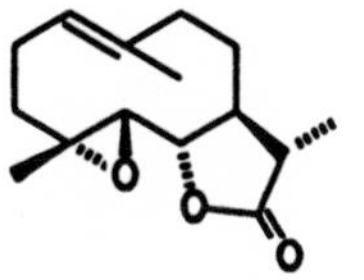 **2** = >399.5 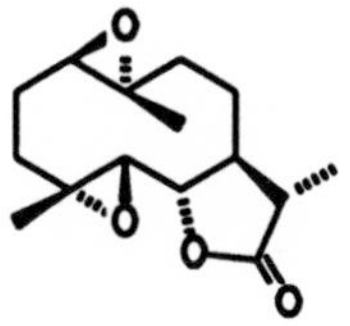 **3** = >413.0

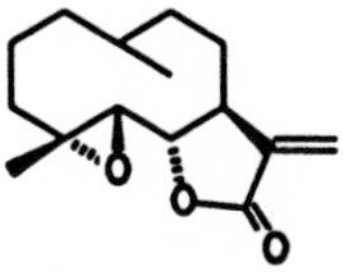

4 = 40.58 (24.73 - 85.20) 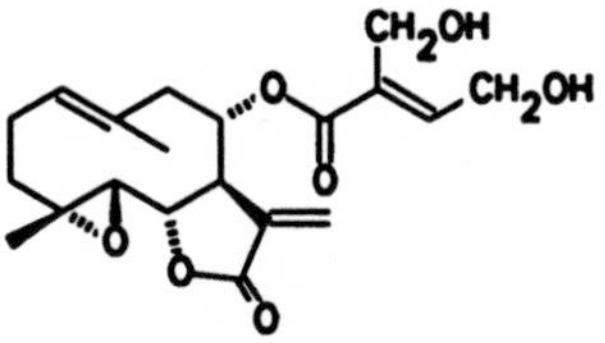 **5** = 5.82 (4.20 - 7.96) 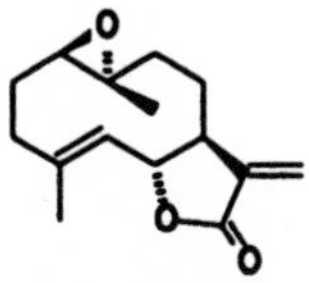 **6** = 121.3 (85.42 - 179.3)

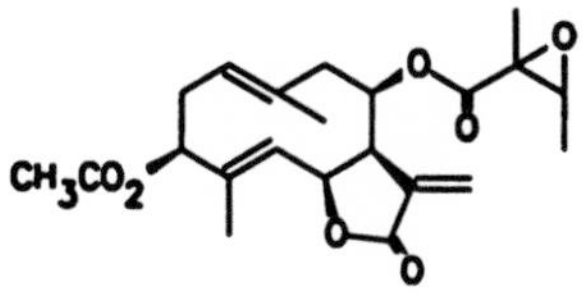

7 = 1.78 (1.11 - 2.73) 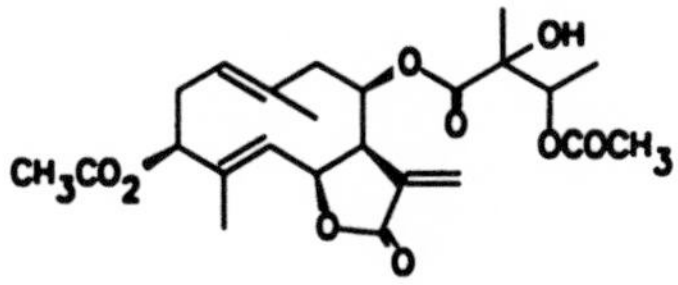 **8** = 5.30 (3.52 - 7.81)

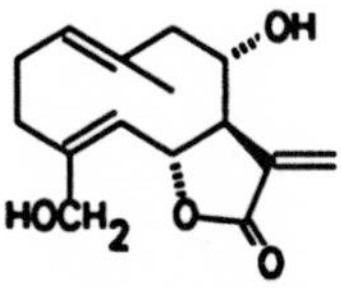

9 = 10.15 (7.30 - 13.95) 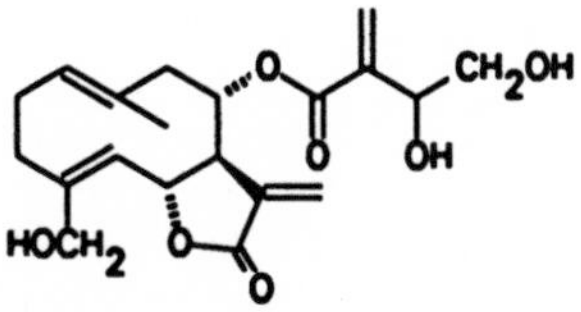 **10** = 3.52 (2.49 - 4.87) 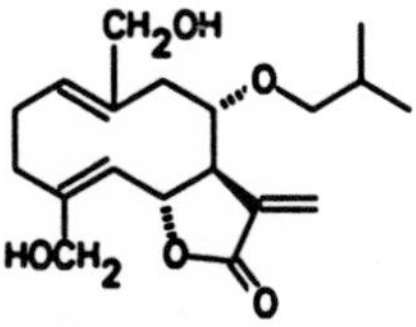 **11** = 5.77 (4.45 - 7.46)

Figure 1. Germacranolides and their 5-HT release inhibition IC_{50} in μM (95% confidence limits). Bold numbers refer to compounds named in Table 2.

all cases (**2, 3, 20, 24, 37,** and **38**) completely destroyed activity, as was expected from the previous findings of structure activity studies done for sesquiterpene lactone cytotoxicity.[17-24]

The mechanism of action established by these authors for sesquiterpene lactone cytotoxicity involves the functional group $O=C-C=CH_2$, which is most

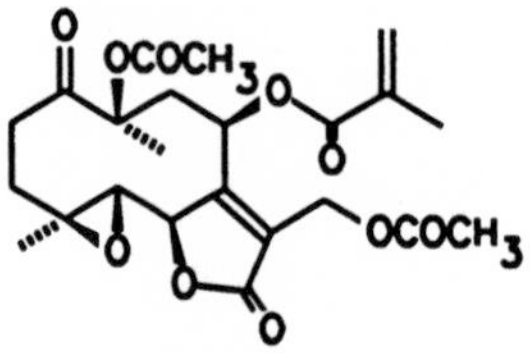

12 = 20.92 (16.57 - 25.92) 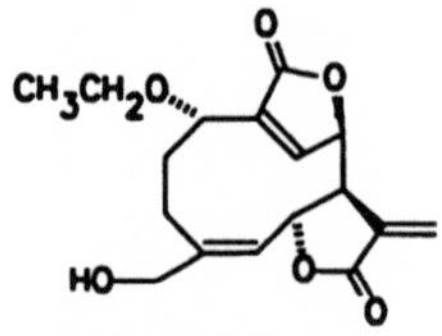 **13** = 3.52 (2.55 - 4.77) 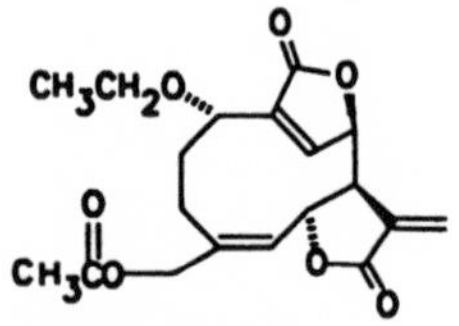 **14** = 2.04 (1.53 - 2.69)

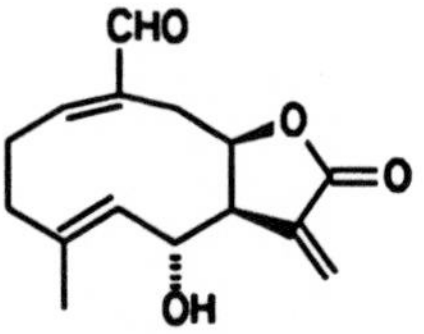

15 = 277.7 (228.8 - 342.3) 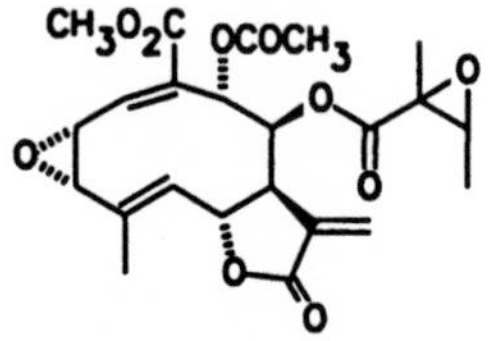 **16** = 4.68 (3.28 - 6.49) 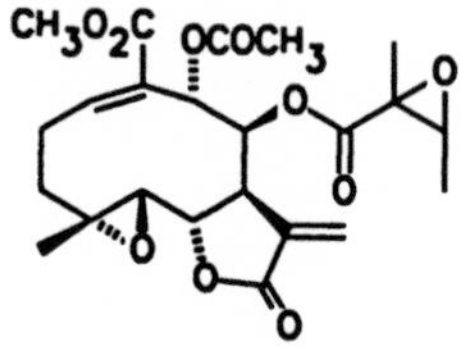 **17** = 8.76 (6.79 - 11.12)

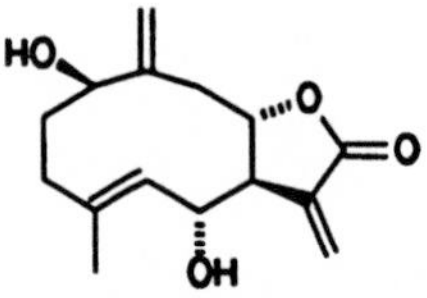

18 = 26.97 (19.05 - 38.70)

Figure 1. *Continued.*

commonly part of the α-methylene-γ-lactone, but may be present in a β-unsubstituted cyclopentenone or other ester or ketone, reacting by a Michael-type addition to a biological nucleophile, particularly the sulfhydryl groups of reduced glutathione and L-cysteine. Thus, these compounds inhibit cellular enzyme activities and metabolism leading to cell death. The same structural requirements also have been reported for sesquiterpene lactone contact hypersensitivity,[25-28] antiinflammatory,[29,30] and antihyperlipidemic[31] activities. In each case thiol-bearing enzymes or metabolically important proteins appear to be the primary targets through alkylation by a Michael addition, with the exocyclic α-methylene of the γ-lactone being the most active form of this functional group, although other forms such as the endocyclic methylene of cyclopentenone contribute to the activity.

We too have observed this enhancement of activity with 5-HT release inhibition, e.g. for germacrolides **5, 7 - 12**, and melampolides **16 - 18** having

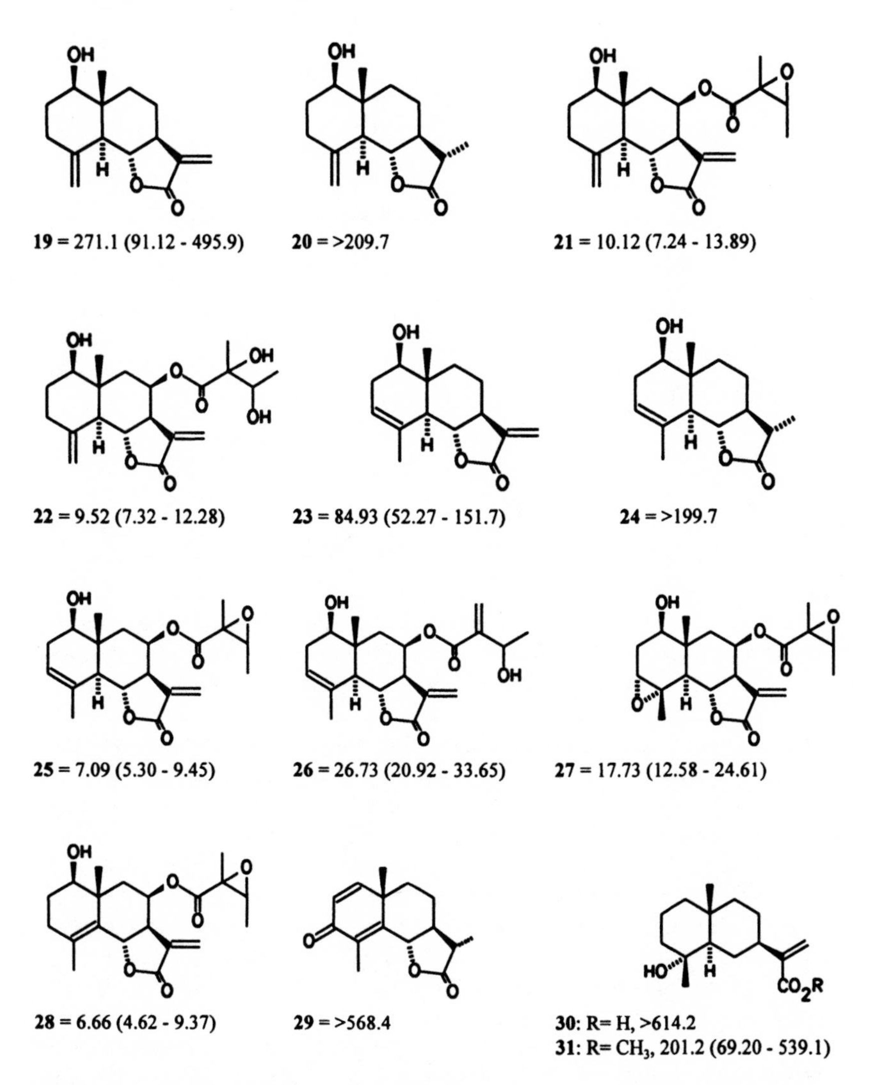

Figure 2. Eudesmanolides and their 5-HT release inhibition IC_{50} in μM (95% confidence limits). Bold numbers refer to compounds named in Table 2.

oxygenated substituents at the 8- and/or 15-positions, and for leucantholides **13** and **14** having an additional unsaturated lactone . In the eudesmanolide series it is interesting to note that two functional groups were necessary for activity: the α-methylene-γ-lactone and also a hydroxy group in the 1β position, e.g. compounds capable of Michael addition but lacking a hydroxyl group (**32**)

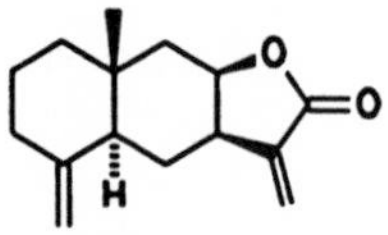

32 = >516.5

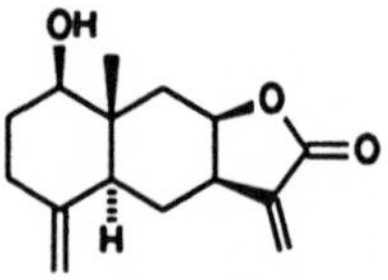

33 = 57.62 (42.85 - 76.80)

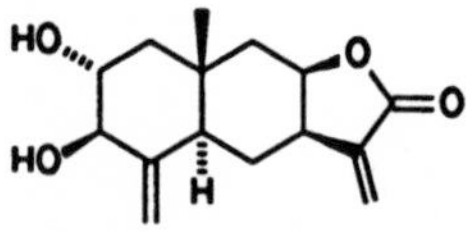

34 = >416.2

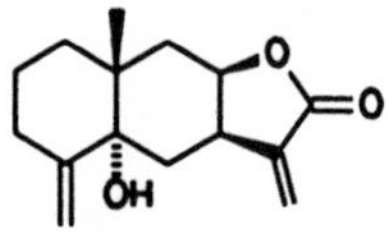

35 = >563.8

Figure 2. *Continued.*

or with hydroxyl groups in other positions (**34, 35**) were inactive. This may relate to hydrogen bonding requirements for appropriate fit into a receptor molecule. In the guaianolide series, oxygenated substituents in the 8-position clearly enhanced activity, e.g. compounds **39** - **41** versus **42**. In the pseudoguaianolide series, the presence of an additional potentially alkylating group, the cyclopentenone, clearly enhanced the activity of **44** over that of **47** - **49**, although an oxygenated substituent at the 8-position (**53**) also provided a good level of activity.

Notably, possessing an α-methylene-γ-lactone is not sufficient for 5-HT release inhibition, as shown by the eudesmanolides **32, 34,** and **35**, and guaianolide **42**. Neither is an α,β-unsaturated cyclopentenone, as in eudesmanolide **29**. Other structural features clearly play a role in the observed range of activities. It is the role of these other structural features that is of particular interest to us.

Kupchan et al.[18] found a direct linear relationship between the level of cytotoxicity of α-methylene-γ-lactone containing sesquiterpene lactones and their hydrophobicity, as determined experimentally from octanol/water partitioning. Thus, more fat-soluble analogues were more cytotoxic than closely related but more water-soluble analogues. This did not correlate with reactivity toward cysteine but rather with the ability of the compound to reach its receptor compartment, and the authors suggested that probably steric factors, and perhaps also electronic factors, contributed to the remaining variation in cytotoxicity.

We have previously examined the correlation between HPLC-determined hydrophobicity and the toxicity of sesquiterpene lactones to mosquito larvae, for which the relationship was not strong.[32] We have had greater success correlating photoactivated thiophene cytotoxicity, antiviral activity, and toxicity to insect and

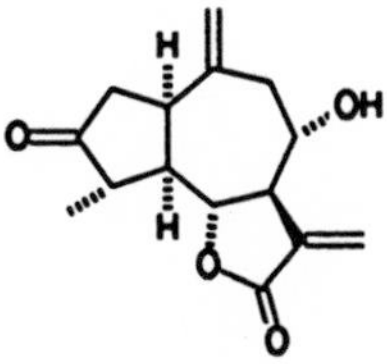

36 = 22.18 (15.88 - 31.12) 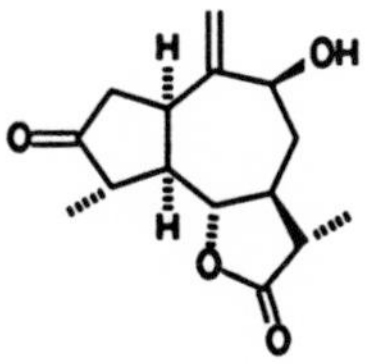 **37** = >454.0 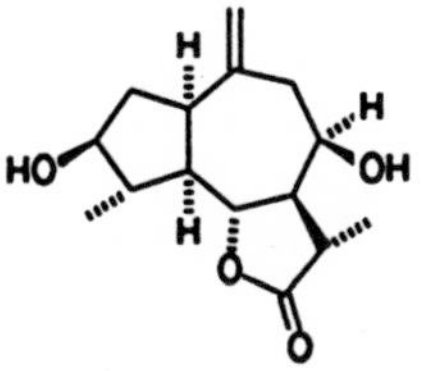 **38** = >488.1

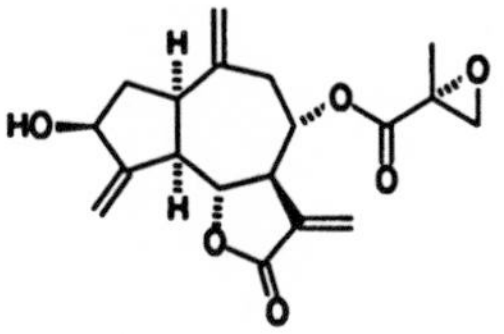

39 = 6.32 (4.65 - 8.47) 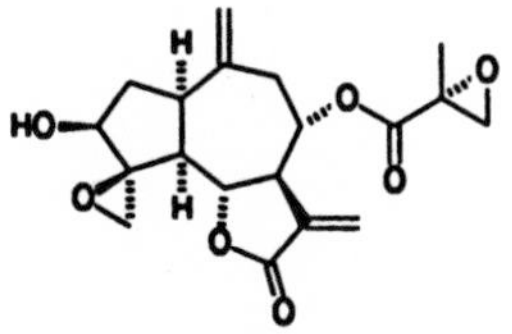 **40** = 9.24 (7.14 - 11.80) 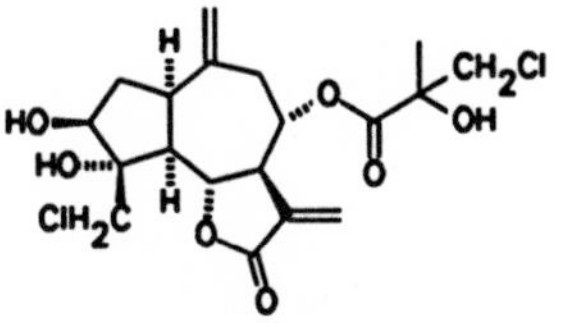 **41** = 6.06 (4.62 - 7.91)

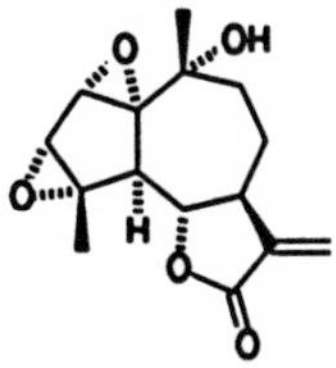

42 = >134.7 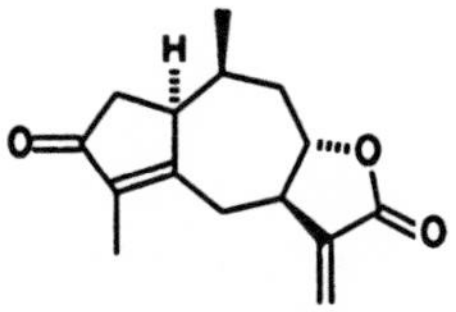 **43** = 101.4 (81.26 - 124.4)

Figure 3. Guaianolides and their 5-HT release inhibition IC_{50} in μM (95% confidence limits). Bold numbers refer to compounds named in Table 2.

crustacean larvae with their hydrophobicity and rate of singlet oxygen production.[33-35] We used a computer program that calculated hydrophobicities on the basis of well-established rules of the additivity of substituents,[36] and showed that the values thus obtained correlated well with experimentally determined values.[33] Table 2 lists the calculated CLOGP values for each of the sesquiterpene lactones tested, as a measure of their hydrophobic properties. While recognizing that experimentally determined log P values would probably be more reliable if measured in a reproducible manner, we hoped to achieve some success with the calculated values for sesquiterpene lactones. The correlation between the few available reported experimental log P values[18, 32] and our CLOGP values (Table 3) was highly significant ($r = 0.95$, $P < 0.0005$).

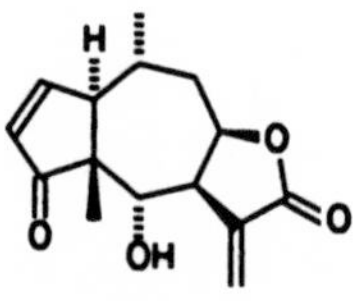

44 = 4.28 (3.20 - 5.64) 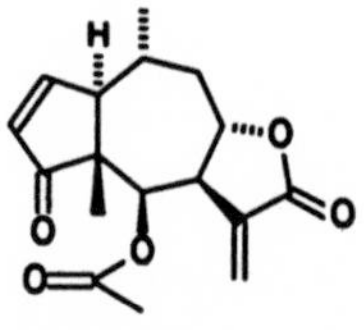 **45** = 11.44 (9.05 - 14.36) 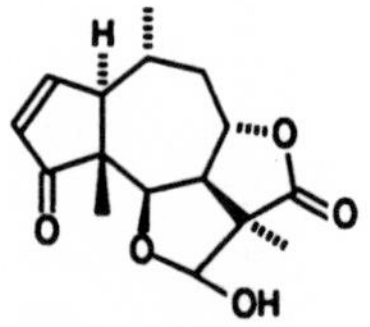 **46** = 361.3 (78.74 - ∞)

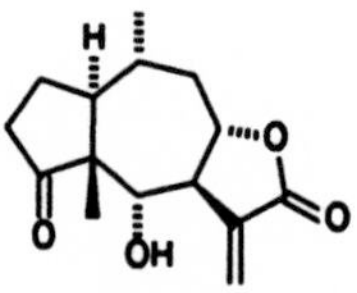

47 = 60.08 (38.52 - 96.47) 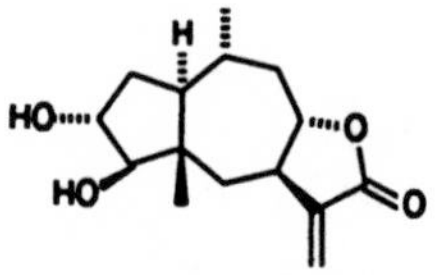 **48** = 73.43 (52.52 - 104.1) 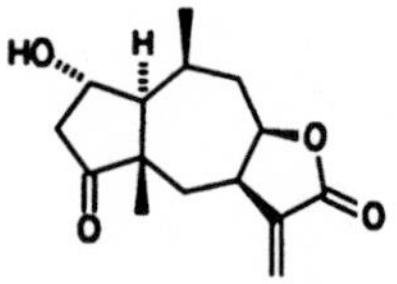 **49** = 251.4 (193.1 - 345.9)

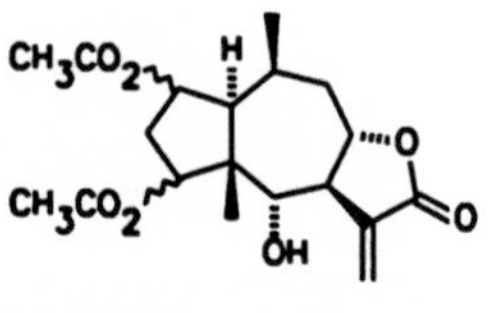

50 = 33.47 (24.62 - 45.28) 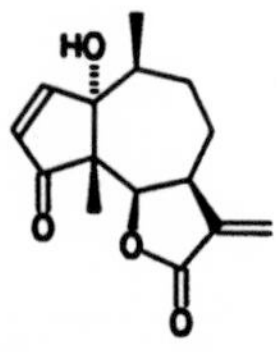 **51** = 129.3 (81.41 - 225.8) 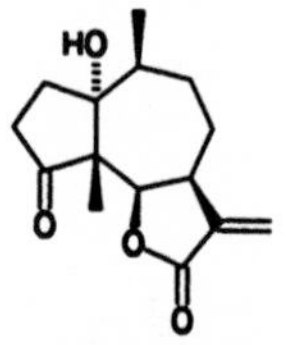 **52** = 248.9 (157.7 - 442.0)

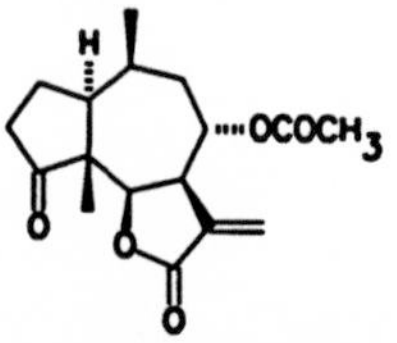

53 = 8.88 (6.39 - 12.25) 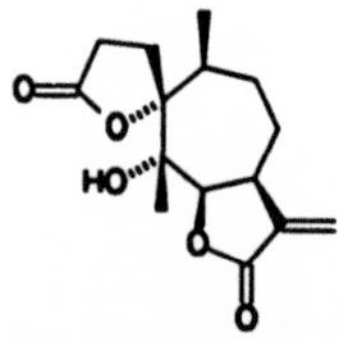 **54** = 44.06 (30.72 - 63.69)

Figure 4. Pseudoguaianolides and their 5-HT release inhibition IC_{50} in μM (95% confidence limits). Bold numbers refer to compounds named in Table 2.

One problem that was recognized in the calculation of CLOGP values for some of the sesquiterpene lactones was the possibility for hydrophilic overlap, as illustrated for stizolicin (**5**) in Figure 5. The long side chain at the 8-position can rotate about the ester bond and possibly overlap the ring structure to form hydrogen bonds between its hydroxy groups and the electronegative carbonyl or

Table 3. Comparison of calculated versus experimentally determined octanol/water partition coefficients

Compound Name	CLOGP	Exper. log P
Allantolactone	4.050	3.380[32]
Allantolactone, iso-	4.050	3.420[32]
Bipinnatin	1.188	1.620[32]
Coronopilin	1.188	0.830[18]
Helenalin	0.536	0.870[18]
Hymenolin	0.909	1.700[32]
Mexicanin I	0.536	0.360[18]
Parthenin	0.892	0.770[18]

(superscripts refer to literature source of experimental log P value)

epoxide oxygens. This would affect the hydrophobicity of the molecule if this were an energetically favorable conformation in solution. Germacranolides **7, 8, 10** and **12** are other examples of where this might be a source of error.

Examining the sesquiterpene lactones by subclass, class, or as a whole revealed no significant correlation ($P < 0.05$) of 5-HT release inhibition (as log $1/IC_{50}$) versus CLOGP. Still unconvinced that hydrophobicity plays no role in this bioactivity, we next used the MolCad molecular modelling program to examine the distribution of hydrophobic potential over the surface of some of the germacranolide molecules. Computer-generated and colored three-dimensional models revealed that molecules such as parthenolide (**1**), ursiniolide A (**7**), and salonitenolide (**9**) have a concave hydrophilic region adjacent to the lactone and functional groups such as the 8-epoxyangelate, 3-acetate, 8-hydroxy, and 15-hydroxymethyl groups. There is a convex hydrophobic region on the opposite side of the molecule (1, 2 and 9, 10, 14 positions). Saturation of the 11β,13 exocyclic methylene not only eliminates Michael addition capability but also introduces hydrophobicity into the previously hydrophilic region of the molecule, which might prevent compensatory

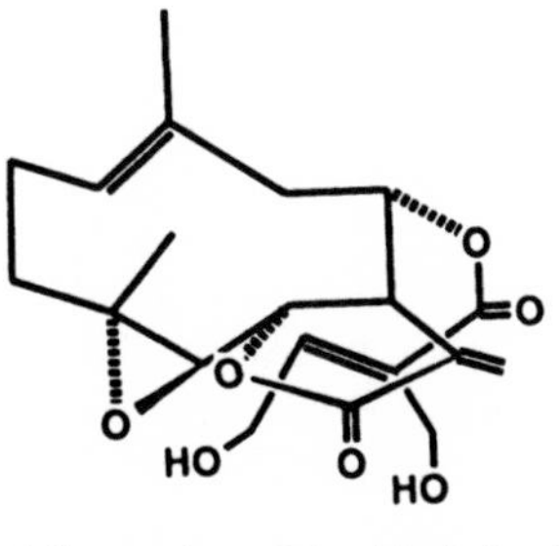

Figure 5. Hydrophilic overlap of the sidechain of stizolicin (**5**).

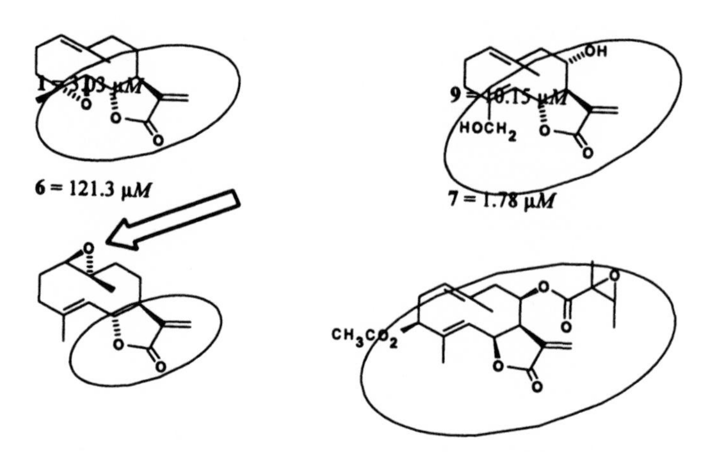

Figure 6. The effect on 5-HT release inhibition IC_{50} of having a hydrophilic region on the same (circled) or opposite (arrow) side of the molecule to the α,β-unsaturated lactone.

activity of other reactive groups such as a cyclopentenone (e.g. **29**). Introduction of a 1,10-epoxide disrupts the hydrophobic side of the molecule, possibly explaining the significantly lower activity of 1,10-epoxycostunolide (**6**). Thus, hydrophobicity does appear to play a significant role in 5-HT release inhibition by affecting possible interactions with a receptor molecule. Figure 6 illustrates in a two-dimensional, black and white representation, some of these principles.

These results suggest that the addition of hydrophilic substituents to the "lower" (as drawn here) portion of the germacranolide skeleton, e.g. positions 3, 4, 5, 8, or 15 on parthenolide, might lead to enhanced 5-HT release inhibition.

In order to examine other steric aspects of sesquiterpene lactone 5-HT release inhibition, we calculated the molar refractivity (CMR) for each of the compounds, as shown in Table 2. Regression analysis at the subclass, class, and overall level indicated that the eudesmanolides were the only class to be directly correlated with CMR ($n = 8$, $r = 0.89$, $P < 0.005$). These steric effects may be coincidental to the hydrophilic nature and overlap potential of the angelate-derived side chains at position 8. To examine the effect of substitution and derivatization on the ring conformation in the germacrolide subclass, the energy-minimized skeleta of compounds **1 - 4**, **6**, **9**, and **11** were overlaid (Fig. 7).

Figure 7 shows the remarkable stability of the germacrolide ring structure in the C6 to C2 region of the molecule. Only reduction of the 1,10-double bond (**4**) significantly changes the conformation of the molecule, and by relieving ring strain tends to deactivate the possibly reactive 4,5-epoxide, perhaps explaining the reduced activity of **4**.

To study further the three-dimensional aspects of sesquiterpene lactone activity, we next subjected our germacrolide series to Comparative Molecular

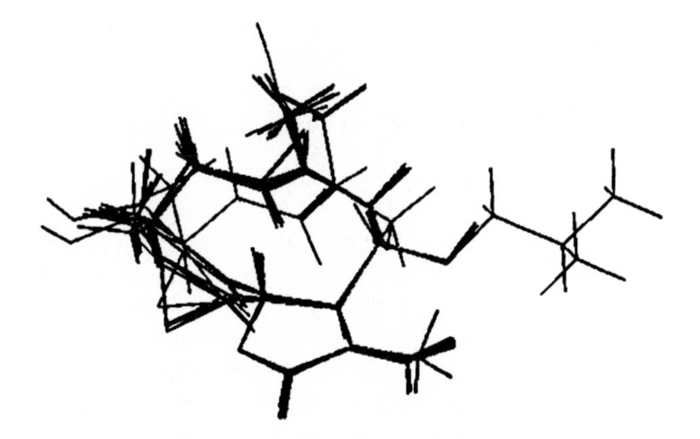

Figure 7. Overlay of the energy-minimized conformations for skeleta of germacrolides **1 - 4, 6, 9,** and **11**.

Field Analysis (CoMFA). The activity of parthenolide and its analogues in inhibiting 5-HT release most likely involves both covalent binding and noncovalent interactions with the target molecule. The CoMFA model provides a numerical estimate of the effects on binding and bioactivity of changes in steric bulk and electrostatic potential fields over the surface of the molecule. It also displays a qualitative graphical view of the most important three-dimensional aspects of the problem. The effects of changes in related structures are assessed by aligning the analogues atop each other by means of fitting several atoms along the common rings. The detailed quantitative analysis is not presented here, but trends suggested by the modelling will be described.

The CoMFA model of germacrolides suggests that their 5-HT release inhibition might be enhanced by significantly increasing steric bulk in the region of the 8α position, and to a lesser extent increasing steric bulk in the 7α region below the plane of the lactone. Steric bulk should be minimized in the regions of the 13-exocyclic methylene, 15-methyl, and 2α - 3α positions. Activity might be enhanced by significantly increasing the positive electrostatic potential in the regions of the 13-exocyclic methylene and 4,5-epoxide, and to a lesser extent in the 10-position. More negative electrostatic potential might be beneficial in the region of the 12- carbonyl and lactone oxygens, and significantly in the 2β region, to a lesser extent in the 2α, 3α, and 14-methyl positions. CoMFA thus provides a predictive model on which synthetic schemes may be based for optimal drug design, and also generates some understanding of the high degree of efficiency which evolution through natural selection has brought to the sesquiterpene lactone class of allelochemicals. For example, it confirms both the beneficial effects on activity of electron withdrawal from the exocyclic methylene through conjugation to the lactone carbonyl (**1**), and of appropriate substitution at the 15 (e.g. **9** - **11**) and 8 (e.g. **5**, **8** - **12**) positions.

Serotonin Release Inhibition versus Cytotoxicity

In the 1970's and early 1980's a great deal of attention was paid to the antitumor potential of sesquiterpene lactones. Interest lessened significantly with the discovery of the lack of predictable selectivity, since alkylation of reduced glutathione will kill any cell, and since many of these compounds have a significant allergenic potential. With the registration in Canada of Tanacet 125® as a *nonprescription* migraine prophylactic drug that must contain at least 0.2% parthenolide in 125 mg of standardized, certified feverfew leaf powder, interest has once again risen with regard to their therapeutic potential and possible adverse effects. The toxicology of standardized feverfew preparations has been well studied, and there have been remarkably few side effects, e.g. mild ulceration of the tongue in 11% of the patients participating in the clinical trial.[5] This may be a systemic effect related to the allergenic potential of sesquiterpene lactones, well known for compounds such as parthenolide (**1**), parthenin (**51**), helenalin (**44**) and coronopilin (**52**).[25-28] In tests for contact allergenicity of sesquiterpene lactones such as parthenolide, specifically sensitized T-lymphocytes were found to detect primarily the α-methylene-γ-lactone group, while structural differences between basic skeletons played only a secondary role.[28] Gastrointestinal upset also was noted in 7% of patients, with symptoms disappearing promptly when dosing was discontinued. Conversely, nausea associated with migraine attacks generally was significantly decreased.[5] These observations suggest that although parthenolide is a known allergen and cytotoxin, its adverse effects may be modified by the presence of other constituents in the crude drug or other factors yet to be established.

We have assessed parthenolide at concentrations up to 800 μM in a forward mutation assay utilizing *Salmonella typhimurium* and found it to be nonmutagenic. At the highest concentration (200 μg/ml) it showed significant antibacterial activity against the test organism. A comparison of migraine patients chronically using feverfew versus matched non-users found no significant change in the frequency of chromosomal aberrations or the frequency of sister chromatid exchange in circulating peripheral lymphocytes, and the mutagenicity of urine from feverfew-using patients was unaffected compared to that of non-users.[37] Despite these reassuring results, if pure sesquiterpene lactones are to be developed as therapeutic agents, care should be taken to minimize side-effects such as cytotoxicity.

We have assessed the cytotoxicity to the KB-3 human cell line of 20 sesquiterpene lactones, and contrasted their KB ED_{50} with the 5-HT release inhibition IC_{50}. The results are presented in Table 4, arranged in descending order of activity in the 5-HT assay. There was no correlation between cytotoxicity and 5-HT release inhibition, even when the extremes were omitted from the regression

(n = 17, r = 0.09). This suggested that a handle on selectivity could be found. Given that pure compounds must have a KB ED_{50} of 4.0 μg/ml or less to be considered active,[38] compounds **2, 5, 9,** and **12** would be considered inactive, and **13** marginal. Many of these compounds, including ursiniolide A (**7**) and B (**8**), parthenolide (**1**), helenalin (**44**), parthenin (**51**), alatolide (**11**), grossheimin (**36**), and salonitenolide (**9**), have previously been reported to be cytotoxic,[17-24] but it was important for comparative purposes to have them all done simultaneously under identical conditions.

Thirteen of the 20 compounds were more cytotoxic than parthenolide, notably alatolide (**11**) and helenalin (**44**), both of which are more hydrophilic than parthenolide. Due to the structural diversity of the series of compounds tested here, it was not surprising that no correlation was found between KB cytotoxicity and CLOGP for this series. However, this relationship was demonstrated previously by Kupchan et al.[18] Hansch et al.[39] stated that without convincing evidence to the contrary, drugs should be made as hydrophilic as possible without loss of efficacy to minimize side effects such as CNS neurotoxicity. Thus, to increase the selectivity of potential antimigraine drugs and to minimize toxicity, sesquiterpene lactones with hydrophilic substituents in the 8- and 15-positions would seem to be a desirable goal, although other structural and physicochemical aspects also will have to be considered. Salonitenolide (**9**) in particular shows less

Table 4. A comparison of sesquiterpene lactones' serotonin (5-HT) release inhibition with their cytotoxic activity against the KB cell line

No.	Compound name	5-HT IC_{50} μM	KB ED_{50} μM
7	Ursiniolide A	1.78	4.70
14	Cinerenin acetate	2.04	4.69
1	Parthenolide	3.03	12.08
13	Cinerenin	3.52	13.74
44	Helenalin	4.28	1.62
16	Melampodin A	4.68	4.32
8	Ursiniolide B	5.30	3.87
11	Alatolide	5.77	1.49
5	Stizolicin	5.82	13.48
17	Enhydrin	8.76	3.88
53	Confertiflorin	8.88	4.24
9	Salonitenolide	10.15	43.13
12	Glaucolide A	20.92	12.06
36	Grossheimin	22.18	6.10
18	Tatridin B	26.97	5.30
33	Asperilin	57.62	6.04
6	Costunolide, 1,10-epoxy-	121.2	8.46
51	Parthenin	129.3	6.86
15	Schkuhriolide	277.7	13.72
2	Parthenolide, 11β,13-dihydro-	>399.5	>79.89

cytotoxicity while maintaining a respectable level (within one order of magnitude of parthenolide) of 5-HT release inhibition, and may therefore provide a lead to future drug development.

Molecular Targets of Serotonin Release Inhibition

Although we have discussed a great many molecular targets for sesquiterpene lactones and have described physicochemical features with the potential of making them more selective migraine prophylactics, the receptor molecule for 5-HT release inhibition is not yet known. Michael addition to thiols is clearly important to the inhibitory effects of parthenolide or feverfew leaf extract on platelets. Addition of cysteine or other sulfhydryl-containing molecules to *in vitro* assays prevented parthenolide's inhibition of platelet secretory activity. Parthenolide also dramatically decreased the number of acid-soluble sulfhydryl groups (mainly reduced glutathione) in platelets. Formation of disulfide-linked protein polymers was observed in feverfew-treated platelets exposed to aggregating agents, although feverfew did not directly induce disulfide linking.[40] However, the glutathione-depleting agents iodoacetamide and 1-chloro-2,4-dinitrobenzene had little effect on platelet behaviour,[40] so we hypothesize that alkylation of reduced glutathione is an undesirable side-reaction leading to cytotoxicity and not the primary mechanism of parthenolide's therapeutically useful 5-HT release inhibition.

Platelet secretion induced by phorbol-12-myristate-13-acetate is more sensitive to inhibition by feverfew extracts and parthenolide than secretion induced by arachidonic acid.[40] Granule secretion in platelets is accompanied by phospholipid turnover resulting in diacylglycerol formation and activation of protein kinase C, accompanied by increases in the concentration of intracellular free calcium. Feverfew extracts have been shown not to affect calcium-mediated aggregation and secretion.[41] This led to the hypothesis that parthenolide may be interacting with the protein kinase C pathway.[42] We have now established that parthenolide neither activates nor inhibits protein kinase C type I from bovine brain preparation at concentrations of up to 200 μg/ml (805 μ*M*), nor does it affect the activity of membrane-associated protein kinase C in isolated S49T-lymphoma cell membranes.

It has been reported previously that extracts of feverfew do not inhibit cyclooxygenase acitivty,[43] and we have confirmed that parthenolide does not inhibit cyclooxygenase activity with enzyme derived from sheep seminal vesicles. Current evidence suggests that sesquiterpene lactones inhibit the release of arachadonic acid from membrane phospholipid stores rather than its conversion into thromboxane B_2 via the cyclooxygenase pathway. Prostaglandin synthetase has been shown to be inhibited by constituents of feverfew, including parthenolide.[44] Human phospholipase A_2 activity has been shown to be strongly inhibited *in vitro* by feverfew leaf extracts.[45,46] This activity is probably due to sesquiterpene lactone alkylation of sulfhydryl groups essential for the regulation of phospholipase A_2 activity in platelet membranes.[47] However, Ysrael and Croft[48] found that the α-methylene-γ-lactone containing germacrolide, scandenolide, showed only a weak inhibitory effect on phospholipase A_2 but showed a pronounced dose-dependent inhibition of 5-lipoxygenase, resulting in inhibition of the formation of its products leukotriene B_4, 5-hydroxyeicosatetraenoic acid (5-HETE) and platelet activating factor (PAF). In contrast, Sumner et al.[49] found that parthenolide and feverfew leaf extracts both inhibited 5-lipoxygenase and phospholipase A_2, and that the crude extract contained non-α-methylene-γ-lactone containing substituents that demonstrated potent dose-dependent inhibition of both thromboxane B_2 and leukotriene B_4 synthesis. Feverfew extracts from fresh leaves and parthenolide have been shown to cause irreversible time- and dose-dependent inhibition of the contractile response of aortic rings to a variety of receptor-acting agonists, but dried feverfew extracts devoid of parthenolide or other α-methylenebutyrolactones elicited potent and sustained aortic smooth muscle contractions not blocked by the specific 5-HT_2 receptor antagonist ketanserin.[50] These observations provide further arguments for the use of bioassays in addition to phytochemical assays to standardize registered crude drugs such as Tanacet 125®.

Sesquiterpene lactones have thus been shown to interact with a number of thiol-bearing enzymes in the platelet membrane. Platelets have been widely used as models for neurons in pharmacological research due to a number of similarities in functions and reactivity including 5-HT receptors, uptake and turnover, and their greater ease of isolation and use *in vitro*.[51] Platelets are the source of most of the circulating plasma 5-HT, levels of which are significantly elevated in migraine sufferers, who have continuously activated platelets.[52] However, while the role of 5-HT as a vasoactive agent and neurotransmitter in migraine etiology is fairly well understood[53,54] and current antimigraine therapeutic agents such as sumatriptan interact with 5-HT receptors,[55] the exact relationship between effects of sesquiterpene lactones on platelets *in vitro* and their functions in migraine prophylaxis *in vivo* are not clear. Further research on the mechanism of 5-HT release inhibition by parthenolide and analogues is underway.

CONCLUSIONS

Sesquiterpene lactones from every structural class possess serotonin release inhibition activity. The α-methylene-γ-lactone functional group is essential but not sufficient for activity, and other functional groups contribute to the level of activity, particularly when substituted in an appropriate region of the molecule. Measurements of the total hydrophobicity and steric bulk do not correlate well with activity. However, molecular modelling of the distribution of hydrophobic potential over the surface of the molecule reveals a significant bifacial hydrophobic/hydrophilic character in some of the most active molecules. For example, it was found that by increasing hydrophilicity in particular regions of the germacranolide skeleton, good activity was obtained while minimizing the cytotoxicity of the analogues. Cytotoxicity does not correlate with 5-HT release inhibition. Comparative Molecular Field Analysis helps to explain how steric bulk and electrostatic potential at particular points on the skeleton of the germacranolides affect their activity. Parthenolide also was found to be nonmutagenic and coincidentally antibacterial. It is inactive as a modulator of protein kinase C and cyclooxygenase activity. The molecular target for sesquiterpene lactone inhibition of 5-HT release may be a platelet membrane lipase, but this is not known for certain.

Structure-activity relationship modelling of sesquiterpene lactones has helped us to make predictions about structural features that will benefit the design of more selective migraine prophylactic compounds. Specifically, we predict that optimal activity should be seen from germacrolides with a hydrophilic surface adjacent to the α-methylene-γ-lactone and a hydrophobic surface around the 2, 1-10 double bond, and 14 positions, with contributions from a bulky hydrophilic group at the 8α-position and a hydroxy group in the 3β or 15 positions. Michael addition type reactions are responsible in part for both the desirable 5-HT release reaction involving a membrane enzyme and for the undesirable alkylation of reduced glutathione and other thiol-containing molecules in the cytoplasm leading to cell death. Therefore the compound should be optimally and regiospecifically hydrophilic in order to maximize interaction with the membrane receptor and minimize passage of the compound through the membrane and into the cell.

ACKNOWLEDGMENTS

This work is the product of an extensive international collaborative network. The authors would like to thank the following contributors of natural or semisythetic sesquiterpene lactones: B. Drożdż, G. Nowak, and H. Grabarczyk of the Department of Medicinal Plants, University of Medical Sciences, Poznań, Poland; W. Kisiel, Division of Phytochemistry, Institute of Pharmacology, Polish Academy of Sciences, Kraków, Poland, G. Appendino of the Università degli Studi di Torino, Dipartimento di Scienza e Tecnologia del Farmaco, Torino, Italy; A. Romo de Vivar, Instituto de Química, Universidad Nacional Autonoma de México, Mexico; N. Fischer, Department of Chemistry, Louisiana State University, Baton Rouge, Louisiana, USA; T. Waddell, Department of Chemistry, University of Tennessee at Chattanooga, Chattanooga, Tennessee, USA; and J. Kaminski, C. Soucy-Breau, and T. Durst, Department of Chemistry, University of Ottawa. We would like to acknowledge the assistance of: R.L. Compadre, Department of Biopharmaceutical Sciences, University of Arkansas for Medical Sciences, Little Rock, Arkansas, USA, for assistance with structure-activity relationship analysis; L. Shamon, N. Suh, H. Chai, and M.-S. Jang, Program for Collaborative Research in the Pharmaceutical Sciences, Department of Medicinal Chemistry and Pharmacognosy, College of Pharmacy, University of Illinois at Chicago, Chicago, Illinois, U.S.A., for performing enzyme and cytotoxicity assays; J. Durkin, Cell Systems Section, Institute of Biological Sciences, National Research Council of Canada, Ottawa, Canada, for assaying parthenolide for membrane-associated protein kinase C activity; S. Heptinstall and J. May, Department of Medicine, University Hospital, Queen's Medical Centre, Nottingham, UK, for assistance with developing the platelet bioassay; D.V.C. Awang, MediPlant Natural Products Consulting Services, Ottawa, Canada, formerly of the Natural Products Section, Bureau of Drug Research, Health and Welfare Canada, for provision of the initial research contract; and Darryl Kindack, Natural Products Section, Bureau of Drug Research, Health and Welfare Canada, for hplc analysis of the compounds; C.H. Thomas Ltd. for provision of bovine blood; and J. Bormanis and D. Bond, Haematology Department, Ottawa Civic Hospital, for provision of a thrombocounter.

This work was funded by a contract (#4001-9-CZ53/01-SZ) from the Bureau of Drug Research, Health and Welfare Canada, and by a grant from the Brandon University Research Committee.

REFERENCES

1. PICKMAN, A.K. 1986. Biological activities of sesquiterpene lactones. Biochem. Syst. & Ecol. 14: 255-281.
2. IVIE, G.W., WITZEL, D.A. 1983. Sesquiterpene lactones: structure, biological action, and toxicological significance. In: Handbook of Natural Toxins Volume 1: Plant and Fungal Toxins (R.F. Keeler, A.T. Tu, eds.). Marcel Dekker, Inc., N.Y. pp. 543-584.
3. RODRIGUEZ, E., TOWERS, G.H.N., MITCHELL, J.C. 1976. Biological activities of sesquiterpene lactones. Phytochemistry 15: 1573-1580.
4. JOHNSON, E.S., KADAM, N.P., HYLANDS, D.M., HYLANDS, P.J. 1985. Efficacy of feverfew as prophylactic treatment of migraine. Brit. Med. J. 291: 569-573.
5. MURPHY, J.J., HEPTINSTALL, S., MITCHELL, J.R.A. 1988. Randomised double-blind placebo-controlled trial of feverfew in migraine prevention. Lancet 1988: 189-192.
6. MARLES, R. J., KAMINSKI, J., ARNASON, J.T., PAZOS-SANOU, L., HEPTINSTALL, S., FISCHER, N.H., CROMPTON, C.W., KINDACK, D.G., AWANG, D.V.C. 1992. A bioassay for inhibition of serotonin release from bovine platelets. J. Nat. Prod. 55: 1044-1056.
7. BOHLMANN, F., ZDERO, C. 1982. Sesquiterpene lactones and other constituents from *Tanacetum parthenium*. Phytochemistry 21: 2543-2549.
8. LIKHITWITAYAWUID, K., ANGERHOFER, C.K., CORDELL, G.A., PEZZUTO, J.M., RUANGRUNGSI, N. 1993. Cytotoxic and antimalarial bisbenzylisoquinoline alkaloids from *Stephania erecta*. J. Nat. Prod. 56: 30-38.
9. PEZZUTO, J.M., SWANSON, S.W., FARNSWORTH, N.R. 1984. Evaluation of the mutagenic potential of endod (*Phytolacca dodecandra*), a molluscicide of potential value for the control of schistosomiasis. Toxicol. Lett. 22: 15-20.
10. KULMACZ, R.J., LANDS, E.M. 1987. Cyclooxygenase: measurement, purification and properties. In: Prostaglandins and Related Substances. A Practical Approach. (C. Benedetto, R.G. McDonald-Gibson, S. Nigam, T.F. Slater, eds.). IRL Press, Oxford. pp. 209-227.
11. DE VRIES, D.J., HERALD, C.L., PETTIT, G.R., BLUMBERG, P.M. 1988. Demonstration of sub-nanomolar affinity of bryostatin 1 for the phorbol ester receptor in rat brain. Biochem. Pharmacol. 37: 4069-4073.

12. STRICKLAND, J.E., GREENHALGH, D.A., KOCEVA-CHYLA, A., HENNINGS, H., RESTREPO, C., BALASCAHK, M., YUSPA, S.H. 1988. Development of murine epidermal cell lines which contain an activated *ras*[Ha] oncogene and form papillomas in skin grafts on athymine nude mouse hosts. Cancer Res. 48: 165-169.
13. LICHTI, U., GOTTESMAN, M.M. 1982. Genetic evidence that a phorbol ester tumor promotor stimulates ornithine decarboxylase activity by a pathway that is independent of cyclic AMP-dependent protein kinases in CHO cells. J. Cell Physiol. 113: 433-439.
14. CHAKRAVARTHY, B.R., BUSSEY, A., WHITFIELD, J.F., SIKORSKA, M., WILLIAMS, R.E., DURKIN, J.P. 1991. The direct measurement of protein kinase C (PKC) activity in isolated membranes using a selective peptide substrate. Anal. Biochem. 196: 144-150.
15. QUICK, A., ROGERS, D. 1976. Crystal and molecular structure of parthenolide [4,5-epoxy-germacra-1(10),11(13)-dien-12,6-olactone]. J.C.S. Perkin II: 465-469.
16. DIAMOND, S. 1991. Migraine Headaches. Med. Clinics N. Amer. 75: 545-566.
17. KUPCHAN, S.M., FESSLER, D.C., EAKIN, M.A., GIACOBBE, T.J. 1970. Reactions of alpha methylene lactone tumor inhibitors with model biological nucleophiles. Science 168: 376-378.
18. KUPCHAN, S.M., EAKIN, M.A., THOMAS, A.M. 1971. Tumor inhibitors. 69. Structure-cytotoxicity relationships among the sesquiterpene lactones. J. Med. Chem. 14: 1147-1152.
19. LEE, K-H., HUANG, E-S., PIANTADOSI, C., PAGANO, J.S., GEISSMAN, T.A. 1971. Cytotoxicity of sesquiterpene lactones. Cancer Res. 31: 1649-1654.
20. LEE, K-H., HALL, I.H., MAR, E-C., STARNES, C.O., ELGEBALY, S.A., WADDELL, T.G., HADGRAFT, R.I., RUFFNER, C.G., WEIDNER, I. 1977. Sesquiterpene antitumor agents: inhibitors of cellular metabolism. Science 196: 533-536.
21. HALL, I.H., LEE, K-H., MAR, E-C., STARNES, C.O., WADDELL, T.G. 1977. Antitumor agents. 21. A proposed mechanism for inhibition of cancer growth by tenulin and helenalin and related cyclopentenones. J. Med. Chem. 20: 333-337.
22. WADDELL, T.G., AUSTIN, A-M., COCHRAN, J.W., GERHART, K.G., HALL, I.H., LEE, K-H. 1979. Antitumor agents: structure-activity relationships in tenulin series. J. Pharm. Sci. 68: 715-718.
23. WADDELL, T.G., GEBERT, P.H., TAIT, D.L. 1983. Michael-type reactions of tenulin, a biologically active sesquiterpene lactone. J. Pharm. Sci. 72: 1474-1476.
24. MERRILL, J.C., KIM, H.L., SAFE, S., MURRAY, C.A., HAYES, M.A. 1988. Role of glutathione in the toxicity of the sesquiterpene lactones hymenoxon and helenalin. J. Toxicol. Envir. Health 23: 159-169.
25. MITCHELL, J.C., DUPUIS, G. 1971. Allergic contact dermatitis from sesquiterpenoids of the Compositae family of plants. Br. J. Derm. 84: 139-150.
26. MITCHELL, J.C., GEISSMAN, T.A., DUPUIS, G., TOWERS, G.H.N. 1971. Allergic contact dermatitis caused by *Artemisia* and *Chrysanthemum* species. The role of sesquiterpene lactones. J. Invest. Derm. 56: 98-101.
27. RODRIGUEZ, E., EPSTEIN, W.L., MITCHELL, J.C. 1977. The role of sesquiterpene lactones in contact hypersensitivity to some North and South American species of feverfew (*Parthenium* - Compositae). Contact Dermatitis 3: 155-162.
28. HAUSEN, B.M., OSMUNDSEN, P.E. 1983. Contact allergy to parthenolide in *Tanacetum parthenium* (L.) Schulz-Bip. (Feverfew, Asteraceae) and cross-reactions to related sesquiterpene lactone containing Compositae species. Acta Derm. Venereol. 63: 308-314.
29. HALL, I.H., LEE, K.H., STARNES, C.O., SUMIDA, Y., WU, R.Y., WADDELL, T.G., COCHRAN, J.W., GERHART, K.G. 1979. Anti-inflammatory activity of sesquiterpene lactones and related compounds. J. Pharm. Sci. 68: 537-542.
30. HALL, I.H., STARNES, C.O.JR., LEE, K.H., WADDELL, T.G. 1980. Mode of action of sesquiterpene lactones as anti-inflammatory agents. J. Pharm. Sci. 69: 537-543.

31. HALL, I.H., LEE, K.H., STARNES, C.O., MURAOKA, O., SUMIDA, Y., WADDELL, T.G. 1980. Antihyperlipidemic activity of sesquiterpene lactones and related compounds. J. Pharm. Sci. 69: 694-697.
32. ARNASON, J.T., PHILOGÈNE, B.J.R., DUVAL, F., MCLACHLAN, D., PICMAN, A.K., TOWERS, G.H.N., BALZA, F. 1985. Effects of sesquiterpene lactones on development of *Aedes atropalpus and relation to partition coefficient. J. Nat. Prod. 48: 581-584.*
33. MARLES, R.J., COMPADRE, R.L., COMPADRE, C.M., SOUCY-BREAU, C., REDMOND, R.W., DUVAL, F., MEHTA, B., MORAND, P., SCAIANO, J.C., ARNASON, J.T.. 1991. Thiophenes as mosquito larvicides: structure-toxicity relationship analysis. Pestic. Biochem. Physiol. 41: 89-100.
34. MARLES, R.J., ARNASON, J.T., COMPADRE R.L., COMPADRE, C.M., SOUCY-BREAU, C., MORAND, P., MEHTA, B., REDMOND, R.W., SCAIANO, J.C. 1991. Quantitative structure-activity relationship analysis of natural products: phototoxic thiophenes. In: Modern Phytochemical Methods (N.H. Fischer, M.B. Isman, H.A. Stafford, eds). Plenum Press, N.Y., Rec. Adv.Phytochemistry 25: 371-395.
35. MARLES, R.J., HUDSON, J.B., GRAHAM, E.A., SOUCY-BREAU, C., MORAND, P., COMPADRE, R.L., COMPADRE, C.M., TOWERS, G.H.N., ARNASON, J.T. 1992. Structure-activity studies of photoactivated antiviral and cytotoxic tricyclic thiophenes. Photochem. Photobiol. 56: 479-487.
36. HANSCH, C., LEO, A. 1979. Substituent Constants for Correlation Analysis in Chemistry and Biology. John Wiley & Sons, N.Y. pp. 1-64.
37. ANDERSON, D., JENKINSON, P.C., DEWDNEY, R.S., BLOWERS, S.D., JOHNSON, E.S., KADAM, N.P. 1988. Chromosomal aberrations and sister chromatid exchanges in lymphocytes and urine mutagenicity of migraine patients: a comparison of chronic feverfew users and matched non-users. Human Toxicol. 7: 145-152.
38. GERAN, R.I., GREENBERG, N.H., MACDONALD, M.M., SCHUMACHER, A.M., ABBOTT, B.J. 1972. Protocols for screening chemical agents and natural products against animal tumors and other biological systems (3rd ed.). Cancer Chemother. Rep. Part 3, 3: 1-5.
39. HANSCH, C., BJÖRKROTH, J.P., LEO, A. 1987. Hydrophobicity and central nervous system agents: on the principle of minimal hydrophobicity in drug design. J. Pharm. Sci. 76: 663-687.
40. HEPTINSTALL, S., GROENEWEGEN, W.A., SPANGENBERG, P., LOESCHE, W. 1987. Extracts of feverfew may inhibit platelet behaviour via neutralization of sulphydryl groups. J. Pharm. Pharmacol. 39: 459-465.
41. HEPTINSTALL, S., WHITE, A., WILLIAMSON, L., MITCHELL, J.R.A. 1985. Extracts of feverfew inhibit granule secretion in blood platelets and polymorphonuclear leucocytes. Lancet 1985: 1071-1074.
42. GROENEWEGEN, W.A., HEPTINSTALL, S. 1990. A comparison of the effects of an extract of feverfew and parthenolide, a component of feverfew, on human platelet activity in-vitro. J. Pharm. Pharmacol. 42: 553-557.
43. COLLIER, H.O.J., BUTT, N.M., MCDONALD-GIBSON, W.J., SAEED, S.A. 1980. Extract of feverfew inhibits prostaglandin biosynthesis. Lancet 1980: 922.
44. PUGH, W.J., SAMBO, K. 1988. Prostaglandin synthetase inhibitors in feverfew. J. Pharm. Pharmacol. 40: 743-745.
45. MAKHEJA, A.N., BAILEY, J.M. 1981. The active principle in feverfew. Lancet 1981: 1054.
46. MAKHEJA, A.N., BAILEY, J.M. 1982. A platelet phospholipase inhibitor from the medicinal herb feverfew (Tanacetum parthenium). Prost. Leuk. Med. 8: 653-660.
47. SCHRÖDER, H., LÖSCHE, W., STROBACH, H., LEVEN, W., WILLUHN, G., TILL, U., SCHRÖR, K. 1990. Helenalin and 11α,13-dihydrohelenalin, two constituents from Arnica

montana L., inhibit human platelet function via thiol-dependent pathways. Thrombosis Res. 57: 839-845.

48. YSRAEL, M.C., CROFT, K.D. 1990. Inhibition of leukotriene and platelet activating factor synthesis in leukocytes by the sesquiterpene lactone scandenolide. Planta Med. 56: 268-270.
49. SUMNER, H., SALAN, U., KNIGHT, D.W., HOULT, J.R.S. 1992. Inhibition of 5-lipoxygenase and cyclo-oxygenase in leukocytes by feverfew. Involvement of sesquiterpene lactones and other components. Biochem. Pharmacol. 43: 2313-2320.
50. BARSBY, R.W.J., SALAN, U., KNIGHT, D.W., HOULT, J.R.S. 1993. Feverfew and vascular smooth muscle: extracts from fresh and dried plants show opposing pharmacological profiles, dependent on sesquiterpene lactone content. Planta Med. 59: 20-25.
51. BARRADAS, M.A., MIKHAILIDIS, D.P. 1993. The use of platelets as models for neurons: possible applications to the investigation of eating disorders. Biomed. & Pharmacother. 47: 11-18.
52. TAKESHIMA, T., SHIMOMURA, T., TAKAHASHI, K. 1987. Platelet activation in muscle contraction headache and migraine. Cephalalgia 7: 239-243.
53. APPENZELLER, O. 1991. Pathogenesis of migraine. Med. Clinics of N. Amer. 75: 763-789.
54. BLAU, J.N. 1992. Migraine: theories of pathogenesis. Lancet 1992: 1202-1209.
55. PEROUTKA, S.J. 1990. Developments in 5-hydroxytryptamine receptor pharmacology in migraine. Neurologic Clinics 8: 829-839.

INDEX

LaVergne, TN USA
18 March 2011
220807LV00002B/14/A